CYTOKINES AND MENTAL HEALTH

NEUROBIOLOGICAL FOUNDATION OF ABERRANT BEHAVIORS

CYTOKINES AND MENTAL HEALTH

edited by

Ziad Kronfol
University of Michigan Health System
Ann Arbor, Michigan

KLUWER ACADEMIC PUBLISHERS
Boston / Dordrecht / New York / London

Distributors for North, Central and South America:
Kluwer Academic Publishers
101 Philip Drive
Assinippi Park
Norwell, Massachusetts 02061 USA
Telephone (781) 871-6600
Fax (781) 681-9045
E-Mail: kluwer@wkap.com

Distributors for all other countries:
Kluwer Academic Publishers Group
Post Office Box 322
3300 AH Dordrecht, THE NETHERLANDS
Telephone 31 786 576 000
Fax 31 786 576 254
E-Mail: services@wkap.nl

Electronic Services <http://www.wkap.nl>

Library of Congress Cataloging-in-Publication Data

Title: CYTOKINES AND MENTAL HEALTH
Editor: Ziad Kronfol
ISBN: 1-4020-7353-4

A C.I.P. Catalogue record for this book is available
from the Library of Congress.

Printed on acid-free paper.

Printed in the United States of America.

To Rima, Zeina and Sara

Contents

List of Contributors xi

Preface xv

Acknowledgments xvii

Section I: Brain Effects of Cytokines: Relevance to Mental Illness

1. Cytokines and Cytokine Receptors: Principles of Action 1
Daniel G. Remick

2. Immunosensory Signaling: Role of Cytokines 15
Lisa E. Goehler

3. Cytokine Pathways in the Brain 39
Julio Licinio and Ma-Li Wong

4. Effects of Cytokines on Cerebral Neurotransmission and Potential Relationships to Function 55
Adrian J. Dunn

5. Cytokines and HPA Axis Regulation 85
Marni N. Silverman, Brad D. Pearce, and Andrew H. Miller

6. Cytokines and Neuropeptides 123
Nicholas P. Plotnikoff, Robert E. Faith, and Anthony J. Murgo

Section II: Behavioral Effects of Cytokines

7. Cytokines and Sickness Behavior 129
Robert Dantzer

8. Cytokines and Sleep Regulation 147
James M. Krueger and Lynn Churchill

9. Cytokines and Appetite 167
Wolfgang Langhans and Brian J. Hrupka

10. Cytokines Effects on Learning and Memory 211
John M. Petitto, Martin J. Repetto, Zhi Huang, Ray D. Beck, Jr. and David A. Hartemink

11. Cytokine-elicited Sensitization: Neuroendocrine, Neurotransmitter and Behavioral Responses 225
Shawn Hayley, Zul Merali and Hymie Anisman

Section III: Cytokines and Neuropsychiatric Disorders

12. Cytokine Regulation in Major Depression 259
Ziad Kronfol

13. Cytokine Regulation in Anxiety Disorders 281
Francesca Brambilla

14. Cytokines and Post Traumatic Stress Disorder 301
Dewleen G. Baker, Thomas D. Geracioti, Jr., John W. Kasckow, Emmanouil Zoumakis and George P. Chrousos

15. Cytokines and Schizophrenia 339
Thomas Pollmächer, Andreas Schuld and Dunja Hinze-Selch

16. Cytokine Regulation in Autism 369
Vijendra K. Singh

17. Cytokine Regulation in Alzheimer's Disease 385
Robert E. Mrak

18. Cytokines, Immunity and Disordered Sleep 403
Michael Irwin

Index 423

Contributors

Hymie Anisman, Institute of Neuroscience, Carleton University, Ottawa, Ontario, Canada: *Chapter 11*

Dewleen G. Baker, Cincinnati Veterans Affairs Medical Center and Department of Psychiatry, University of Cincinnati College of Medicine, Cincinnati, OH: *Chapter 14*

Ray D. Beck, Jr., McKnight Brain Institute, University of Florida, College of Medicine, Gainesville, FL: *Chapter 10*

Francesca Brambilla, Dipartimento Di Salute Mentale-Ospedale Sacco-Milano-Italy: *Chapter 13*

George P. Chrousos, Pediatric and Reproductive Endocrinology Branch, NICHD, NIH, Bethesda, MD: *Chapter 14*

Lynn Churchill, Washington State University, Pullman, WA: *Chapter 8*

Robert Dantzer, Inegrative Neurobiology, INRA-INSERM U394, Bordeaux, France: *Chapter 7*

Adrian J. Dunn, Department of Pharmacology and Therapeutics, Louisiana State University Health Sciences Center, Shreveport, LA: *Chapter 4*

Robert E. Faith, Center for Comprehensive Medicine, Baylor College of Medicine, Houston, TX: *Chapter 6*

Thomas D. Geracioti, Jr., Cincinnati Veterans Affairs Medical Center and Department of Psychiatry, University of Cincinnati College of Medicine, Cincinnati, OH: *Chapter 14*

Lisa E. Goehler, Department of Psychology, University of Virginia, Charlottesville, VA: *Chapter 2*

David A. Hartemink, McKnight Brain Institute, University of Florida, College of Medicine, Gainesville, FL: *Chapter 10*

Shawn Hayley, Institute of Neuroscience, Carleton University, Ottawa, Ontario, Canada: *Chapter 11*

Dunja Hinze-Selch, Department of Psychiatry and Psychotherapy, Christian Albrechts University, Kiel, Germany: *Chapter 15*

Brian J. Hrupka, Institute of Animal Sciences, Swiss Federal Institute of Technology (ETH), Zurich, Switzerland: *Chapter 9*

Zhi Huang, McKnight Brain Institute, University of Florida, College of Medicine, Gainesville, FL: *Chapter 10*

Michael Irwin, Cousins Center for Psychoneuroimmunology, Neuropsychiatric Institute, University of California, Los Angeles, CA: *Chapter 18*

John W. Kasckow, Cincinnati Veterans Affairs Medical Center and Department of Psychiatry, University of Cincinnati College of Medicine, Cincinnati, OH: *Chapter 14*

Ziad Kronfol, Department of Psychiatry, University of Michigan Health System, Ann Arbor, MI: *Chapter 12*

James M. Krueger, Washington State University, Pullman, WA: *Chapter 8*

Wolfgang Langhans, Institute of Animal Sciences, Swiss Federal Institute of Technology (ETH), Zurich, Switzerland: *Chapter 9*

Julio Licinio, Laboratory of Pharmacogenomics, Neuropsychiatric Institute, David Geffen School of Medicine at UCLA, Los Angeles, CA: *Chapter 3*

Zul Merali, Neuroscience Research Institute, OHRI, University of Ottawa, Ottawa, Ontario, Canada: *Chapter 11*

Andrew H. Miller, Department of Psychiatry and Behavioral Sciences, Emory University School of Medicine, Atlanta, GA: *Chapter 5*

Robert E. Mrak, University of Arkansas for Medical Sciences, Little Rock, AK: *Chapter 17*

Anthony J. Murgo, National Cancer Institute, Rockville, MD: *Chapter 6*

Brad D. Pearce, Department of Psychiatry and Behavioral Sciences, Emory University School of Medicine, Atlanta, GA: *Chapter 5*

John M. Petitto, McKnight Brain Institute, University of Florida, College of Medicine, Gainesville, FL: *Chapter 10*

Nicholas P. Plotnikoff, College of Pharmacy, University of Illinois, Chicago, IL: *Chapter 6*

Thomas Pollmächer, Max Planck Institute of Psychiatry, Munich, Germany: *Chapter 15*

Daniel G. Remick, Department of Pathology, University of Michigan Health System, Ann Arbor, MI: *Chapter 1*

Martin J. Repetto, McKnight Brain Institute, University of Florida, College of Medicine, Gainesville, FL: *Chapter 10*

Andreas Schuld, Max Planck Institute of Psychiatry, Munich, Germany: *Chapter 15*

Marni N. Silverman, Department of Psychiatry and Behavioral Sciences, Emory University School of Medicine, Atlanta, GA: *Chapter 5*

Vijendra K. Singh, Department of Biology & Biotechnology Center, Utah State University, Logan, UT: *Chapter 16*

Ma-Li Wong, Laboratory of Pharmacogenomics, Neuropsychiatric Institute, David Gaffen School of Medicine at UCLA, Los Angeles, CA: *Chapter 3*

Emmanouil Zoumakis, Pediatric and Reproductive Endocrinology Branch, NICHD, NIH, Bethesda, MD: *Chapter 14*

Preface

Cytokines have been known for many years as mediators between cells of the immune system. They were also known to communicate with non-immune cells such as fibroblasts and endothelial cells. Functions outside the immune system or in relation to inflammation, however, were thought to be limited. The fields of neuroscience and psychiatry were far removed from developments in this area. Then came several discoveries over the last few years showing that peripheral cytokines communicate directly or indirectly with brain cells, that genes for specific cytokines and their receptors can be expressed in different areas of the brain under certain conditions and that cytokines are associated with numerous brain effects resulting in neurochemical, neuroendocrine, neuroimmune and behavioral changes. This series of discoveries has led to a burgeoning of the field of psychoneuroimmunology, particularly among neuroscientists. The clinical applications of these activities, however, has remained somewhat limited. Psychiatrists became aware of the neurotoxic and behavioral adverse effects of treatment with cytokines when some patients who were receiving interferon for malignant melanoma or hepatitis C were showing signs of depression and cognitive impairment. Several investigators were also assessing the potential role of cytokines in the pathophysiology of psychiatric disorders such as schizophrenia and major depression. However, in spite of these efforts, the gap between bench and bedside has remained wide and as a result most psychiatrists have not paid too much attention to the exciting developments in psychoneuroimmunology, nor have they been able to access the potential impact of these developments on their field.

It is with this background in mind that the idea for this book has been formulated. While a great deal of new and exciting data in the basic science of psychoneuroimmunology has recently been generated, translational research has been relatively slow. We therefore decided to assemble under the same cover state-of-the-art reviews of selected topics in mental health that span the spectrum from the molecular biology of brain cytokines to their potential role in various psychiatric disorders. We hope this book will help give the neuroscientist working with cytokines a sense of "connectedness" with clinical medicine and particularly psychiatry and give the clinician/psychiatirst a "taste" of the basic research in this field. We expect such a dialogue will result in new discoveries and applications that would integrate cytokines and their receptors into the design of the drugs of the future for the treatment and prevention of mental illness.

Ziad Kronfol
Ann Arbor, Michigan

Acknowledgements

First, I want to thank the many contributors who have made this volume possible. I also wish to thank Ila Atwood and Rima Kronfol who helped in the preparation of the manuscript. Last, I wish to thank the University of Michigan Department of Psychiatry and Depression Center for providing the intellectual stimulation for this work to be accomplished.

Chapter 1

CYTOKINES AND CYTOKINE RECEPTORS: *PRINCIPLES OF ACTION*

Daniel G. Remick
Department of Pathology, University of Michigan Health System, Ann Arbor, Michigan

1. INTRODUCTION

Cytokines have become recognized as important mediators of the several aspects of inflammation. This introductory chapter will provide some basic aspects of cytokine biology including the functions of cytokines and their receptors. Given the brief nature of the chapter it is not possible to exhaustively review all of the cytokines and their biological activities. The major purpose of this short, introductory chapter is to provide a reader with the basic underpinnings of what are cytokines and how they work.

There are multiple methods that can be used to demonstrate the importance of cytokines in health and disease, some more quantitative than others. At the health food store, one can purchase a cytokine oral spray to improve one's health. A simple internet search using the word cytokine yields a multitude of websites with varying degrees of accuracy and timeliness. Scores of books have been published which are exclusively devoted to the subject of cytokines, some of which are already in their third edition. An International Cytokine Society has been formed to further the study of cytokines and this society holds annual meetings. The National Institutes of Health in the United States is funding over 3000 grants in the year 2002 which study some aspect of cytokines.

The clearest indication that cytokines have an important impact on human health may be found in the studies where modulation of cytokines has dramatically improved the lives of humans. Elevation of cytokines and inhibition of cytokines have both been done with dramatic success stories.

2. WHAT ARE CYTOKINES?

A simple definition is: Cytokines are peptide mediators which modulate aspects of immunity and inflammation. While such a short definition lacks sensitivity and specificity, it does allow one to focus on the unifying principles of these remarkable molecules. Cytokines have been compared to the endocrine system and have been called the hormones of the immune system. The cytokines share similarities with hormones since both are made in one organ/cell and move to exert a biological effect on another organ/cell. Insulin represents a classic example of a typical hormone which is produced by the islet cells in the pancreas and then regulates metabolism in several different tissues. In a similar manner, erythropoietin is primarily produced in the kidney and then stimulates the production of new red blood cells in the bone marrow. However, there are also important differences between the cytokine system and the hormone system. Many cytokines work locally and not at distant sites. In this regard they exert an autocrine or paracrine mode of action rather than an endocrine type of response. Many classic hormones serve to maintain homeostasis within the body and are constitutively produced. In contrast, cytokines are typically synthesized when the host senses danger and are upregulated in response to specific stimuli.

The difference between hormones and cytokines may be highlighted by comparing the circulating levels of hormones and cytokines. A typical hormone has a normal circulating level, for example insulin has a normal level of 1 – 21 microunits/ml in an adult, parathyroid hormone has a level of 8.6 to 10.2 mg/dl, also in the adult. These levels change depending on different disease states, and also change with different physiological states. For example, the insulin levels that are listed are typical fasting levels and do not reflect the levels that would be found after a large meal when blood glucose levels increase. In contrast, most of the cytokines do not have normal circulating levels. Therefore, the baseline is near zero, or below detection limit of the assay. It is not appropriate to consider increases above baseline for most of the cytokines, because the baseline is zero.

Cytokines are typically small proteins with a molecular weight generally less than 30 kilodaltons. Some of the cytokines are glycosylated which accounts for some of their variation in molecular weight. The glycoslyation is generally not necessary for the biological activity of the cytokine. Significant structural homology exists between species with regards to both the DNA organization and the amino acid sequence of a specific cytokine. For example, the cDNA for IL-6 has 65% homology between man and mouse and the predicted amino acid sequence has 42% homology between the man and mouse (1). Cytokines may exist alone in solution or exist as dimers (2) or trimers in solution (3).

The immune system has been divided into the innate immune response and the adaptive immune response. Cytokines participate in both arms of the immune reaction. For example, the cytokine interleukin 8 helps to recruit neutrophils to sites of bacterial infection to help clear this infectious challenge in a classic example of the innate immune response. An example of the adaptive immune response is found where the cytokine eotaxin will help recruit eosinophils to sites of allergic inflammation.

Cytokines are pleotropic in that they exert multiple biological activities. In fact, most of the cytokines have several biological activities both *in vitro* and *in vivo*. Tumor necrosis factor will induce necrosis of tumors, but it will also stimulate synthesis of several other cytokines and participates in the acute inflammatory condition observed in animals after injection of Gram negative bacteria (4). Cytokines are also overlapping and more than one cytokine will exert the same biological effect. Both tumor necrosis factor and interleukin 1 will induce the synthesis of interleukin 8 (5).

3. CYTOKINE NOMENCLATURE

The names for the cytokines are confusing and chaotic. At one point in time, cytokines were termed lymphokines if they primarily secreted by lymphocytes and monokines if they were secreted by monocytes/mononuclear cells. This terminology has largely fallen by the wayside since the cell of origin frequently depends on the stimulus used and the state of cellular activation. Historically, the names for virtually all cytokines were based on their biological activity. Tumor necrosis factor (TNF) serves as a specific example. This cytokine was originally defined as a factor present in the serum of mice after lipopolysaccharide injection that would induce necrosis of tumors (6). A significant problem with this approach is that several groups could be examining different biological activities of the same molecule. Such an event occurred with TNF where another group was investigating the serum factor that induces cachexia (7).

Many cytokines retain their original names. Included in this large group are tumor necrosis factor, the interferons, the colony stimulating factors and several others. There are some merits in the approach of naming a cytokine for it's biological activity. First, these names are easier to remember compared to an arbitrarily assigned number. Second, such a naming protocol helps to focus attention on at least one of cytokine's biological activities. Third, even if this biological activity is not the major function of the cytokine it will help to provide an historical context for the protein.

An attempt to bring order to this chaos has been initiated by the interleukin system. With this, each cytokine is given a designated interleukin number. The protocol is that the protein in question must have a unique amino acid sequence and functional activity with leukocytes. The protein is proposed as a cytokine to the Nomenclature and Standardization Committee of the International Cytokine Society and the Union of Immunological Societies. After appropriate evaluation, a recommendation is made to the World Health Organization. Even with these guidelines there have been mistakes and it is not clear if interleukin 14 (IL-14) exists at all.

4. CLASSES OF CYTOKINES

Cytokines may be grouped based on a variety of factors such as common biological activities, shared cell surface receptors or similarity of signal transduction mechanisms. An even broader description divides the cytokines into those with pro-inflammatory activity, anti-inflammatory activity, and the hematopoietic stimulatory factors. Some investigators include growth factors as cytokines, these would include molecules such as platelet derived growth factor (PDGF), epidermal growth factor (EGF) and transforming growth factor (TGF). Of these, TGF is clearly considered to be a cytokine.

4.1 Pro-inflammatory cytokines

Many of the cytokines promote active inflammation within the host. This increased inflammation serves to help eliminate pathogens and also speed the resolution of the inflammatory process. A partial listing of these cytokines is provided in Table 1. There are multiple mechanisms whereby cytokines will exert a pro-inflammatory effect. Individual cells may be activated and increase their capacity to function more efficiently. Macrophages and neutrophils may be energized to better kill invading infectious agents. NK cells and T cells will become switched on to more effectively lyse their targets. Cytokines will also provide signals to T cells to augment their proliferation, and B cells may be induced to proliferate or begin to secrete immunoglobulins. Numerous cytokines may be considered to be pro-inflammatory. Among these are tumor necrosis factor, interferon gamma, interleukin 1 and interleukin 2.

The pro-inflammatory effects of the cytokines extend beyond activation of specific cells. Several aspects of physiology may be altered to a more pro-inflammatory state. At a systemic level, cytokines have been shown to induce fever in both experimental animals and man (8). Cytokines, in particular interleukin 6, will increase synthesis of acute phase proteins within the liver. At the local level, there is enhanced recruitment of inflammatory cells to sites of inflammation

Cytokine	**Size (kD)**	**Cellular Sources**	**Typical Biological Activities** ***in vivo actions listed in italics***
IL-1α IL-1β	17.5 17.5	Macrophage	Activates T, B cells, endothelial cells, *causes fever, induces hepatic acute phase proteins*
IL-2	15-20	T cells	Activates T, B, and NK cells, *anti-tumor*
IL-6	21-28	Most cells	IgG production, T cell activation, *induces hepatic acute phase proteins, causes fever*
IL-8	8-10	Monocytes, other cells	Neutrophil and T cell chemotaxis, *neutrophil recruitment*
IL-12	35, 40	T cells	NK growth factor, decreases IgE, *decreases parasite load*
IL-18	24	Macrophages	Induces γ-IFN production, *accelerates clearance of intracellular pathogens*
TNF	17	Macrophages	Induces other cytokines, *causes cachexia, induces tissue injury and shock*
γ-IFN	20-25	T and NK cells	Increases lysis of virally infected cells, *augments macrophage function*

Table 1. List of pro-inflammatory cytokines. This table lists some typical pro-inflammatory cytokines. The size is the molecular weight of the cytokine and represents the size of the mature, secreted protein. The cellular source is the typical cell that would secrete the cytokine after stimulation. The biological activities include those found after incubation of the cytokines with target cells or when it is administered *in vivo*.

It must be mentioned that the cytokine response has evolved to confer a survival benefit to the host. As such, the pro-inflammatory effects of the cytokines need to be viewed as primarily beneficial. Inhibition of cytokine action, or decreasing the synthesis of cytokines, is tantamount to immunosuppression. While there are certain situations where it is beneficial to decrease inflammation this carries the risk of a potential increase in infectious diseases (9).

There is a specific class of pro-inflammatory cytokines termed the chemokines. These are chemotactic cytokines (chemokines) which serve to recruit, activate, and retain leukocytes to local sites of inflammation. The chemokines represent a subgroup of cytokines. The best characterized of the chemokines is interleukin 8 which acts primarily on neutrophils. Other chemokines act on other inflammatory cells with relative, although not absolute specificity. Eotaxin is a chemokine recruits eosinophils.

4.2 Anti-inflammatory cytokines

Some cytokines decrease the inflammatory response by inhibiting immune responses. These anti-inflammatory cytokines are listed in Table 2. As an example, IL-10 was originally named cytokine synthesis inhibitory factor(10) since it would specifically decrease the production of pro-inflammatory cytokines such as TNF. Other examples of cytokines which exert an anti-inflammatory effect include IL-4, IL-13, and TGF-β. The number of anti-inflammatory cytokines is smaller compared to those with pro-inflammatory effects.

Cytokine	Size (kD)	Cellular Sources	Typical Biological Activities *in vivo actions listed in italics*
IL-4	15-19	T cells, BM stromal cells	T, B, NK cell activation, IgG and IgE class switching, suppresses monocytes, *anti-tumor*
IL-10	19	T cells	Inhibition of cytokine synthesis, proliferation of T cells, *inhibition of endotoxin shock*
IL-13	10	T cells	Suppress macrophage cytotoxicity, decreases production of pro-inflammatory cytokines, *anti-tumor*
TGF-β	25	Platelets	Inhibits growth of fibroblasts, T and B cell proliferation, *improves wound healing, bone repair*

Table 2. Anti-inflammatory cytokines. This table lists some typical anti-inflammatory cytokines. The size is the molecular weight of the cytokine and represents the size of the mature, secreted protein. The cellular source is the typical cell that would secrete the cytokine after stimulation. The biological activities include those found after incubation of the cytokines with target cells or when it is administered *in vivo*.

It is sometimes difficult to precisely define the biological repertoire of a particular cytokine as strictly pro-inflammatory or anti-inflammatory. This relates to the arbitrary nature of attempting to place a molecule into a narrowly limited scope of biological activity. For example, IL-8 will serve to recruit neutrophils to local sites of inflammation (11). However, high intravascular levels of IL-8 serve to limit neutrophil extravasation to sites of inflammation (12).

Pathogenic organisms have developed their own versions of cytokines or cytokine receptors to dampen the immune response and permit enhanced survival. These have been termed decoy receptors and they represent viral gene products with significant structural homology to the human proteins. One of the best examples is interleukin 10 where the viruses have a protein which shares biological activity with mammalian IL-10. When this protein is secreted by a virally infected cell, it decreases production of pro-inflammatory cytokines. With less inflammation, there is decreased clearance of the pathogen.

4.3 Hematopoietic cytokines

A significant function of cytokines is to increase the production of white blood cells and red blood cells from the bone marrow. These cytokines may be given interleukin numbers or may be defined by the type of response they induce and a partial listing is given in Table 3. The biological activity of this group of cytokines has been defined by adding the cytokine to bone marrow precursor cells and then determining which colonies have been induced. This name is applied by looking at the formation of groups of cells (colonies) which form in soft agar cultures (13). For example, if a cytokine induces the formation of granulocytes it is called granulocyte colony stimulating factor (G-CSF). Some cytokines induce more than type of colony, such as granulocyte monocyte colony stimulating factor (GM-CSF). Note that some of these cytokines are approved for use in humans. The hematopoietic cytokines differ from many of the other cytokines since they may be constitutively produced. This is in keeping with their biological activity, since they serve to maintain ongoing production of the cells in the peripheral blood which need to be continuously replenished.

Cytokine	Size (kD)	Cellular/Organ Sources	Cells Formed Following Stimulation
Erythropoietin	30	Kidney	Erythroid cells, red blood cells
Granulocyte-CSF	25-30	Many cells	Granulocytes
Granulocyte-monocyte CSF	22	Monocytes, T cells	Granulocytes, monocytes
Monocyte CSF	45-70	T cells	Monocytes
IL-3	14-30	T cells	Multi-lineage
IL-5	21.5	T cells	Eosinophils
IL-9	32-39	T cells	Erythroid cells
IL-11	23	Stromal cells	Megakaryocytes

Table 3. Hematopoietic cytokines. The column listing the cells which respond refers to the types of cells which are induced in cultures of bone marrow cells. CSF, colony stimulating factor

4.4 Growth factors

There are several growth factors that can be considered to be cytokines. The growth factors may be named for the cellular source, such as platelet derived growth factor (PDGF). Alternatively, growth factors can be named for their biological activity, such as nerve growth factor (NGF). NGF will be of interest to neuroscientists since it induces the growth of peripheral nerves and will also increase wound healing. NGF binds to both the NGF receptor as well as the receptor for tumor necrosis factor. Since the growth factors are not extensively considered in this book they will not be further discussed, other than TGFβ.

5. CYTOKINE METABOLISM

Cytokine metabolism includes both the production and clearance of cytokines. Except for the hematopoietic cytokines, most of these mediators are produced when the host determines that there is a need to modulate the inflammatory response. In response to an invading pathogen the host will increase production of several pro-inflammatory cytokines to help orchestrate the immune response and speed clearance of the pathogen. The production of the cytokines is extremely complex, non-linear and multi-dimensional. In the not-too-distant past, cytokines were considered to be made in a simple cascade model where one cytokine would induce the production of the next cytokine. It is now apparent that the production is better defined as a network with complex inter-relationships.

Multiple stimuli will induce the synthesis of cytokines. Many of these have been defined using *in vitro* cell culture techniques. However, extensive work has also been performed *in vivo* using both animal models and human volunteers. It is not an oversimplification to state that virtually any perturbation of the host has the potential to induce synthesis of cytokines. For example, physical exercise can increase the circulating levels of IL-6 (14).

While multiple, diverse stimuli may increase the synthesis of cytokines there are classic, well defined inducers of cytokine synthesis that are frequently employed to dissect the immune response. Lipopolysaccharide (LPS) derived from the cell wall of Gram negative bacteria strongly stimulates cells to produce pro-inflammatory cytokines. LPS works effectively both *in vitro* (15) as well as *in vivo* (16). Other stimulators of cytokine synthesis include viral infections, trauma, organ or tissue transplantation and ischemia/reperfusion injury.

Once the cytokines have been produced they are generally rapidly cleared. The half life of most cytokines which have been injected intravenously is usually measured in minutes (17). There are significant differences between clearance of cytokines from tissue culture experiments and clearance of the intravenously injected cytokines. The tissue culture conditions do not have hepatic or renal clearance mechanisms and the half life of the cytokines is substantially greater than that observed *in vivo.* During cell culture conditions the only clearance occurs when the cytokines bind to cytokine receptors.

6. CYTOKINE RECEPTORS

No cytokine story is complete without a discussion of the cytokine receptors. For a cytokine to exert biological activity, both the cytokine and a cell surface receptor need to be present and functional. All of the cytokines exert their biological effects by binding to receptors on the surface of cells (18). Once the cytokine has bound to its receptor, cellular signal transduction mechanisms are initiated which direct the cell to proceed to the next step in the inflammation pathway. Most cytokine receptors are composed of three portions. The first is the extracellular domain which is responsible for binding to the cytokine. The specificity of the response is dictated by the specificity of the extracellular domain for the cytokine. The second portion is the transmembrane spanning portion. This is the part of the polypeptide chain that actually crosses the cell membrane. The last segment of the receptor is the intracellular signal transduction region responsible for initiating the sequence of signal transduction events that culminate in cell activation.

Cytokine receptors may be classified into different categories based on their biochemical profile (19). Some cytokine receptors are composed of multiple subunits or chains. For example, the IL-2 receptor has several chains including the alpha, beta and gamma chains. Some of these chains are shared with other cyotkine receptors, such that IL-2, IL-7 and IL-15 share a common gamma subunit. The functional unit, i.e. the assembly of cell surface proteins necessary for biological activity, varies between cytokines.

There is one notable exception to the traditional ligand-receptor interaction which typifies most cytokines. Interleukin 1 has a naturally occurring structural homolog termed the interleukin 1 receptor antagonist (IL-1RA) (20). This protein shares significant structural homology with IL-1. However, when IL-1RA binds to the IL-1 receptor it does not stimulate the cell and thereby serves as an antagonist of IL-1. During either *in vivo* inflammatory conditions or *in vitro* cell stimulation events, IL-1RA blocks the biological activity of IL-1. Since this is a naturally occurring protein in humans, it has emerged as a potential inhibitor of inflammation.

The receptors for the cytokines may be shed from the surface of the cell and enter the circulation of the host. These soluble, shed receptors will bind to their specific cytokines and in this extracellular form serve to inactivate the cytokine. If the cytokine is already complexed to a receptor, then it is generally not available to bind to a cell surface receptor. In effect, these soluble receptors act as endogenous inhibitors of cytokines. In some situations, the soluble receptor represents the binding portion of the same receptor responsible for activating the cell, as in the case of the TNF receptors (21). In other situations, the soluble receptor represents a form that may be expressed on the surface of cells but is not involved in signal transduction, such as with IL-1 and IL-18. These soluble forms of the shed receptors have been exploited as therapeutic agents to treat diseases, as described below.

An important biological concept has emerged concerning the ratio of the soluble receptors and their cytokines. The final activity of the cytokine is not dictated by the actual concentration of the cytokine, but rather by the ratio of the soluble receptor to the cytokine (22). If the soluble receptor concentrations are high, they will inhibit the activity of the cytokine. In effect, this is similar to the free level of the cytokine rather than the total level.

7. MEASURING CYTOKINES

Since cytokines are important, and since cytokines and cytokine inhibitors are being used to treat human diseases, there is considerable interest in precise quantitation of these inflammatory mediators. Whatever method used to measure the cytokines is dictated by the source of the material, which may not reflect the actual level of cytokine in the local environment. As a specific example, if a patient has rheumatoid arthritis, the cytokine concentrations within the inflamed joint are most likely responsible for dictating the progression of disease. Many studies have only examined blood levels of the cytokines since that is the sample that is most frequently available. Even if the blood levels of the cytokine in question are measured with the utmost precision and accuracy, the blood levels may not truly reflect the levels in the inflamed joint. Clearly, repeated access to peripheral blood is easier to obtain than samples of synovial fluid. Given the topic under consideration in this book, cytokines and mental health, this caveat takes on added significance.

The complexity of measuring cytokines extends beyond the mere intricacies of how to properly perform an assay for a cytokine. As previously mentioned, cytokines frequently exert their principal biological effect in the local environment rather than acting at the systemic level. Therefore, measuring plasma levels of the cytokines may not accurately reflect the true level where the cytokines are exerting their biological activity. One way to circumvent this problem is to perform *ex vivo* stimulation studies. With this type of a protocol, blood or isolated blood cells are taken from the patient, stimulated *ex vivo* and the production of cytokines measured. This type of measurement allows one to understand the capacity of the individual patient to mount an inflammatory response which may provide some insight into disease. For example, in patients with multiple sclerosis, it has been shown that the production of tumor necrosis factor following *ex vivo* stimulation of whole blood predicts relapses (23).

The potential to produce cytokines has other applications in the understanding of human disease. If there are no detectable circulating cytokines it may indicate that the patient is normal, or severely immunosuppressed and unable to mount an appropriate inflammatory response. Clearly, the implications for potential treatment of the individual patient are very different. This has been studied in patients with severe bacterial infections where patients from an intensive care unit with bacteremia had decreased production of cytokines following endotoxin stimulation of whole blood but augmented production when the blood was challenged with heat killed *Staphylococcus aureus* (24).

There are several different methods to measure cytokines. Cytokines may be detected at either the messenger RNA or the protein level. Measurement of only messenger RNA coding for specific cytokines will not tell you if this mRNA has ever been translated into protein. However, mRNA analysis can be done an extremely sensitive level. Methods for measuring specific cytokine mRNAs include, in increasing order of sensitivity, Northern blot analysis, ribonuclease protection assay, and reverse transcriptase polymerase chain reaction (RT-PCR). To precisely localize the cell of origin for a specific cytokine it is possible to perform *in situ* hybridization.

The two most frequently used methods to measure cytokine proteins are either bioassays or immunodetection methods. There are advantages and pitfalls to both of these. A bioassay will tell you if the cytokine is present, and also if it is still biologically active, which may be a significant issue. Unfortunately, bioassays are plagued with lack of specificity since the cell responding to the particular cytokine may respond to several other cytokines. In some situations, the bioassays may be either more sensitive or less sensitive than immunodetection methodologies. The most widely used immuno detection method is an ELISA format. For this method, an antibody specific for the cytokine in question is coated in a 96 well plate. This permits detection of up to 48 samples at once if the assays are performed in duplicate. Radioactivity is not involved and only a chromogenic substrate is used for detecting the cytokine. The advantages of the ELISA format is that several different antibody pairs are available for detection of specific cytokines. The antibodies have tremendous sensitivity and they are carefully checked for cross-reactivity to ensure that they exhibit appropriate specificity. A disadvantage is that the ELISA may detect cytokines bound to soluble receptors that are biologically inactive. Another disadvantage is that the antibodies may detect degraded fragments of the cytokine which also lack biological relevancy. The vast majority of the current literature measures cytokines by using immunodetection methods such as the ELISA. Cytokines within individual cells may also be visualized using immunohistochemistry.

8. CYTOKINE FAILURES AND SUCCESSES

One of the reasons for the intense interest in the cytokines has been the potential to alter human disease by modulating the activity of different cytokines. A notable failure of cytokine inhibition for the treatment of human diseases is the use of specific inhibitors of TNF and IL-1 for the treatment of sepsis. Despite several, large scale clinical trials enrolling thousands of patients, the inhibitors were not effective (25).

In contrast, inhibitors of TNF have proven remarkable efficacious for the treatment of at least 2 chronic inflammatory conditions. TNF inhibitors were approved for the treatment of both rheumatoid arthritis and Crohn's disease by the Food and Drug Administration in the United States. Augmentation of cytokines has also been effective and Table 4 lists hematopoietic cytokines which are presently approved for use in humans.

Name	Biological Activity	Year Approved for Use
Erythropoietin	Increases red blood cells	1989
Granulocyte Colony Stimulating Factor	Increases neutrophils	1991
Interleukin 11	Increases platelets	1997

Table 4. Partial list of hematopoietic cytokines currently approved for use in humans. This is a partial listing of hematopoietic cytokines which are used to treat human diseases or correct deficiencies such as anemia.

There are other success stories with the cytokines. Interferon has been successfully used to treat hepatitis C and multiple sclerosis. Interleukin 2 has been shown to be effective against some types of cancer, and even better efficacy with lower toxicity is achieved when the patients' cells are stimulated *ex vivo* and then given back to the patient. IL-2 has demonstrated particular promise against melanoma and renal cell carcinoma. Tumor necrosis factor was not effective against tumors because of dose limiting toxicity, but it has proven beneficial as an anti-cancer drug when used in an isolated limb perfusion protocol. Cytokine gene therapy has also been proposed for the potential treatment of human diseases (26).

Hopefully, this brief overview has provided the fundamentals of cytokine biology and will serve as a framework for understanding other portions of this book. Not all cytokines have been discussed and not all cytokine biological activities have been presented. The major purpose of this introduction was to deliver the basics of cytokine biology. Another important component was to impart to the reader a sense of the importance of the cytokines in health and disease.

ACKNOWLEDGEMENTS

This work is supported in part by NIH grant GM 44918 and GM 50501.

REFERENCES

1. Northemann, W., Braciak, T.A., Hattori, M., et al. Structure of the rat interleukin 6 gene and its expression in macrophage-derived cells. *J.Biol.Chem.* 1989;264: 16072-16082
2. Clore, G.M., Appella, E., Yamada, M., et al. Three-dimensional structure of interleukin 8 in solution. *Biochem.* 1990;29: 1689-1696
3. Smith, R.A. and Baglioni, C. The active form of tumor necrosis factor is a trimer. *J.Biol.Chem.* 1987;262: 6951-6954
4. Tracey, K.J., Fong, Y., Hesse, D.G., et al. Anti-cachectin/TNF monoclonal antibodies prevent septic shock during lethal bacteraemia. *Nature* 1987;330: 662-664
5. DeForge, L.E., Kenney, J.S., Jones, M.L., et al. Biphasic production of IL-8 in lipopolysaccharide (LPS)- stimulated human whole blood. Separation of LPS- and cytokine- stimulated components using anti-tumor necrosis factor and anti- IL-1 antibodies. *J.Immunol.* 1992;148: 2133-2141
6. Carswell, E.A., Old, L.J., Kassel, R.L., et al. An endotoxin-induced serum factor that causes necrosis of tumors. *Proc. Natl. Acad. Sci.* U.S.A. 1975;72: 3666-3670
7. Beutler, B., Greenwald, D., Hulmes, J.D., et al. Identity of tumour necrosis factor and the macrophage-secreted factor cachectin. *Nature* 1985;316: 552-554
8. Kettelhut, I.C. and Goldberg, A.L. Tumor necrosis factor can induce fever in rats without activating protein breakdown in muscle or lipolysis in adipose tissue. *J.Clin.Invest.* 1988;81: 1384-1389
9. Keane, J., Gershon, S., Wise, R.P., et al. Tuberculosis associated with infliximab, a tumor necrosis factor alpha-neutralizing agent. *New Engl. J. Med.* 2001;345: 1098-104

10. Howard, M. andO'Garra, A. Biological properties of interleukin 10. *Immunol. Today* 1992;13: 198-200
11. Baggiolini, M., Walz, A. and Kunkel, S.L. Neutrophil-activating peptide-1/interleukin 8, a novel cytokine that activates neutrophils. *J.Clin.Invest.* 1989;84: 1045-1049
12. Gimbrone, M.A., Jr., Obin, M.S., Brock, A.F., et al Endothelial interleukin-8: a novel inhibitor of leukocyte- endothelial interactions. *Science* 1989;246: 1601-1603
13. Ogawa, M. Differentiation and proliferation of hematopoietic stem cells. *Blood* 1993;81: 2844-2853
14. Drenth, J.P., Van Uum, S.H., Van Deuren, M., et al. Endurance run increases circulating IL-6 and IL-1ra but downregulates ex vivo TNF-alpha and IL-1 beta production. *J. Appl. Physiol.* 1995;79: 1497-503
15. Wollenberg, G.K., DeForge, L.E., Bolgos, G., et al. Differential expression of tumor necrosis factor and interleukin- 6 by peritoneal macrophages in vivo and in culture. *Am.J.Pathol.* 1993;143: 1121-1130
16. Remick, D.G., Strieter, R.M., Eskandari, M.K., et al. Role of tumor necrosis factor-alpha in lipopolysaccharide- induced pathologic alterations. *Am. J. Pathol.* 1990;136: 49-60
17. Beutler, B.A., Milsark, I.W. and Cerami, A. Cachectin/tumor necrosis factor: production, distribution, and metabolic fate in vivo. *J.Immunol.* 1985;135: 3972-3977
18. Foxwell, B. and Barrett, K. Introduction to cytokine receptors: structure and signal transduction. *Int.Rev.Exp.Pathol.* 1993;34 Pt B: 105-121
19. Shepherd, V.L. and Abdolrasulnia, R. Cytokine receptors. In: *Cytokines in Health and Disease.* Remick DG, Friedland JS, eds., 2nd ed. New York, New York: Marcel Dekker, Inc, 1997, pp. 263-280
20. Carter, D.B., Deibel, M.R., Jr., Dunn, C.J., et al. Purification, cloning, expression and biological characterization of an interleukin-1 receptor antagonist protein [see comments]. *Nature* 1990;344: 633-638
21. Tracey, K.J. and Cerami, A. Tumor necrosis factor: an updated review of its biology. *Crit.Care Med.* 1993;21: S415-S422
22. Girardin, E., Roux-Lombard, P., Grau, G.E., et al. Imbalance between tumour necrosis factor-alpha and soluble TNF receptor concentrations in severe meningococcaemia. The J5 Study Group. *Immunol.* 1992;76: 20-23
23. Chofflon, M., Juillard, C., Juillard, P., et al. Tumor necrosis factor alpha production as a possible predictor of relapse in patients with multiple sclerosis. *Eur.Cytokine.Netw.* 1992;3: 523-531
24. Mitov, I.G., Kropec, A., Benzing, A., et al. Differential cytokine production in stimulated blood cultures from intensive care patients with bacterial infections. *Infection* 1997;25: 206-12
25. Abraham, E. Why immunomodulatory therapies have not worked in sepsis. *Intensive Care Med.* 1999;25: 556-66
26. Davis, I.D. and Lotze, M.T. Cytokine Gene Therapy. In: *The Cytokine Handbook.* Thomson AW, ed.. Third ed. New York: Academic Press, 1998:823-853

Chapter 2

IMMUNOSENSORY SIGNALING: *ROLE OF CYTOKINES*

Lisa E. Goehler
Department of Psychology, University of Virginia, Charlottesville, Virginia

1. INTRODUCTION

Traditional conceptualizations of neural-immune interactions focused on mechanisms by which the brain modulates immune functions, for instance how stress, and/or personality variables can affect immune function. However, it is now well established that the converse is also true: immune activation, following infection or during neoplastic or autoimmune conditions, exerts profound influences upon brain function. Thus, neural-immune interactions are *bi-directional*.

This bi-directional relationship allows both the brain and immune system to respond to fluctuations or changes in concert to physiologically relevant immune parameters. Host defense is a homeostatic process, involving the detection of dangerous micro-organisms, and leading to the initiation of responses that eliminate the infection and restore a more optimal physiological state. In this way, the neuro-immune axis functions as a physiological subsystem (immunophysiology), comprised of *immunosensory signaling pathways* that activate networks of nuclei in the brain that are ultimately responsible for producing *physiological, neuroendocrine, and behavioral responses* that both support the efforts of the immune system and modulate its function. Importantly, immune influences on the brain include effects not previously associated with host defense, notably the regulation of cognition and affect. For example, low doses of an immune stimulant (in humans) and a subclinical bacterial infection (in mice) can produce changes in affective states and cognitive function or anxiety-like behavior, in the absence of overt illness symptoms (1,2). These findings suggest that subtle changes in immune functions, which are not necessarily perceived as illness, may induce or modulate changes in ongoing mental functions.

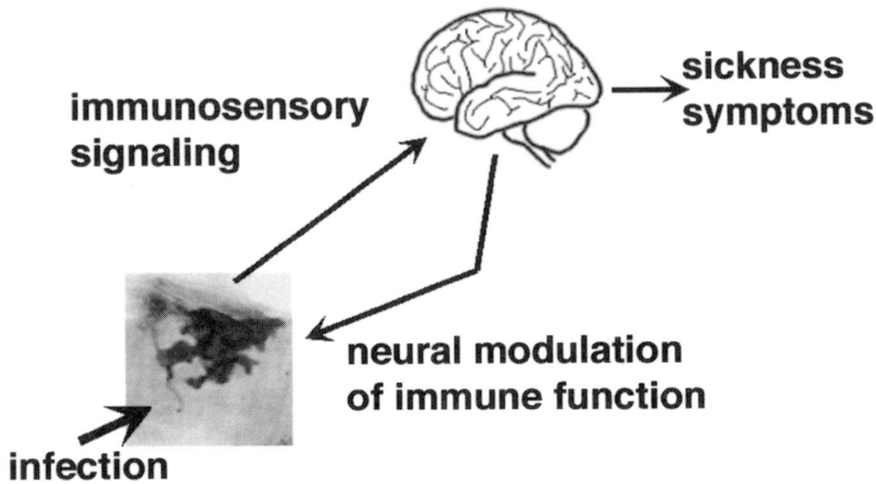

Figure 1. Schematic representation of bi-directional communication between the immune and nervous systems.

One of the most active areas of research into neural-immune interactions centers around the specific mechanisms by which immune activation can induce the constellation of illness, (also called brain-mediated acute phase) responses. These responses serve generally to support the efforts of the immune system and include activation of the hypothalamic-pituitary-adrenal (HPA) axis, increased sleep, and changes in cardiovascular and behavioral parameters. The question is: how is information about immune activation conveyed to the nervous system? Taken together, the findings from several lines of investigation into this issue support the idea that immunosensory pathways are organized in a manner similar to other viscerosensory systems, such as "satiety". The sensory signaling pathways (see below) share common features, and the brain neurocircuitry that processes viscerosensory and immunosensory information, and ultimately mediates the appropriate responses, overlap extensively, if not entirely (3).

Because the physical immunosensory stimuli (pathogens) are detected by mobile non-neuronal cells (immune cells), a critical issue concerns the mechanisms by which these immune cells interface with neural elements. Activated immune cells release a variety of chemical mediators that can act in both paracrine- and endocrine-like manners (4). These mediators serve both to initiate and modulate immune and brain responses necessary for effective defense. Of these, certain cytokines appear to play a major role in signaling the brain (5). Thus, cytokines form an important link between peripheral immune activation and brain processing leading to illness symptoms.

It is important to note that many cytokines are expressed within the brain, especially following brain infection, trauma, neoplastic disease or stroke (5). In addition, cytokines generated in the periphery induce the expression of cytokines in the brain during normal immune responses (6), and also participate in the regulation of homeostatic functions, such as sleep (7), under conditions not related to illness. However, the focus of this chapter principally concerns the organization of sensory signaling pathways activated during infection, and mechanisms by which peripherally generated cytokines can influence the brain.

2. CYTOKINES, IMMUNE STIMULI, AND THE RECEPTORS THAT DETECT THEM

Cytokines constitute a large group of protein molecules that are derived primarily from immune-type cells, including neuroglia (4,5). The members of this group perform a wide variety of functions, ranging from host defense to the regulation of normal homeostatic functions, to the regulation of growth, development and cell differentiation (4,5). A given cytokine may participate in more than one of these types of functions, and several cytokines contribute to each kind of function. Thus, attempts to organize cytokines into coherent categories that accurately reflect their biological actions can be extremely difficult, if possible at all. However, such conceptualizations can be useful, as long as their inherent limitations are recognized. The following represents a (highly) generalized description of features shared by cytokines implicated immunosensory signaling.

Cytokines most consistently implicated in the induction of brain mediated illness responses, interleukin-1 beta (IL-1β), IL-6, and tumor necrosis factor alpha (TNFα), belong to a group often called "pro-inflammatory" (4). These cytokines, along with two others, IL-2 and gamma interferon (IFNγ), are expressed early in immune activation, and serve to induce and co-ordinate cell-mediated host defense responses, and the concomitant inflammation. It is important to note that although "pro-inflammatory" cytokines seem to exert somewhat similar actions, they can be induced following different kinds of immune challenge. For instance, bacterial stimuli (see below) induce IL-1β and TNFα, whereas viral stimuli potently induce interferons (4). Similarly, stimuli that require the activation of T cells (see below) induce IL-2 , TNFα, and IFNγ. This differential expression of cytokines to different immune stimuli may support a tailoring of immune responses to specific pathogens. In addition, it seems that that these different cytokines induce slightly different patterns of brain responses, suggesting that immunosensory signaling may involve parallel pathways conveying information regarding type of pathogen, and that this information is relevant to the induction of different patterns of illness responses.

In addition to "pro-inflammatory" cytokines, another set of cytokines can be categorized as "anti-inflammatory", based on their ability to antagonize and limit the actions of pro-inflammatory cytokines. This is a critical function, because prolonged or unrestrained inflammation is extremely deleterious, and/or fatal. These cytokines, which include IL-4, IL-5, IL-10, IL-13 as well as beta interferon, are expressed slightly later in the course of immune activation, and in addition to opposing the actions of "pro-inflammatory" cytokines, tend to encourage the development of humoral immunity (8). Some of these cytokines, notably IL-4, IL-10, IL-13, and the IL-1 receptor antagonist (IL-1ra; an endogenous cytokine that binds IL-1 receptors but does not activate them) exert opposing effects on brain functions such as sleep and fever induced by pro-inflammatory cytokines such as IL-1 (5,9). In this way, cytokines homeostatically regulate brain - mediated illness responses. That is: pro-inflammatory cytokines signal peripheral immune activation and activate appropriate responses. These responses are then modulated and terminated by anti-inflammatory cytokines subsequently released.

2.1 Pathogen signals that induce cytokines: what features of pathogens do immune cells detect?

The immune system needs to be able to detect dangerous micro-organisms (or transformed human cells), and be able to distinguish between them and normal cells of our bodies. A major way that immune cells do this is by detecting chemical constituents of pathogens or abnormal cells that are unique to these kinds of cells (4,10,11,12). This endeavor is complicated by the fact that microorganisms replicate, and thus evolve, rapidly. It is as necessary for them to evade detection as it is for our immune system to detect them. Thus, the immune cells that function as sentinels or immunosensory receptor cells detect critical structural or metabolic components of micro-organisms that cannot be modified. So, for example, cell wall constituents such as proteoglycans (muramyl dipeptide is an example of one used experimentally) and lipopolysaccharides (LPS) provide evoluntionarily stable signals regarding gram positive and gram negative bacteria (12). The challenge for the immune system is to recognize cellular constituents found in pathogens, but not in ourselves. Mistakes can induce pathological conditions, such as the cell-mediated autoimmune disease, multiple sclerosis. Thus, mechanisms of pathogen detection are significant clinical interest.

2.2 Which immune cells function as immunosensory receptors?

The cytokines that appear to play important roles in activating and influencing the brain are derived from two major sources: immune sentinel cells of the monocyte/macrophage lineage and T lymphocytes (see below). These immunosesnory receptor cells can respond to somewhat different types of pathogenic stimuli, and contribute differently to the activation of peripheral immune and brain-mediated illness responses. Another important cell type, the mast cell, can respond to inflammatory signals and release neuroactive substances such as histamine, serotonin, and substance P, but these cells are not known as significant sources of cytokines and will not be discussed in depth here.

2.3 Immune sentinels release pro-inflamatory cytokines

Dendritic cells and macrophages are derived from circulating monocytes, and function as immune sentinels for the detection of infectious or inflammatory conditions (10,13,14,15), although their close relatives the microglia are key players in the initiation of responses to infection or injury in the brain. Thus, these cells provide the earliest signals derived from immune interaction with pathogens. Dendritic cells and macrophages are distributed throughout the body (16). They express receptors for bacterial and viral constituents, including CD14 (the best characterized LPS receptor), and TOLL-like receptors (TLRs; 11). TLRs are mammalian homologs of drosophila proteins that in adult flies mediate pathogen recognition (11). Interestingly, each member of the TOLL family in flies induces defense responses to specific types of pathogens. Humans express many more TLRs than do flies, and it is unknown if this apparent receptor selectivity obtains in vertebrates. However, expression of different, but related receptor proteins can provide a mechanism for the immune system to tailor its responses to specific types of pathogens (12).

When dendritic cells and macrophages are activated they release IL-1β and TNFα (10,16) both implicated in immunosensory signaling. In addition, dendritic cells are the most important immune cells for the activation (via antigen presentation) of T lymphocytes, and in this way they are able to signal the presence of pathogens directly via releasing cytokines and indirectly by activating T cells.

2.4 Activated T cells release pro-inflamatory cytokines

T cells are lymphoid cells that play a critical role in the orchestration of host defense. In addition to activating other cells to produce antibody (humoral defense) or directly destroy pathogens (cell mediated defense), T cells themselves (some subsets) can destroy infected cells (4,8). When activated via antigen presentation, T cells produce both TNF and interleukin-2 (IL-2), two cytokines implicated in immune to brain signaling.

T-cells are conventionally divided into three categories, based on the type of co-receptors they express, the types of cytokines they release, and their function (4,8). Cytotoxic T cells are able to recognize cells containing pathogens, and destroy them, whereas the other two types, Th1 and Th-2 are critical for the induction of cell-mediated and humoral types of defense responses, respectively. The two types of "helper" (hence the h) T cells are the most relevant source of cytokines that signal the brain.

Th1 cells are sometimes called "inflammatory" T cells, because when they are activated they release cytokines that activate macrophages and other cells that destroy infected cells (4). These cytokines also produce signs of inflammation, including swelling and redness. The characteristic Th1 type of cytokine pattern involves the release of TNF, IL-2, and interferons.

Whereas Th-1 cytokines activate cytotoxic cells (e.g. natural killer cells, cytoktoxic T cells, macrophages and neutrophils) that participate in the destruction of infected cells, Th-2 cells enable the activation of lymphocytes that express antibodies, which are critical components of humoral (blood-borne) responses (8). In this way the two types of immune responses co-ordinate to exert effective host defense. Like Th-1 cells, Th-2 cells preferentially release a characteristic pattern of cytokines: IL-4, IL-5, IL-10 and IL-13, that are often called "anti-inflammatory" (4), which are released somewhat later during immune activation that are the "pro-inflammatory" cytokines (8). A summary of immune cells and their characteristic cytokine expression is shown in Figure 2. It is important to note, however, that each type of T cell can be induced to release "uncharacteristic" cytokines (8). Nevertheless, the Th-1 and Th-2 categories provide a structure with which to conceptualize likely actions and outcomes of differential T cell activation. In addition, the pattern of cytokines generated during immune activation and/or inflammatory conditions has implications for the type of responses generated in the brain, and may control or modulate the type of affective or cognitive changes seen in these conditions.

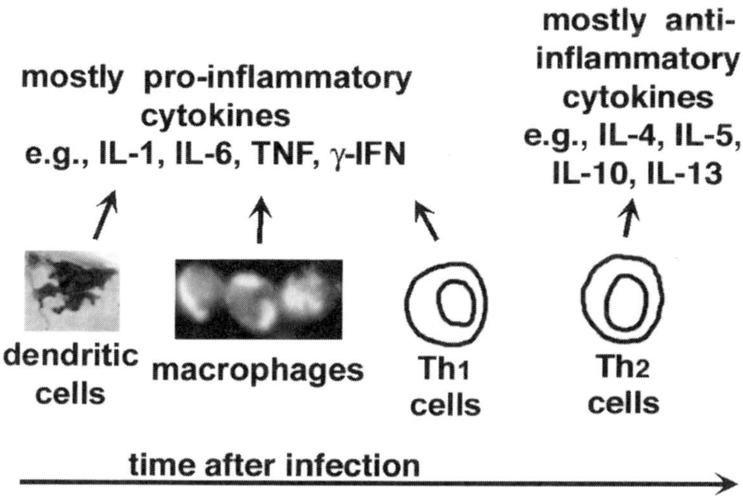

Figure 2. Schematic representation of various types of immune cells that both respond early to infections and produce cytokines relevant to immunosensory signaling.

2.5 Experimental approaches typically used to study immune effects on the brain

Studies investigating mechanisms by which the immune system influences brain function typically involve either the exogenous administration of immune stimulants or immune derived mediators such as cytokines, or inoculation of laboratory animals with live viral or bacterial cultures. In addition, local inflammation can be modeled using injections of immune stimulants injected into air pouches in the skin, or subcutaneously in a limb, such as the hindpaw.

Immune stimulants or cytokines are most commonly administered via three routes: intraperitoneal (i.p.; into the abdomen), intravenous (i.v.; into the systemic circulation), or intracerebroventricular (i.c.v.; into the fluid-filled spaces of the brain). The route of administration is usually chosen based on assumptions about the site of action conveyed by each route. For example, i.p. injections presumably stimulate receptors in the abdomen, whereas i.c.v. injections are presumed to stimulate receptors inside the brain. However, it is extremely important to recognize that although these routes of administration probably do lead to activation of receptors at these sites, substances injected either i.p. or i.c.v reach the systemic circulation fairly rapidly (17,18,19), and higher doses the concentration of these substances in the circulation can be substantial. Further, a major route by which i.p. and i.c.v. injections reach the systemic circulation is through the lymphatics (16,17,18), which are active immunosensory interfaces in their own right.

Thus, conclusions about the site of action of exogenously administered substances must be made very cautiously.

3. IMMUNOSENSORY INTERFACES: SPECIALIZED BRAIN TISSUES AND PERIPHERAL NERVES

The results from studies utilizing several different approaches have demonstrated two general types of immunosensory mechanisms. Because immune cells release cytokines into the systemic circulation, initial interest focused on how these cytokines exert their influences on the brain. This work has demonstrated that specialized tissues associated with specific areas of the brain likely serve as critical interfaces between cytokines (notably IL-1β, IL-6 and IFNγ) and the brain neurons that initiate illness responses. Because cytokines are relatively large, lipophobic proteins, they cannot diffuse passively into the brain. Consequently, most of the pathways identified to date that support direct action of peripherally generated cytokines on the brain involve interactions of cytokines with receptors on cells outside of the blood brain barrier. Such brain-associated immunosensory tissues include cells, such as endothelium and choroid plexus (3,15,20,21), associated with the blood brain barrier, as well as in specialized brain regions called circumventricular organs (CVO's; 22,23) and the ventricular ependyma (24). In addition, subsequent research has revealed a role for peripheral nerves as pathways by which cytokines may exert their effects on the brain (3).

3.1 Immunosensory signaling via meninges, choroid plexus and endothelium-

Cells associated with the blood-brain barrier provide a critical interface allowing circulating cytokines, as well as those released into the cerebrospinal fluid (CSF), to interact with the brain. Because these immunosensory brain barrier cells are not obviously innervated, they must, upon activation by cytokines, release diffusable mediators capable of interacting with neurons into the brain parenchyma. The most important of potential mediators are prostaglandins (see below). The meninges, choroid plexus, and endothelium each possess unique features that likely contribute to respond to different circumstances of immune activation.

3.1.1 Meninges and choroid plexus-

The meninges form the outer coverings of the brain, as well as provide the outer structure of sinuses and the fourth ventricle. This location ensures that they are in contact with cytokines or other immune mediators present in the CSF. Meninges contain immune cells, primarily dendritic cells and macrophages (Figure 3A). As described above, these cells can respond directly to pathogens, via their expression of receptors for pathogen products or structural components, or indirectly via their expression of receptors for cytokines (24,25). Studies have described receptors for both IL-1 and IL-6 on meningeal cells, and these cells respond to both central and peripheral immune activation with the expression of cytokines, notably IL-1 (27,28). The choroid plexus is best known for its role in the production of CSF, but it also contains abundant immune cells (Figure 3B). Like those within the meninges, these cells are primarily dendritic cells and macrophages, and express both cytokines and cytokine receptors (25,28). Choroid plexus is distributed throughout the ventricular system, and with the meninges, provide a significant contribution to the production of cytokines within the brain.

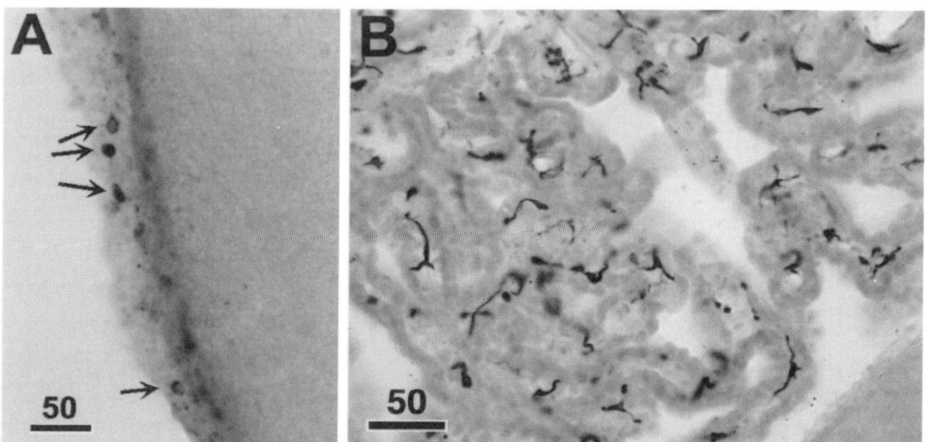

Figure 3. Macrophages and dendritic cells in meninges (A) and choroid plexus (B) express immunoreactivity for IL-1β after i.p. administration of lipopolysaccharide (400 μg/kg) and contribute to immunosensory signaling. In the meninges, macrophages predominate (arrows in A) whereas the choroid plexus is enriched with dendritic-type cells (B). Scale bars are calibrated in micrometers.

3.1.2 Endothelium-

Endothelium consists of the luminal lining of cells within blood vessels. Thus, endothelial cells are in contact with virtually every circulating substance, including those derived from immune activation such as cytokines. These cells have been reported to express a wide variety of substances and receptors, including receptors for IL-1, IL-6 and IFNγ (25,29) and they also express enzymes necessary for producing prostaglandins (29,30,31).

Prostaglandin E2 (PGE2) is strongly implicated in the mediation of several illness responses, including fever and activation of the HPA axis (23,32,30), based in part on the observations that drugs that prevent the synthesis of prostaglandins (e.g. indomethacin, ketorolac, and aspirin) prevent these responses to systemic cytokine treatment. Further, intracerebroventricular administration of PGE2 induces fever (33,34), and direct injection into the ventrolateral medulla activates the HPA axis (30). The actions of PGE2 are mediated by several receptor subtypes that are distributed selectively in different regions of the neuraxis (29, 35). These different receptor types may mediate specific effects of PGE2.

Compelling evidence indicates that PGE2 produced in endothelial cells mediates actions of circulating IL-1. Ek et al.(20,29) have shown that IL-1 receptors are co-distributed in endothelial cells with synthetic enzymes for PGE2. In addition, this co-distribution occurs in areas of the brain in which nearby neurons respond to IL-1 (as assessed by induction of activation markers), express PGE2 receptors (EP3 & 4), and have been previously implicated in the mediation of illness responses. These regions include the area postrema, nucleus of the solitary tract, and ventrolateral medulla of the caudal brainstem, as well as midline thalamic nuclei and the preoptic area in the forebrain (29) Thus, immunosensory endothelial cells that express IL-1 receptors can release PGE2 to into the adjacent brain parenchyma to activate neurons that mediate brain responses to immune activation. In this way, prostaglandins represent important diffusible substances releasing via the actions of cytokines circulating within the brain vasculature, which can then activate neurons that mediate illness responses such as HPA axis activation and fever.

3.1.3 Circumventricular organs-

Circumventicular organs [the area postrema, organum vasculosum lamina terminalis (OVLT), subfornical organ, and median eminence] are situated along the brain ventricles (hence the name) at different locations throughout the neuraxis. The blood brain barrier is weak within these structures, allowing access to circulating substances excluded from the brain parenchyma. Like the meninges and choroid plexus, CVOs contain numerous immune cells (Figure 4A,B). In addition, interspersed among the immune cells are neurons. Interestingly, immunosensitive dendritic cells appear to selectively contact specific neurons in the area postrema (Figure 4B). This arrangement suggests the possibility that immune cells in CVOs can respond to circulating substances not accessible to neurons in the brain parenchyma, signal adjacent neurons via a local, or paracrine kind of mechanism, and those neurons could then activate brain neurocircuitry subserving illness responses. This type of interaction would allow for a more specific kind of activation; that is activated immune cells could interface with specific neurons within the CVO that project to and activate specific regions of the brain. In this way CVOs could activate selected illness responses. An additional advantage of immune cells as receptors is that some cytokines (e.g. TNF) can induce apoptosis. Because immune cells are more easily regenerated than neurons, this arrangement would allow for a replaceable pool of receptor cells similar to those of other chemical senses, notably the olfactory and gustatory systems.

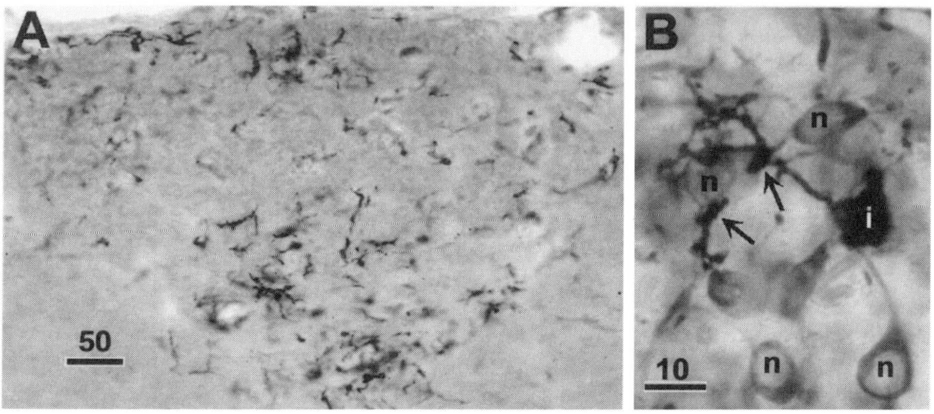

Figure 4. Immune cells in the area postrema, a circumventricular organ lining the caudal pole of the fourth ventricle in the lower brainstem. A: Dendritic-like cells express IL-1b immunoreactivity after i.p. administration of lipopolysaccharide (400 μg/kg). B Close association of dendritic-like immune cells and their processes (arrows) with neurons within the area postrema in an untreated rat. The immune cells (i) are stained for MHC class II immunoreactivity (black), whereas the neurons are stained for tyrosine hydroxylase (n, lighter grey). Scale bars are calibrated in micrometers

3.1.4 Ventricular ependyma-

The ventricular ependyma consists of cells lining the walls of the ventricular system. As such they are in contact with substances, including cytokines, present within the cerebrospinal fluid, anywhere in the brain or spinal cord. The cells possess cilia, a typical characteristic of sensory cells (e.g. hair cells, photoreceptors, taste cells and olfactory receptors). They express IL-1receptors (25), and respond to intracerebroventricular injection of LPS or IL-1β with the expression of c-Fos protein (24), an activation marker for many cells (Figure 5). These observations are consistent some kind of immunosensory role for these cells, but the mechanisms by which they interface with neural, or glial, elements is unknown. Interestingly, in catfish, ventricular ependymal cells of the fourth ventricle are coupled, either synaptically or via gap junctions, with vagal motor neurons (L. Goehler, unpublished observations). It is possible that in mammals these cells may also interact with neurons in structures adjacent to the ventricles.

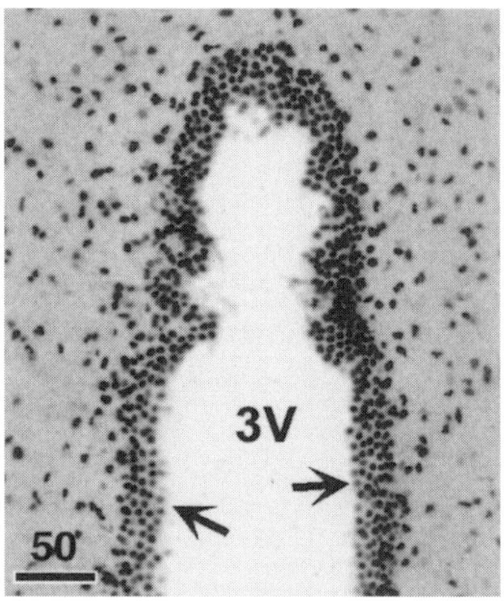

Figure 5. Ependymal cells lining the wall of the third ventricle (3V), indicated with arrows, are strongly stained for the protein c-Fos, an activation marker, sixty minutes following intracerebroventricular administration of IL-1β (50 ng). These cells are highly responsive to this cytokine.

3.1.5 Active transport of cytokines into the brain-

Whereas, as mentioned above, cytokines can not passively diffuse into the brain, evidence exists for an active transport mechanism, that at high concentrations of circulating cytokines carries them across the blood-brain barrier (36,37). Although circulating IL-1 or TNF may not reach concentrations for which transport might lead to significant brain activation, levels of IL-6 do (5). Thus active transport constitutes an additional mechanism for signal transduction of circulating signals, and because its relevance may be selective for certain cytokines, such as IL-6, may provide one mechanism by which different cytokines exert different effects on the brain.

3.2 Immunosensory signaling via peripheral nerves

Because pathogens multiply rapidly, it is important that host defense responses are generated promptly. Whereas detection by the brain is clearly critical in generating these responses to systemic or brain infection, the detection of infection locally at other sites of infection or inflammation is also a feature of immunosensory systems. To date, two cranial nerves have been demonstrated to induce brain-mediated illness responses following exposure to pathogen products or cytokines: the glossopharyngeal (38) and the vagus (3). Together these two nerves innervate the alimentary canal, an important point of entry for many pathogens, as well as many other important visceral tissues including lung and lymph nodes (3,39). In addition, the trigeminal nerve (fifth cranial nerve, that innervates nasal and oral cavities, and face) responds to LPS and IL-1 (40), although whether these responses leads induction of illness responses has not been reported. The contribution of other nerves to the induction of brain-mediated host defense responses is largely unexplored. Although many peripheral nerves have been shown to respond to cytokines or other immune mediators when damaged (indeed it is likely they all do) to date it has not been demonstrated that they convey immunosensory information to the brain. However it is possible that other nerves, notably other visceral nerves, or C fibers in somatic nerves, do in fact signal some features of inflammation in the tissues that they innervate. Alternatively, such signals may be conveyed to either the systemic circulation or the vagus nerve (see below) via the lymphatics. Clearly this is an issue for which more data is necessary in order to be resolved.

3.2.1 Glossopharyngeal nerve

The glossopharyngeal nerve (the ninth cranial nerve) innervates the posterior two-thirds of the tongue as well as other posterior oral structures.

Given that important immune structures such as the tonsils are located in this region, the glossopharyngeal nerve is well positioned for a role in immunosensory surveillance. In support of this idea, Romeo et al. (38) have recently shown that application of either LPS or IL-1 into the soft palate induces a fever that can be blocked by prior section of the glossopharyngeal nerve. Section of the glossopharyngeal nerve had no effect, however, on fevers induced by LPS or IL-1 injected into the peritoneal cavity (i.p.), showing that this nerve is indeed signaling local immune activation.

3.2.2 Vagus nerve

Like the glossopharyngeal nerve, the vagus nerve (the tenth cranial nerve) is well positioned to interact with pathogen products and cytokines. This nerve innervates nearly every internal structure (40), from the larynx to the colon and thus deserves its name (vagus means wanderer). In addition, like other viscerosensory nerves, the vagus innervates lymph nodes (3, 41,42,43). Thus, internal tissues commonly in contact with pathogens, notably the lungs, gastrointestinal tract, and liver are richly supplied with vagal afferents potentially capable of signaling immune activation in these tissues. Abundant immune type tissue and cells are found throughout the gastrointestinal tract, which is not surprising as this is a barrier site for infectious agents. Specialized immune tissue, including lymphoid nodules (similar to lymph nodes) and Peyers patches of the small intestine, reside directly beneath the epithelium. In addition, macrophages and dendritic cells line the epithelium, and overlie the Peyers patches (44). Vagal sensory neurons that innervate the submucosal and epithelial regions of the intestine are closely associated with a cell type described as possessing several long dendrite-like processes that may be dentritic cells (39). Thus these sensory neurons occupy a position in which they might be sensitive to cytokines produced by immune cells responding to local infection. In this way, vagal sensory fibers could be rapidly activated in response to pathogens in the gastrointestinal system.

In addition to the fortuitous location of sensory nerve fibers, the abdominal vagus contains abundant immune cells, including dendritic cells, macrophages, and mast cells (16). Numerous dendritic cells are interspersed themselves vagal nerve fibers, and within the paraganglia (specialized sensory structures that are innervated by vagal sensory neurons) their processes encircle adjacent chemosensory (glomus) cells (Figure 6). When treated with intraperitoneal LPS, these dendritic cells and mast cells express IL-1 (16), potentially providing a source of cytokine signal to vagal afferents that convey information about immune activation in the gut.

Experimental evidence supporting a role for the vagus in immunosensory signaling initially relied on studies that involved cutting the vagus nerve in the abdomen, below the diaphragm (subdiaphragmatic vagotomy). When animals recovered, they were challenged with different immune stimulants and the effects of the surgery on illness responses were observed. The results from these studies showed that vagotomy can block a wide range of illness

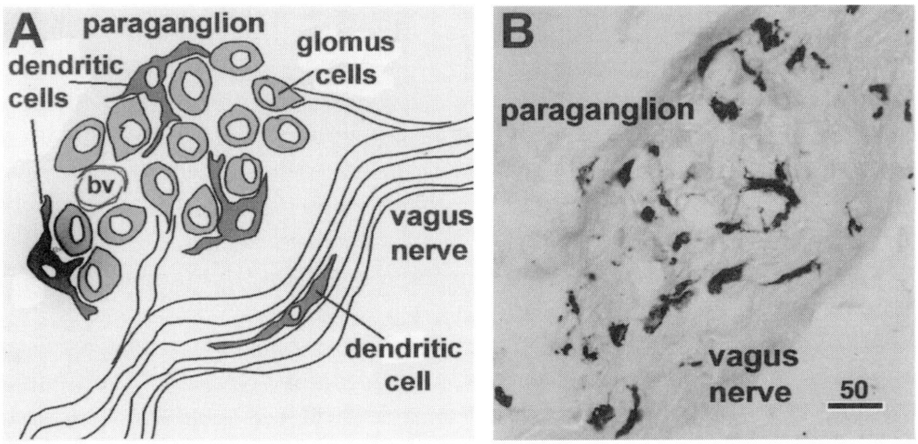

Figure 6. Abundance of immune cells in the abdominal vagus nerve and associated paraganglia, depicted both in a schematic drawing (A) and in a photomicrograph of a section (16 μm thickness) through these structures in an untreated rat. Dendritic-like immune cells, immunoreactive for MHC class II (black) are intricately associated with chemosensory glomus cells and nerves.

responses, including hyperalgesia (45,46), fever (47,48,49), hypersomnelence (50), HPA activation (51,52,53), conditioned taste aversion (54), and social withdrawal (55). The effects of vagotomy were most pronounced when the dose of stimulant was low (56,57). These findings suggest that the vagus nerve may contribute to the signaling of immune activity locally in visceral tissues (where the concentration of cytokines is likely to be low, at least early on during an infection), and that higher doses of immune stimulants such as cytokines recruit of the immunosensory pathways associated with the brain.

Although the results from vagotomy studies support a role for this nerve in immunosensory signaling, the conclusions from these studies are complicated by the fact that the vagus carries both sensory and motor nerve fibers. Thus, cutting the vagus may inhibit illness responses not because it interrupts sensory signaling but because it produces side effects or impairs immune functioning. These alternative explanations for the effects of vagotomy have, for the most part, been ruled out (58) Vagotomy does not affect the ability of animals to, for example, mount fevers (33,34). Vagotomized animals have been shown to respond identically to controls when the thermogenic stimulus is not associated with cytokine treatment. In addition, vagotomy does not impair either cytokine expression following LPS treatment nor does it impair the entry of cytokines or LPS into the systemic circulation (59). In fact, vagotomy blocks fever responses to low doses of i.p. injected IL-1, even when the injected IL-1 reaches the systemic circulation (60). Taken together, these findings support a role for vagal sensory transmission of cytokine signals.

In support of the idea that in fact, vagal sensory neurons signal immune activation, peripheral administration of IL-1 induces c-fos mRNA (61) and c-Fos protein (62) in vagal sensory neurons, and increases electrically recorded neural firing in hepatic vagal sensory fibers (63). Thus, like other sensory pathways, vagal immunosensation involves the activation of primary sensory neurons as an initial stimulus-to-nervous system interface. In addition, both i.p. and i.v. injections of LPS activate vagal sensory neurons (64) as does i.p. staphylococcus enterotoxin B (SEB; 3), a product of gram positive bacteria. Activation of illness responses by SEB is a T-cell mediated process (65) that likely takes place in lymph nodes, and leads to the relatively rapid (1-2 hours) induction of IL-2 and TNF (66). The finding that both LPS (T-cell independent) and SEB (T-cell dependent) induce activation in vagal sensory neurons demonstrate that this immunosensitive pathway likely carries information regarding immune activation to a wide variety of pathogens.

Vagal sensory neurons express receptors for immune-derived mediators, including IL-1 and prostaglandins (61). In addition, vagal sensory neurons contain one of the synthetic enzymes for producing prostaglandins, cyclooxegenase 1 (31), suggesting that one source of prostaglandins signaling immune activation (see above) may be vagal afferents.

In addition to activating vagal afferents directly, cytokines may activate vagal immunosensitive pathways via the chemoreceptive cells located in the vagal paraganglia (67) and/or similar, vagally innervated structures, the neuroepithelial bodies (68), which are found in lung airways. Specialized cells in the paraganglia called glomus cells, express binding sites for IL-1. Glomus cells are chemoreceptive when located in other similar tructures, (69), and are innervated by vagal afferents (40). Vagal paraganglia are penetrated by blood and lymph vessels, suggesting that these structures are likely monitoring substances circulating in body fluids. Because IL-1 immunoreactivity is induced in paraganglionic dendritic cells following immune challenge with bacterial endotoxin (16) that co-distribution with cells expressing IL-1 binding sites, it seems likely that vagal paraganglia provide a mechanism by which the vagus may monitor immune-related stimuli circulating in blood or lymph.

Sensory fibers associated with the vagal ganglia collect sensory signals from the tissues that they innervate, and then convey this information to brainstem relay nuclei: the nucleus of the solitary tract and the area postrema (which is also a circumventricular organ, see above). Following i.p. administration of either LPS or IL-1β, the vagal sensory neurotransmitter glutamate (70), is released into these structures (71), demonstrating that immune activation leads to functional activation of vagal sensory neurons. Both of these brainstem structures, the nucleus of the solitary tract and area postrema, express activation markers (.e.g. c-Fos) following i.p., i.v., and i.c.v. adminstration of immune stimulants, and can activate both local, protective reflexes, such as emesis and gastric retention, and activate brain neurocircuitry that mediate illness responses (3,24,30).

4. FURTHER ISSUES AND SUMMARY

Within the last ten years great progress has been made toward our understanding of mechanisms by which signals arising from peripheral infection or inflammation can access and influence the brain. There are still, however, a number of important issues yet to be resolved. For instance:

4.1 What is the relative contribution of each of the identified immunosensory pathways?

One striking feature of the organization of immunosensory signaling is the multiple, potentially parallel pathways by which immune mediators can activate the brain. In addition to several immunosensitive tissues associated with the brain itself, peripheral nerves also contribute. Why are there so many different pathways? An obvious answer would be that infection or inflammation can occur anywhere in the body and thus these different pathways are signaling something about the location of the immune activation. For instance vagal signals may be most relevant to signals arising from the viscera, whereas endothelial and CVO receptors signal systemic infection, and ependymal or meningeal receptors are tuned to brain insult or meningeal infection. Multiple pathways may also contribute to a phenomenon in other sensory systems termed "spatial summation" whereby recruitment of multiple receptors induce enhanced responding, and could in this system signal intensity, or severity of an infection. However, evidence for these ideas is lacking.

4.2 What is the relative contribution of each of the cytokines reported to influence the brain?

In addition to the multiple immunosensitive signaling pathways, an impressive array of cytokines have been reported to participate in either the activation or modulation of many brain-mediated illness responses. Why so many? Given that a somewhat unique pattern of cytokine expression is associated with different pathogens and different signal transduction mechanisms (e.g. viral vs. bacterial vs. parasites; T-cell dependent vs. independent), it seems possible that the different cytokines signal specific features of an infection (e.g. type of pathogen) and induce situation-specific illness responses. This idea awaits careful comparison of the effects of different pathogens and different cytokines on pattern of effects on the brain.

4.3 What is the array of cytokines that actually signal infection and activate illness responses?

Whereas numerous cytokines when administered exogenously affect particular brain functions, their roles in naturalistic infections are not well understood. For instance, which cytokines are involved in the initiation of illness responses? Reports vary regarding the types and concentrations of cytokines during natural infection. It is possible that differences and assay sensitivity across studies accounts for some variability, but the fact of the matter is that in non-experimental illness it is not possible to determine the earliest signals, as they occur prior to interaction with the medical community. Nonetheless, at least one criterion can be advanced in order to support the role of a particular cytokine in early signaling of immune activation: receptors for the cytokine must be constitutively expressed in immunosensory tissues. For a cytokine to effect the earliest signals, the receptor must be already present. Of the cytokines known to be released early in immune activation, constitutive receptors in immunosensory tissues have been reported for IL-1, IL-6, and IFNγ. (25,26,73,74)

However, other cytokines, such as alpha and beta interferons, IL-4, -10, and -13 also modulate specific brain functions, including sleep and fever. These cytokines seem to be induced slightly later in the course of immune activation (8), and serve to limit or modulate the inflammatory responses. Indeed, they have been shown to induce opposite effects on, for example, sleep and fever (75,76,77,78), from pro-inflamatory cytokines such as IL-1. Do these cytokines access the nervous system, either directly via interaction with their own receptors, or indirectly via the induction of other mediators? Or are they generated by neuroglia within the brain? Because they are generally expressed later in infection (8), "anti-inflammatory" or Th2 cytokines are unlikely to signal the presence, per se, of infection or inflammation. However, because they seem to be more sensitive to increased doses (8), they may provide signals relevant to behavioral changes such as social withdrawal and behavioral depression, seen either later during infection (or following immune stimulant administration) or at higher doses.

4.4 What is the role of peripherally generated cytokines in the regulation of affect and cognition?

It is becoming evident that changes in affect and cognition, notably learning and memory, comprise a normal component of illness responses. These changes may follow the induction of cytokines such as IL-1, within the brain or peripherally, as a normal response to infection or inflammation. It is also possible that affective and cognitive changes, especially anxiety and memory impairment, are a hallmark of early or subclinical stages of infection in peripheral tissues such as lung and gut.

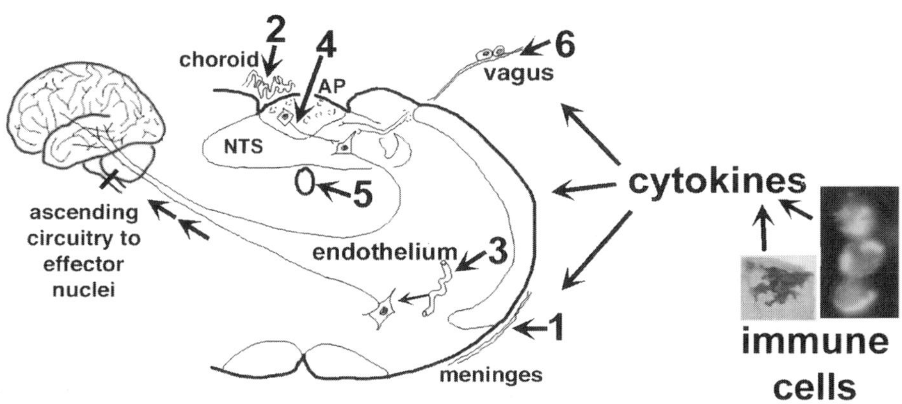

Figure 7. Mechanisms of immunosensory signaling. Cytokines released from immune cells can interact with the nervous system at the level of (1) meninges, (2) choroid plexus present in the brain's ventricular space, (3) brain endothelium, (4) the circumventricular organs, (5) the ependyma lining the ventricular space, and (6) primary afferents of the peripheral nerves such as the vagus. The schematic diagram depicts a cross-section of the lower brainstem (indicated with short thick bar in the human brain drawing) as an important site of "primary" immunosensory processing, which gives rise to engagement of effector systems for the mediation of sickness responses. However, immunosensory interfaces can also be found in other areas of the brain .

4.5 Summary

Immune cells, following interactions with pathogens or abnormal cells, release chemical mediators such as cytokines, which then interact with immunosensory tissues associated with the brain or peripheral nerves. Cytokines activate these tissues directly via specific receptors, or via diffusible mediators such as prostaglandins. Immunosensory structures responding to cytokines then activate brain neurocircuitry responsible for the induction of illness responses (Figure 7). In addition, peripherally generated cytokines may also utilize these pathways to influence brain function after illness symptoms have been induced, or during chronic inflammatory conditions.

REFERENCES

1. Reichenberg, A., R., Yirmiya, A., Schuld, T., et al. Cytokine-associated emotional and cognitive disturbances in humans. *Arch. Gen. Psychiatry* 2001; 58:445-452.
2. Lyte, M., Varcoe, J.J. and Bailey M.T. Anxiogenic effect of subclinical bacterial infection in mice in the absence of overt immune activation. *Physiol. Behav.* 1998; 65:63-69.
3. Goehler, L.E., Gaykema, R.P.A., Hansen, M.K., et al. Vagally mediated fever: a visceral chemosensory modality. *Auton. Neurosci.*, 2000; 85:49-59
4. Janeway, C.A. and Travers, P. *Immunobiology: The Immune System in Health and Disease.* London: Current Biology Ltd./Garland Publishing, 1997, 7:26.
5. Rothwell, N.J., Luheshi, G. and Toulmond, S. Cytokines and their receptors in the central nervous sytem: Physiology, pharmacology, and pathology. *Pharmacol. Ther.* 1996; 69:85-95.
6. Laye, S., Gheusi, G., Cremona, S., et al. Endogenous brain IL-1 mediates LPS-induced anorexia and hypothalamic cytokine expression. *Am. J. Physiol.* 2000; 279:R93-R98.
7. Krueger, J.M., Fang, J., Taishi, P., et al. Sleep: a physiologic role for IL-1β and TNF-a. *Ann. N.Y. Acad. Sci.* 1998; 856:148-159
8. Abbas, A.K., Murphy, K.M. and Sher, A. Functional diversity of helper T lymphocytes. *Nature*, 1996; 383:787-793.
9. Cartmell, T., Luheshi, G.N., Hopkins, S.J., et al. Role of endogenous interleukin-1 receptor antagonist in regulating fever induced by localised inflammation in the rat. *J. Physiol.* 2001; 531:171-180.
10. Banchereau ,J. and Steinman, R.M. Dendritic cells and the control of immunity, *Nature* 1998; 392:245-252.
11. Kopp, E.B. and Medzhitov, R. The Toll-receptor family and control of innate immunity. *Curr. Opin. Immunol.* 1999; 11:13-18.
12. Medzhitov, R. and Janeway, C.A. Innate immunity: impact of the adaptive immune response. *Curr. Opin. Immunol.* 1997; 9:4-9.
13. Reis e Sousa, C., Sher, A. and Kaye, P. The role of dendritic cells in the induction and regulation of immunity to microbial infection. *Curr. Opin. Immunol.* 1999; 11:392-399.
14. Steinman, R.M. The dendritic cell and its role in immunogenicity. *Ann. Rev. Immunol.* 1991; 9:271-296.

15. Van Dam, A.M.W., Brouns, M., Louisse, S., et al. Appearance of interleukin-1 in macrophages and ramified microglia in the brain of endotoxin-treated rats: a pathway for the induction of non-specific symptoms of sickness? *Brain Res.* 1992; 588:291-296.
16. Goehler, L.E., Gaykema, R.P.A., Nguyen, K.T., et al. Interleukin-1β in immune cells of the abdominal vagus nerve: an immune to nervous system link? *J. Neurosci.* 1999b; 17: 2799-2806.
17. Chen, G. and Reichlin, S. Clearance of [^{125}I]-tumot necrosis factor-a from the brain into the blood after intracerebroventricular injection in rats. *NeuroImmunoModulation* 1998; 5:261-269.
18. Dickstein, J.B., Moldovsky, H., Lue F.A., et al. Intracerebroventricular injection of TNF-α promotes sleep and is recovered in cervical lymph. *Am. J. Physiol.* 1999; 276:R1018-R1022.
19. Di Santo, E., Benigni, F., Agnello, D., et al. Peripheral effects of centrally administered interleukin-1b in mice in relation to its clearance from the brain into the blood and tissue distribution. *Neuroimmunomodulation* 1999; 6:300-304.
20. Ek, M., Engblom, D., Saha, S., et al. Inflammatory response: pathway across the blood-brain barrier. *Nature* 2001; 410:430-431.
21. Herkenham, M., Lee, H. Y. and Baker, R.A. Temporal and spatial patterns of c-fos mRNA induced by intravenous interleukin-1: A cascade of non-neuronal cellular activation at the blood brain barrier. *J. Comp. Neurol.* 1998; 400:175-196.
22. Lee, H.Y., Whiteside, M.B. and Herkenham, M. Area postrema removal abolishes stimulatory effects of intravenous interleukin-1β on hypothalamic-pituitary-adrenal axis activity and *c-fos* mRNA in the hypothalamic paraventricular nucleus. *Brain Res. Bull.* 1998; 46:495-503.
23. Saper, C.B. and Breder, C.D. The neurological basis of fever. *N. Engl. J. Med.* 1994; 330:1880-1886.
24. Day, H.E.W. and Akil, H. Differential pattern of c-fos mRNA in rat brain following central and systemic administration of interleukin-1-beta: Implications for mechanism of action. *Neuroendocrinol.* 1996; 63:207-218.
25. Ericsson, A., Kovacs, K. J. and Sawchenko, P.E. A functional neuroanatomical analysis of central pathways subserving the effects of interleukin-1 on stress-related neuroendocrine neurons. *J. Neurosci.* 1994; 14: 897-813.
26. Valleires, L. and Rivest, S. Regulation of the genes encoding interleukin-6, its receptor, and gp130 in the rat brain in response to the immune activator lipopolysaccharide and the pro-inflamatory cytokine interleukin-1beta. *J. Neurochem.* 1997; 69:1668-1683.
27. Vernet-der Gerabedian, B., Lemaigre-Dubreuil, Y. and Mariani, J. Central origin of IL-1β produced during peripheral inflammtion: role of meninges. *Mol. Brain Res.* 2000; 75:259-263
28. Quan, N., Stern, E.L., Whiteside, M.B., et al. Induction of pro-inflammatory cytokine mRNAs in the brain afterperipheral injection of subseptic doses of lipopolysaccharide in the rat. *J. Neuroimmunol.* 1999; 93: 72-80
29. Ek, M., Arias, C., Sawchenko, P., et al. Distribution of the EP3 prostaglandin E2 receptor subtype in the rat brain: relationship to sites of interleukin-1-induced cellular responsiveness. *J. Comp. Neurol.* 2000; 428:5-20.
30. Ericsson, A., Arias, C. and Sawchenko, P.E. Evidence for an intramedullary prostaglandin-dependent mechanism in the activation of stress-related neuroendocrine circuitry by intravenous interleukin-1. *J. Neurosci.* 1997; 17:7166-7179.
31. Matsumura, K., Kaihatsu, S., Imai, H., et al. Cyclooxygenase in the vagal afferents: is it involved in the brain prostaglandin response evoked by lipopolysaccharide? *Auton. Neurosci.* 2000; 85:88-92

32. Scammell, T.E., Elmquist, J.K., Griffin, J.D., et al. Ventromedial preoptic prostaglandin E2 activates fever-producing autonomic pathways. *J. Neurosci.* 1996; 16:6246-6254.
33. Milligan, E., McGorry, M.M., Fleshner, M., et al. Subdiaphragmatic vagotomy does not prevent fever following intracerebroventricular prostaglandin: further evidence for the importance of vagal afferents in immune-to-brain communication. *Brain Res.* 1997; 766:240-243
34. Sugimoto, N., Simons, C.T. and Romanovsky, A.A. Vagotomy does not affect thermal responsiveness to intrabrain prostaglandin E2 and cholecystokinin octapeptide. *Brain Res.* 1999; 844:157-163.
35. Zhang, J. and Rivest, S. A functional analysis of EP4 receptor-expressing neurons in mediating the action of prostaglandin E2 within specific nuclei of the brain in response to circulating interleukin-1β. *J. Neurochem.* 2000; 74:2134-2145.
36. Banks, W.A., Ortiz, L., Lotkin, S.R., et al. Human interleukin (IL) 1α, murine, IL-1 α, and murine IL-1β are transorted from the blood to the brain by a shared saturable mechanism. *J. Pharmacol. Exp. Therap.* 1994; 259:988-996.
37. Luheshi, G.N., Gay, J. and Rothwell, N.J. Circulating IL-6 is transported into the brain via a saturable transport mechanismin the rat. *Br. J. Pharm.* 1996; 111:146P.
38. Romeo, H.E., Tio, D.L., Rahman, S.U., et al. The glossopharyngeal nerve as a novel pathway in immune-to-brain communication: relevance to neuroimmune surveillance of the oral cavity. *J. Neuroimmunol.* 2001; 115:91-100.
39. Kobierski, L. A., Srivastava, S. and Borsook, D. Systemic lipopolysaccharide and interleukin-1β activate the interleukin 6: STAT intracellular signaling pathway in neurons of mouse trigeminal ganglion. *Neurosci. Lett.* 2000; 281:61-64
40. Berthoud, H.-R. and Neuhuber, W.L Functional and chemical anatomy of the afferent vagal system. *Autonom. Neurosci. Basic Clin.* 2000; 85:1-17.
41. Felton, D.L., Livnat, S., Felton, S.Y., et al. Sympathetic innervation of lymph nodes in mice. *Brain Res. Bull* 1984; 13:693-696.
42. Fink, T. and Weihe, E. Multiple neuropeptides in nerves supplying mammalian lymph nodes: messenger candidates for sensory and autonomic neuroimmunomodulation. *Neurosci. Lett.* 1988; 19:39-44.
43. Popper, P., Mantyh, C.R., Vigna, S.R., et al. The localization of sensory nerve fibers and receptor binding sites for sensory neuropeptides in canine lymph nodes. *Peptides* 1988; 9:257-267.
44. Reudl, C., Reiser, C., Bock, G., et al. Phenotypic and functional characterization of CD11c+ dendritic cell population in mouse Peyer's patches. *Eur. J. Immunol.* 1996; 26:1801-1806.
45. Watkins, L.R., Weirtelak, E.P., Goehler, L.E., et al. Characterization of cytokine induced hyperalgesia. *Brain Res.* 1994; 654:15-26
46. Watkins, L.R., Goehler, L.E., Relton, J., et al. Immune-to-brain communication: Systemic tumour necrosis factor-alpha (TNF-alpha) produces behavioral hyperalgesia via vagal afferents. *Brain Res.* 1995; 692:244-250.
47. Watkins, L.R., Goehler, L.E., Relton, J.K., et al. Blockade of interleukin-1-induced fever by subdiaphragmatic vagotomy: evidence for vagal mediation of immune-brain communication. *Neurosci. Lett.* 1995; 183:27-31.
48. Romanovsky, A.A., Simons, C.T., Szekely, M., et al. The vagus nerve in the thermoregulatory response to systemic inflammation. *Am. J. Physiol.* 1997; 273:R407-R413.
49. Sehic, E. and Blatteis, C.M. Blockade of lipopolysaccharide-induced fever by subdiaphragmatic vagotomy in guinea pigs. *Brain Res.* 1996; 726:160-166.

50. Hansen, M.K. and Kreuger, J.M. Subdiaphragmatic vagotomy blocks the sleep- and fever- promoting effects of interleukin-1β. *Am. J. Physiol.* 1997; 273:R1246-R1253.
51. Fleshner, M., Goehler, L.E., Hermann, J., et al. Interleukin 1β induced corticosterone elevation and hypothalamic NE depletion is vagally mediated. *Brain Res. Bull.* 1995; 37:605-610.
52. Gaykema, R.P.A., Dijkstra, I. and Tilders, F.J.H. Subdiaphragmatic vagotomy suppresses endotoxin-induced activation of the hypothalamic corticotropin-releasing hormones neurons and ACTH secretion. *Endocrinol.* 1995; 136:4717-4720.
53. Kapcala, L.P., He, J.R., Gao, Y., et al. Subdiaphragmatic vagotomy inhibits intra-abdominal interleukin-1β stimulation of adrenocorticotropin secretion. *Brain Res.* 1996; 728:247-254
54. Goehler, L.E., Busch, C.R., Tartaglia, N., et al. Blockade of cytokine induced conditioned taste aversion by subdiaphramatic vagotomy: further evidence for vagal mediation of immune-brain interactions. *Neurosci. Lett.* 1995; 185:163-166.
55. Bluthe, R.-M., Walter, V., Parnet, P., et al. Lipopolysaccharide induces sickness behavior in rats by a vagal mediated mechanism. *CR. Acad. Sci. Paris* 1994; 317:499-503.
56. Hansen, M.K., O'Conner, K.A., Goehler, L.E., et al. The role of the vagus nerve in interleukin-1β-induced fever is dependent on dose. *Am. J. Physiol.* 2001; 280:R929-R934.
57. Van Dam, A.M.W., Bol, J.G.J.M., Gaykema, R.P.A., et al. Vagotomy does not inhibit high dose LPS-induced interleukin-1β immunoreactivity in rat brain and pituitary gland. *Neurosc. Lett.* 2000; 285:169-172.
58. Romanovsky, A.A. Thermoregulatory manifestations of system inflammation: lessons from vagotomy. *Auton. Neurosci. Basic Clin.* 2000; 85:39-48.
59. Hansen, M.K., Nguyen, K.T., Fleshner, M., et al. Effects of vagotomy on circulating levels of endotoxin, pro-inflammatory cytokines, and corticosterone following intraperitoneal lipoplysaccharide. *Am. J. Physiol.* 2000; 278:R331-R336.
60. Gaykema, R.P.A., Goehler, L.E., Hansen, M.K., et al. Subdiaphragmatic vagotomy blocks interleukin-1β-induced fever but does not reduce interleukin-1β levels in the circulation. *Auton. Neurosci. Basic Clin.* 2000; 85:72-77.
61. Ek, M., Kurosawa, M., Lundeberg, T., et al. Activation of vagal afferents after intravenous injection of interleukin-1β: role of endogenous prostaglandins. *J. Neurosci.* 1998; 18:9471-9479.
62. Goehler, L.E., Gaykema, R.P.A., Hammack, S.E., et al. Interleukin-1 induces c-Fos immunoreactivity in primary afferent neurons of the vagus nerve. *Brain Res.* 1998; 804:306-310.
63. Niijima, A. The afferent discharges from sensors for interleukin-1β in the hepatoportal system in the anesthetized rat. *J. Auton. Nerv. Syst.* 1996; 61:287-291
64. Gaykema, R.P.A., Goehler, L.E., Tilders, F. J. H., et al. Bacterial endotoxin induces Fos immunoreactivity in primary afferent neurons of the vagus nerve. *NeuroImmunoMod.* 1998; 5:234-240.
65. Shurin, G., Shanks, N., Nelson, L., et al. Hypothalamic-pituitary-adrenal activation by the bacterial superantigen staphylococcal enterotoxin B: role of macrophages and T cells. *Neuroendocrinol.* 1997; 65:18-28.
66. Litton, M.J., Sander, B., Murphy, E., et al. Early expression of cytokines in lymph nodes after treatment in vivo with Staphyolococcus enterotoxin B. *J. Immunol. Meth.* 1994; 175:47-58.

67. Goehler, L.E., Relton, J.K., Dripps, D., et al. Vagal paraganglia bind biotinylated interleukin-1 receptor antagonist: A possible mechanism for immune-to-brain communication. *Brain Res. Bull.* 1997; 43:357-364.
68. Adriaensen, D., Timmermans, J.-P., Brouns, I., et al. Pulmonary intraepithelial vagal nodose afferent nerve terminals are confined to neuroepithelial bodies: an anterograde tracing and confocal microscopy study in adult rats. *Cell. Tiss. Res.* 1998; 293:395-405.
69. Matsuura, S. Chemoreceptor properties of glomus tissue found in the carotid region of the cat. *J. Physiol. Lond.* 1973; 235:57-73.
70. Schaffar, N., Roa, H., Kessler, J.P., et al. Immunohistochemical detection of glutamate in rat vagal sensory neurons. *Brain Res.* 1997; 778:302-308.
71. Mascarucci, P., Perego, C., Terrazzino, S., et al. Glutamate release in the nucleus tractus solitarius induced by peripheral lipopolysaccharide and interleukin-1β. *Neurosci.* 1998; 86:1285-1290.
72. Wan, W., Wetmore, L., Sorensen, C. M., et al. Neural and biochemical mediators of endotoxin and stress-induced c-fos expression in the rat brain. *Brain Res. Bull.* 1994; 34:7-14.
73. Robertson, B., Kong, G., Peng, Z., et al. Interferon-gamma-responsive neuronal sites in the normal rat brain: receptor protein distribution and cell activation revealed by Fos induction. *Brain Res. Bull.* 2000; 52:61-74.
74. Wei, Y.P., Kita, M., Shimura, K., et al. Expression of IFN-gamma in cerebrovascular endothelial cells from aged mice. *J. Interferon Cytokine Res.* 2000; 20:403-409.
75. Leon, L.R., Kozak, W., Rudolph, K., et al. An antipyretic role for interleukin-10 in LPS fever in mice. *Am. J. Physiol.* 1999; 276:R81-R89.
76. Kushikata, T., Fang, J., Wang, Y., et al. Interleukin-4 inhibits spontaneous sleep in rabbits. *Am. J. Physiol.* 1998; 275:R1185-R1191.
77. Kushikata, T., Fang, J. and Krueger, J.M. Interleukin-10 inhibits spontaneous sleep in rabbits. *J. Interferon Cytokine Res.* 1999; 19:1025-1030.
78. Kubota, T., Fang, J., Kushikata, T., et al. Interleukin-13 and transforming growth factor-beta1 inhibit spontaneous sleep in rabbits. *Am. J. Physiol.* 2000; 279:R786-R792.

Chapter 3

CYTOKINE PATHWAYS IN THE BRAIN

Julio Licinio and Ma-Li Wong
Laboratory of Pharmacogenomics, Neuropsychiatric Institute, Brain Research Institute, David Geffen School of Medicine at UCLA. Los Angeles, California

> "For a long time, ever since Hippocrates, it has been observed that every now and then, mental patients were healed or favourably influenced, when they were attacked by a fever."
>
> Presentation speech by Professor W. Wernstedt, Dean of the Royal Caroline Institute, at the award of the Nobel Prize in Physiology or Medicine to Julius Wagner-Jauregg in 1927.

1. INTRODUCTION

This book is focused on the role of cytokines in mental health and mental disorders. Work in this area has been conducted for over a century: The first Nobel Prize awarded to a psychiatrist was given in 1927 to Julius Wagner-Jauregg, who in 1887 was already systematically investigating the effects of febrile diseases on psychoses (1).

There has been in contemporary science a confluence of evidence indicating that elements of the immune system may affect key functions of the brain. The most obvious line of reasoning stems from the observations that when the immune system is activated (in conditions such as infection or inflammation) there may be clinical improvement in psychiatric patients, while in previously healthy individuals immune system activation causes behavioral changes that are reminiscent of the features of psychiatric disorders. Most of the work in this field was limited to observational and descriptive clinical studies until the 80's, when the genes encoding cytokines and their receptors were cloned (2-11). Such advances made it possible for our group and others to develop lines of investigation aimed at determining whether cytokine circuits exist within the brain. As an example of that type of work, we showed over 10 years ago that the genes encoding several cytokines where expressed in the brain (12-15). Specifically, we showed that gene encoding a key member of the interleukin 1 (IL-1) system, namely

IL-1 receptor antagonist (IL-1ra), is expressed in areas of the mammalian brain that are relevant to biological systems known to be dysregulated in depression (16). This led us to propose that cytokines synthesized within brain circuits relevant to psychiatric disorders might have a role in the biology of those disorders.

Genetic engineering methods made it possible to synthesize recombinant cytokines from various species, permitting the study of the effects of each cytokine on various organs and systems, including the brain and the central nervous system (CNS). Until then it had been difficult to study the effects of cytokines because hallmarks of cytokine biology are pleiotropism and redundancy. Studies conducted initially with endotoxin, and subsequently with specific recombinant cytokines showed that exogenous administration of inflammatory stimuli resulted in changes in mental function not only in experimental animals but also in humans who received cytokines as part of chemotherapy treatment. In 1996 Yirmiya showed that systemic endotoxin administration to rats caused reduction in the preference for and consumption of saccharine solutions and suppression of sexual behavior, which the author interpreted as evidence of anhedonia, a key feature of depression (17). Endotoxin also resulted in decreases of body weight, social interaction, and activity in the open-field test. These where likewise interpreted by the author to be reminiscent of features of depression. Moreover, chronic, but not acute, treatment with a tricyclic antidepressant suppressed such effects.

Robert Dantzer's group has demonstrated that systemic or central infusion of recombinant cytokines induces the full-blown syndrome of sickness behavior (18). Clinical research has shown that the administration of cytokines to humans during chemotherapy can result in symptoms of psychosis, depression, or cognitive dysfunction that are similar to the presentations of major psychiatric disorders. Those symptoms can be successfully treated with psychotropic medication (19-27).

Where is this field now? First, there is much misunderstanding about the role of inflammatory mediators in psychiatric disorders. A key concept in the field, which we advanced in 1997 (28), is that there are distinct central and peripheral cytokines compartments, which are integrated but differentially regulated. Cytokines acting in the periphery mediate the classic inflammatory response, which is characterized by the accumulation of fluid and white blood cells. In contrast, cytokines in the brain can be expressed in the absence of local inflammation and rather than promote accumulation of fluid and white blood cells, they trigger neuronal and neuroendocrine events that contribute favorably to the organism's overall response to an inflammatory challenge. Those include activation of the hypothalamic-pituitary-adrenal (HPA) axis, induction of sleep, fever, and suppression of locomotion, exploratory behavior, and sexual and reproductive function. Therefore, in sensu stricto, the expression of cytokines in the brain is not necessarily an indication of inflammation at that site; it is rather one element of a complex cascade that can be elicited (not only) by the response to inflammation, but which also involves complex networks of neural, endocrine, and behavioral responses that contribute to successful adaptation during the course of inflammation and infection.

We believe that in psychiatric disorders these central cytokine networks are activated. That does not imply that psychiatric disorders are caused by inflammation: central cytokines can be activated by stress, nutritional signals, neurodegeneration, ischemia, autoimmunity, infectious agents, and possibly by genetic predisposition. It has been proposed that central cytokine activation is not, in itself evidence of inflammation (29, 30). Therefore, documentation of brain cytokine activation in psychiatric disorders may not be indicative of inflammatory, autoimmune, or infectious etiologies. The ongoing investigation on the roles of immune activation and central cytokines in depression are not necessarily synonymous. In other words, activation of central cytokine circuits is not a sine qua non for an immune or autoimmune etiology for psychiatric disorders.

2. CENTRAL CYTOKINE CIRCUITS

2.1 Cytokine gene expression in the brain

2.1.1 Interleukin 1

How is a cytokine message transduced in the brain? Even though cytokines can be induced in the brain by a variety of mechanisms, the pathways for cytokine activation of CNS tissue have been dissected during the course of inflammatory stimuli, usually by eliciting an inflammatory response in the periphery. We showed in 1997 that peripheral or systemic inflammation causes induction of the expression of genes encoding inflammatory mediators in brain parenchyma, even though there is no histological evidence of inflammation within the brain. We showed that soon (2 h) after peritoneal lipopolyssacharide (LPS) administration there was induction of the RNA encoding IL-1ß (figure 1) (28) and iNOS (or NOS2) (inducible nitric oxide synthase or nitric oxide synthase type 2) (15) in areas of blood-brain interface. By 6 h after LPS injection there was expression of

IL-1ß and iNOS within brain parenchyma. In the periphery, induction of IL-1 is rapidly followed by induction of IL-1ra in levels that far exceed those of IL-1ß. This is necessary because a large excess of IL-1ra is required for its

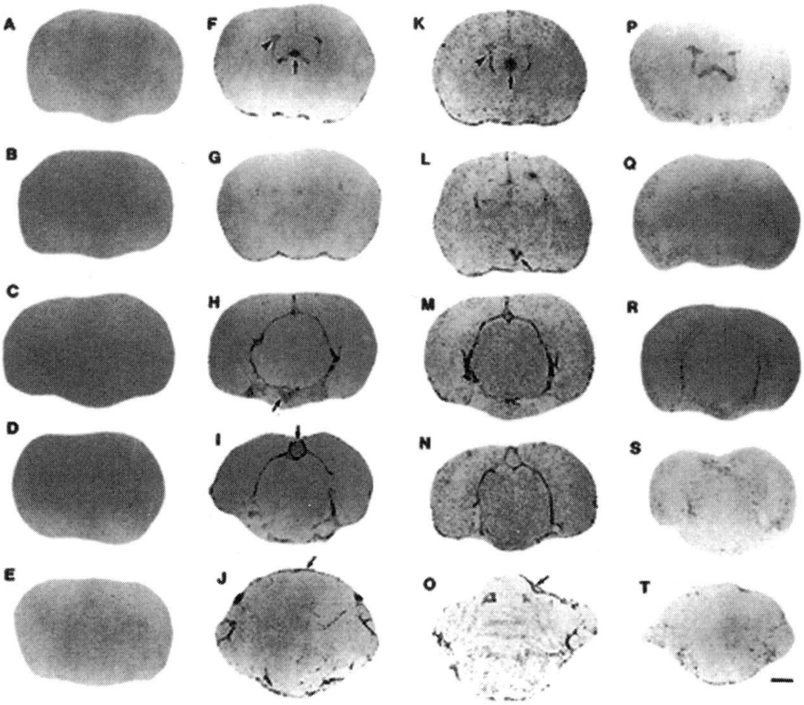

Figure 1. Localization of IL-1ß mRNA in the rat brain by ISHH after treatment with LPS. In Figs. 1, 2, 3, 4, 5 images are representative of findings in six animals/group. A series of film autoradiographs is arranged from rostral to caudal (top to bottom), showing the regional pattern of IL-1ß gene expression. Brain slices are shown in the first column (*A-E*) represent the hybridization of IL-1 antisense riboprobe in the brain of control rats, showing no detectable IL-1ß mRNA. Two hours after a single LPS injection i.p. (5.0 mg/animal), the induction of IL-1ß mRNA the brain is shown in the second column (*F-J*). There was induction of IL-1ß mRNA in the choroid plexus (arrowhead in *F*) and subfornical organ (arrow in *F*), posterior pituitary (arrow in *H*), pineal (arrow in *I*), and meninges (arrow in *J*). Six hours after a single LPS injection, the induction of IL-1 throughout the brain is shown in the third column (*K-O*). There was a remarkable induction of IL-1ß mRNA in the paraventricular nucleus of the hypothalamus (arrow in *L*); the induction in the choroid plexus (arrowhead in *K*), meninges (arrow in *O*), and in the subfornical organ (arrow in *K*) persists. Twenty-four hours after a single LPS injection the levels of IL-1ß mRNA throughout the rat brain were considerably decreased (fourth column, *P-T*). (Bar = 1.3 cm.) Reproduced from ref. (28) with permission.

receptor antagonist properties to be manifest. We found that the CNS, in contrast to peripheral tissues, exhibited IL-1ra (and IL-10) anti-inflammatory responses that were far more modest than those observed in peripheral tissues (28). This finding led us to develop the concept that the central and peripheral cytokine compartments are integrated, but differentially regulated. Our findings were subsequently replicated by Herkenham's group (31).

2.1.2 Tumor necrosis factor α (TNF-α)

Saper's group showed in 1994 that following peripheral LPS administration, TNF-α mRNA is induced in murine brain in patterns that were analogous to those that we later observed for IL-1ß. That consisted of initial expression in areas of blood-brain interface, such as perivascular cells and circumventricular organs, including the vascular organ of the lamina terminalis, median eminence, and area postrema, as well as the ventral surface of the medulla. Subsequently, TNF-α mRNA was in brain parenchyma, including neurons of in the pericircumventricular nuclei, such as the anteroventral periventricular and arcuate nuclei of the hypothalamus, and the nucleus of the solitary tract. Those authors proposed that TNF-α synthesized by a cascade of neurons within the brain may participate in the autonomic, neuroendocrine, metabolic, and behavioral responses to infection and inflammation.

Once it was established that peripheral cytokine induction caused central cytokine induction in widespread but discrete patterns, there was considerable interest to identify the secondary signals that promote the effects of brain cytokines, because neither IL-1 receptor gene expression nor IL-1 binding have been identified in significant levels in key areas that respond to IL-1ß.

2.2 Molecules that mediate cytokine signaling.

2.2.1 Nitric oxide

We hypothesized that perivascular nitric oxide (NO) might modulate the effects of systemic IL-1ß in the brain, and studied the expression of the genes encoding for IL-1ß, the signal-transducing IL-1 receptor type I (IL-1RI) and iNOS (or NOS2) constitutively and during systemic inflammation in vascular and perivascular regions of the rat brain (32). We showed that IL-1RI is constitutively expressed at the interface of the vascular wall and perivascular glia. During systemic inflammation there is induction of IL-1ß gene expression in the vascular wall, accompanied by perivascular induction of iNOS mRNA. We conclude that during systemic inflammation vascular IL-1ß, binding to vascular and perivascular IL-1RI receptors, may induce perivascular iNOS gene expression, leading to the production of NO and modulation of the effects of IL-1ß in the brain. Those data led us to suggest that the vascular and peri-vascular induction of iNOS mRNA by IL-1ß might represent a mechanism for the modulation of the central nervous system effects of peripheral inflammatory mediators. The idea that a signal originating from areas of blood-brain interface would disseminate into brain parenchyma was further supported by the findings from Herkenham's group showing that after intravenous injection of IL-1ß c-fos mRNA was induced initially in areas of blood-brain interface, with a general spread of activation directed away from the circumventricular organs over time (33).

2.2.2 Inhibitory factor κ Bα (IκBα)

Quan et al. studied IκBα mRNA expression in cells of the rat brain following intraperitoneal (i.p.) injection of LPS (31). IκBα controls the activity of nuclear factor κB (NF-κB), which is in turn a key regulatory component of intracellular signal pathways in cells of the immune system. This factor and related family of transcription factors responds to numerous immune challenges by activating gene transcription for a wide variety of immune signal molecules. The expression of IκBα mRNA parallels both the NF-κB activity and the duration of the activating extracellular stimulation. Quan et al. traced the activation of IκBα gene expression in the CNS after peripheral immune challenges to identify the extent and neuroanatomical location of the processes that are involved in responding to peripheral immune signals and, in turn, generating centrally derived immune molecules. The findings and identified pathways were very similar to those we had identified for iNOS (32): it starts in cells lining the blood side of the blood-brain barrier and progresses to cells inside brain (see figure 2). These findings further substantiated our hypothesis that cells of the blood-brain barrier synthesize immune signals to activate cells inside the CNS in response to peripheral inflammation.

2.2.3 Cyclooxygenase 2 (COX 2)

Ericsson et al. examined the means by which circulating IL-1 might access medullary aminergic neurons, including mechanisms involving sensory components of the vagus, the area postrema, or perivascular cells bearing IL-1 receptors. They showed that neither abdominal vagotomy nor area postrema lesions reliably altered c-fos expression induced in the

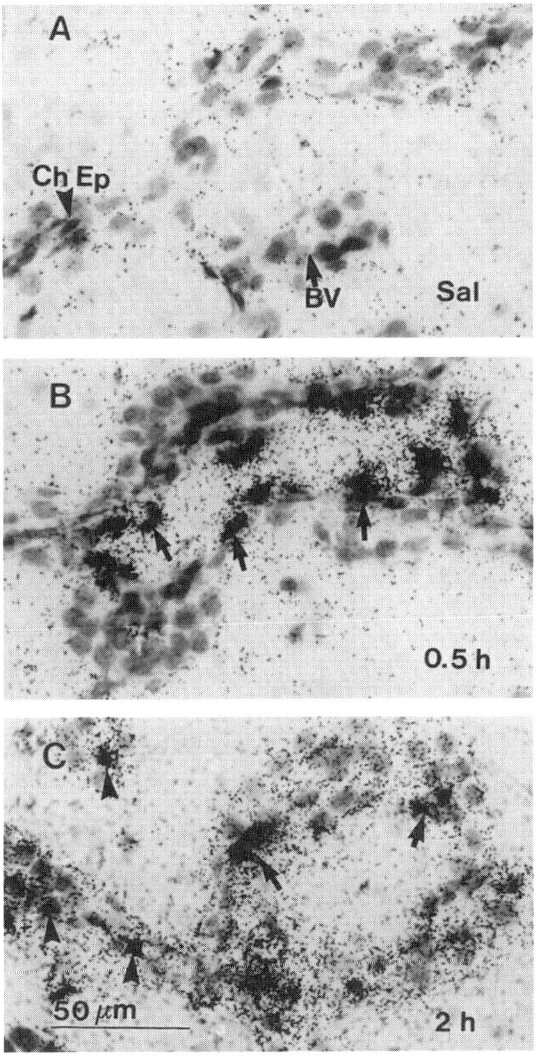

Figure 2. High-magnification bright-field photomicrographs show labeling of IκBα mRNA in the choroid plexus in animals sacrificed after saline injection (Sal; *A*) and at 0.5 hr (*B*) and 2 hr (*C*) after the LPS injection. Arrows point to cells of the blood vessels (BV), and arrowheads point to cells in the choroidal ependyma (Ch Ep). Reproduced from ref. (31) with permission.

medulla or the paraventricular nucleus of the hypothalamus (PVN) in response to a moderate dose of IL-1. Cytokine-stimulated increases in corticotropin-releasing hormone (CRH) mRNA in the PVN were also unaffected by either ablation. By contrast, systemic administration of the cyclooxygenase inhibitor indomethacin resulted in parallel dose-related attenuations of IL-1 effects in hypothalamus and medulla (see figure 3). Those effects include c-fos expression and CRH mRNA expression (in PVN). The authors also showed that microinjections of prostaglandin E2 (PGE2; 10 ng) in rostral ventrolateral medulla, the principal seat of IL-1-sensitive neurons that project to the PVN, provoked discrete patterns of cellular activation in hypothalamus and medulla that mimicked those seen in response to intravenous IL-1. They suggested that paracrine effects of PGE2 released from perivascular cells in the medulla as a consequence of IL-1

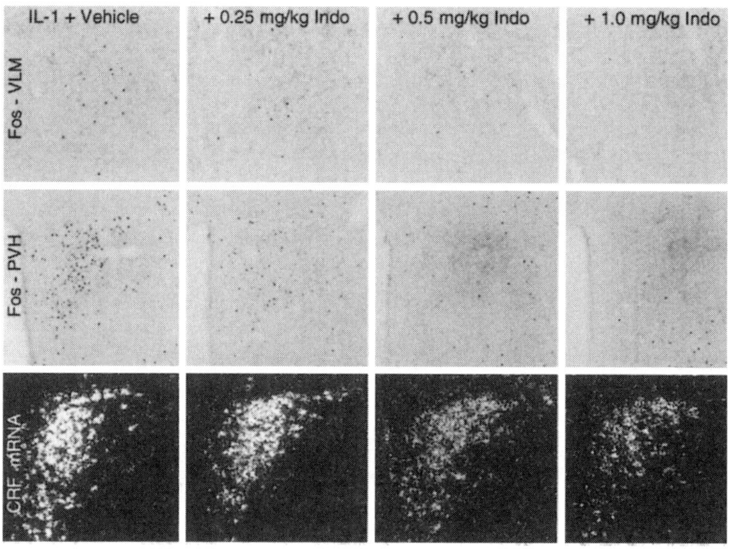

Figure 3. Effects of graded levels of prostaglandin synthesis blockade on IL-1-stimulated Fos-ir and CRF mRNA expression. Bright-field photomicrographs show sections through similar levels of the PVH (*middle*) and rostral ventrolateral medulla (*top*) stained for Fos-ir, and dark-field photomicrographs show PVH sections hybridized with probes for CRF mRNA (*bottom*) from animals pretreated intravenously with either vehicle of varying doses of indomethacin (*Indo*) 15 min before an intravenous challenge with 1.87 μg/kg IL-1. Indomethacin pretreatment produces a dose-related decrease in the expression of Fos-ir and CRF mRNA in the PVH and of Fos-ir in the region of the C1 catecholamine cell group. The patterns and strength of expression of both markers in the PVH and of Fos-ir in the VLM at the higher indomethacin doses are not distinguishable from those seen in rats injected with vehicle in lieu of IL-1. Note that Fos-ir expression in the regions immediately adjoining the PVH is ostensibly unaffected by any dosage of indomethacin. Magnification, 60X. Reproduced from ref. (41) with permission.

stimulation and, acting through prostanoid receptors on or near local aminergic neurons that project to the PVN, contribute to the stimulatory effects of increased circulating IL-1 on neurons constituting the central limb of the HPA axis.

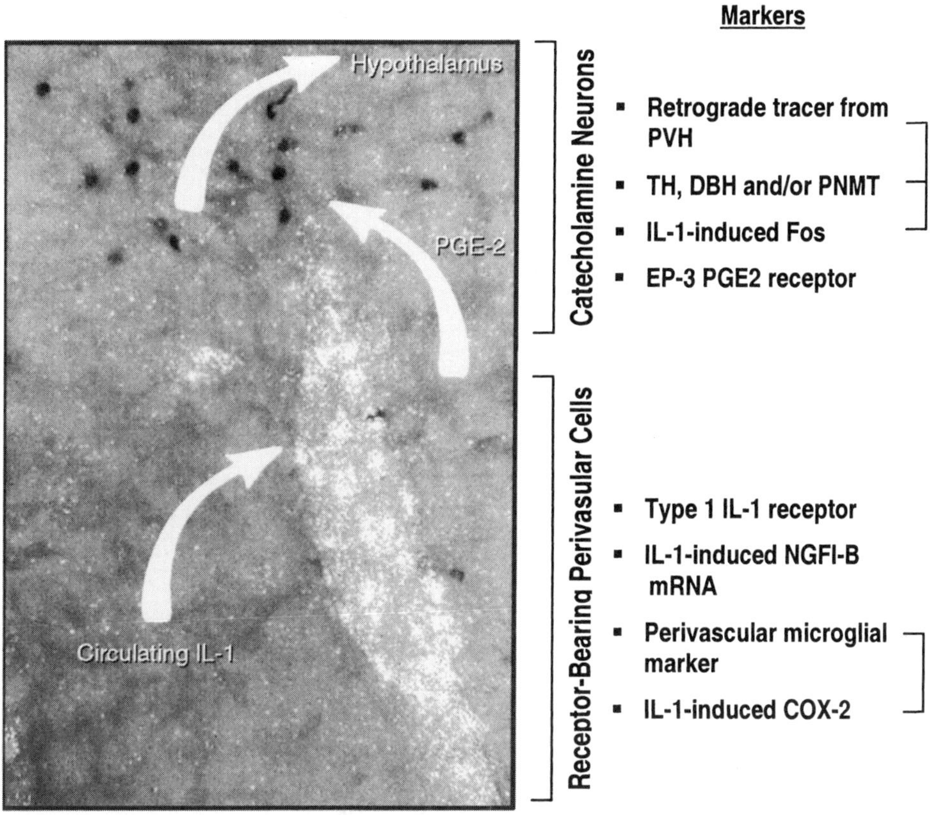

Figure 4. Possible mechanism for intravenous IL-1-mediated stimulation of central HPA control systems. A polarized epifluorescence illumination image of a section through the rostral ventrolateral medulla shows combined hybridization histochemical localization of perivascular cells displaying IL-1R1 mRNA and immunoperoxidase detection of nuclear Fos-ir in the region of the C1 catecholamine cell group. Ericsson *et al.* suggested that circulating IL-1 binds its cognate receptor on perivascular cells in the region, inducing them to synthesize PGE2, which in turn diffuses through the extracellular space to (directly or indirectly) stimulate nearby aminergic neurons and, consequently, CRF-expressing targets of their axonal projections in the endocrine hypothalamus. Listed at the *right* are markers of potentially relevant components of this signaling cascade that have been localized in the requisite regions, either under basal conditions or in response to intravenous IL-1 or endotoxin. Markers that have been co-localized to date are *bracketed.* It remains to be determined how the others are distributed with respect to the key perivascular (i.e., IL-1R1-expressing) and neuronal (IL-1-sensitive, hypothalamically projecting, and catecholaminergic) cell types. Reproduced from ref. (41) with permission.

This line of reasoning is summarized in figure 4 and is further supported by the work of Quan et al. who used in situ hybridization histochemistry following intraperitoneal or intravenous LPS administration to assess the COX 2 mRNA expression. Constitutive expression of COX 2 mRNA was found in neurons of cortex, hippocampus, and amygdala, but not in cells of the blood vessels. COX 2 mRNA levels were not altered in saline-injected animals as compared to non-injected controls. In LPS-injected animals, no consistent changes of neuronal COX 2 mRNA expression were observed. COX 2 mRNA expression was induced in cells closely associated with blood vessels; new labeling increased to a peak at 2 h and subsided gradually to basal levels by 24 h. The authors showed that initially, labeling was observed in cells comprising major surface-lying blood vessels and meninges. Later, vascular and perivascular cells associated with smaller penetrating blood vessels were labeled. These results support Ericsson et al.'s conclusion that peripheral LPS affects the neuronal activity of widespread brain regions by elevating the levels of prostaglandins (34).

3. RECENT ADVANCES: THE ROLE OF TOLL-LIKE RECEPTOR 4 IN BRAIN CYTOKINE PATHWAYS

A missing link for the transduction events linking cytokine bioactivity, and pro-inflamatory gene transcription in the brain is now attributed to the mammalian homologs of Toll. The Toll receptor was initially identified in Drosophila melanogaster. It is known that the cytokine-induced activation cascade of NF-κB in mammals and the activation of the morphogen dorsal in Drosophila embryos show striking structural and functional similarities (Toll/IL-1, Cactus/I-κB, and dorsal/NF-κB). Lemaitre et al. demonstrated that these parallels extended to the immune response of Drosophila, and that the Toll-receptor system has been shown to induce the innate immune response in adult Drosophila. First they showed the intracellular components of the dorsoventral signaling pathway and the extracellular Toll ligand, spatzle, control expression of the antifungal peptide gene drosomycin in adults. They also showed that mutations in the Toll signaling pathway dramatically reduced survival after fungal infection (35). Medzhitov et al. (36) discovered the human homologue of the Drosophila toll protein (Toll). Like Drosophila Toll, human Toll is a type I transmembrane protein with an extracellular domain consisting of a leucine-rich repeat (LRR) domain, and a cytoplasmic domain homologous to the cytoplasmic domain of the human IL-1 receptor (also called Toll/IL-1receptor motif). Both Drosophila Toll and the IL-1 receptor are known to signal through the NF-κB pathway. The authors show that a constitutively active mutant of human Toll transfected into human cell lines can induce the activation of NF-kB and the expression of NF-kB-controlled genes for the inflammatory cytokines IL-1, IL-6 and IL-8, as well as the expression of the co-stimulatory molecule B7.1, which is required for the activation of naive T cells. They conclude that a human homolog of the Drosophila Toll protein (Toll-like receptor, TLR) signals activation of adaptive immunity. Of the ten distinct TLRs, TLR4 appears to be the most crucial, because mice which are knockout for TLR4 (but not for other TLRs) are resistant to LPS (37, 38).

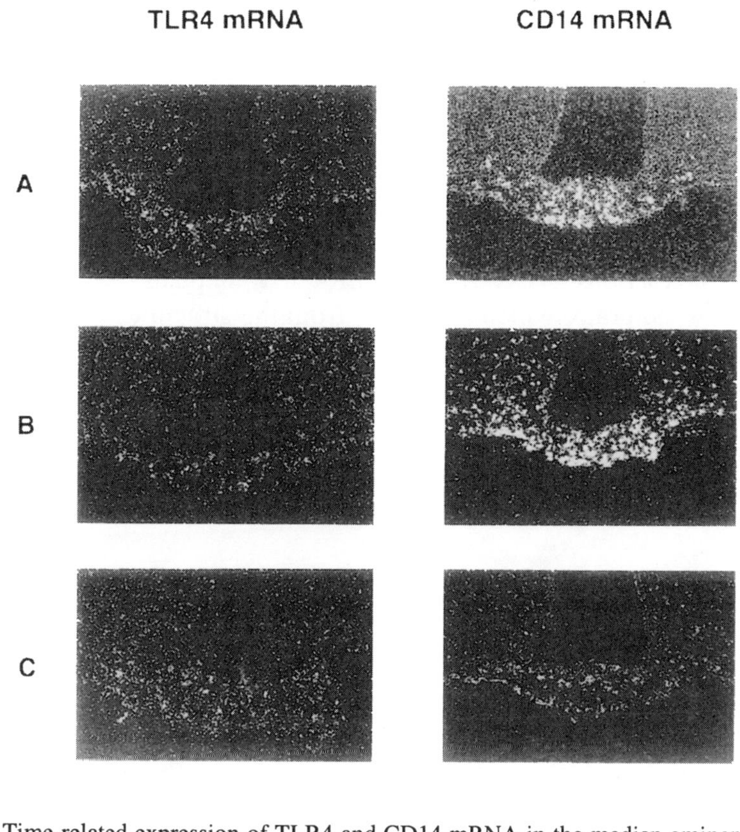

Figure 5. Time-related expression of TLR4 and CD14 mRNA in the median eminence (ME) and its surrounding parenchymal structures in response to a single i.v. bolus of lipopolysaccharide (LPS). These darkfield photomicrographs of 30 μm brain sections dipped into NTB-2 emulsion milk depict a robust CD14 signal that is localized within the ME 1 h after the LPS challenge (*A*), but gradually spreads over the anatomical boundaries of the organ at time 3 h postinjection (*B*). The signal for CD14 mRNA decreased 6 h after the treatment (*C*, right column), whereas such an expression pattern was not observed for the gene encoding TLR4. As shown by the left column, TLR4 hybridization signal was barely detectable in the ME of endotoxin-challenged rats. x25. Reproduced from ref. (39) with permission.

These findings led Laflamme & Rivest to investigate the interactions between TLR4 and CD14, a glycoprotein which serves as the membrane LPS receptor. Laflamme & Rivest cloned the rat TLR4 cDNA and used it as probe for in situ hybridization experiments that demonstrated TLR4 gene expression in the brain, in the following areas: leptomeninges, choroid plexus, subfornical organ, organum vasculosum of the lamina terminalis, median eminence, and area postrema (39). Scattered small cells also displayed a convincing hybridization signal within the brain parenchyma. Few well-defined nuclei exhibited positive TLR4 transcript: the supramamillary nucleus, cochlear nucleus, and the lateral reticular nucleus. The circumventricular organs, the leptomeninges, and choroids plexus also exhibited constitutive expression of the LPS receptor CD14. The authors examined the effects of peripheral LPS administration on TLR4 and CD14 gene expression. They found that in contrast to the marked up-regulation of the gene encoding CD14 during endotoxemia, neither LPS nor IL-1ß caused an increase in the TLR4 mRNA levels across the CNS (figure 5). A down-regulation of the gene encoding TLR4 was found in the cerebral tissue of immune-challenged animals. They concluded that the expression and possible interactions of CD14 and TLR4 may explain the innate immune response in the brain, which originates from the structures devoid of blood–brain barrier in presence of circulating LPS. Other toll-like receptors may be of relevance to brain-immune interactions. As an example, TLR2 is another toll-like receptor that is important in the transduction of signals initiated by bacterial peptidoglycan or lipopeptides, which are common constituents of both Gram-negative and Gram-positive bacteria.

4. DIRECTIONS FOR FUTURE WORK

Multiple cytokines (IL-1ß, TNF-α, IL-6, and others) trigger several signals such as nitric oxide, cyclooxygenase, and the NF-κB system to activate CNS circuits. This further confirms the redundancy of mechanisms and pathways by which cytokines activate the brain. Evidence reviewed here shows that cytokines affect CNS function through widespread, discrete, and redundant pathways that include gene expression of pro-inflammatory cytokines and molecules involved in signal transduction. This body of work was mostly conducted during the course of experimentally-induced endotoxemia. However, inflammation, infection and auto-immunity are only some of multiple pathways that can result in central cytokine activation. The pathways subserving other mechanisms for activation of cytokine pathways in the brain remain to be elucidated. These alternative mechanisms include stress, neurodegeneration, ischemia, genetic predisposition, and gene-environment interactions (e.g., nutritional factors, drugs). Further investigation in this field should dissect the pathways by which cytokines are stimulated by these other mechanisms and then compare and contrast those pathways with the ones that are utilized in the context of inflammation or infection. Such work would permit us to determine whether brain cytokine pathways are a non-specific marker of CNS pathology as suggested by Vitkovic (30, 40) or whether they may have a specific role in the fundamental biology of psychiatric disorders.

ACKNOWLEDGEMENTS

J.L. is supported by NIH grants GM61394, HL04526, DK58851, RR16996, RR017611, RR000865, HG002500, and by an award from the Dana Foundations. M.-L. W. is supported by NIH grants MH062777, RR000865, GM61394, HL04526, and an award from NARSAD.

REFERENCES

1. http://www.nobel.se/medicine/laureates/1927/press.html.
2. Fujita, T., Takaoka, C., Matsui, H., et al. Structure of the human interleukin 2 gene. *Proc Natl Acad Sci USA* 1983; 80: 7437-7441.
3. Mita, S., Maeda, S., Obaru, K., et al. Isolation and characterization of a human interleukin 2 gene. *Biochem Biophys Res Commun* 1983; 117: 114-121.
4. Maeda, S., Nishino, N., Obaru, K., et al. Cloning of interleukin 2 mRNAs from human tonsils. *Biochem Biophys Res Commun* 1983; 115: 1040-1047.
5. Devos, R., Plaetinck, G., Cheroutre, H., et al. Molecular cloning of human interleukin 2 cDNA and its expression in E. coli. *Nucleic Acids Res* 1983; 11: 4307-4323.
6. Taniguchi, T., Matsui, H., Fujita, T., et al. Structure and expression of a cloned cDNA for human interleukin-2. *Nature* 1983; 302: 305-310.
7. Degrave, W., Tavernier, J., Duerinck, F., et al. Cloning and structure of the human interleukin 2 chromosomal gene. *Embo J* 1983 ;2: 2349-2353.
8. Cosman, D., Cerretti, D.P., Larsen, A., et al. Cloning, sequence and expression of human interleukin-2 receptor. *Nature* 1984; 312: 768-771.
9. Fung, M.C., Hapel, A.J., Ymer, S., et al. Molecular cloning of cDNA for murine interleukin-3. *Nature* 1984; 307: 233-237.
10. Leonard, W.J., Depper, J.M., Kanehisa, M., et al. Structure of the human interleukin-2 receptor gene. *Science* 1985; 230: 633-639.
11. March, C.J., Mosley, B., Larsen, A., et al. Cloning, sequence and expression of two distinct human interleukin-1 complementary DNAs. *Nature* 1985; 315: 641-647.
12. Wong, M.L., Licinio, J. Localization of stem cell factor mRNA in adult rat hippocampus. *Neuroimmunomodulation* 1994; 1: 181-187.
13. Wong, M.L., Licinio, J. Localization of interleukin 1 type I receptor mRNA in rat brain. *Neuroimmunomodulation* 1994; 1: 110-115.
14. Wong, M.L., Bongiorno, P.B., Gold, P.W., et al. Localization of interleukin-1 beta converting enzyme mRNA in rat brain vasculature: evidence that the genes encoding

the interleukin-1 system are constitutively expressed in brain blood vessels. Pathophysiological implications. *Neuroimmunomodulation* 1995; 2: 141-148.

15. Wong, M.L., Rettori, V., Al-Shekhlee, A., et al. Inducible nitric oxide synthase gene expression in the brain during systemic inflammation. *Nature Med* 1996; 2: 581-584.
16. Licinio, J., Wong, M.L., Gold, P.W.. Localization of interleukin-1 receptor antagonist mRNA in rat brain. *Endocrinology* 1991; 129: 562-564.
17. Yirmiya, R. Endotoxin produces a depressive-like episode in rats. *Brain Res* 1996; 711: 163-174.
18. Dantzer, R. Cytokine-induced sickness behavior: mechanisms and implications. *Ann N Y Acad Sci* 2001; 933: 222-234.
19. Yirmiya, R., Pollak, Y., Morag, M., et al. Illness, cytokines, and depression. *Ann N Y Acad Sci* 2000; 917: 478-487.
20. Maes, M., Bonaccorso, S., Marino, V., et al. Treatment with interferon-alpha (IFN alpha) of hepatitis C patients induces lower serum dipeptidyl peptidase IV activity, which is related to IFN alpha-induced depressive and anxiety symptoms and immune activation. *Mol Psychiatry* 2001; 6: 475-480.
21. Valentine, A.D., Meyers, C.A., Kling, M.A., et al. Mood and cognitive side effects of interferon-alpha therapy. *Semin Oncol* 1998; 25: 39-47.
22. Licinio, J., Kling, M.A., Hauser, P. Cytokines and brain function: relevance to interferon-alpha-induced mood and cognitive changes. *Semin Oncol* 1998; 25: 30-38.
23. Neilley, L.K., Goodin, D.S., Goodkin, D.E., et al. Side effect profile of interferon beta-1b in MS: results of an open label trial. *Neurology* 1996; 46: 552-554.
24. Okada, F. Interferon-induced depression: just one of Bonhoeffer's exogene Reaktionstypen or a clue to understanding psychoimmunological aspects of depression? *J Mol Med* 1995; 73: 99-100.
25. Levenson, J.L., Fallon, H.J. Fluoxetine treatment of depression caused by interferon-alpha. *Am J Gastroenterol* 1993; 88: 760-1.
26. Musselman, D.L., Lawson, D.H., Gumnick, J.F., et al. Paroxetine for the prevention of depression induced by high-dose interferon alfa. *N Engl J Med* 2001; 344: 961-966.
27. Renault, P.F., Hoofnagle, J.H., Park, Y., et al. Psychiatric complications of long-term interferon alfa therapy. *Arch Intern Med* 1987; 147: 1577-1580.
28. Wong, M.L., Bongiorno, P.B., Rettori, V., et al. Interleukin (IL) 1beta, IL-1 receptor antagonist, IL-10, and IL-13 gene expression in the central nervous system and anterior pituitary during systemic inflammation: pathophysiological implications. *Proc Natl Acad Sci USA* 1997; 94: 227-232.
29. Licinio, J., Wong, M.L. The role of inflammatory mediators in the biology of major depression: central nervous system cytokines modulate the biological substrate of depressive symptoms, regulate stress-responsive systems, and contribute to neurotoxicity and neuroprotection. *Mol Psychiatry* 1999; 4: 317-327.
30. Vitkovic, L., Maeda, S., Sternberg, E. Anti-inflammatory cytokines: expression and action in the brain. *Neuroimmunomodulation* 2001; 9: 295-312.
31. Quan, N., Whiteside, M., Kim, L., et al. Induction of inhibitory factor kBa mRNA in the central nervous system after peripheral lipopolysaccharide administration: An in situ hybridization study in the rat. *Proc Natl Acad Sci USA* 1997; 94: 10985-10990.
32. Wong, M.L., Bongiorno, P.B., al-Shekhlee, A., et al. IL-1 beta, IL-1 receptor type I and iNOS gene expression in rat brain vasculature and perivascular areas. *Neuroreport* 1996; 7: 2445-2448.
33. Brady, L.S., Lynn, A.B., Herkenham, M., et al. Systemic interleukin-1 induces early and late patterns of c-fos mRNA expression in brain. *J Neurosci* 1994; 14: 4951-4964.

34. Quan, N., Whiteside, M., Herkenham, M. Cyclooxygenase 2 mRNA expression in rat brain after peripheral injection of lipopolysaccharide. *Brain Res* 1998; 802: 189-97.
35. Lemaitre, B., Nicolas, E., Michaut, L., et al. The dorsoventral regulatory gene cassette spatzle/Toll/cactus controls the potent antifungal response in Drosophila adults. *Cell* 1996; 86: 973-983.
36. Medzhitov, R., Preston-Hurlburt, P., Janeway, C.A., Jr. A human homologue of the Drosophila Toll protein signals activation of adaptive immunity. *Nature* 1997; 388: 394-397.
37. Takeuchi, O., Hoshino, K., Kawai, T., et al. Differential roles of TLR2 and TLR4 in recognition of gram-negative and gram-positive bacterial cell wall components. *Immunity* 1999; 11: 443-451.
38. Hoshino, K., Takeuchi, O., Kawai, T., et al. Cutting edge: Toll-like receptor 4 (TLR4)-deficient mice are hyporesponsive to lipopolysaccharide: evidence for TLR4 as the LPS gene product. *J Immunol* 1999; 162: 3749-3752.
39. Laflamme, N., Rivest, S. Toll-like receptor 4: the missing link of the cerebral innate immune response triggered by circulating gram-negative bacterial cell wall components. *FASEB J* 2001; 15: 155-163.
40. Vitkovic, L., Bockaert, J., Jacque, C. "Inflammatory" cytokines: neuromodulators in normal brain? *J Neurochem* 2000; 74: 457-471.
41. Ericsson, A., Arias, C., Sawchenko, P.E. Evidence for an intramedullary prostaglandin-dependent mechanism in the activation of stress-related neuroendocrine circuitry by intravenous interleukin-1. *J Neurosci* 1997; 17: 7166-7179.

Chapter 4

EFFECTS OF CYTOKINES ON CEREBRAL NEUROTRANSMISSION AND POTENTIAL RELATIONSHIPS TO FUNCTION

Adrian J. Dunn
Department of Pharmacology and Therapeutics, Louisiana State University Health Sciences Center, Shreveport, Louisiana

1. INTRODUCTION

A relationship between immune function and neurotransmission in the brain has been suspected for some time. In 1972, Pohorecky *et al.* reported that endotoxin (lipopolysaccharide, LPS) increased the turnover of norepinephrine (NE), but not serotonin (5-hydroxytryptamine, 5–HT) in the brain (1). Although LPS is a powerful stimulator of the immune system, it has many non-immune effects. A decade later, Besedovsky *et al.* showed that administration of sheep red blood cells (SRBC) to rats altered the turnover of NE in the hypothalamus (assessed by their disappearance following inhibition of synthesis) and that this response correlated with the presence of an immune response (2). Subsequently SRBC administration was shown to decrease the NE content of the hypothalamic paraventricular nucleus (PVN) and the 5-HT content of the hypothalamic paraventricular and supraoptic nuclei, but no changes were observed in the other hypothalamic or extrahypothalamic regions studied (3). These changes occurred only at the peak of the immune response to SRBC. The mechanism of these responses was unknown, but because immune activation was known to result in the synthesis and secretion of cytokines, it was postulated that they may function as immune cell-to-brain messengers.

Subsequent studies have confirmed that peripheral administration of purified cytokines can indeed affect brain neurochemistry, in particular the metabolism of neurotransmitters.

2. NEUROCHEMICAL RESPONSES TO IMMUNE ACTIVATION

2.1 Responses to Bacterial and Viral Infections

The literature contains several indications that plasma concentrations of glucocorticoids are elevated in sick animals. It has been a common observation in our own laboratory that animals sick from a variety of causes (mostly undetermined) exhibit elevations of plasma corticosterone. Bacterial infections have long been known to increase plasma concentration, indicating an activation of the hypothalamo-pituitary-adrenocortical (HPA) axis (4,5). More recently Smith *et al.* found that administration of Newcastle disease virus (NDV) to mice elevated plasma concentrations of corticosterone (6). We subsequently showed that NDV administration also elicited neurochemical changes. The NE catabolite, 3–methoxy,4–hydroxyphenylethyleneglycol (MHPG), the dopamine (DA) catabolite, 3,4–dihydroxyphenylacetic acid (DOPAC), and the 5-HT catabolite, 5-hydroxyindoleacetic acid (5-HIAA) were all increased in a number of brain regions (7,8). Free tryptophan was also elevated throughout the brain. The HPA and neurochemical changes lasted only a few hours.

Because NDV does not truly infect mice in the sense of producing live virus, we studied infection of mice with influenza virus. Infusion of the virus into the lungs (the natural site of influenza virus infection) induced a chronic elevation of plasma corticosterone (9), in contrast to the transient elevation seen with most commonly studied stressors. The changes in corticosterone were accompanied by neurochemical ones. MHPG and MHPG:NE ratios were elevated in all brain regions studied, but the magnitude of the response was greater in the hypothalamus than in the other brain regions (9). DOPAC and DOPAC:DA ratios, and homovanillic acid (HVA) and HVA:DA ratios were not significantly altered. Tryptophan concentrations were elevated in all regions studied, as were 5-HIAA and 5-HIAA:5-HT ratios. All of these changes appeared around 36 h after influenza virus infection and were present as long as the animals appeared sick.

Similar changes in plasma corticosterone have been associated with infection with other bacteria and viruses, such as *Mycoplasma fermentans* (10), Pichinde virus (11), Herpes virus (12), and lymphocytic choriomeningitis virus and murine cytomegalovirus (13) and changes in brain catecholamines and indoleamines. It seems likely that infections are rather generally associated with activation of the HPA axis, and activations of brain noradrenergic and serotonergic systems, and increases in brain tryptophan.

The similarity between the neurochemical and physiological responses to infections and other stressors commonly studied in the laboratory (e.g., footshock or restraint) is striking (see Table 1). In the CNS, a major stress-related response occurs in noradrenergic neurons, but responses occur in dopaminergic and serotonergic neurons too (14,15). The NE response is widespread and appears to affect both the locus coeruleus (A6) system which innervates dorsal structures, such as the cortex, hippocampus and cerebellum via the dorsal noradrenergic ascending bundle (DNAB), as well as the nucleus tractus solitarius (A1/A2) system innervating ventral structures, such as the hypothalamus, via the ventral noradrenergic ascending bundle (VNAB). The DA response is also widespread, such that all the major neuronal systems show responses (nigrostriatal, mesolimbic, mesocortical), but the magnitude of the response appears much greater in the prefrontal and cingulate cortices compared to other regions, like the neostriatum. There is also a robust elevation of tryptophan concentrations in all regions of the brain. This increase is quite uniform and does not appear to be related in any obvious way to the extent of the serotonergic innervation of a region (16,17).

	Electric Footshock	Viral Infection
HPA Axis (Plasma Corticosterone)	+++	+++
Sympathetic Nervous System	+++	+
Adrenal Medulla	+++	+
Brain Norepinephrine	++	+++
Brain Dopamine	+	0
Brain Tryptophan	++	++
Brain 5-Hydroxytryptamine	+	+

Table 1. Physiological responses to laboratory stressors and viral infection (immune stress).

+ indicates an increase in activity; +++ a large increase in activity; 0 no change. Modified from Ref. 154.

Nevertheless, the responses associated with what are sometimes regarded as "immune stressors" differ from those associated with physical and psychological stressors. The noradrenergic responses in the hypothalamus (the A1/A2, VNAB system) to infections and illness are much greater than in other brain regions, whereas the responses to footshock and restraint are relatively uniform on a regional basis. DA responses are not commonly associated with infections. When they occur they are meager and not focused on the prefrontal cortex as they are after most stressors studied. The chronicity of the HPA activation is also important, it is continuous not transient as it is to electric shock and other stressors commonly used in the laboratory.

2.2 Responses to Endotoxin (LPS)

The ability of LPS to activate the HPA axis has been known for half a century. Administration of high doses of LPS significantly decreased the brain content of NE, but did not alter that of 5-HT (1). This occurred whether the LPS was injected intraperitoneally (ip) or intracerebroventricularly (icv). LPS also accelerated the disappearance of icv administered [^{3}H]NE. In mice, low ip doses of LPS (e.g., 30 µg/kg) rapidly induce elevations of plasma ACTH and corticosterone reaching peaks around 2 h (18). MHPG concentrations and MHPG:NE ratios were also increased throughout the brain (18-22). As with influenza virus infections, the response was greatest in the hypothalamus, suggesting a relatively greater activation of the VNAB compared to the DNAB. LPS also induced small increases of DOPAC in most brain regions, including prefrontal cortex, hypothalamus and brain stem (18-23). The peak responses for both DA NE occurs around 2 h (18). Tryptophan and 5-HIAA were also increased, but in a regionally nonselective manner (18,19,22-24). However, these indoleamine responses reach a peak much later at around 8 h (18). Very similar changes were observed following icv LPS (18).

That these neurochemical changes reflect increased neurotransmitter release is suggested by *in vivo* microdialysis studies which have indicated increased extracellular concentrations of DA, DOPAC, NE, DHPG, MHPG and 5-HIAA in the medial prefrontal cortex and hypothalamus of rats following LPS administration (15 µg/kg ip) (25). Other studies have indicated increased 5-HT in microdialysates of the hippocampus (26,27), and of NE, MHPG, 5–HT and 5-HIAA from the preoptic area (28). LPS injection (100 µg ip) increased DA and 5–HIAA in microdialysates from the nucleus accumbens (29). Thus LPS administration induces a pattern of neuroendocrine and neurochemical responses quite similar to those described above for bacterial and viral infections, but with the addition of some changes in cerebral DA that are not regionally selective.

2.3 Responses to Cytokines

2.3.1 Interleukin-1 (IL-1)

2.3.1.1 Effects on Catecholamines

Peripheral administration of recombinant IL-1 substantially increased brain concentrations of MHPG and MHPG:NE ratios in mice (Figure 1)

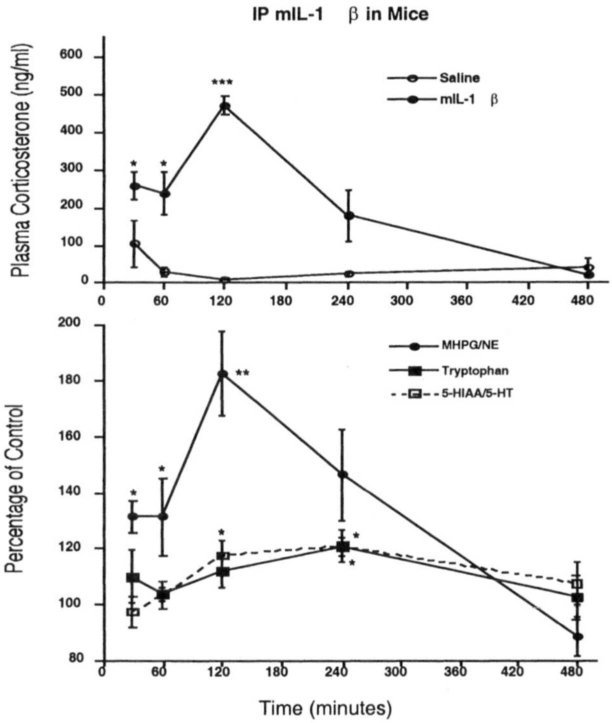

Figure 1. Time course of the neurochemical responses to IL-1β. Mice were injected ip with 100 ng mIL-1β. Upper panel - plasma corticosterone: open circles – saline; filled circles - mIL-1β. Lower panel - hypothalamic amines: filled circles – MHPG:NE ratios; filled squares – tryptophan; open squares – 5-HIAA:5-HT ratios. *Significantly different from saline-treated mice at the same time after injection (*$p < 0.01$,***$p < 0.01$)

(18,30,31) and rats (32). The responses in the hypothalamus were larger than in the cortex and other brain regions studied. Within the hypothalamus, the greatest response was found medially (30). The selectivity for the hypothalamus suggested a preferential activation of the ventral noradrenergic system which originates largely from NTS, although the medial hypothalamus also receives a small NE input from the locus coeruleus. An IL-1-induced increase in hypothalamic NE turnover was indicated by the accelerated loss of cerebral NE following inhibition of synthesis by metyrosine in rats (33). This effect was also greatest in the hypothalamus. IL-1 also decreased the hypothalamic content of NE in rats (34), suggesting that NE was turned over faster than it could be replaced by synthesis. The different forms of IL-1 (IL-1α and IL-1β) elicited similar responses as did IL-1 from different species (human, rat, mouse) consistent with the known affinities of these forms of IL-1 for the type I receptor (35). Similar noradrenergic responses are observed in response to intravenous (iv) (36) or subcutaneous injections, and to icv administration at considerably lower doses (18). These neurochemical effects of IL-1 are not due to contamination with endotoxin because heat-denatured IL-1 is inactive (30).

DA metabolism is not substantially affected by IL-1 as indicated by changes in DOPAC or HVA (30,32). However, Zalcman *et al.* reported increased DOPAC in mouse prefrontal cortex and certain other brain regions (31). We have occasionally observed statistically significant increases of DOPAC or DOPAC:DA ratios following ip IL-1, but such responses were unusual and not observed consistently. In the few experiments that showed increases in DOPAC, the anatomical pattern was not typical of responses to physical stressors, which exhibit in the medial prefrontal cortex much greater than in other brain regions. Zalcman *et al.* reported a similar anatomical pattern (31). Interestingly, administration of low doses of IL-1β to rats in the first few days of life, was reported to decrease DA permanently in the hypothalamus and superior cervical ganglion (37).

The effect of IL-1 on NE most likely reflects increased synaptic release, because *in vivo* microdialysis studies have indicated increased extracellular concentrations of NE in the medial hypothalamus (38-40). One study with push-pull cannulation also suggested increased hypothalamic PVN NE release (41). DA was not altered in dialysates from prefrontal cortex, nucleus accumbens or hippocampus by ip IL-1β (40). IL-1β was also reported to increase tyrosine hydroxylase activity in the median eminence of rats (42).

IL-1 also activates the adrenal medulla and the sympathetic nervous system, although this response is weaker than that induced by commonly used laboratory stressors (43-45). Intravenous IL-1β increases the electrical activity of vagal nerves innervating the thymus (46). NE turnover in the spleen and pancreas was also increased by ip IL-1β, but not by IL-6 (47,48).

NE release was also increased in a splenic dialysis study (49). In contrast, *in vitro* studies have shown IL-1β-induced decreases of NE release from the rat myenteric nerves (50), from the perfused spleen (51) and from the heart (52).

2.3.1.2 Effects on Tryptophan and Serotonin (5-HT)

Brain concentrations of the amino acid tryptophan, the natural precursor to 5-HT, are increased in all regions of mouse brain studied following peripheral IL-1 administration (Figure 1) (19,30). These changes are followed by increases in 5-HIAA, and are most easily observed by increases in 5-HIAA:5-HT ratios (Figure 1) (19,30,31). Increases in 5-HIAA in push-pull samples from the hypothalamus were detected after intrahypothalamic injection of IL-1β (53). Peripheral administration of IL-1 did not alter microdialysate concentrations of 5-HT from the hippocampus, but 5-HIAA was increased in nucleus accumbens and hippocampus (40). Icv IL–1β increased 5-HT in dialysates from the hippocampus (27). Studies with voltammetry found increases in 5-HIAA in the hypothalamus (54) and the dorsal raphe nucleus after peripheral or icv IL-1β (55).

The brain content of tryptophan is increased in response to a variety of stimuli, including a number of psychotropic drugs, increases in body temperature and many stressors (16,17,56,57). The experience in our laboratory has been that illness associated with infection, wounds or surgery or food or water deprivation are almost always associated with increases of brain tryptophan. These increases are often accompanied by increases in 5-HIAA. This change may be driven by the increased tryptophan because peripheral administration of high doses of tryptophan, increases brain tryptophan and 5-HIAA (58 and Dunn, unpublished observations), but it is not clear that the increased 5-HIAA necessarily reflects increased synaptic release (59,60). The increases in tryptophan and 5-HIAA in response to footshock, restraint, IL-1 and LPS appear to depend upon sympathetic activation, because they can be blocked by pretreatment with the ganglionic blocker, chlorisondamine (57). The stress-associated increases in tryptophan are largely prevented by the β-adrenergic receptor antagonist, propranolol, but not by the α–adrenergic receptor antagonist, phentolamine, or the muscarinic receptor antagonist, scopolamine (57). They are mimicked by β_2-adrenergic agonists (61,62).

The noradrenergic and indoleamine responses to IL-1 and LPS appear to involve different mechanisms. The noradrenergic response to ip IL-1 reaches a peak at around 2 hours paralleling the increases in plasma ACTH and corticosterone (Figure 1); whereas the peak responses in tryptophan and 5-HIAA appear later at around 4 hours (18). The peak noradrenergic responses to LPS are similar, around 2 hours, but the tryptophan and 5-HIAA responses occurred around 8 hours (18). Moreover, the anatomic patterns are distinct; the noradrenergic responses are far greater in the hypothalamus than in the cortex and brain stem, whereas the magnitudes of the increases in tryptophan and 5-HIAA are very similar from region to region. The NE and 5-HT responses can be dissociated. Mice of the endotoxin-resistant, C3H/HeJ, strain exhibit very small plasma corticosterone responses to LPS and no elevation of MHPG, but the increases in tryptophan and 5-HIAA are very similar to those in the control strains (63). The corticosterone and neurochemical responses to IL-1 are similar in both endotoxin-resistant and normal strains. Conversely, the nitric oxide synthase (NOS) inhibitor, L–NAME, can prevent the tryptophan and 5-HIAA response to both IL-1 and LPS, without obvious alterations in the plasma corticosterone and NE responses (64). Thus the NE and indoleamine responses appear to be activated by distinct mechanisms.

2.3.1.3 Effects on Acetylcholine, Histamine and the Amino Acids

Ip IL-1β decreased the secretion of acetylcholine from the hippocampus, as measured by microdialysis in freely moving rats (65). This effect occurred only at relatively high doses of 20 or 50 μg/kg. It is to be noted that iv IL-1β increased the activity of vagal nerves (46). Histamine turnover (assessed by accumulation following inhibition of degradation by the monoamine oxidase inhibitor, pargyline) was increased in the rat hypothalamus by 25 ng IL-1β icv (66), and increased release of histamine from the rat hypothalamus was observed following intrahypothalamic injection of IL-1β in a microdialysis study (67). Ip injection of hIL-1β (20 μg/kg, but not 10 μg/kg) decreased hippocampal concentrations of glutamate, glutamine and GABA one hour later (68).

2.3.1.4 Other Neurochemical Responses

Peripheral administration of IL-1 increases Fos protein in several brain regions (69-71). The most consistent and obvious effect is in the hypothalamic PVN, which contains the perikarya of neurons that synthesize the CRF involved in the activation of the HPA axis. Increases in Fos protein are also observed in the central amygdaloid nucleus, the medial preoptic area, the bed nucleus of the stria terminalis, and the NTS. The Fos response in the PVN appears to depend on activation by noradrenergic neurons because it was markedly attenuated in mice depleted of NE with 6-hydroxydopamine (6-OHDA) (72). Protein synthesis in the brain was altered by subcutaneous injection of hIL-1β in the rat (73). Such increases were observed in the subfornical organ, choroid plexus, medial habenula, dentate gyrus and the anterior and posterior lobes of the pituitary, whereas decreases were observed in the cingulate cortex and the pineal gland.

2.3.2 Interleukin-2 (IL-2)

IL-2 (200 ng ip) increased MHPG concentrations and MHPG:NE ratios in the hypothalamus of BALB/c mice 2 hours after injection (31). However, there were no changes in plasma corticosterone concentrations. DOPAC:DA ratios were also increased in the prefrontal cortex. A study from the same laboratory found that peripheral IL-2 reversed the decreases in brain NE and 5-HT in olfactory bulbectomized rats as well as the behavioral and immunological deficits (74). Surprisingly, peripheral IL-2 administration induced decreased DA in microdialysates from the nucleus accumbens (75). Icv IL-2 (500, but not 50 ng) was reported to increase 5-HT and 5-HIAA in hippocampal dialysates (76). Ip injection of human IL-2 (5 μg/kg) induced small increases in glutamine in the cortex and hippocampus of mice one hour later (68).

2.3.3 Interleukin-6 (IL-6)

Administration of IL-6 has frequently been reported to elevate plasma concentrations ACTH and corticosterone. Human IL-6 (200 ng, ip) injected into mice elevated 5-HIAA in the hippocampus, and 5-HIAA and DOPAC in the prefrontal cortex (31). In our laboratory, injection of mIL-6 into mice iv or ip elicited modest increases in plasma ACTH and corticosterone; such that maximal concentrations were much lower than observed with IL-1 (77). Corticosterone concentrations reached a peak 30-60 minutes after injection and returned to baseline by 2 hours. Iv or ip mouse IL-6 consistently elevated tryptophan in all brain regions at around 2 hours, and 5-HIAA in the brain stem at the same time, but had no effect on MHPG or DOPAC in any brain region (77). Icv mIL-6 had similar endocrine and neurochemical effects at lower doses. Our results are consistent with the lack of effect of IL-6 on NE turnover (using metyrosine to block catecholamine synthesis) in rats (33). The effect of IL-6 on serotonin was confirmed by a microdialysis study that showed increases in dialysate 5-HT and 5-HIAA from the hypothalamus of rats injected ip with rat IL-6 (2 μg per rat) (78). A similar response to rat IL-6 was observed in microdialysates from the striatum (79). This result is consistent with an amperometric study in which it was found that similar doses of IL-6 increased the apparent release of 5-HT from the striatum following stimulation of the dorsal raphe nucleus (79). These results strongly suggest an effect of IL-6 on cerebral serotonin metabolism. Because IL-1 and LPS administration both stimulate the synthesis and secretion of IL-6, it is possible that IL-6 mediates the indoleamine responses to IL-1 and LPS.

Changes in Fos protein have also been reported to peripheral administration of IL-6, but the results have varied. The PVN was affected in some studies (80), but not others (81). Other structures involved, include the central amygdala, the bed nucleus of stria terminalis, the superoptic nuclei, the NTS, and the superchiasmatic nucleus (82).

2.3.4 Tumor Necrosis Factor-α (TNFα)

The human forms of IL-1 and IL-6 are active on the respective cytokine receptors in the mouse and the rat so that the responses to the human and rodent cytokines are similar, but this is not the case for TNFα. Although amino acid sequence homology between mouse and human TNFα is 79% (83), mouse TNFα is a glycosylated dimer whereas hTNFα is not glycosylated (84). Human TNFα does not bind to mouse type 2 TNF receptors (mTNF-R2), whereas mouse TNFα binds to both mTNF-R1 and mTNF-R2 (85). hTNFα essentially lacks some actions of mTNFα in mice (86), so that physiologically relevant effects should only be studied using the homologous cytokines. In an early study, we observed no effect of injection of either hTNFα or mTNFα into mice on cerebral catecholamines or indoleamines (87). However, mTNFα elevated plasma corticosterone, whereas hTNFα did not. Consistent with this, Terao et al.(33) found no effect of hTNFα on NE turnover in rats. In a more recent study in mice, mTNFα increased brain MHPG and tryptophan, but only at relatively high doses (1 μg or more). Icv mTNFα (50 or 100 ng) did not alter hippocampal dialysate concentrations of 5-HT or 5-HIAA, but did elevate body temperature and plasma corticosterone (76). A recent report indicated that hTNFα (4 μg ip) increased hypothalamic MHPG in mice, and that chronic treatment sensitized this response (88).

By contrast, TNFα inhibited NE release from the median eminence (89). Similarly TNFα inhibited NE release from hippocampal slices (90). Interestingly, chronic treatment with the antidepressant, desmethyl-imipramine, reversed this effect, such that TNFα stimulated NE release. In the periphery, TNFα inhibited NE release from the rat myenteric plexus (91), and the mouse heart (52).

2.3.5 Interferons

No significant changes in plasma corticosterone or cerebral biogenic amines were observed following ip injection of human interferon-α (hIFNα:

1000 or 10000 Units, (87). Mouse IFNα administration at doses and times that were behaviorally active (92) did not result in significant changes in plasma corticosterone, or catecholamine or serotonin metabolism in any of five brain regions 2 h following ip administration of 400, 800 or 1600 Units (Leenhouwers, Crnic & Dunn, unpublished observations). However, chronic treatment of mice with interferon-γ increased brain concentrations of quinolinic acid and the activity of the enzyme indoleamine-2,3-dioxygenase (IDO), a key enzyme in indoleamine degradation (93). Peripheral administration of IFNγ to rats did not alter 5-HIAA in the dorsal raphe, but local administration increased 5-HT activity assessed by voltammetry (55). In cultured immune cells, IFNγ stimulates the activity of GTP cyclohydrolase, a critical enzyme for the synthesis of tetrahydrobiopterin (94). Tetrahydrobiopterin is an essential co-factor for NO synthase and also for the biosynthesis of catecholamines and serotonin.

2.3.6 Other Cytokines

Ip administration of granulocyte/macrophage colony-stimulating factor (GM-CSF) to rats (5 and 10 μg) significantly reduced hypothalamic glutamate, glutamine, aspartate and GABA, as well as NE and 5-HT, but not DA (95).

2.3.7 Effects of Intracerebral Administration of Cytokines

Many investigators have studied the responses to intracerebral administration of cytokines. Such studies have often been performed to demonstrate direct actions of cytokines on the brain. However, the results are difficult to interpret because of the tissue damage inherent in the introduction of the injectors, the microdialysis probes, the push-pull cannulae, and/or the electrodes used. The lesions induce macrophage and lymphocyte invasion, and microglial and astroglial proliferation. Needless to say there will be substantial local production of cytokines and other factors. This presumably elevates the basal concentrations of cytokines and creates the potential for interaction of the injected substances with any of the endogenous factors.

There are important implications for those who wish to promote an intracerebral function for IL-1. For example, Shintani *et al.* reported that anterior hypothalamic infusion of IL-1β increased microdialysate concentrations of DA, NE and 5-HT from the anterior hypothalamus (96). Subsequently, they showed that infusion of the IL-1-receptor antagonist (IL-1ra) into this same region attenuated the increases in plasma ACTH and microdialysate DA, NE and 5-HT in response to immobilization (97). They concluded that hypothalamic IL-1 plays a role in the HPA and neurochemical responses to the stressor. The problem is that the insertion of microdialysis probes creates a massive lesion and thus a site for the local production of cytokines, such as IL-1 and IL-6 as demonstrated by Woodroofe *et al.* (98). Thus the net effect on plasma ACTH and dialysate amines reflects the sum of the effects of the stressor and the local cytokine activity. Attenuation of the response to immobilization may indicate only the removal of the contribution of local IL-1, rather than an attenuation of the response to the stressor. In simple terms, the intracerebral antagonists may simply antagonize the artifacts of the cannulae and probe insertions. We have not observed any attenuation by icv IL-1ra of the neurochemical responses to footshock or restraint in mice (99).

3. MECHANISM(S) OF THE EFFECTS OF THE CYTOKINES

The foregoing has described a variety of effects of peripherally administered cytokines on brain neurochemistry. Cytokines are too large to readily pass the blood-brain barrier. They could act on one or other of the brain regions that lack a blood-brain barrier, the circumventricular organs, such as the median eminence, organum vasculosum laminae terminalis (OVLT), or the area postrema. Some researchers believe that IL-1, especially when administered iv, may act directly on CRF-containing terminals in the median eminence to initiate HPA axis activation (100). Local application of IL-1 in the median eminence elevates plasma ACTH and corticosterone in rats, and these responses were prevented by local administration of α- and β-adrenergic receptor antagonists (101) or indomethacin (102).

An alternative possibility is that cytokines may cross the blood-brain barrier using specific uptake systems. Banks *et al.* have demonstrated specific uptake of many cytokines from the blood to the brain (103). Uptake has been demonstrated for IL-1α, IL-1β, IL-1ra, IL-2, IL-6, and TNFα, as well as the soluble receptors for IL-1 and TNFα. However, the capacity of these uptake systems is quite low, and it is not known whether the uptake enables active concentrations of the cytokines to be attained in the appropriate brain regions (104).

An important mechanism for cytokine actions on the brain involves the vagus nerve, and possibly other neural afferents. Initially it was demonstrated that a subdiaphragmatic vagotomy prevented the cerebral Fos response to ip LPS (105). Subsequently, several groups demonstrated that vagal lesions could prevent the CNS effects of ip IL-1 or LPS on behavior (106,107). Subdiaphragmatic vagotomy also attenuated the HPA response to IL-1 and prevented the decreases in hypothalamic NE (34). Abdominal vagotomy attenuated the IL-1β-induced increase in hypothalamic dialysate NE (39; for a review, see 108). It is believed that IL-1 binds to receptors on paraganglion cells associated with the vagus (109). The vagal afferents terminate in the NTS in the brain stem, which also contains the cell bodies of the A1/A2 noradrenergic projection system, thus providing an obvious pathway for activation of the VNAB and thence the PVN.

Yet another possibility is that cytokines act on cerebral blood vessels. Receptors for IL–1 and LPS are known to be present on endothelial cells. These receptors appear to be coupled to cyclooxygenases which enable production of prostaglandins, leukotrienes, thromboxanes and other lipid mediators that can easily pass into and through the brain. Despite the remarkable sensitivity of the brain to IL-1, the brain contains very few receptors for IL-1. Binding sites for IL-1 have not been clearly demonstrated in the rat, and in the mouse they appear only in the hippocampus (110,111). mRNA for IL-1 receptors has also been difficult to demonstrate in the parenchyma of normal brain. However, IL-1-receptors are common in the capillary endothelium and choroid plexus. It is likely that these distinct mechanisms operate in parallel, so that when cytokines are administered, more than one mechanism operates.

The cyclooxygenase inhibitor, indomethacin, prevented the increases in MHPG (36) and NE turnover (33,112). Indomethacin also prevented the IL-1-induced increase in microdialysate NE from rat medial hypothalamus (Filippov & Dunn, unpublished observations). The opiate receptor antagonist, naloxone had no effect on NE turnover, but an antibody to CRF attenuated the response (33). However, we have observed normal neurochemical responses to ip mIL-1β in mice lacking the gene for CRF (Dunn, unpublished observations).

4. PARTICIPATION OF CYTOKINES IN THE RESPONSES TO LPS AND INFECTION

There are remarkable similarities in the neurochemical, behavioral and HPA responses to IL-1 and those to LPS and to bacterial and viral infections. Thus an obvious question is to what extent does IL-1 account for the HPA and neurochemical responses. Surprisingly, studies with antibodies to IL-1 and IL-1ra suggest that IL-1 is not critical for the responses to LPS (99,113). Although LPS is a potent stimulator of IL-1 production, little IL-1 appears in the first hour, and the peak response in the plasma is around 2 hours. Therefore, if IL-1 were the mediator of responses to LPS, the endocrine, neurochemical and behavioral responses should be delayed more than is actually observed (18,99). Studies with antibodies to IL-1 and with IL-1ra failed to show blockade of either the increases in plasma ACTH and corticosterone, the behavioral responses, or those in MHPG or tryptophan (87,99,113), although there was a trend towards an attenuation of the HPA activation at the later times (99) as might be expected from the time considerations mentioned above. These results are consistent with observations on the responses to LPS in IL-1β-knockout mice (114). Because LPS is a potent stimulator of IL-1β synthesis and secretion, it is unlikely that IL-1 does not contribute to the HPA, behavioral and neurochemical responses to LPS (99,113). However, IL-1β is clearly not the only factor involved.

Experiments with antibodies to IL-6 have suggested a role for IL-6 in the HPA responses to both the administration of LPS (115,116) and IL-1 (116). In our experiments, the sensitivity to IL-1 antibody was confined to the later phases of the HPA response, consistent with the delay necessary for the production and secretion of IL-6. Pretreatment with a neutralizing monoclonal antibody to mouse IL-6 also attenuated the increases in tryptophan and 5-HIAA following LPS administration to mice, but not that to IL-1. This suggests that IL-6 contributes only to the indoleamine responses to LPS, and is not involved in those to IL-1 (116). Treatment with a neutralizing antibody to TNFα also failed to prevent the HPA and neurochemical responses to LPS, even when supplemented with IL-1ra (87).

5. THE FUNCTIONAL SIGNIFICANCE OF THE NEUROCHEMICAL RESPONSES TO CYTOKINES

A critical question is whether any of the neurochemical responses observed following cytokine administration can be linked to a functional response. The most obvious candidates are the HPA responses, and various behavioral responses.

5.1 Neurotransmitter Involvement in the HPA Response to IL-1

NE and 5-HT have both been implicated in HPA activation (117), and are thus obvious candidates for mediating the responses to IL-1 (118). The responses in hypothalamic NE response and those in plasma ACTH and corticosterone are closely linked in time, and in our experiments, have been very highly correlated over a large number of experiments involving a large number of different manipulations. This association was confirmed in a microdialysis study in freely moving rats in which a very close temporal relationship between extracellular concentrations of NE in the hypothalamus and plasma concentrations of corticosterone was observed following both iv and ip IL-1β (38). The association between hypothalamic NE and HPA activation is also supported by some other studies. For example, when the noradrenergic neurotoxin 6-OHDA was injected into the VNAB or the PVN of rats, it resulted in depletions of PVN NE of 75% or more, and the plasma corticosterone response to ip IL-1 was largely prevented (119). A similar result was obtained using icv IL-1 in 6-OHDA-treated rats (120). Similarly, 6-OHDA treatment prevented the increase in plasma corticosterone following infection with Herpes simplex virus, whereas a lesion of serotonergic systems with 5,7-dihydroxytryptamine (5,7-DHT) did not (12). Surprisingly however, when whole brain NE was depleted by around 98% using 6-OHDA in mice, no impairment of the plasma corticosterone response to IL-1β was observed (72), and other unpublished observations). Also, α- and β-adrenergic receptor antagonists had little effect on the response to IL-1 in rats (100) and mice (121), and the α_1-adrenergic receptor antagonist, prazosin, attenuated the response in mice only at high doses (121). Thus it is likely that brain NE is involved in the HPA response to IL-1, but it does not appear to be essential.

The role of corticotropin-releasing factor (CRF) in the HPA response is clearer. The original studies on IL-1 action on the HPA axis, implicated CRF in this response, because an antibody to CRF largely prevented the increases in plasma ACTH and corticosterone to IL-1 (122-124) and IL-1 increased secretion of CRF into portal blood (123). Studies in CRF knockout mice showed that although several behavioral responses were normal, the plasma corticosterone response to IL-1 was dramatically reduced but not completely absent (125). This very small response to IL-1 could reflcct a limited direct effect of IL-1 on the pituitary or the adrenal cortex, but this is clearly not the major mechanism involved in the HPA response to IL-1.

5.2 Neurotransmitter Involvement in Behavioral Responses to Cytokines

Illness is associated with familiar alterations in behavior known as sickness behavior (126). Sickness behaviors include decreases in activity, in feeding, in exploration and sexual activity, and increases in sleep. Similar behaviors are elicited by IL-1 and LPS (113,126,127). We have studied the drinking of sweetened milk by mice as a simple model of sickness behavior (127). The reductions in ingestive behavior induced by IL-1 are largely prevented by cyclooxygenase inhibitors, such as aspirin, indomethacin and ibuprofen (128,129). However, such inhibitors are less effective on LPS-induced behaviors (127,129), and have only small effects on the behavioral changes associated with influenza virus infection (127). The responses appear to be largely mediated by the COX2 isozyme (129,130). However, COX1 is involved in the early phase of the response (131).

We have used the milk drinking model to study the potential involvement of cerebral NE and 5-HT. Both neurotransmitters are excellent candidates because they have previously been implicated in hypophagia (132). Norepinephrine appeared more likely to be involved because it responds more rapidly to IL-1 than serotonin, and the time course of the hypophagia paralleled its response more closely. Thus we tested the ability of adrenergic receptor antagonists to reverse the hypophagia. However, neither α_1-, α_2-, nor nonselective α- or β-adrenergic receptor antagonists, alone or in combination induced significant reductions in IL-1- or LPS-induced hypophagia in mice (133). Likewise, pretreatment with 6-OHDA or DSP-4 to deplete cerebral NE failed to alter the hypophagic response to IL-1 or LPS (133).

However, studies of other cytokine-induced behaviors have produced some evidence implicating noradrenergic mechanisms. For example, Ovadia *et al.* found that pretreatment with 6-OHDA, the β-adrenoreceptor blocker, propranolol, or the α_2-adrenoreceptor blocker, yohimbine, prevented IL-1-induced fever (134). Also, 6-OHDA pretreatment or prazosin prevented the antinociceptive effect of icv hIL-1α determined in the hot-plate test (135). These findings may reflect the different neural pathways involved in these responses.

Surprisingly, we have also failed to find any effects of cerebral 5-HT depletion (with 5,7-DHT) or pretreatment with a variety of 5-HT-receptor antagonists (136). However, this is consistent with the lack of effect of NOS to IL-1 and LPS (64), and the observation that administration of IL-6 which induces increases in brain tryptophan and 5-HIAA like those to IL-1 (77), yet IL-6 fails to alter milk drinking in mice (113). We also examined the role of histamine in the IL-1- and LPS-induced hypophagia. Pretreatment with histamine H_1, H_2 and H_3 antagonists and the histamine synthesis inhibitor, α-fluoromethylhistidine, all failed to alter the feeding responses to IL-1β (137).

A number of different peptides are known to affect ingestive behavior. A role for the endorphins can be excluded, because naloxone failed to alter the

IL-1- and LPS-induced reductions in milk intake. Similar findings were obtained for the substance P antagonist, L703,606, the (NP)Y1-receptor antagonist (BIBP 3226), and the MC4-receptor antagonist, SHU9119. CRF and urocortin do not appear to be involved, because high icv doses of the mixed antagonist, alpha-helical $CRF_{9\text{-}41}$ failed to alter the IL-1- and LPS-induced reductions in milk intake. Moreover, CRF knockout mice responded normally to IL-1 and LPS (138). However, Bianchi, et al. (135) found icv alpha-helical $CRF_{9\text{-}41}$ prevented the antinociceptive effects of icv IL-1 in the hot-plate test.

The effects of IL-6 on cerebral 5-HT are interesting in that mice lacking the gene for IL-6 (IL-6 knockout mice) show a high degree of aggressive behavior, whereas mice overexpressing IL-6 show more social interactions (139). IL-6 administration has no dramatic effects on behavior in mice, and does not reduce food intake (113). However, Zalcman, et al. (140) reported modest increases in locomotor activity, exploratory activity and grooming when hIL-6 was administered to mice. Whether or not these activities are related to the indoleamine changes is not known.

6. RELATIONSHIPS OF CYTOKINE-INDUCED EFFECTS ON NEUROTRANSMISSION TO MENTAL HEALTH

The potential relationships of the cytokine-induced changes in cerebral neurotransmission to mental health can only be speculated upon. The strongest links are between cytokines and depression, although schizophrenia has often been associated with immune abnormalities (141). Both norepinephrine and serotonin have been implicated in clinical depression, primarily because the major therapeutic drugs inhibit reuptake of NE or 5-HT or both. There is also some evidence for a hyperactivity of cerebral noradrenergic systems in depressed patients (e.g., 142) and substantial evidence that a significant majority of depressed patients hypersecrete cortisol (143). These facts suggest that IL-1 which activates the HPA axis, as well as cerebral NE and 5–HT metabolism is a potential mediator of depression.

Several authors have noted the similarities between sickness behavior and depression (144-146). Thus it has been argued that sickness behavior in animals may be a model for depression. Yirmiya showed that LPS administration to rats reduced their consumption of saccharin solution and argued that this represented anhedonia, like the anhedonia which is a hallmark of depression (146). In a creative experiment, he showed that chronic treatment with the antidepressant, imipramine, reversed the reduction of saccharin ingestion by LPS, reinforcing the concept that LPS treatment could model depression (146). The ability of antidepressant drugs to prevent the reduction in pleasurable activities by LPS has been replicated by some (147-149), but not others (147,150). Yirmiya, et al. (145) subsequently reported that the protective effect of antidepressants was evident only with LPS and not with IL-1. They speculated that the effect of the antidepressants was associated with LPS-induced IL-1 production. However, it is not easy to explain why inhibitors of neurotransmitter reuptake should exert such an effect. A study from another laboratory showed that chronic treatment of rats with the atypical antidepressant, tianeptine, attenuated sickness behavior induced by IL-1 or LPS, but only when these agents were administered peripherally, not icv (148). It has been postulated that the sickness behavior induced by peripheral IL-1 or LPS is mediated by induction of IL-1 centrally because icv IL-1 and LPS can induce some sickness behaviors and IL-1 antagonists can attenuate the effects of peripheral IL-1 and LPS (126). However, if cerebral IL-1 is responsible for the sickness behaviors, then tianeptine treatment should have attenuated the responses to icv IL-1, unless its site of action is peripheral which would conflict with most theories of antidepressant action.

The similarities between the responses to IL-1 and depression do not hold up well to close scrutiny. Although depression is frequently associated with decreased activity, anorexia, and loss of libido, it can also be associated with hyperactivity and over-eating. Proponents of the cytokine theory have pointed out that both depression and IL-1 administration can be associated with abnormalities in sleep (144,145,151). However, IL-1 increases the duration of slow-wave sleep (152), whereas by far the most common sleep abnormality in depressed patients is insomnia. Administration of cytokines to humans can induce some symptoms of sickness, but this is as true of IL-2 and the interferons as it is of IL-1. The interferons have not been shown to induce sickness behavior, or for that matter HPA or cerebral noradrenergic activation. Although elevated circulating concentrations of IL-1 have been reported in some depressed patients, this has not been a universal finding (153). Thus, cytokine administration and depression may share certain aspects, perhaps because they are both stress-like states. Cytokines may contribute to depressive symptoms, but are unlikely to be the major cause of depression.

7. SUMMARY

Administration of certain cytokines to animals can alter CNS neurotransmission. Table 2 summarizes the effects of cytokines reported on catecholamines and indoleamines, as well as those on the HPA axis (154).

Stimulus	Corticosterone	NE	DA	Tryptophan	5-HT
Influenza virus	+	+	0	+	+
LPS	+	+	+	+	+
IL-1α/IL-1β	+	+	0	+	+
IL-2	0	+	+	Nd	0
IL-6	+	0	0	+	+
TNFα	+	(+)	0	(+)	0
IFNα	0	0	0	0	0

Table 2. Comparison of HPA and brain neurochemical responses to viral infection, LPS and various cytokines

+ increased; 0 no change; (+) indicates increases only at the highest doses of TNFα (1 μg or more); nd not determined. Modified from Ref. 154.

Cytokine administration can also elicit a number of effects on the brain, including neuroendocrine and behavioral effects. The most well documented effect is the activation by IL-1 of the HPA axis, which is accompanied by a stimulation of hypothalamic NE metabolism, probably reflecting increased NE release. IL-1 also increases brain concentrations of tryptophan, and the metabolism of serotonin. IL-6 induces effects on tryptophan and 5-HT similar to those of IL-1. TNFα has effects on the HPA axis similar to those of IL-6, but affects NE and tryptophan only at high doses. Interferon α had no effects on the parameters studied. The effects of IL-1 are remarkably similar to those observed following administration of endotoxin and infections, such as influenza virus. They also resemble quite closely the responses that are observed to stressors commonly studied in laboratory animals, such as electric shock or restraint (Table 1). The major differences are that the NE response to shock or restraint is very uniform throughout the brain, whereas that to IL-1, LPS or infection is significantly greater in the hypothalamus. Also, responses of DA systems are normally observed to shock or restraint, with especially prominent responses in the limbic cortex, whereas DA responses are rarely observed in response to IL-1 and immune stimuli, and when they do occur, the response in the limbic cortex is similar to that in other brain regions. A small DA response occurs throughout the brain to peripheral administration of LPS. The neurochemical responses to cytokines are likely to underlie the endocrine and behavioral responses. The NE response to IL-1 appears to be related to the HPA activation. Neither the noradrenergic nor the serotonergic systems appear to be involved in the hypophagic responses. The significance of the indoleaminergic responses is not known.

ACKNOWLEDGEMENTS

The author's research described here was supported by the Office of Naval Research (N0001-4-85K-0300), the National Institute of Mental Health (MH46261) and the National Institute of Neurological Diseases and Stroke (NS25370). I am grateful for the excellent technical assistance of Bunney Powell, Sandra Vickers, Michael Adamkiewicz, Lynn Pittman-Cooley and Glenn Farrar, and the clerical assistance of Sharon Farrar, Karen Scott and Julie Droddy.

REFERENCES

1. Pohorecky, L.A., Wurtman, R.J., Taam, D. et al. Effects of endotoxin on monoamine metabolism in the rat. *Proc. Soc. Exptl. Biol. Med.* 1972; 140:739-46.
2. Besedovsky, H.O., del Rey, A.E., Sorkin, E. et al. The immune response evokes changes in brain noradrenergic neurons. *Science* 1983; 221:564-6.
3. Carlson, S.L., Felten, D.L., Livnat, S. et al. Alterations of monoamines in specific central autonomic nuclei following immunization in mice. *Brain Behav. Immun.* 1987; 1:52-63.
4. Beisel, W.R. "Alterations in hormone production and utilization during infection." In *Infections: The Physiologic and Metabolic Responses of the Host*, M.C. Powanda and P.G. Canonico, eds. Elsevier/North-Holland Biomedical Press, 1981, pp.147-72.
5. Kass, E.H. and Finland, M. Corticosteroids and infections. *Adv. Intl. Med.* 1958; 9:45-80.
6. Smith, E.M., Meyer, W.J. and Blalock, J.E. Virus-induced corticosterone in hypophysectomized mice: a possible lymphoid adrenal axis. *Science* 1982; 218:1311-2.
7. Dunn, A.J., Powell, M.L., Moreshead, W.V. et al. Effects of Newcastle disease virus administration to mice on the metabolism of cerebral biogenic amines, plasma corticosterone, and lymphocyte proliferation. *Brain Behav. Immun.* 1987; 1:216-30.
8. Dunn, A.J. and Vickers, S.L. Neurochemical and neuroendocrine responses to Newcastle disease virus administration in mice. *Brain Res.* 1994; 645:103-12.

9. Dunn, A.J., Powell, M.L., Meitin, C. et al. Virus infection as a stressor: influenza virus elevates plasma concentrations of corticosterone, and brain concentrations of MHPG and tryptophan. *Physiol. Behav.* 1989; 45:591-4.
10. Weidenfeld, J., Wohlman, A. and Gallily, R. Mycoplasma fermentans activates the hypothalamo-pituitary adrenal axis in the rat. *NeuroReport* 1995; 6:910-2.
11. Guo, Z.M., Qian, C.G., Peters, C.J. et al. Changes in platelet-activating factor, catecholamine, and serotonin concentrations in brain, cerebrospinal fluid, and plasma of Pichinde virus-infected guinea pigs. *Lab. Animal Sci.* 1993; 43:569-74.
12. Ben Hur, T., Rosenthal, J., Itzik, A. et al. Adrenocortical activation by herpes virus: involvement of IL-1β and central noradrenergic system. *NeuroReport* 1996; 7:927-31.
13. Miller, A.H., Spencer, R.L., Pearce, B.D. et al. Effects of viral infection on corticosterone secretion and glucocorticoid receptor binding in immune tissues. *Psychoneuroendocrinol.* 1997; 22:455-74.
14. Stone, E.A. "Stress and catecholamines." In *Catecholamines and Behavior. Vol. 2 Neuropsychopharmacology*, A.J. Friedhoff, ed. New York: Plenum Press, 1975, pp.31-72.
15. Dunn, A.J. and Kramarcy, N.R. "Neurochemical responses in stress: relationships between the hypothalamic-pituitary-adrenal and catecholamine systems." In *Handbook of Psychopharmacology*, L.L. Iversen, S.D. Iversen and S.H. Snyder, eds. New York: Plenum Press, 1984, pp.455-515.
16. Curzon, G., Joseph, M.H. and Knott, P.J. Effects of immobilization and food deprivation on rat brain tryptophan metabolism. *J. Neurochem.* 1972; 19:1967-74.
17. Dunn, A.J. Changes in plasma and brain tryptophan and brain serotonin and 5-hydroxyindoleacetic acid after footshock stress. *Life Sci.* 1988; 42:1847-53.
18. Dunn, A.J. Endotoxin-induced activation of cerebral catecholamine and serotonin metabolism: comparison with interleukin-1. J. *Pharmacol. Exptl. Therapeut.* 1992; 261:964-9.
19. Mefford, I.N. and Heyes, M.P. Increased biogenic amine release in mouse hypothalamus following immunological challenge: antagonism by indomethacin. *J. Neuroimmunol.* 1990; 27:55-61.
20. Masana, M.I., Heyes, M.P. and Mefford, I.N. Indomethacin prevents increased catecholamine turnover in rat brain following systemic endotoxin challenge. *Prog. Neuro-Psychopharmacol. Biol. Psychiat.* 1990; 14:609-21.
21. Lacosta, S., Merali, Z. and Anisman, H. Behavioral and neurochemical consequences of lipopolysaccharide in mice: anxiogenic-like effects. *Brain Res.* 1999; 818:291-303.
22. Delrue, C., Deleplanque, B., Rouge-Pont, F. et al. Brain monoaminergic, neuroendocrine, and immune responses to an immune challenge in relation to brain and behavioral lateralization. *Brain Behav. Immun.* 1994; 8:137-52.
23. Molina-Holgado, F. and Guaza, C. Endotoxin administration induced differential neurochemical activation of the rat brain stem nuclei. *Brain Res. Bull.* 1996; 40:151-6.
24. Heyes, M.P., Quearry, B.J. and Markey, S.P. Systemic endotoxin increases L-tryptophan, 5–hydroxyindoleacetic acid, 3-hydroxykynurenine and quinolinic acid content of mouse cerebral cortex. *Brain Res.* 1989; 491:173-9.
25. Lavicky, J. and Dunn, A.J. Endotoxin administration stimulates cerebral catecholamine release in freely moving rats as assessed by microdialysis. *J. Neurosci. Res.* 1995; 40:407-13.
26. Linthorst, A.C.E., Flachskamm, C., Holsboer, F. et al. Activation of serotonergic and noradrenergic neurotransmission in the rat hippocampus after peripheral administration of bacterial endotoxin: involvement of the cyclo-oxygenase pathway. *Neuroscience* 1996; 72:989–97.

27. Linthorst, A.C.E., Flachskamm, C., Müller-Preuss, P. et al. Effect of bacterial endotoxin and interleukin-1β on hippocampal serotonergic neurotransmission, behavioral activity, and free corticosterone levels: an *in vivo* microdialysis study. *J. Neurosci.* 1995; 15:2920-34.
28. Linthorst, A.C.E., Flachskamm, C., Holsboer, F. et al. Intraperitoneal administration of bacterial endotoxin enhances noradrenergic neurotransmission in the rat preoptic area: relationship with body temperature and hypothalamic-pituitary-adrenocortical axis activity. *Europ. J. Neurosci.* 1995; 7:2418-30.
29. Borowski, T., Kokkinidis, L., Merali, Z. et al. Lipopolysaccharide, central *in vivo* amine variations, and anhedonia. *NeuroReport* 1998; 9:3797-802.
30. Dunn, A.J. Systemic interleukin-1 administration stimulates hypothalamic norepinephrine metabolism parallelling the increased plasma corticosterone. *Life Sci.* 1988; 43:429-35.
31. Zalcman, S., Green-Johnson, J.M., Murray, L. et al. Cytokine-specific central monoamine alterations induced by interleukin-1, -2 and -6. *Brain Res.* 1994; 643:40-9.
32. Kabiersch, A., del Rey, A., Honegger, C.G. et al. Interleukin-1 induces changes in norepinephrine metabolism in the rat brain. *Brain Behav. Immun.* 1988; 2:267-74.
33. Terao, A., Oikawa, M. and Saito, M. Cytokine-induced change in hypothalamic norepinephrine turnover: involvement of corticotropin-releasing hormone and prostaglandins. *Brain Res.* 1993; 622:257-61.
34. Fleshner. M., Goehler. L.E., Hermann. J. et al. Interleukin-1β induced corticosterone elevation and hypothalamic NE depletion is vagally mediated. *Brain Res. Bull.* 1995; 37:605-10.
35. Liu. C., Takao. T., Hashimoto. K. et al. "Interleukin-1 receptors in the nervous system." In *Cytokines in the Nervous System*, N.J. Rothwell, ed. R. G. Landes Company, 1996, pp.21-40.
36. Dunn, A.J. and Chuluyan, H. The role of cyclo-oxygenase and lipoxygenase in the interleukin-1-induced activation of the HPA axis: dependence on the route of injection. *Life Sci.* 1992; 51:219-25.
37. Kabiersch, A., Furukawa, H., del Rey, A. et al. Administration of interleukin-1 at birth affects dopaminergic neurons in adult mice. *Ann. N.Y. Acad. Sci.* 1998; 840:123-7.
38. Smagin, G.N., Swiergiel, A.H. and Dunn, A.J. Peripheral administration of interleukin-1 increases extracellular concentrations of norepinephrine in rat hypothalamus: comparison with plasma corticosterone. *Psychoneuroendocrinol.* 1996; 21:83-93.
39. Ishizuka, Y., Ishida, Y., Kunitake, T. et al. Effects of area postrema lesion and vagotomy on interleukin-1b-induced norepinephrine release in the hypothalamic paraventricular nucleus region in the rat. *Neurosci. Letts.* 1997; 223:57-60.
40. Merali, Z., Lacosta, S. and Anisman, H. Effects of interleukin-1β and mild stress on alterations of norepinephrine, dopamine and serotonin neurotransmission: a regional microdialysis study. *Brain Res.* 1997; 761:225-35.
41. MohanKumar, P.S. and Quadri, S.K. Systemic administration of interleukin-1 stimulates norepinephrine release in the paraventricular nucleus. *Life Sci.* 1993; 52:1961-7.
42. Abreu, P., Llorente, E., Hernández, M.M. et al. Interleukin-1β stimulates tyrosine hydroxylase activity in the median eminence. *NeuroReport* 1994; 5:1356-8.
43. Berkenbosch, F., De Goeij, D.E.C., del Rey, A. et al. Neuroendocrine, sympathetic and metabolic responses induced by interleukin-1. *Neuroendocrinol.* 1989; 50:570-6.

44. Niijima, A., Hori, T., Aou, S. et al. The effects of interleukin-1β on the activity of adrenal, splenic and renal sympathetic nerves in the rat. J. *Auton. Nerv. System.* 1991; 36:183-92.
45. Kannan, H., Tanaka, Y., Kunitake, T. et al. Activation of sympathetic outflow by recombinant human interleukin-1β in conscious rats. *Amer. J. Physiol.* 1996; 270:R479-R85.
46. Niijima, A., Hori, T., Katafuchi, T. et al. The effect of interleukin-1β on the efferent activity of the vagus nerve to the thymus. *J. Autonomic Nervous System* 1995; 54:137-44.
47. Akiyoshi, M., Shimizu, Y. and Saito, M. Interleukin-1 increases norepinephrine turnover in the spleen and lung in rats. *Biochem. Biophys. Res. Commun.* 1990; 173:1266-70.
48. Terao, A., Oikawa, M. and Saito, M. Tissue-specific increase in norepinephrine turnover by central interleukin-1, but not by interleukin-6, in rats. *Amer. J. Physiol.* 1994; 266:R400-4.
49. Shimizu, N., Hori, T. and Nakane H. An interleukin-1b-induced noradrenaline release in the spleen is mediated by brain corticotropin-releasing factor: an *in vivo* microdialysis study in conscious rats. *Brain Behav. Immun.* 1994; 8:14-23.
50. Hurst, S. and Collins, S.M. Interleukin-1b modulation of norepinephrine release from rat myenteric nerves. *Amer. J. Physiol.* 1993; 264:G30-5.
51. Bognar, I.T., Albrecht, S.-A., Farasaty, M. et al. Effects of human recombinant interleukins on stimulation-evoked noradrenaline overflow from the rat perfused spleen. *Naunyn-Schmiedeberg's Arch. Pharmacol.* 1994; 349:497-502.
52. Foucart, S. and Abadie, C. Interleukin-1β and tumor necrosis factor-α inhibit the release of [^{3}H]-noradrenaline from mice isolated atria. *N.S. Arch. Pharmacol.* 1996; 354:1-6.
53. Mohankumar, P.S., Thyagarajan, S. and Quadri, S.K. Interleukin-1β increases 5–hydroxyindoleacetic acid release in the hypothalamus in vivo. *Brain Res. Bull.* 1993; 31:745-8.
54. Gemma, C., De Luigi, A. and De Simoni, M.G. Permissive role of glucocorticoids on interleukin-1 activation of the hypothalamic serotonergic system. *Brain Res.* 1994; 651:169-73.
55. Clement, H.W., Buschmann, J., Rex, S. et al. Effects of interferon-γ, interleukin-1β, and tumor necrosis factor-a on the serotonin metabolism in the nucleus raphe dorsalis of the rat. *J. Neural. Transm.* 1997; 104:981-91.
56. Chaouloff, F. Physiopharmacological interactions between stress hormones and central serotonergic systems. *Brain Res. Rev.* 1993; 18:1-32.
57. Dunn, A.J. and Welch, J. Stress- and endotoxin-induced increases in brain tryptophan and serotonin metabolism depend on sympathetic nervous system activation. *J. Neurochem.* 1991; 57:1615-22.
58. Lookingland, K.J., Shannon, N.J., Chapin, D.S. et al. Exogenous tryptophan increases synthesis, storage, and intraneuronal metabolism of 5-hydroxytryptamine in the rat hypothalamus. *J. Neurochem.* 1986; 47:205-12.
59. Joseph, M.H. and Kennett, G.A. Stress-induced release of 5-HT in the hippocampus and its dependence on increased tryptophan availability: an in vivo electrochemical study. *Brain Res.* 1983; 270:251-7.
60. De Simoni, M.G., Sokola, A., Fodritto, F. et al. Functional meaning of tryptophan-induced increase of 5-HT metabolism as clarified by in vivo voltammetry. *Brain Res.* 1987; 411:89-94.

61. Edwards, D.J., Sorisio, D.A. and Knopf, S. Effects of the b_2-adrenoceptor agonist clenbuterol on tyrosine and tryptophan in plasma and brain of the rat. *Biochem. Pharmacol.* 1989; 38:2957-65.
62. Lenard, N., O'Donnell, J.M. and Dunn, A.J. Clenbuterol-induced elevation of brain tryptophan is not related to its antidepressant effects. Soc. Neurosci. Abstr. 2000; 26:1768.
63. Dunn, A.J. and Chuluyan, H.E. Endotoxin elicits normal tryptophan and indolamine responses but impaired catecholamine and pituitary-adrenal responses in endotoxin-resistant mice. *Life Sci.* 1994; 54:847-53.
64. Dunn, A.J. Nitric oxide synthase inhibitors prevent the cerebral tryptophan and serotonergic responses to endotoxin and interleukin-1. *Neurosci. Res. Commun.* 1993; 13:149-56.
65. Rada, P., Mark, G.P., Vitek, M.P. et al. Interleukin-1β decreases acetylcholine measured by microdialysis in the hippocampus of freely moving rats. *Brain Res.* 1991; 550:287-90.
66. Kang, M., Yoshimatsu, H., Ogawa, R. et al. Thermoregulation and hypothalamic histamine turnover modulated by interleukin-1 beta in rats. *Brain Res. Bull.* 1994; 35:299-301.
67. Niimi, M., Mochizuki, T., Yamamoto, Y. et al. Interleukin-1 beta induces histamine release in the rat hypothalamus in vivo. *Neurosci. Lett.* 1994; 181:87-90.
68. Bianchi, M., Ferrario, P., Zonta, N. et al. Effects of interleukin-1β and interleukin-2 on amino acids levels in mouse cortex and hippocampus. *NeuroReport* 1995; 6:1689-92.
69. Veening, J.G., van der Meer, M.J.M., Joosten, H. et al. Intravenous administration of interleukin-1β induces fos-like immunoreactivity in corticotropin-releasing hormone neurons in the paraventricular hypothalamic nucleus of the rat. *J. Chem. Neuroanatomy* 1993; 6:391-7.
70. Chang, S.L., Ren, T. and Zadina, J.E. Interleukin-1 activation of Fos proto-oncogene protein in the rat hypothalamus. *Brain Res.* 1993; 617:123-30.
71. Ericsson, A., Kovács, K.J. and Sawchenko, P.E. A functional anatomical analysis of central pathways subserving the effects of interleukin-1 on stress-related neuroendocrine neurons. *J. Neurosci.* 1994; 14:897-913.
72. Swiergiel, A.H., Dunn, A.J. and Stone E.A. The role of cerebral noradrenergic systems in the Fos response to interleukin-1. *Brain Res. Bull.* 1996; 41:61-4.
73. Williams, L.M., Ballmer, P.E., Hannah, L.T. et al. Changes in regional protein synthesis in rat brain and pituitary after systemic interleukin-1b administration. *Amer. J. Physiol.* 1994; 267:E915-20.
74. Song, C. and Leonard, B.E. Interleukin-2-induced changes in behavioural, neurotransmitter, and immunological parameters in the olfactory bulbectomized rat. *Neuroimmunomodulation* 1995; 2:263-73.
75. Anisman, H., Kokkinidis, L. and Merali, Z. Interleukin-2 decreases accumbal dopamine efflux and responding for rewarding lateral hypothalamic stimulation. *Brain Res.* 1996; 731:1-11.
76. Pauli, S., Linthorst, A.C. and Reul, J.M. Tumour necrosis factor-α and interleukin-2 differentially affect hippocampal serotonergic neurotransmission, behavioural activity, body temperature and hypothalamic-pituitary-adrenocortical axis activity in the rat. *Europ. J. Neurosci.* 1998; 10:868-78.
77. Wang, J.P. and Dunn, A.J. Mouse interleukin-6 stimulates the HPA axis and increases brain tryptophan and serotonin metabolism. *Neurochem. Intl.* 1998; 33:143-54.

78. Barkhudaryan, N. and Dunn, A.J. Molecular mechanisms of actions of interleukin-6 on the brain, with special reference to serotonin and the hypothalamo-pituitary-adrenocortical axis. *Neurochem. Res.* 1999; 24:1169-80.
79. Zhang, J.-J., Terreni, L., De Simoni, M.-G. et al. Peripheral interleukin-6 administration increases extracellular concentrations of serotonin and the evoked release of serotonin in the rat striatum. *Neurochem. Intl.* 2001; 38:303-8.
80. Niimi, M., Wada, Y., Sato, M. et al. Effect of continuous intravenous injection of interleukin-6 and pretreatment with cyclooxygenase inhibitor on brain c-fos expression in the rat. *Neuroendocrinol.* 1997; 66:47-53.
81. Callahan, T.A. and Piekut, D.T. Differential Fos expression induced by IL-1β and IL-6 in rat hypothalamus and pituitary gland. *J. Neuroimmunol.* 1997; 73:207-11.
82. Tinsley, S.L., Knight, D. and Dunn, A.J. The effects of interleukin-6 on Fos expression in the rat brain. *Soc. Neurosci. Abstr.* 2001; 27:1749.
83. Fransen, L., Müller, R., Marmenout, A. et al. Molecular cloning of mouse tumor necrosis factor cDNA and its eukaryotic expression. *Nucleic Acids Res.* 1985; 13:4417-29.
84. Sherry, B., Jue, D.M., Zentella, A. et al. Characterization of high molecular weight glycosylated forms of murine tumor necrosis factor. *Biochem. Biophys. Rsch. Comm.* 1990; 173:1072-8.
85. Lewis, M., Tartaglia, L.A., Lee, A. et al. Cloning and expression of cDNAs for two distinct murine tumor necrosis factor receptors demonstrate one receptor is species specific. *Proc. Natl. Acad. Sci.* USA 1991; 88:2830-4.
86. Brouckaert, P., Libert, C., Everaerdt, B. et al. Selective species specificity of tumor necrosis factor for toxicity in the mouse. *Lymphokine Cytokine Res.* 1992; 11:193-6.
87. Dunn, A.J. The role of interleukin-1 and tumor necrosis factor α in the neurochemical and neuroendocrine responses to endotoxin. *Brain Res. Bull.* 1992; 29:807-12.
88. Hayley, S., Brebner, K., Lacosta, S. et al. Sensitization to the effects of tumor necrosis factor-a: neuroendocrine, central monoamine, and behavioral variations. *J. Neurosci.* 1999; 19:5654-65.
89. Elenkov, I.J., Kovacs, K., Duda, E. et al. Presynaptic inhibitory effect of TNF-a on the release of noradrenaline in isolated median eminence. *J. Neuroimmunol.* 1992; 413:117-20.
90. Ignatowski, T.A. and Spengler, R.N. Tumor necrosis factor-a: presynaptic sensitivity is modified after antidepressant drug adminstration. *Brain Res.* 1994; 665:293-9.
91. Hurst, S.M. and Collins, S.M. Mechanism underlying tumor necrosis factor-a suppression of norepinephrine release from rat myenteric plexus. *Amer. J. Physiol.* 1994; 266:G1123-9.
92. Crnic, L.S. and Segall, M.A. Behavioral effects of mouse interferon-a and interferon-γ and human interferon-a in mice. *Brain Res.* 1992; 590:277-84.
93. Saito, K., Markey, S.P. and Heyes, M.P. Chronic effects of γ-interferon on quinolinic acid and indoleamine-2,3-dioxygenase in brain of C57BL6 mice. *Brain Res.* 1991; 546:151-4.
94. Werner, E.R., Werner-Felmayer, G. and Wachter, H. Tetrahydrobiopterin and cytokines. *Proc. Soc. Exptl. Biol. Med.* 1993; 203:1-12.
95. Bianchi, M., Clavenna, A., Bondiolotti, G.P. et al. GM-CSF affects hypothalamic neurotransmitter levels in mice: involvement of interleukin-1. *NeuroReport* 1997; 10:3587-90.
96. Shintani, F., Kanba, S., Nakaki, T. et al. Interleukin-1β augments release of norepinephrine, dopamine, and serotonin in the rat anterior hypothalamus. *J. Neurosci.* 1993; 13:3574-81.

97. Shintani, F., Nakaki, T., Kanba, S. et al. Involvement of interleukin-1 in immobilization stress-induced increase in plasma adrenocorticotropic hormone and in release of hypothalamic monoamines in the rat. *J. Neurosci.* 1995; 15:1961-70.
98. Woodroofe, M.N., Sarna, G.S., Wadhwa, M. et al. Detection of interleukin-1 and interleukin-6 in adult rat brain, following mechanical injury, by in vivo microdialysis: evidence of a role for microglia in cytokine production. *J. Neuroimmunol.* 1991; 33:227-36.
99. Dunn, A.J. Effects of the interleukin-1 (IL-1) receptor antagonist on the IL-1- and endotoxin-induced activation of the HPA axis and cerebral biogenic amines in mice. *Neuroimmunomodulation* 2000; 7:36-45.
100. Rivier, C. Influence of immune signals on the hypothalamic-pituitary axis of the rodent. *Front Neuroendocrinol.* 1995; 16:151-82.
101. Matta, S.G., Singh, J., Newton, R. et al. The adrenocorticotropin response to interleukin-1b instilled into the rat median eminence depends on the local release of catecholamines. *Endocrinol.* 1990; 127:2175-82.
102. McCoy, J.G., Matta, S.G., Sharp, B. Prostaglandins mediate the ACTH response to interleukin-1-beta instilled into the hypothalamic median eminence. *Neuroendocrinol.* 1994; 60:426-35.
103. Banks, W.A., Ortiz, L., Plotkin, S.R. et al. Human interleukin (IL)1a, murine IL-1α and murine IL-1β are transported from blood to brain in the mouse by a shared saturable mechanism. *J. Pharmacol. Exptl. Therapeut.* 1991; 259:988-96.
104. Banks, W.A., Kastin, A.J. and Broadwell, R.D. Passage of cytokines across the blood-brain barrier. *Neuroimmunomodulation* 1995; 2:241-8.
105. Wan, W., Wetmore, L., Sorensen, C.M. et al. Neural and biochemical mediators of endotoxin and stress-induced *c-fos* expression in the rat brain. *Brain Res. Bull.* 1994; 34:7-14.
106. Bret-Dibat, J.-L., Bluthé, R.-M., Kent, S. et al. Lipopolysaccharide and interleukin-1 depress food-motivated behavior in mice by a vagal-mediated mechanism. *Brain, Behav. Immun.* 1995; 9:242-6.
107. Watkins, L.R., Wiertelak, E.P., Goehler L.E. et al. Neurocircuitry of illness-induced hyperalgesia. *Brain Res.* 1994; 639:283-99.
108. Watkins, L.R., Maier, S.F. and Goehler, L.E. Cytokine-to-brain communication: a review & analysis of alternative mechanisms. *Life Sci.* 1995; 57:1011-26.
109. Goehler, L.E., Gaykema, R.P.A., Nguyen, K.T. et al. Interleukin-1β in immune cells of the abdominal vagus nerve: a link between the immune and nervous systems? *J. Neurosci.* 1999; 19:2799-806.
110. Haour, F., Ban, E., Marquette, C. et al. "Brain Interleukin-1 receptors: mapping, characterization and modulation." In *Interleukin-1 in the Brain*, N.J. Rothwell and R.D. Dantzer, eds. Oxford: Pergamon Press, 1992, pp.13-25.
111. Takao, T., Tracey, D.E., Mitchell, W.M. et al. Interleukin-1 receptors in mouse brain: characterization and neuronal localization. *Endocrinol.* 1990; 127:3070-8.
112. Terao, A., Kitamura, H., Asano, A. et al. Roles of prostaglandins D_2 and E_2 in interleukin-1-induced activation of norepinephrine turnover in the brain and peripheral organs of rats. *J. Neurochem.* 1995; 65:2742-7.
113. Swiergiel, A.H., Smagin, G.N., Johnson, L.J. et al. The role of cytokines in the behavioral responses to endotoxin and influenza virus infection in mice: effects of acute and chronic administration of the interleukin-1-receptor antagonist (IL-1ra). *Brain Res.* 1997; 776:96-104.

114. Fantuzzi, G., Zheng, H., Faggioni, R. et al. Effect of endotoxin in IL-1β-deficient mice. *J. Immunol.* 1996; 157:291-6.
115. Perlstein, R.S., Whitnall, M.H., Abrams, J.S. et al. Synergistic roles of interleukin-6, interleukin-1, and tumor necrosis factor in the adrenocorticotropin response to bacterial lipopolysaccharide *in vivo*. *Endocrinol.* 1993; 132:946-52.
116. Wang, J.P. and Dunn, A.J. The role of Interleukin-6 in the activation of the hypothalamo-pituitary-adrenocortical axis induced by endotoxin and interleukin-1β. *Brain Res.* 1999; 815:337-48.
117. Plotsky, P.M., Cunningham, E.T. and Widmaier, E.P. Catecholaminergic modulation of corticotropin-releasing factor and adrenocorticotropin secretion. *Endocrine Rev.* 1989; 10:437-58.
118. Turnbull, A.V. and Rivier, C. Regulation of the HPA axis by cytokines. *Brain Behav. Immun.* 1995; 9:253-75.
119. Chuluyan, H., Saphier, D., Rohn, W.M. et al. Noradrenergic innervation of the hypothalamus participates in the adrenocortical responses to interleukin-1. *Neuroendocrinol.* 1992; 56:106-11.
120. Weidenfeld, J., Abramsky, O. and Ovadia, H. Evidence for the involvement of the central adrenergic system in interleukin 1-induced adrenocortical response. *Neuropharmacol.* 1989; 28:1411-4.
121. Dunn, A.J. Cytokine activation of the HPA axis. *Ann. N.Y. Acad. Sci.* 2000; 917:608-17.
122. Uehara, A., Gottschall, P.E., Dahl, R.R. et al. Interleukin-1 stimulates ACTH release by an indirect action which requires endogenous corticotropin releasing factor. *Endocrinol.* 1987; 121:1580-2.
123. Sapolsky, R., Rivier, C., Yamamoto, G. et al. Interleukin-1 stimulates the secretion of hypothalamic corticotropin-releasing factor. *Science* 1987; 238:522-4.
124. Berkenbosch, F., van Oers, J., del Rey, A. et al. Corticotropin-releasing factor-producing neurons in the rat activated by interleukin-1. *Science* 1987; 238:524-6.
125. Dunn, A.J. and Swiergiel, A.H. Behavioral responses to stress are intact in CRF-deficient mice. *Brain Res.* 1999; 845:14-20.
126. Dantzer, R., Bluthé, R.-M., Castanon, N. et al. "Cytokine effects on behavior." In *Psychoneuroimmunology*, R. Ader, D. Felten and N. Cohen, eds. San Diego, CA: Academic Press, 2001, pp.703-27.
127. Swiergiel, A.H., Smagin, G.N. and Dunn, A.J. Influenza virus infection of mice induces anorexia: comparison with endotoxin and interleukin-1 and the effects of indomethacin. *Pharmacol. Biochem. Behav.* 1997; 57:389-96.
128. Hellerstein, M.K., Meydani, S.N., Meydani, M. et al. Interleukin-1-induced anorexia in the rat. Influence of prostaglandins. *J. Clin. Invest.* 1989; 84:228-35.
129. Dunn, A.J. and Swiergiel, A.H. The role of cyclooxygenases in endotoxin- and interleukin-1-induced hypophagia. *Brain Behav. Immun.* 2000; 14:141-52.
130. Swiergiel, A.H. and Dunn, A.J. Cyclooxygenase 1 is not essential for hypophagic responses to interleukin-1 and endotoxin in mice. *Pharmacol. Biochem. Behav.* 2001; 69:659-63.
131. Swiergiel, A.H. and Dunn, A.J. Distinct roles for COX1 and COX2 in interleukin-1-induced behavioral changes. *J. Pharmacol Exptl. Therap.* 2002; in press.
132. Cooper, S.J. and Clifton P.G. *Drug receptors subtypes and ingestive behavior.* London: Academic Press, 1996.
133. Swiergiel, A.H., Burunda, T., Peterson, B. et al. Endotoxin- and interleukin-1-induced hypophagia are not affected by noradrenergic, dopaminergic, histaminergic and muscarinic antagonists. *Pharmacol. Biochem. Behav.* 1999; 63:629-37.

134. Ovadia, H., Abramsky, O. and Weidenfeld J. Evidence for the involvement of the central adrenergic system in the febrile response induced by interleukin-1 in rats. *J. Neuroimmunol.* 1989; 25:109-16.
135. Bianchi, M. and Panerai, A.E. CRH and the noradrenergic system mediate the antinociceptive effects of central interleukin-1α in the rat. *Brain Res. Bull.* 1995; 36:113-7.
136. Swiergiel, A.H. and Dunn, A.J. Lack of evidence for a role of serotonin in interleukin-1-induced hypophagia. *Pharmacol. Biochem. Behav.* 2000; 65:531-7.
137. Swiergiel, A.H., Burunda, T., Patterson, B. et al. Endotoxin- and interleukin-1-induced hypophagia are not affected by noradrenergic, dopaminergic, histaminergic and muscarinic antagonists. *Pharmacol. Biochem. Behav.* 1999; 63:629-37.
138. Swiergiel, A.H. and Dunn, A.J. CRF-deficient mice respond like wild type mice to hypophagic stimuli. *Pharmacol. Biochem. Behav.* 1999; 64:59-64.
139. Alleva, E., Cirulli, F., Bianchi, M. et al. Behavioural characterization of interleukin-6 overexpressing or deficient mice during agonistic encounters. *Europ. J. Neurosci.* 1998; 10:3664-72.
140. Zalcman, S., Murray, L., Dyck, D.G. et al. Interleukin-2 and -6 induce behavioral-activating effects in mice. *Brain Res.* 1998; 811:111-21.
141. Hinze-Selch, D. and Pollmächer, T. In *vitro* cytokine secretion in individuals with schizophrenia: results, confounding factors, and implications for further research. *Brain Behav. Immun.* 2001; 15:282-318.
142. Wong, M.L., Kling, M.A., Munson, P.J. et al. Pronounced and sustained central hypernoradrenergic function in major depression with melancholic features: relation to hypercortisolism and corticotropin-releasing hormone. *Proc. Natl. Acad. Sci.* USA 2000; 97:325-30.
143. Kasckow, J.W., Baker, D. and Geracioti, T.D. Corticotropin-releasing hormone in depression and post-traumatic stress disorder. *Peptides* 2001; 22:845-51.
144. Charlton, B.G. The malaise theory of depression: major depressive disorder is sickness behavior and antidepressants are analgesic. *Med. Hypotheses* 2000; 54:126-30.
145. Yirmiya, R., Weidenfeld, J., Pollak Y. et al. "Cytokines, "Depression due to a general medical condition," and antidepressant drugs." In *Cytokines, Stress, and Depression*, R. Dantzer, E.E. Wollman and R. Yirmiya, eds. New York: Kluwer Academic/Plenum Publishers, 1999, pp.283-316.
146. Yirmiya, R. Endotoxin produces a depressive-like episode in rats. *Brain Res.* 1996; 711:163-74.
147. Connor, T.J. and Leonard, B.E. Depression, stress and immunological activation: the role of cytokines in depressive disorders. *Life Sci.* 1998; 62:583-606.
148. Castanon, N., Bluthé, R.-M. and Dantzer, R. Chronic treatment with the atypical antidepressant tianeptine attenuates sickness behavior induced by peripheral but not central lipopolysaccharide and interleukin-1b in the rat. *Psychopharmacol.* 2001; 154:50-60.
149. Yirmiya, R., Pollak, Y., Barak, O. et al. Effects of antidepressant drugs on the behavioral and physiological responses to lipopolysaccharide (LPS) in rodents. *Neuropsychopharmacol.* 2001; 24:531-44.
150. Dunn, A.J. and Swiergiel, A.H. The reductions in sweetened milk intake induced by interleukin-1 and endotoxin are not prevented by chronic antidepressant treatment. *Neuroimmunomodulation* 2001; 9:163-9.

151. Dantzer, R., Wollman, E.E., Vitkovic, L. et al. "Cytokines, stress, and depression." In *Cytokines, Stress, and Depression*, R. Dantzer, E.E. Wollman and R. Yirmiya, eds. New York: Kluwer Academic/Plenum Publishers, 1999, pp.317-29.
152. Krueger, J.M. and Majde, J.A. Microbial products and cytokines in sleep and fever regulation. *Crit. Rev. Immunol.* 1995; 14:355-79.
153. Zorrilla, E.P., Luborsky, L., McKay, J.R. et al. The relationship of depression and stressors to immunological assays: a meta-analytic review. *Brain Behav. Immun.* 2001; 15:199-226.
154. Dunn, A.J. "Effects of cytokines and infections on brain neurochemistry." In *Psychoneuroimmunology*, R. Ader, D.L. Felten and N. Cohen, eds. New York: Academic Press, 2001, pp.649-66.

Chapter 5

CYTOKINES AND HPA AXIS REGULATION

Marni N. Silverman, Brad D. Pearce, and Andrew H. Miller
Department of Psychiatry and Behavioral Sciences, Emory University School of Medicine, Atlanta, Georgia

1. INTRODUCTION

Compelling data has been amassed indicating that soluble factors, or cytokines, from the immune system can have profound effects on the neuroendocrine system, in particular the hypothalamic-pituitary-adrenal (HPA) axis. HPA activation by cytokines, in turn, has been found to play a critical role in restraining and shaping immune responses. Thus, cytokine-HPA interactions represent a fundamental consideration regarding the maintenance of homeostasis and the development of disease.

Maintenance of homeostasis during immune challenge involves activation of the immune system, resolution of the challenge, and protection of the host against potentially toxic inflammatory processes. Examples of immune challenges include, infection with viruses, bacteria or parasites, tissue damage and destruction, and inappropriate responses to auto-antigens that may result in the development of autoimmune disease. Upon immune challenge, the immune system is activated to release numerous protein hormones called cytokines. One group of these cytokines is the pro-inflamatory cytokines, which include tumor necrosis factor alpha (TNF-α), interleukin-1 (IL-1), and interleukin-6 (IL-6). These cytokines are released during the early innate immune response from activated immune cells, including macrophages and their CNS counterparts microglia, vascular endothelial cells, and fibroblasts. In addition to contributing to the progression of the immune response, the pro-inflamatory cytokines also activate the HPA axis (1-6) (Figure 1).

Activation of the HPA axis begins with the release of corticotropin-releasing hormone (CRH) from the parvocellular cells of the paraventricular nucleus (PVN) of the hypothalamus into the median eminence (ME) and hypophyseal portal circulation. CRH, in turn, stimulates the release of adrenocorticotropic hormone (ACTH) from the corticotrophs of the anterior pituitary. ACTH then is released into the peripheral circulation and

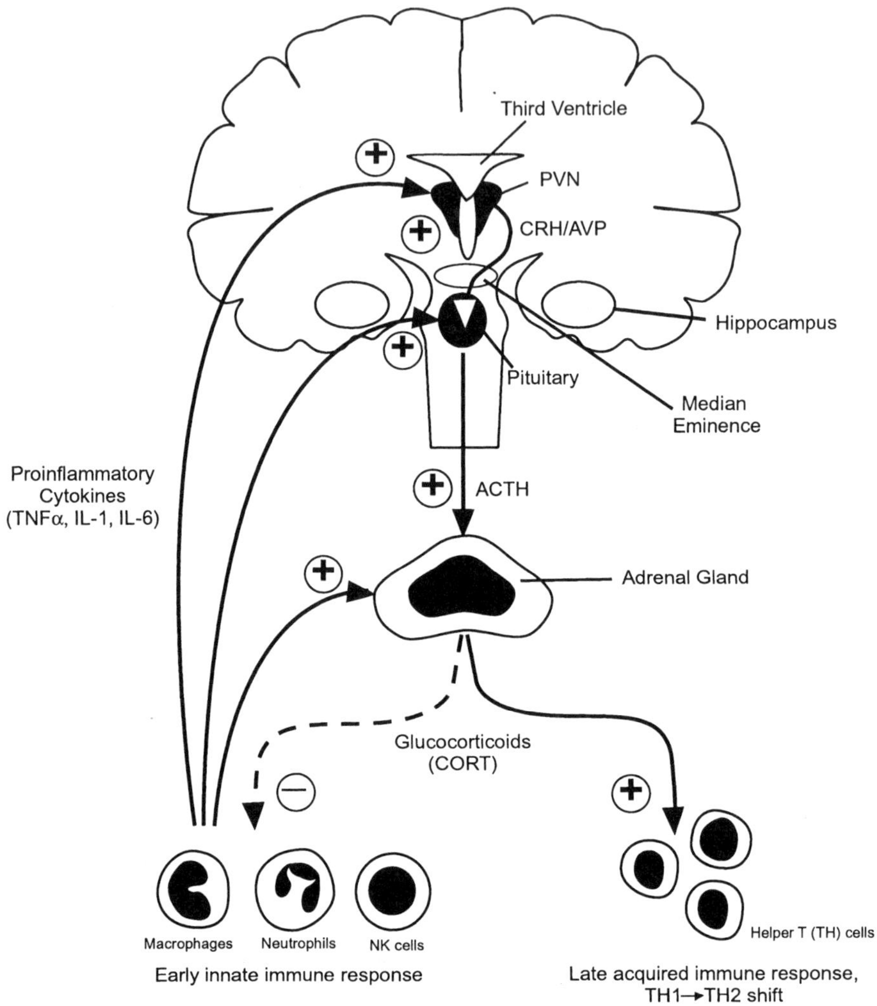

Figure 1. Bidirectional communication between the immune system and the hypothalamic-pituitary-adrenal (HPA) axis (human brain). The immune system, via pro-inflamatory cytokines (TNFα, IL-1, and IL-6), stimulates glucocorticoid release by acting at all three levels of the HPA axis. In turn, glucocorticoids negatively feedback on the immune system to suppress the further synthesis and release of pro-inflamatory cytokines. In addition, glucocorticoids play an important role in shaping downtream acquired immune responses, by causing a shift from cellular (Th1/inflammatory) to humoral (Th2/anti-inflammatory) type immune responses. By doing so, glucocorticoids protect an organism from the detrimental consequences of overactive inflammatory immune responses. PVN, paraventricular nucleus of the hypothalamus; CRH, Corticotropin-releasing hormone; AVP, arginine vasopressin; ACTH, adrenocorticotropic hormone; IL, interleukin; TNF, tumor necrosis factor.

stimulates the release of glucocorticoids (CORT, cortisol in humans and monkeys; corticosterone in rodents) from the adrenal cortex. In turn, glucocorticoids negatively feedback onto immune cells to suppress the further synthesis and release of pro-inflamatory cytokines, thereby protecting the host from the detrimental consequences of an overactive immune response. In addition, glucocorticoids play an important role in shaping downstream acquired immune responses, by causing a shift from cellular (Th1/inflammatory) to humoral (Th2/anti-inflammatory) type immune responses (7).

Although hypothalamic CRH is considered a primary mechanism by which cytokines stimulate glucocorticoid release, increasing evidence supports a direct action of cytokines at the level of the pituitary and adrenal glands. Cytokine receptors have been detected at all HPA axis levels and, therefore, each level can serve as an integration point for immune and neuroendocrine signals. In addition to circulating cytokines being able to act upon all three levels of the HPA axis, TNFα, IL-1, and IL-6 are synthesized in the brain, the anterior pituitary, and the adrenal gland. The production of local cytokines may function in a paracrine manner to amplify and maintain elevated HPA activity during chronic inflammation. Once glucocorticoids are released, maintenance of appropriate glucocorticoid activity is accomplished by a local regulation of glucocorticoid availability and action by factors such as corticosterone binding globulin (CBG), 11β-hydroxysteroid dehydrogenase (11β-HSD), the multidrug resistance transporter (MDR), and ultimately, the glucocorticoid receptor.

Given the critical role of the HPA axis and glucocorticoid responses in maintaining a balance between the beneficial and detrimental effects of pro-inflamatory cytokines, as well as shaping downstream immune responses, it has become increasingly apparent that cytokine-HPA axis interactions are fundamental to immune regulation. Moreover, redundant pathways of glucocorticoid induction, incorporating all three levels of the HPA axis, exist to ensure the survival of the organism during immune challenge.

2. HPA AXIS ACTIVATION DURING SYSTEMIC IMMUNE ACTIVATION

A number of model systems have been used to investigate the complex and redundant pathways through which immune stimuli activate the HPA axis. Early seminal work by Hugo Besedovsky and colleagues demonstrated that glucocorticoids increase during the immune response to foreign antigens, such as sheep red blood cells, when administered to rats (8). Complementary studies demonstrated that viral infection was associated with increased glucocorticoid levels (9-12) and identified the pro-inflamatory cytokine IL-1 as being the critical mediator of the HPA axis response during viral infection (9,13). Subsequent to this early landmark work, a plethora of studies have examined the direct impact of various cytokines alone or in combination on HPA axis function.

Some general properties of cytokines are that 1) they are pleiotropic (one cytokine can exert many actions), 2) they are redundant (different cytokines can exert the same action), 3) they often influence the synthesis of other cytokines (TNF→IL-1→IL-6, while IL-6 inhibits TNF and IL-1 synthesis), and 4) they often influence the action of other cytokines (TNF, IL-1, and IL-6 can act synergistically). The importance of synergistic actions of cytokines has been demonstrated by several studies. In an *in vitro* preparation of isolated rat hypothalami, Buckingham et al (2) showed that the release of CRH by conditioned media from lipopolysaccharide (LPS)-stimulated peritoneal macrophages (containing multiple cytokines) was much greater than that observed in response to either TNFα, IL-1, or IL-6 alone. In addition, there is *in vivo* evidence that supports the role of synergistic actions of these pro-inflamatory cytokines in stimulating a complete, LPS-induced ACTH response (14) and in stimulating greater HPA axis activity together compared to their individual effect (15). By studying CORT release that is induced by a complex inflammatory stimulus, such as bacteria or viruses rather than a bolus of purified cytokine, one can study a more natural situation of immune – neuroendocrine interaction. This gives the advantage of eliciting immune and neuroendocrine responses that evolve over time and occur in the physiological milieu of an *in vivo* immune response (release and action of multiple cytokines at physiologically relevant concentrations).

To mimic a bacterial induced immune response, the administration of LPS (or endotoxin) is commonly used. LPS is a constituent of a Gram-negative bacterial cell membrane that is released upon bacterial lysis and recognized by the immune system. However, upon injecting an animal with LPS, the animal is not actively infected. Peak ACTH and CORT responses to systemic administration of LPS occur about 1-2h post-injection. IL-1, IL-6 and TNFα seem to play a role in LPS-induced HPA axis activation (6). HPA axis activity in response to viral infections has received less attention than bacterial stimuli. To investigate viral-induced immune-neuroendocrine interactions, investigators have employed Newcastle disease virus (NDV), which does not actively infect the host, and polyinosinic polcytidilic acid (PolyI:C), a synthetic double-stranded RNA used to mimic viral exposure. Both NDV and PolyI:C produce a marked HPA response about 1-2h post-injection, in which NDV stimulates IL-1-dependent (9) and PolyI:C stimulates IL-6-dependent (16) ACTH and CORT release. Our lab has been using a murine model of cytomegalovirus (MCMV), a replicating virus, to help elucidate the mechanisms of immune-HPA axis interactions during viral infection.

MCMV is a cytopathic herpes virus that induces an early natural killer (NK) cell-mediated, anti-viral defense. The anti-viral immune response is characterized by high levels of IL-12 and NK cell–produced IFN-γ (17), and NK cell–mediated liver pathology (18). In addition to the induction of anti-viral cytokines, the pro-inflamatory cytokines, TNFα, IL-1, and IL-6, are induced during the innate immune response to MCMV infection. Peak serum cytokine levels occur around 36 - 44 hours after infection and are paralleled by peak neuroendocrine (ACTH and CORT) responses, which occur at 36 hours after MCMV infection (16) (Figure 2). Previous studies have shown

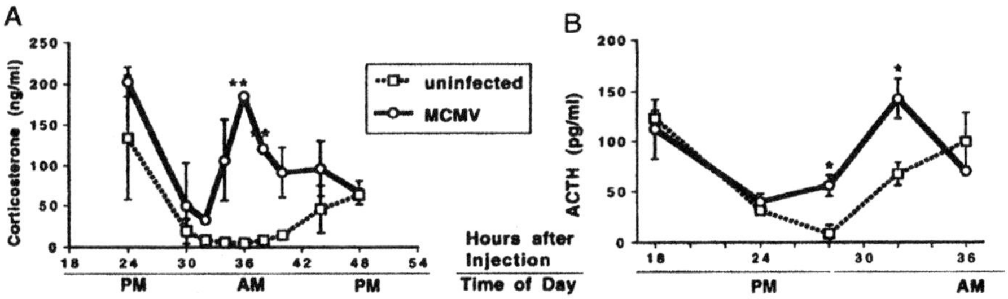

Figure 2. Plasma corticosterone and ACTH levels after MCMV infection. Corticosterone and ACTH levels were measured from plasma samples collected from C57BL/6 mice under low-stress conditions (mice were bled within 4 min of handling). Corticosterone was measured at 2 to 4 h intervals between 24 and 48 h (A), and ACTH was measured at 4 to 6 h intervals between 18 and 36 h (B) after MCMV ($5x10^4$ PFU/mouse) or vehicle injection. Data are presented as means ± SEM. Results are significant at *p<0.05 and **p<0.01. [Reproduced from Ruzek et al., The Journal of Experimental Medicine, 1997, 185(7), 1185-1192 by permission of the Rockefeller University Press.]

that the MCMV-induced CORT response is dependent on IL-6, with IL-1α contributing to IL-6 production (16) (Figure 3). CORT has been shown to negatively regulate TNFα production during MCMV infection. If CORT is removed by adrenalectomy, splenic TNFα mRNA expression and circulating TNFα protein levels in infected mice remain high (19) and the mice die due to septic shock. CORT replacement or the administration of TNF anti-sera restores survival (19) (Figure 4). Therefore, the glucocorticoid response during MCMV infection is essential for protection against TNFα-mediated lethality. This essential role of glucocorticoids in protecting the host against

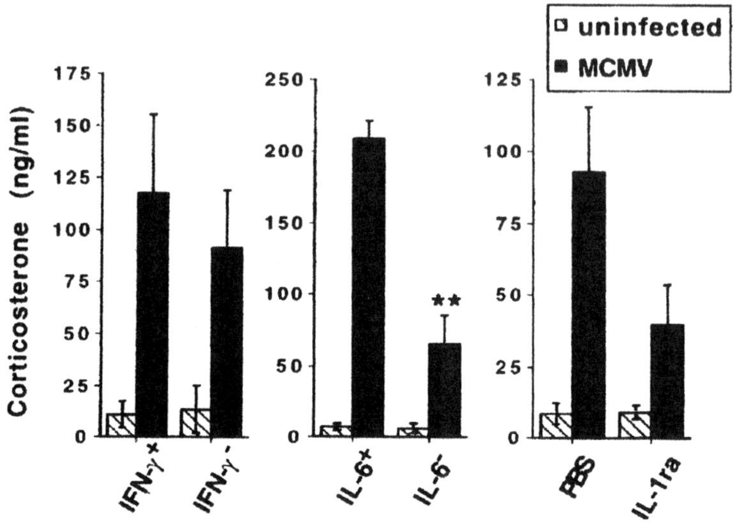

Figure 3. Plasma corticosterone in mice with neutralized cytokine function. Corticosterone levels in MCMV-infected ($5x10^4$ PFU/mouse) (filled bar) or vehicle-injected (striped bar) in IFN-γ deficient (IFN-γ^-), IL-6 deficient (IL-6^-), IL-1ra-treated (IL-1ra) mice or respective control mice (IFN-γ^+, IL-6^+, PBS). Corticosterone levels were measured from plasma samples collected from C57BL/6 mice under low-stress conditions (mice were bled within 4 min of handling) 36 h after injection. Data are presented as means ± SEM. Results are significant at $**p<0.01$. [Reproduced from Ruzek et al., The Journal of Experimental Medicine, 1997, 185(7), 1185-1192 by permission of the Rockefeller University Press.]

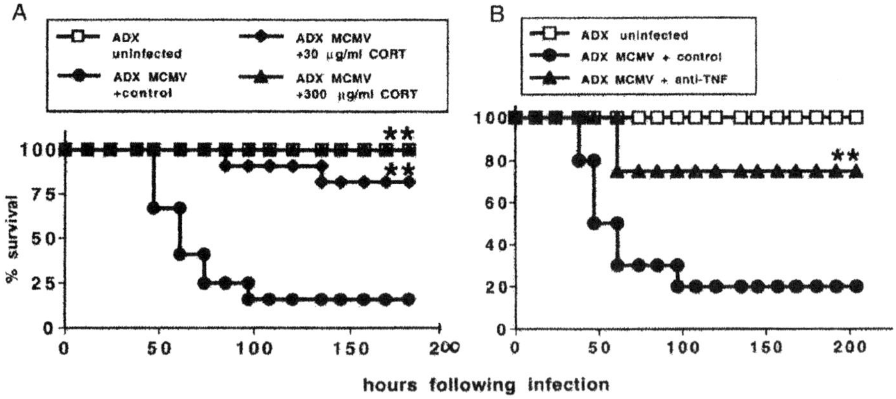

Figure 4. Role of glucocorticoids in protection against cytokine (TNF)-mediated lethality. A) Adrenalectomized (ADX) mice were vehicle injected (squares) or infected with MCMV ($1x10^5$ PFU/mouse) 5 days following surgery and either not given (circles) or given corticosterone in 0.9% saline as drinking water at either 30 ug/ml (diamonds) or 300 ug/ml (triangles). Control mice received vehicle-treated water. Denoted p value represents significant differences between corticosterone and vehicle-treated MCMV-infected ADX mice. B) Control (circles) or anti-TNF (triangles) antibodies (Abs) were administered 8-10 h before infection of ADX mice. Denoted p value represents significant differences between TNF and control Ab-treated MCMV-infected ADX mice.

a lethal over-activation of inflammatory responses has been demonstrated in a number of animal model systems (20-22) and is relevant to clinical conditions in which glucocorticoids are a mainstay in the treatment of endotoxic shock.

3. MECHANISMS OF CYTOKINE EFFECTS ON THE HPA AXIS

Given the importance of the glucocorticoid response in restraining potentially detrimental immune responses, redundant or complementary pathways of glucocorticoid induction are activated by pro-inflammatory cytokines at all three levels of the HPA axis, including the hypothalamus, pituitary, and adrenal gland.

3.1 Cytokine – HPA Interactions: Hypothalamus / ME

The majority of evidence indicates that activation of hypothalamic CRH is the primary means by which cytokines stimulate the release of ACTH and CORT. Since cytokines are large, soluble peptides, significant consideration has been given as to how cytokines cross the blood-brain barrier (BBB) and activate CRH-producing neurons in the PVN. Several mechanisms have been identified. Cytokines may 1) stimulate visceral (vagal) afferents that project to the nucleus tractus solitarius (NTS) in the brainstem, activating the release of (norepinephrine) NE from catecholaminergic terminals in the PVN, 2) passively cross the BBB at "leaky" regions where the BBB in not intact, such as the circumventricular organs (CVO's) [i.e., median eminence (ME), organum vascularis of lateral terminalis (OVLT), and area postrema (AP)], and directly activate neurons that project to the hypothalamus, 3) exert a direct effect on CRH nerve terminals in the ME, 4) act on endothelial cells of brain vasculature or glial cells in the CVO's, inducing the synthesis/release of secondary messengers such as central cytokines, prostaglandins (PG's), or nitric oxide (NO), which in turn activate hypothalamic neurons, and 5) cross the BBB themselves via active transport. Evidence to support all of these pathways exists, depending on the cytokine dose, time interval, and route of administration (6,23,24) (Figure 5).

Figure 5

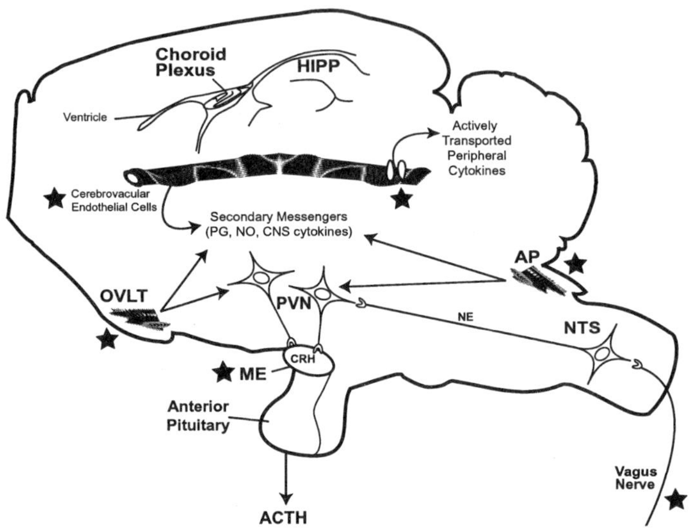

Figure 5. Proposed mechanisms by which peripheral cytokines activate CRH-producing neurons of the PVN in the hypothalamus (see text) (rat brain). ACTH, adrenocorticotropic hormone; AP, area postrema; CRH, corticotropin-releasing hormone; HIPP, hippocampus; ME, median eminence; NE, norepenephrine; NO, nitric oxide; NTS, nucleus tractus solitarius; OVLT, organum vascularis of lateral terminalis; PG, prostaglandins; PVN, paraventricular nucleus. ★ = site of entry of peripheral cytokine signals.

3.1.1 Cytokine Receptor Expression

Once the peripheral immune signal reaches the brain, cytokines can interact with their receptors in multiple brain regions involved in HPA axis regulation, including the hypothalamus (PVN) and the hippocampus, which provides negative feedback regulation on PVN activity (25). In addition, many of the aforementioned brain regions involved in relaying the circulating cytokine message to the PVN have been shown to express pro-inflamatory cytokine receptors.

Of the two types of IL-1 receptors, both IL-1RI (activating) and IL-IRII (decoy) mRNA and protein have been detected in the mouse brain (26,27). In the mouse, high levels of IL-1RI mRNA expression are found in the hippocampus (dentate gyrus), the midline raphe system, the choroid plexus, meninges, and endothelial cells of postcapillary venules throughout the brain (28,29). However, only a scattered signal for IL-1RI mRNA has been detected in the ME and no signal in the hypothalamus has been found. In contrast, IL-1RI mRNA in the rat brain is predominantly expressed over blood-brain barrier-related cells and not strongly expressed in certain neuronal populations as seen in the mouse brain (29). More specifically, constitutive expression of IL-1RI mRNA has been detected at the interface of the vascular wall and perivascular glia (30) and on endothelial cells (31) of the rat brain. In addition, IL-1 binding sites are detected in vagus nerve paraganglia (32). Peripheral administration of LPS has been shown to downregulate IL-1 receptors in the mouse brain (33,34).

Binding of IL-6 to IL-6R is followed by the rapid association of the gp130 molecule, the signal transducing component of the IL-6R complex, which increases the affinity of the IL-6R for IL-6. In the untreated rat, IL-6R mRNA has been detected in numerous brain regions that have been implicated in playing a role in HPA axis regulation, including the hippocampus, hypothalamus, and circumventricular organs such as the median eminence (35,36). However, the PVN displays no specific hybridization signal for IL-6R mRNA (35,36). Valliere et al (36) have demonstrated that intraperitoneal (ip) injection of LPS in rats is able to upregulate the expression of IL-6R mRNA in selective regions of the brain that constitutively express IL-6R mRNA. Interestingly, LPS also induces IL-6R mRNA expression in some areas in which no IL-6R mRNA is detected under basal conditions, such as the PVN and endothelial cells of cerebral blood vessels (36). IL-6R mRNA has been localized to glial cells, ependymal cells, and neurons (35,37). Gp130 immunoreactivity also has been observed in both glial and neural cells, and its distribution in the rat brain overlaps with that of IL-6R (38). Gp130 mRNA expression is detected throughout the rat brain (including the PVN) and is upregulated in the OVLT and in endothelial cells of cerebral microvasculature in response to peripheral administration of LPS (36).

TNF exerts its functions through two structurally related, but functionally distinct receptors, TNF-R1 (p55) and TNF-R2 (p75). Whereas TNF-R1 mRNA is constitutively expressed in the CVO's (including the ME), choroid plexus, meninges, and microvasculature, and to a lesser extent in the PVN, TNF-RII mRNA is barely detectable in the rat brain under basal conditions (39). Systemic treatment with LPS induces an increase TNF-RI mRNA expression in cells of barrier-associated structures (39). Specific binding sites for TNF in mouse brain homogenates have been shown to be widespread, but maximal in the brainstem (40). Both types of TNF receptors have been localized to cerebrovascular endothelial cells (41), glial cells (42), and neurons (43).

3.1.2 Cytokine Induction of CRH in the PVN and ME

Release of CRH from PVN neurons into the ME is established as a primary pathway by which cytokines stimulate the HPA axis. Early groundbreaking studies by Sapolsky et al (44) and Berkenbosch et al (45) concluded that ip administration of IL-1 (in rats) induces a CRH-dependent ACTH response, based on effective neutralization of the response by prior administration of CRH antisera and detection of increased CRH release, as assessed by an increase of CRH in the hyposhyseal portal blood (Sapolsky) and a decrease in CRH content in the median eminence (Berkenbosch). Furthermore, participation of CRH-producing neurons of the PVN is implicated by the findings of IL-1-induced expression of c-fos mRNA (a cellular marker of neuronal activation) and an increase in CRH mRNA in the PVN (46). Similarly, it has been demonstrated that PVN lesions completely prevent the increase in ACTH induced by intravenous (iv) administration of IL-6 and TNFα (47), and that immunoneutralization of CRH blocks the stimulatory effects of IL-6 and TNFα on ACTH and/or CORT release (48-51). IL-1, IL-6 and TNFα also have been shown to stimulate the release of CRH from rat hypothalamic explants (48,52,53). Moreover, both IL-1 and IL-6 are capable of acting directly on CRH nerve terminals in the ME (53-56), thereby increasing CRH release without necessarily increasing CRH synthesis in the PVN. However, a direct action of TNF at the level of the ME has not been demonstrated. Although iv administration of TNFα results in an increase in CRH release from the ME (57), Sharp et al (56) failed to show an effect of TNFα on ACTH secretion when injected directly above the ME.

Some studies have reported differential effects of pro-inflamatory cytokines on PVN activity. While IL-1 and IL-6 were able to stimulate ACTH secretion after intra-ME delivery, only IL-1 was able to stimulate ACTH secretion after injection into the PVN (54). Callahan et al. (58) also reported an increase in Fos expression in CRH-producing neurons of the PVN in response to ip administration of IL-1β, but not to IL-6. Moreover, an ip injection of IL-1β, in addition to eliciting ACTH and CORT responses, produced an increase in CRH mRNA expression in the PVN, while IL-6 did not induce an increase in CRH mRNA expression, despite an increase in ACTH and CORT (59). Initially, Vallieres et al. (60) reported that a single systemic injection of IL-6 was insufficient by itself to induce c-fos and CRH mRNA in the PVN. However, if IL-6 receptors were first induced in the PVN by prior LPS (iv) exposure (IL-6R was barely expressed in the PVN under basal conditions), then IL-6 (iv) was able to induce an increase in c-fos mRNA and CRH hnRNA in the PVN, concordant with a significant

CORT response (61). In addition, continuous infusion of IL-6 has been shown to induce Fos expression in the PVN (62). Vallieres et al (61) also demonstrated that although IL-6 did not appear to be essential during the early phases of endotoxemia, IL-6 was required during the later phases to prolong the activation of neurons throughout the brain and to maintain CRH expression in the PVN. In addition, iv administration of TNFα induces c-fos mRNA and CRH hnRNA expression in the PVN of rats, concordant with elevated plasma CORT levels (39). Taken together, these data suggest that IL-1 and TNFα may be responsible for activating CRH-producing neurons of the PVN in response to an acute immunological challenge, while IL-6 may be responsible for the PVN response to chronic inflammation or in instances where the PVN has been previously sensitized to IL-6 by prior exposure to other cytokines (i.e., TNFα, IL-1).

Aside from CRH, other hypothalamic secretagogues may contribute to cytokine-HPA axis interactions. Arginine vasopressin (AVP) is one such peptide, which plays a synergistic role with CRH in stimulating ACTH release from the anterior pituitary. IL-1, IL-6 and TNFα have been shown in one study or another to influence AVP synthesis or release (2,6). However, a portion of this AVP response may originate from the magnocellular, rather than the parvocellular, cells of the PVN and be destined for release from the posterior pituitary and ultimately the general circulation. There is strong evidence that under conditions of chronic inflammation, such as rheumatoid arthritis, cytokine effects on HPA axis activity may be driven primarily by AVP. In arthritic rats, Harbuz et al (63) have demonstrated an increase in pro-opiomelanocortin (POMC) mRNA expression (the ACTH precursor) in the anterior pituitary concordant with a paradoxical decrease in CRH mRNA in the PVN and CRH release into the hypophyseal portal blood (HPB), while an increase in AVP release into the HPB was detected. In addition, Lewis rats, which are susceptible to developing chronic inflammatory disease due to impaired CRH and CORT secretion, have high plasma levels, hypothalamic content, and *in vitro* release of AVP in comparison with the inflammatory-disease resistant Fisher rats (64). Similar findings have been reported in animal models of other chronic inflammatory disorders, such as lupus (65,66). Taken together, these results suggest that an increase in AVP may attempt to compensate for low CRH drive in HPA axis regulation.

3.1.3 Cytokine Synthesis in the CNS

During systemic inflammation or infection, the central production of TNFα, IL-1, and IL-6 may function in a paracrine manner to amplify the signal of circulating cytokines and maintain elevated HPA axis activity. Evidence for a role in centrally synthesized cytokines comes from the observation that centrally administered IL-1, IL-6, and TNFα stimulate ACTH release at doses much lower than the effective doses administered peripherally (6). In addition, central administration of cytokine antibodies or antagonists attenuate HPA axis activity induced by the peripheral administration of LPS and particular cytokines (67). In healthy brains, cytokines are expressed at low levels or not at all (68). Constitutive expression of cytokines tends to be localized to neurons, whereas induced central cytokines are mainly produced by glial cells, especially microglia (69).

Constitutive expression of IL-1 protein has been detected in the rat hypothalamus and hippocampus (70). Although some studies have failed to detect basal IL-1 expression in the brain (71-73). The discrepancy in results may be due to a difference in sensitivity of the assays used. In response to systemic LPS administration, an increase in IL-1β immunoreactivity or mRNA expression has been detected in perivascular microglia and macrophages in the meninges, choroid plexus, CVO's, and hypothalamus in the rat brain (72,73) and in the mouse hypothalamus and hippocampus (69,74). Ip injection of LPS has been shown to induce a biphasic response in IL-1β mRNA expression in the rat brain, where an early phase of induced transcription (2h post-injection) can be detected in areas without an effective blood-brain barrier, such as the choroid plexus, meninges, and cerebrovasculature, followed by a later phase of expression (6h post-injection) in the parenchyma, including areas such as the hypothalamus (75).

Whereas Schobitz et al (35) have shown that IL-6 mRNA in the rat brain is constitutively expressed in regions such as the hippocampus and hypothalamus (colocalized with IL-6R mRNA), Vallieres et al (36) have reported that IL-6 mRNA expression is undetectable under basal conditions. Upon ip LPS administration, however, IL-6 mRNA is increased in the CVO's (including the ME) and the choroid plexus of the rat (36) and in the brainstem, hippocampus, and hypothalamus of the mouse (69,74). Whereas LPS induces a strong and transient increase in IL-1 and TNFα mRNAs in the hypothalamus and hippocampus, the induction of IL-6 mRNA tends to be progressive over time (69). *In vitro*, IL-6 mRNA and release is detected in rat medial basal hypothalamic fragments under basal conditions and is further stimulated by LPS, presumably from glial cells (76).

While no expression of TNFα mRNA has been detected in the rat brain under basal conditions (77), the mouse brain has been shown to exhibit constitutive levels of TNFα mRNA (74) and TNFα immunoreactivity in neurons of the hypothalamus, caudal raphe system, and the ventral surface of the brainstem (78). Increased TNFα mRNA expression is observed in the rat (79) and the mouse (69) hypothalamus in response to systemic LPS administration. Similar to the pattern of central induction of IL-1mRNA, ip injection of LPS tends to stimulate a biphasic response in TNFα mRNA expression in both the rat (77) and the mouse (80) brain, where an early phase of induced transcription can be detected in barrier-related structures such as the choroid plexus, CVO's, and meninges, followed by a later phase of expression in the parenchyma, including areas such as the hypothalamus and the NTS. Naduau et al (77) have shown that the induced TNFα is mainly a product of microglia and brain macrophages.

3.2 Cytokine – HPA Interactions: Pituitary Gland

Although strong evidence has been provided in support of cytokine action at the level of the CNS to activate the HPA axis, numerous *in vitro* studies indicate that cytokines can exert a direct effect on ACTH release from the anterior pituitary. In addition, *in vivo* work by our lab and others (81-83) have demonstrated that animals whose CRH or CRH-R1 gene has been eliminated are still capable of producing ACTH and CORT responses to immune challenges. These data indicate that pathways are in place for direct cytokine-pituitary interactions.

3.2.1 Cytokine Receptor Expression

Receptors for pro-inflamatory cytokines have been identified in various cell types that reside in the pituitary. Of the two types of IL-1 receptors, both IL-1RI and IL-IRII mRNA's have been detected in the mouse anterior pituitary and in the mouse corticotropic tumor cell line AtT-20 (28,84), and IL-1RI mRNA is constitutively expressed in the rat pituitary gland (85). Furthermore, IL-1RI and IL-1RII immunoreactivity have been localized in somatotrophs in the normal mouse anterior pituitary (86). IL-6R mRNA has been detected in the rat anterior pituitary gland (87) and the expression of IL-6R protein has been localized to mouse corticotrophs (81) and to human gonadotrophs (87). TNF binding sites are found at high concentrations in the anterior lobe of both mouse and rat pituitary glands (88). In addition, the ACTH-producing AtT-20 cells express both Type I and II TNF receptor mRNA and the folliculostellate TtT/GF cells (pituitary source of IL-6, see below) have been found to express TNF-R1 mRNA (89).

Although cytokine receptors have been found on AtT-20 cells, it is unclear to what extent corticotroph tumors accurately represent normal anterior pituitary corticotrophs. The lack of evidence on some cytokine receptor localization to normal corticotrophs may suggest that paracrine interactions are involved between the endocrine cells that express those receptors and the ACTH-producing corticotrophs. Moreover, the presence of cytokine receptors on non-endocrine cell types, suggest that in addition to exerting a direct effect on ACTH release, cytokines (i.e., TNFα) may indirectly modulate ACTH release by first stimulating the release of local cytokines, such as IL-6 from folliculostellate cells.

3.2.2 Cytokine Induction of ACTH

Numerous *in vitro* studies support the contention that pro-inflamatory cytokines can directly stimulate ACTH release from the pituitary gland. IL-1 has been shown to induce POMC transcription and ACTH release from cultured rat anterior pituitary cells (90-92), while IL-6 has been demonstrated to stimulate ACTH secretion from cultured hemipituitaries (93). In addition, both IL-1 and IL-6 have been shown to increase POMC mRNA expression or promoter activity and ACTH release from AtT-20 cells (91,94-96). Moreover, Fukata et al (94) have shown that IL-1β can sensitize AtT-20 cells to release ACTH in response to CRH. TNFα also has the ability to induce ACTH release from cultured rat anterior pituitary cells (48) and to stimulate POMC transcription and ACTH release from AtT-20 cells (89,95). However, according to Gaillard et al (97), TNFα exerts an inhibitory effect on CRH-stimulated ACTH release from rat anterior pituitary cells, while having no effect on basal ACTH secretion.

Nevertheless, there are studies that have failed to demonstrate a direct action of IL-1 (44,45,52), IL-6 (52), and TNFα (52,56,92) on cultured anterior pituitary cells to stimulate ACTH release. The discrepancy between these reports may be explained by differences in the incubation times used, in which short incubation periods (<4h) may not be long enough to observe a direct effect of the cytokine on the pituitary. Therefore, CRH-dependent ACTH release may mediate the rapid onset (<1h) of an acutely induced CORT response, whereas a direct action of cytokines on the anterior pituitary may mediate more latent CORT responses due to chronic inflammation. Another possible reason for the lack of an effect of individual cytokine action on ACTH release at early time points may be the need for syngergistic actions between multiple cytokines. For example, Buckingham et al (2) have demonstrated that rat anterior pituitary cells incubated with conditioned media from macrophages stimulated with LPS (which contains multiple cytokines) increase their ACTH secretion, while no effect was observed when incubated with various cytokines alone.

In vivo, ip administration of either IL-1 or IL-6 has been shown to induce Fos expression in the anterior pituitary (58). However, IL-1, but not IL-6, produces an increase in POMC mRNA in the anterior pituitary with a concomitant increase in plasma ACTH (59). Perhaps IL-6 could be exerting a direct effect on the pituitary to increase ACTH release, without increasing its synthesis. Ip administration of TNFα decreases POMC mRNA in the rat anterior pituitary (98).

Besides IL-1 being able to increase pituitary sensitivity to CRH, CRH has been shown to play a role in increasing pituitary sensitivity to circulating cytokines. Payne et al (99) demonstrated that low levels of exogenously administered (or endogenously produced CRH) may be necessary for IL-1 to induce ACTH release via direct action on the anterior pituitary. Once sensitized, IL-1-induced ACTH release was no longer inhibitable by a CRH antagonist. CRH also has been shown to increase IL-1R expression in AtT-20 cells (100) and increase IL-1α binding in the anterior pituitary (101). Moreover, PVN-lesioned rats injected (ip) with LPS exhibit a delayed ACTH response compared to CRH-intact animals (102). These data support the idea that while the initial ACTH response to cytokines may be dependent on CRH release from the hypothalamus, a more sustained ACTH response may result from a direct action of cyokines on the pituitary gland that has been first sensitized by CRH. However, without the prior sensitization by CRH (possibly by the upregulation of cytokine receptor expression), direct cytokine action on the pituitary may occur only after prolonged exposure to circulating cytokines.

3.2.3 Cytokine Synthesis in the Pituitary

Like the brain, the pituitary can synthesize its own set of local cytokines and serve as an integration point for immune and neuroendocrine signals. IL-1β protein and mRNA have been detected in the rat anterior pituitary and both are increased following ip LPS injections (33,85,103). In addition, Koenig et al (103) showed that IL-1β was localized to cytoplamsic granules in anterior pituitary endocrine cells and colocalized with TSH in thyrotrophs. IL-1α/β mRNA also have been demonstrated in monocytes in both the anterior and posterior lobes of the rat pituitary gland (72). Cultured rat anterior pituitary cells spontaneously secrete large quantities of IL-6, which is enhanced by exposure to LPS (104) and to IL-1α /β (105,106). The source of pituitary IL-6 has been shown to be the folliculostellate (FS) cell (107), which is not a hormone-secreting cell, but a cell of monocytic lineage that is thought to be involved in paracrine regulation of hormone secretion from the pituitary (108,109). Moreover, LPS and TNFα have been shown to directly stimulate IL-6 release from mouse pituitary FS TtT/GF tumor cells (89,110), while LPS produces a similar effect on FS cells of mouse pituitary cell cultures (110). IL-6 immunoreactivity also is expressed in ACTH and FSH/LH-secreting cells in normal human pituitary glands (111) and IL-6 mRNA is induced in the rat pituitary by ip administration of LPS (112). TNFα mRNA is constitutively expressed in the mouse pituitary (74) and is upregulated in response to systemic injection of LPS in the mouse and rat (69,74,77,79). The role of local, pituitary-derived cytokines may be to amplify and maintain elevated pituitary-adrenal activity during chronic inflammation.

3.3 Cytokine – HPA Interactions: Adrenal Gland

There has been a growing body of evidence to support the idea of extrahypothalamic and extrapituitary regulation of adrenal steroidogenesis, whereby circulating cytokines act directly on the adrenal gland. The adrenal cortex is comprised of three histologically distinct layers. From the outermost to innermost (adjacent to the medulla), these layers are the zona granulosa (synthesize mineralocorticoids) and the zonas fasciculata and reticularis (synthesize glucocorticoids and androgens). Although blood flow within the adrenal gland is directed centripetally from the cortex to the medulla, secretory products released by medullary chromaffin cells (catecholamines and various neuropeptides) are able to influence steroidogenesis in the adrenal cortex in a paracrine manner (113). Rays of medullary tissue have been found to extend through the entire cortex of the human adrenal gland and steroid producing cells occur diffusely within the entire adrenal medulla, with the two cell types located in direct apposition (114-116). Systemic cytokine actions at the adrenal gland and their effect on glucocorticoid synthesis/release may be dependent upon intricate intra-adrenal interactions between cortical and medullary cells, the vascular supply, neural input, and locally synthesized cytokines and neuropeptides (i.e., an intra-adrenal CRH-ACTH system) (117,118).

3.3.1 Cytokine Receptor Expression

Although receptors for IL-1 have not been detected in any part of the adrenal gland (28), receptors for both IL-6 and TNFα have been identified in the adrenal gland of numerous species. IL-6R mRNA has been detected primarily in the zona reticularis and the zona fasciculata (to a lesser extent in the zona glomerulosa and the medulla) of the mouse (81) and the human adrenal gland (119), the zona fasciculata of the bovine adrenal gland (120), and more intensely in the medulla than in the cortical regions of the rat adrenal gland (121). Based on immunohistochemical studies, TNF receptors (type I and II) have been localized to the zona fasciculata of the bovine adrenal gland (120). In addition, TNF binding has been demonstrated throughout the rat and bovine adrenal cortex and in the human adrenocortical tumor cell line H295R (122).

3.3.2 Cytokine Induction of Glucocorticoids

Data from numerous *in vitro* studies have demonstrated that pro-inflamatory cytokines can directly affect glucococorticoid release from the adrenal gland. Despite the lack of evidence for the presence of IL-1 receptors in the adrenal gland, IL-1 has been shown to directly stimulate glucocorticoid release. Andreis et al (123) have demonstrated that IL-1β is capable of stimulating the release of CORT from adrenal slices including both cortex and medulla (but not in adrenal cortical cells alone), showing its dependence on medullary chromaffin cells. In addition, IL-1 has been shown to stimulate CORT release on its own from rat (124,125) and bovine (126) adrenal cells in primary culture and to act synergistically with ACTH (125). Possible mechanisms of how IL-1 stimulates CORT release include the involvement of an intra-adrenal CRH/ACTH system (123,127), catecholamines released from medullary chromaffin cells (via α-adrenergic mechanisms) (124,128), and prostaglandins (125,126,129).

IL-6 also has been shown to stimulate the release of CORT from primary cultures of rat, human, and bovine adrenal cells on its own (119-121,130,131) and in synergy with ACTH (120,130). Due to the long incubation time required in the majority of the studies for IL-1 and IL-6 to stimulate CORT release (>12h), it has been suggested that (CRH-dependent) ACTH may mediate early CORT responses to an immune challenge, while a direct action of these cytokines on the adrenal gland may mediate long-term CORT induction. Furthermore, the observation that CORT levels can remain elevated several days after burn trauma (132) or during chronic treatment with IL-6 (133) in humans, or in response to turpentine-induced inflammation in rats (134), despite declining plasma ACTH levels, lends support to the idea that pro-inflamatory cytokines may contribute to a sustained CORT response by directly stimulating CORT release. Nevertheless, there also have been reports in which IL-1 and IL-6 failed to stimulate CORT production by cultured normal rat adrenal cells (44) and by mouse adrenal tumor cells (96), which may be due to the lack of long enough incubation periods.

While IL-1 and IL-6 exert a positive influence on CORT release from the adrenal gland, TNFα has been shown to have an inhibitory effect on CORT synthesis and release. Barney et al (120) demonstrate that TNFα inhibits basal CORT release from bovine adrenal zona fasciculata cells after a 16 incubation period, while inhibiting ACTH-stimulated CORT release at 4h. In a preparation of rat adrenal cells, TNFα also inhibits ACTH-induced CORT release at 6h, but does not have an effect on basal CORT release at that time point (51). In addition to attenuating ACTH-stimulated CORT release, TNFα has been shown to inhibit the ACTH-induced increase in cytochrome P450 oxidase mRNA expression (a group of enzymes involved in glucocorticoid synthesis) in cultured human fetal adrenal cells (135). Interestingly, the direct effect of TNFα on the adrenal gland contradicts the stimulatory effect of circulating TNFα on HPA axis activity seen in an intact animal. Therefore, the role of intra-adrenal TNFα may be different than that of plasma TNFα.

Although *in vitro* studies have demonstrated the ability of cytokines to stimulate CORT release directly from adrenal cells, numerous *in vivo* studies have shown that immune stimuli (i.e., NDV, MCMV, and IL-1) given to hypophysectomized animals can not induce a CORT response independent of pituitary-derived ACTH (10,136-138). In contrast, two studies have reported that NDV (12) and IL-1β (123) can induce a CORT response in hypophysectomized mice.

Studies from our lab (82) have shown that MCMV-infected mice administered CRH anti-sera are able to produce a robust CORT response despite their inability to produce an ACTH response. However, baseline levels of ACTH are still present (unlike hypophysectomized mice) and may serve as a permissive factor to allow for direct cytokine-adrenal interactions. In addition, mice deficient in the CRH or CRH-R1 gene are able to mount a significant CORT response to a variety of immunological stressors, such as carageenan (139), turpentine (83), LPS (81,83,139), and MCMV (82). (Even though they exhibit markedly attenuated CORT responses to physical stressors such as restraint, ether, fasting, and forced swim (140-142).) Of note, the genetically-manipulated animals (and those treated with CRH anti-sera) that receive the immune stimulus, exhibit dramatically greater increases in IL-6 compared to immune challenged control animals. This marked increase in IL-6 may compensate for the lack of CRH function in driving the HPA axis, and become a predominant driving force for glucocorticoid induction by acting directly on the adrenal glands.

3.3.3 Cytokine Synthesis in the Adrenal Gland

Just as the brain and anterior pituitary can synthesize their own cytokines, IL-1, IL-6 and TNFα synthesis by multiple cell types of the adrenal gland have been demonstrated. IL-1α is produced by rat chromaffin cells (143). IL-1β, IL-6 and TNFα mRNA's have been detected in steroid producing cells and macrophages of the zona reticularis in the human adrenal gland (144-146). Macrophages have been localized in all zones of the human and bovine adrenal gland and lie in direct contact with cortical and chromaffin cells (147,148). Nobel and Schultzberg (149) demonstrate that LPS administered to rats induces an increase in IL-1β mRNA expression in both the adrenal cortex and medulla. IL-6 is primarily produced by rat and bovine adrenal zona glomerulosa cells (and to a lesser extent by zona reticularis and fasciculata cells) (150,151). In addition, its release is stimulated by IL-1α/β, LPS, angiotensin II, and ACTH, with IL-1β also potentiating the stimulatory effect of ACTH on IL-6 secretion (151). IL-6 mRNA in the rat adrenal gland also is induced by ip administration of LPS (112). TNFα immunoreactivity is found throughout the bovine adrenal cortex (most prominent in the zona glomerulosa), and while its release is stimulated by IL-1α/β, LPS, and angiotensin II, it is inhibited by ACTH (150). The same has been shown for TNFα release from rat zona glomerulosa cells (152). As with the pituitary, the role of local, adrenal-derived cytokines may be to amplify and maintain elevated adrenal activity in response to chronic inflammation.

4. OTHER CYTOKINES IN IMMUNE-HPA INTERACTIONS

Although this review has focused on the stimulatory effects of the macrophage-produced pro-inflamatory cytokines, TNFα, IL-1, and IL-6, numerous other cytokines and their receptors have been identified in HPA axis tissue and have been shown to modulate immune-HPA axis interactions. Some of these cytokines are briefly described below.

The Type I interferons (IFN) α and β, released during the early innate immune response by virus-infected cells, and the pro-inflamatory Th1-produced cytokines, IL-2 and IFN-γ (Type II interferon) have been reported to have HPA axis-stimulating activity *in vivo*. IL-2 and its receptor have been identified in all three levels of the HPA axis (except for IL-2 expression in the adrenal gland). In addition, IL-2 has been shown to stimulate CRH, ACTH, and CORT release from cultured hypothalamic, pituitary and adrenal cells, respectively. IFN-α and IFN-γ and their receptors also are expressed in the brain. Whereas IFN-α has been shown to stimulate CRH release from hypothalamic cultures and both IFN-α and IFN-γ induce CORT secretion from adrenal cultures, neither of these cytokines has been reported to exert a direct effect on ACTH release from pituitary cells (6).

Another pro-inflamatory cytokine that has recently gained attention for its role in neuroendocrine function is leukemia inhibitory factor (LIF). LIF is member of the IL-6 cytokine superfamily, in which the LIF receptor (LIF-R) shares the gp130 receptor subunit for signal transduction. Serum LIF is increased in a variety of inflammatory conditions, concordant with HPA activation, and systemic injection of LIF has been shown to stimulate pituitary POMC transcription and ACTH release alone and in synergy with

CRH. In addition, the local expression of mouse LIF and LIF-R in the hypothalamus and the pituitary (corticotrophs) is upregulated by various inflammatory stressors. Moreover, LIF KO mice fail to mount an ACTH response and demonstrate blunted POMC and CORT responses to inflammatory stressors (153,154).

Like LIF, macrophage migration inhibitory factor (MIF) is another pro-inflamatory cytokine that is detected in the corticotrophs (and thyrotrophs) of the anterior pituitary and also is expressed (although in lesser amounts) in brain regions such as the hypothalamus, hippocampus, and perivascular epithelial cells. The release of MIF from the pituitary is stimulated by CRH and inflammatory stimuli, such as LPS. Once released into the general circulation, MIF plays a role in opposing the anti-inflammatory actions of glucocorticoids. Interestingly, physiological concentrations of glucocorticoids stimulate macrophage release of MIF, which is in contrast to the typical inhibitory action of glucocorticoids on pro-inflamatory cytokines (67,155,156).

IL-10 is an anti-inflammatory macrophage- and Th2-produced cytokine. Similar to glucocorticoids, it inhibits the production of pro-inflamatory cytokines from macrophages and Th1 cells. IL-10 is produced in pituitary, hypothalamic, and neural tissue. In addition, IL-10 has been shown to stimulate ME fragments to release CRH and AtT-20 pituitary cells to release ACTH and MIF. However, IL-10 seems to negatively regulate glucocorticoid release from the adrenal gland, since cold-swim and PolyI:C-induced glucocorticoid responses are enhanced in IL-10 deficient mice and Y-1 adrenal cells treated with an anti-IL-10 Ab (157).

Finally, IL-18 (also known as interferon-γ-inducing factor (IGIF) or IL-1γ) is a recently identified cytokine that has been detected in the brain, pituitary, and adrenal cortex, specifically in the zona reticularis and fasciculata. Microglial IL-18 mRNA expression is induced by LPS and adrenal IL-18 mRNA is upregulated by acute cold stress and ACTH treatment. (158-160). Although HPA axis stimulating activity has not yet been demonstrated for this cytokine, the localization of IL-18 in the three levels of the HPA axis suggests that it may play a modulatory role in immune-neuroendocrine interactions.

5. LOCAL FACTORS REGULATING GLUCOCORTICOID BIOAVAILABILITY AND ACTION

5.1 Factors Regulating Glucocorticoid Bioavailability

There is no question that circulating levels of glucocorticoid hormones are relevant to steroid action, however, at the cellular level, activity of glucocorticoids is determined by local factors that regulate the access of free hormone to its receptor. Such factors include corticosterone binding globulin (CBG), 11β-hydroxysteroid dehydrogenase (11β-HSD), and the multidrug resistance transporter (MDR), all of which have been shown to be altered under conditions of immune activation (Figure 6).

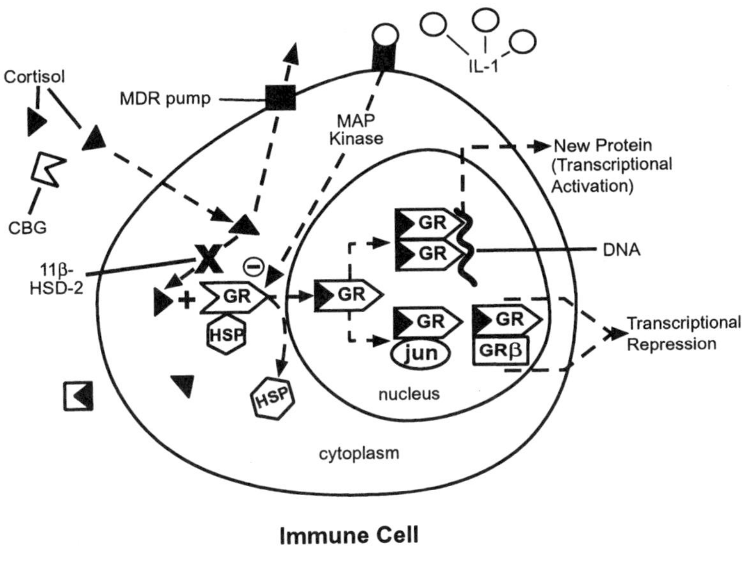

Figure 6. Local factors regulating glucocorticoid bioavailability and action. 1) corticosterone binding globulin (CBG), 2) 11β-hydroxysteroid dehydrogenase (11β-HSD), 3) multidrug resistance transporter (MDR pump), 4) glucocorticoid receptor (GR = GRα) nuclear translocation, 5) GR interaction with other transcription factors (AP-1 [jun/fos], NFκB, MAPK), and 6) ratio of GRα:GRβ isoform expression. HSP, heat shock protein.

5.1.1 Corticosterone Binding Globulin (CBG)

Only free or unbound glucocorticoids are capable of diffusing across the plasma membrane and activating the glucocorticoid receptor (GR). Over 90% of circulating glucocorticoids are bound to corticosteroid-binding globulin (CBG) (161). Therefore, the relative concentration of CBG is an important determinant of "free" and available glucocorticoid. For example, decreases in the circulating level of CBG have been associated with evidence of occupation/activation of GR in the spleen of stressed rats (162). Likewise, several studies have shown that during immune activation in rats, such as turpentine-induced inflammation (163) and endotoxin administration (164), and in a murine model of systemic lupus erythrematosus (66), reduced CBG levels are observed. In addition, a decrease in plasma CBG-binding capacity is associated with human septic shock (165). However, two studies examining plasma CBG during viral infection detected no change of CBG levels in infected subjects compared to controls (166,167). It is important to note that even if a change in plasma CBG levels is not detected, local changes (for instance, within the microenvironment of the various lymphoid compartments) in the concentration and/or affinity of CBG during inflammation/infection remain a consideration.

Since CBG has a primary structure with marked homology to the family of serpins (serine protease inhibitors), it has been suggested that local release of proteases (such as elastase) by activated neutrophils at the site of inflammation may cleave CBG (168). This results in a conformational change in the molecule that prevents its steroid binding activity and therefore reduces CBG's affinity for glucocorticoids. As a result, the local concentration of free glucocorticoid is increased, thereby making the steroid hormone available to interact with GR. It is unlikely that elastase cleavage of CBG occurs in the general circulation due to an excess of other serine protease inhibitors (eg. α_1-proteinase inhibitor/A1-PI) that may form an inhibitory complex with neutrophil elastase, therefore preventing the elastase cleavage of CBG. However, at the local site of inflammation, the production of oxygen-free radicals interacts with the active site of A1-PI and destroys its ability to inhibit elastase produced by activated neutrophils (168).

5.1.2 11β-hydroxysteroid dehydrogenase (11β-HSD)

Another factor regulating the access of glucocorticoids to their receptors in target immune cells and tissues is 11β-hydroxysteroid dehydrogenase (11β-HSD). This enzyme comes in two isoforms, type I and type II. Whereas 11β-HSD-1 is a reversible oxidoreductase, 11β-HSD-2 only exhibits oxidative or dehydrogenase activity (169). 11β-HSD-2 breaks down naturally occurring glucocorticoids (but not synthetic glucocorticoids) as they enter the cell, leaving the hormones in the form of inactive metabolites (i.e., cortisol to cortisone in humans; corticosterone to 11-dehydrocorticosterone in rodents). Interestingly, significant differences in

11β-HSD-2 activity have been found among immune compartments, and there is a direct correlation between 11β-HSD-2 activity and the preferential production of Th1 versus Th2 cytokines by T cells residing in particular lymphoid organs (170). Inhibition of 11β-HSD-2 activity (which would enhance the available amount of hormone to bind to GR) leads to reduced Th1 responses and enhanced Th2 cytokine production by activated T cells (170).

Recent studies suggest that the bi-directional 11β-HSD-1 prefers the reductase direction unless cells are disrupted (169). Therefore, in intact tissues, 11β-HSD-1 reactivates the inactive 11-keto glucocorticoids (corticosterone/11-dehydro-corticosterone) into their active 11-hydroxy glucocorticoid form (cortisol/corticosterone). Inflammatory glomerular cells exposed to TNFα or IL-1β have been shown to exhibit an increase in 11β-HSD-1 expression and reductase activity, thereby favoring the formation of active glucocorticoids to counterbalance the pro-inflamatory effect of such cytokines (171). Therefore, changes in the relative activity of 11β-HSD in immune tissues during inflammation/infection may influence the relative impact of glucocorticoids on immune responses. Thus, the microenvironment of the various immune compartments is a potentially important site for regulation of neuroendocrine-immune interactions.

5.1.3 Multidrug Resistance P-glycoprotein (MDR Pgp)

MDR Pgp is an ATP-dependent multidrug efflux pump that decreases intracellular concentrations of potentially toxic chemicals (drugs and hormones). It is expressed in both human and rodent tissues, including the adrenal gland, kidney, liver, colon, small intestine, and brain and testis capillary endothelial cells (172). In the interest of glucocorticoids, the MDR pump at the BBB, in both mice and humans, transports cortisol and the synthetic glucocorticoid dexamethasone, but not corticosterone out of endothelial cells lining the brain (173). *In vitro* studies have indicated that cytokines, such as IL-1, IL-6 and TNFα, decrease MDR expression and/or function in rodent hepatocytes (174,175), human colon carcinoma cells (176), and human brain endothelial cells (177). In contrast, patients with autoimmune diseases, such as rheumatoid arthritis (178), colitis/Crohn's disease (179), and lupus (180), tend to exhibit high lymphocitic MDR expression and/or activity. Greater MDR function in immune cells may reduce glucocorticoid availability and thereby enhance the release of pro-inflamatory cytokines and exacerbate inflammatory responses.

5.2 The Glucocorticoid Receptor

The ultimate effect of glucocorticoids on immune system regulation is established at the receptor level. The majority of the well-characterized effects of glucocorticoids on target tissue function are mediated through intracellular glucocorticoid receptors (GR). Upon glucocorticoid binding to cytosolic GR's, a conformational change in GR causes heat-shock protein 90 (hsp90) and other ancillary proteins to dissociate from the receptor, and the ligand-bound receptor then translocates into the nucleus. Here, it dimerizes with another ligand-bound GR, thereby allowing it to alter the transcription rates of glucocorticoid–sensitive genes, either through protein-DNA (glucocorticoid response element - GRE) interactions or through protein-protein (other transcription factors) interactions. A tissue's sensitivity to glucocorticoid activity can be influenced by a change in 1) GR number or affinity, or 2) GR function, including its ability for nuclear translocation, its interaction with other signal transduction pathways, and the expression of particular GR isoforms (Figure 6). Pro-inflamatory cytokines have been shown to impact a number of these factors.

5.2.1 GR number and affinity

There is a large body of data on the impact of cytokines on GR number (181). Interestingly, however, the results are split into those studies that find an increase in GR number following cytokine administration and those that find a decrease. The discrepancy in results appears to depend on how the receptors were measured. Studies using whole-cell assay binding techniques tend to find an increase in cytokine-induced GR expression, while those using cytosolic receptor binding techniques tend to find GR number to be decreased. Results are further complicated by the fact that the studies utilizing whole-cell binding techniques tend to use longer incubation times (i.e., 24h), whereas the studies employing cytosolic receptor binding tend to examine shorter incubation periods (i.e., 4-6h). Few studies have documented changes in receptor affinity. However, Kam et al (182) report that lymphocytes simultaneously exposed to Th1 and Th2 cytokines exhibit a reduced affinity of GR for glucocorticoids.

5.2.2 GR Function

Several autoimmune/inflammatory disorders have been associated with impaired GR function, possibly contributing to the excessive inflammation characteristic of these illnesses. Pro-inflamatory cytokines (TNFα, IL-1, and

IL-6) and cytokines that mediate lymphocyte growth and differentiation (IL-2, IL-4) have been found to inhibit GR function (181).

Although the mechanism by which cytokines inhibit GR is unknown, several possibilities have been considered. First, cytokines may influence GR function through their effects on GR translocation from the cytoplasm into the nucleus. For example, TNFα and IL-1 have been shown to block dexamethasone-induced nuclear GR translocation (183,184). Another possible mechanism by which cytokines may influence GR is through cross-talk among signal transduction pathways. For example, downstream in the IL-1 and TNFα signal transduction pathways are the transcription factors NFκB and AP-1 (which consists of jun and fos proteins). GR and NFκ/AP-1 have been shown to mutually antagonize each other's actions by binding to each other in the nucleus and inhibiting the other's binding to their respective recognition sites on DNA (185-187). In addition, IL-1 activates mitogen-activated protein kinase (MAPK) signaling pathways, which have been shown to be inhibitory to GR function (188). Finally, cytokines may affect GR function by altering the ratio of GRα:GRβ isoform expression. Alternative splicing of the human GR primary transcript produces two isoforms (189), GRα and GRβ, the latter of which may negatively regulate GR activity. Oakely et al (190), have demonstrated that the GRα isoform binds hormone and activates glucocorticoid-responsive genes, while the GRβ isoform fails to bind hormone and activate glucocorticoid-responsive genes, and attenuates the trans-activation of the hormone-bound GRα isoform. These findings suggest that, when GRβ is present in excess in various tissues, it might repress glucocorticoid-mediated GRα function. In fact, Webster et al (191) have shown that IL-1 and TNFα lead to the selective accumulation of GRβ protein in cells of lymphoid origin and the development of a glucocorticoid resistant state. Moreover, increases in GRβ levels have been found in immune cells of patients with glucocorticoid-resistant asthma (192) and colitis (193).

6. CLINICAL CONSIDERATIONS

Given the critical role played by glucocorticoids in regulating immune responses, (especially restraining inflammatory responses), patient populations with altered HPA axis activity may be especially vulnerable to the pathological consequences of prolonged exposure to elevated circulating levels of pro-inflamatory cytokines. Two stress-related neuropsychiatric disorders are of special consideration. Both Post Traumatic Stress Disorder (PTSD) and major depression are disorders characterized by deficient glucocorticoid signaling, and both have been found to be associated with an exaggerated inflammatory response system. In PTSD, patients exhibit reduced plasma, urinary and salivary concentrations of CORT (194), and in major depression, patients exhibit reduced glucocorticoid receptor function as manifested by impaired responsiveness to dexamethasone administered either *in vivo* or *in vitro* (195). Patients with PTSD and major depression also have been shown to have elevated levels of pro-inflamatory cytokines, including IL-1 and IL-6, as well as increases in acute phase reactants which are produced in the liver in response to these cytokines (especially IL-6) (196-198). Of note, pro-inflamatory cytokines can have profound effects on cellular function throughout the body and can affect behavior. For example, TNF, IL-1 and IL-6 have been shown to contribute to cell death in multiple regions of the CNS as well as contributing to osteoporosis and insulin resistance; all of which are pathophysiologic features of these disorders (in particular, major depression) (199,200). Finally, given the capacity of pro-inflamatory cytokines to cause changes in behavior, including the induction of sickness behavior (which includes symptoms of depression, fatigue, anhedonia, and cognitive dysfunction), it has been suggested that pro-inflamatory cytokines may contribute to some of the behavioral features of both major depression and PTSD, as well as other disorders, including chronic fatigue syndrome and fibromyalgia, which also are characterized by reduced glucocorticoids and increased inflammatory mediators (201-205).

7. CONCLUSIONS

The immune system, via pro-inflamatory cytokines (TNFα, IL-1, and IL-6), stimulates glucocorticoid release by acting at all three levels of the HPA axis. Cytokine receptors have been detected at all HPA axis levels and therefore, each level can serve as an integration point for immune and neuroendocrine signals. It appears that the initial activation of the HPA axis is mediated via CRH dependent pathways, for example, through the traditional CRH to ACTH to CORT pathway. However, when a sustained CORT response to an immunological stressor is neccessary, complementary, CRH-independent mechanisms of glucocorticoid induction may occur to maintain the initial rise in CORT levels. These alternative pathways include the direct action of cytokines on the pituitary and adrenal glands. The local production of cytokines at each level of the HPA axis also may contribute to the amplification and maintainance of elevated HPA activity during chronic inflammation.

The fact that CRH-independent mechanisms of CORT induction may contribute to the maintenance phase of the CORT response rather than the initiation is supported by the findings that, *in vitro*, cytokines can directly stimulate ACTH release from pituitary tissue and CORT from adrenal tissue, but only after prolonged incubation periods. The direct action of cytokines on the pituitary and adrenal glands, however, may not be completely independent of CRH or ACTH, respectively, where CRH / ACTH may be required as a permissive factor to allow a direct action of cytokines on the lower levels of the HPA axis. This hypothesis is supported by the *in vivo* findings that hypophysectomized animals administered various immunological challenges (in the majority of the studies) are unable to exhibit an increase in glucocorticoids.

Once the pro-inflamatory cytokines have stimulated glucocorticoid release to keep inflammatory responses in check, it is also important to regulate the resultant activity of glucocorticoids themselves. Either too much or too little activity of this steroid hormone may result in pathological consequences, due to the inappropriate shaping of an immune response. Therefore, several local factors are in place to maintain appropriate glucocorticoid activity at the level of the immune cell. In conclusion, a closed loop, regulatory feedback circuit exists, with redundant pathways for cytokines to stimulate glucocorticoid release and regulate glucocorticoid activity, which in turn, is crucial to prevent the overshoot of inflammatory responses that may be detrimental to the organism.

REFERENCES

1. Besedovsky, H.O. and del Rey, A. Immune-neuro-endocrine interactions: facts and hypotheses. *Endoc. Rev.* 1996; 17: 64-102.
2. Buckingham, J.C., Loxley, H.D., Christian, H.C., et al. Activation of the HPA axis by immune insults: roles and interactions of cytokines, eicosanoids, glucocorticoids. *Pharmacol., Biochem. Behav.* 1996; 54: 285-298.
3. Harbuz, M.S. and Lightman, S.L.. Stress and the hypothalamo-pituitary-adrenal axis: acute, chronic and immunological activation. *J. Endocrinol.* 1992; 134: 327-339.
4. McEwen, B.S., Biron, C.A., Brunson, K.W., et al. The role of adrenocorticoids as modulators of immune function in health and disease: neural, endocrine and immune interactions. *Brain Res. - Brain Res. Rev.*1997; 23: 79-133.
5. Schobitz, B., Reul, J.M. and Holsboer, F. The role of the hypothalamic-pituitary-adrenocortical system during inflammatory conditions. *Critical Rev. Neurobiol.* 1994; 8: 263-291.
6. Turnbull, A.V. and Rivier, C.L. Regulation of the hypothalamic-pituitary-adrenal axis by cytokines: actions and mechanisms of action. *Physiol. Rev.* 1999; 79: 1-71.
7. Chrousos, G.P. The stress response and immune function: clinical implications. The 1999 Novera H. Spector Lecture. *Ann. N.Y. Acad. Sci.* 2000; 917: 38-67.
8. Besedovsky, H., Sorkin, E., Keller, M., et al. Changes in blood hormone levels during the immune response. *Proc. Soc. Exp.Biol. Med.*1975; 150: 466-470.
9. Besedovsky, H.O. and del Rey, A. Mechanism of virus-induced stimulation of the hypothalamus-pituitary-adrenal axis. *J. Steroid Biochem.* 1989; 34: 235-239.

10. Dunn, A.J., Powell, M.L. and Gaskin, J.M. Virus-induced increases in plasma corticosterone. *Science* 1987; 238: 1423-1425.
11. Dunn, A.J., Powell, M.L., Meitin, C., et al. Virus infection as a stressor: influenza virus elevates plasma concentrations of corticosterone, and brain concentrations of MHPG and tryptophan. *Physiol. Behav.*1989; 45: 591-594.
12. Smith, E.M., Meyer, W.J. and Blalock, J.E. Virus-induced corticosterone in hypophysectomized mice: a possible lymphoid adrenal axis. *Science* 1982; 218: 1311-1312.
13. Besedovsky, H., del Rey, A., Sorkin, E., et al. Immunoregulatory feedback between interleukin-1 and glucocorticoid hormones. *Science* 1986; 233: 652-654.
14. Perlstein, R.S., Whitnall, M.H., Abrams, J.S., et al. Synergistic roles of interleukin-6, interleukin-1, and tumor necrosis factor in the adrenocorticotropin response to bacterial lipopolysaccharide in vivo. *Endocrinol.*1993; 132: 946-952.
15. Van der Meer, M.J., Sweep, C.G., Pesman, G.J., et al. Synergism between IL-1 beta and TNF-alpha on the activity of the pituitary-adrenal axis and on food intake of rats. *Am. J. Physiol.* 1995; 268: E551-E557.
16. Ruzek, M.C., Miller, A.H., Opal, S.M., et al. Characterization of early cytokine responses and an interleukin (IL)-6-dependent pathway of endogenous glucocorticoid induction during murine cytomegalovirus infection. *J. Exp. Med.*1997; 185: 1185-1192.
17. Orange, J.S., Wang, B., Terhorst, C., et al. Requirement for natural killer cell-produced interferon gamma in defense against murine cytomegalovirus infection and enhancement of this defense pathway by interleukin 12 administration. *J. Exp. Med.*1995; 182: 1045-1056.
18. Salazar-Mather, T.P., Orange, J.S. and Biron, C.A. Early murine cytomegalovirus (MCMV) infection induces liver natural killer (NK) cell inflammation and protection through macrophage inflammatory protein 1alpha (MIP-1alpha)-dependent pathways. *J. Exp. Med.* 1998; 187: 1-14.
19. Ruzek, M.C., Pearce, B.D., Miller, A.H., et al. Endogenous glucocorticoids protect against cytokine-mediated lethality during viral infection. *J. Immunol.*1999; 162: 3527-3533.
20. Kapcala, L.P., Chautard, T. and Eskay, R.L. The protective role of the hypothalamic-pituitary-adrenal axis against lethality produced by immune, infectious, and inflammatory stress. *Ann. N.Y. Acad. Sci.*1995; 771: 419-437.
21. Price, P., Olver, S.D., Silich, M., et al. Adrenalitis and the adrenocortical response of resistant and susceptible mice to acute murine cytomegalovirus infection. *Eur. J. Clin. Invest.* 1996; 26: 811-819.
22. Semenkov, V.F., Levina, L.V., Lavrov, V.F., et al. The influence of endogenous hypocorticism on the replication of influenza virus in the mouse lungs and the production of specific antibodies. *Acta Virolog.*1990; 34: 37-43.
23. Rivest, S., Lacroix, S., Vallieres, L., et al. How the blood talks to the brain parenchyma and the paraventricular nucleus of the hypothalamus during systemic inflammatory and infectious stimuli. *Proc. Soc. Exp.Biol. Med.* 2000; 223: 22-38.
24. Watkins, L.R., Maier, S.F. and Goehler, L.E. Cytokine-to-brain communication: a review & analysis of alternative mechanisms. *Life Sci.* 1995; 57: 1011-1026.
25. Herman, J.P., Schafer, M.K. Young, E.A, et al. Evidence for hippocampal regulation of neuroendocrine neurons of the hypothalamo-pituitary-adrenocortical axis. *J. Neurosci.* 1989; 9: 3072-3082.
26. French, R.A., VanHoy, R.W., Chizzonite, R., et al. Expression and localization of p80 and p68 interleukin-1 receptor proteins in the brain of adult mice. *J. Neuroimmunol.*1999; 93: 194-202.

27. Parnet, P., Amindari, S., Wu, C., et al. Expression of type I and type II interleukin-1 receptors in mouse brain. *Brain Res.- Molec. Brain Res.*1994; 27: 63-70.
28. Cunningham, E.T., Wada, E., Carter, D.B., et al. In situ histochemical localization of type I interleukin-1 receptor messenger RNA in the central nervous system, pituitary, and adrenal gland of the mouse. *J. Neurosci.*1992; 12: 1101-1114.
29. Loddick, S.A., Liu, C., Takao, T., et al. Interleukin-1 receptors: cloning studies and role in central nervous system disorders. *Brain Res. - Brain Res. Rev.*1998; 26: 306-319.
30. Wong, M.L., Bongiorno, P.B., al-Shekhlee, A., et al. IL-1 beta, IL-1 receptor type I and iNOS gene expression in rat brain vasculature and perivascular areas. *Neuroreport* 1996; 7: 2445-2448.
31. Van Dam, A.M., De Vries, H.E., Kuiper, J., et al. Interleukin-1 receptors on rat brain endothelial cells: a role in neuroimmune interaction? [erratum appears in *FASEB* J 1996; 10: 660]. *FASEB J.*1996; 10: 351-356.
32. Goehler, L.E., Relton, J.K., Dripps, D., et al. Vagal paraganglia bind biotinylated interleukin-1 receptor antagonist: a possible mechanism for immune-to-brain communication. *Brain Res. Bull.* 1997; 43: 357-364.
33. Takao, T., Culp, S.G. and De Souza, E.B. Reciprocal modulation of interleukin-1 beta (IL-1 beta) and IL-1 receptors by lipopolysaccharide (endotoxin) treatment in the mouse brain-endocrine-immune axis. *Endocrinol.* 1993; 132: 1497-1504.
34. Marquette, C., Van Dam, A.M., Van Rooijen, N., et al. Peripheral macrophage depletion prevents down regulation of central interleukin-1 receptors in mice after endotoxin administration. *Psychoneuroendocrinol.* 1994; 19: 189-196.
35. Schobitz, B., de Kloet, E.R., Sutanto, W., et al. Cellular localization of interleukin 6 mRNA and interleukin 6 receptor mRNA in rat brain. *Eur. J. Neurosci.* 1993; 5: 1426-1435.
36. Vallieres, L. and Rivest, S. Regulation of the genes encoding interleukin-6, its receptor, and gp130 in the rat brain in response to the immune activator lipopolysaccharide and the pro-inflamatory cytokine interleukin-1beta. *J. Neurochem.*1997; 69: 1668-1683.
37. Sawada, M., Itoh, Y., Suzumura, A., et al. Expression of cytokine receptors in cultured neuronal and glial cells. *Neurosci. Lett.* 1993; 160: 131-134.
38. Watanabe, D., Yoshimura, R., Khalil, M., et al. Characteristic localization of gp130 (the signal-transducing receptor component used in common for IL-6/IL-11/CNTF/LIF/OSM) in the rat brain. *Eur.J. Neurosci.*1996; 8: 1630-1640.
39. Nadeau, S. and Rivest, S. Effects of circulating tumor necrosis factor on the neuronal activity and expression of the genes encoding the tumor necrosis factor receptors (p55 and p75) in the rat brain: a view from the blood-brain barrier. *Neurosci.* 1999; 93: 1449-1464.
40. Kinouchi, K., Brown, G., Pasternak, G., et al. Identification and characterization of receptors for tumor necrosis factor-alpha in the brain. *Biochem. Biophys.Res. Comm.* 1991; 181: 1532-1538.
41. Bebo, B.F. and Linthicum, D.S. Expression of mRNA for 55-kDa and 75-kDa tumor necrosis factor (TNF) receptors in mouse cerebrovascular endothelium: effects of interleukin-1 beta, interferon-gamma and TNF-alpha on cultured cells. *J. Neuroimmunol.* 1995; 62: 161-167.
42. Dopp, J.M., Mackenzie-Graham, A., Otero, G.C., et al. Differential expression, cytokine modulation, and specific functions of type-1 and type-2 tumor necrosis factor receptors in rat glia. *J. Neuroimmunol.*1997; 75: 104-112.
43. Boka, G., Anglade, P., Wallach, D., et al. Immunocytochemical analysis of tumor necrosis factor and its receptors in Parkinson's disease. *Neurosci. Lett.* 1994; 172: 151-154.

44. Sapolsky, R., Rivier, C., Yamamoto, G., et al. Interleukin-1 stimulates the secretion of hypothalamic corticotropin-releasing factor. *Science* 1987; 238: 522-524.
45. Berkenbosch, F., van Oers, J., del Rey, A., et al. Corticotropin-releasing factor-producing neurons in the rat activated by interleukin-1. *Science* 1987; 238: 524-526.
46. Ericsson, A., Kovacs, K.J., Sawchenko, P.E. A functional anatomical analysis of central pathways subserving the effects of interleukin-1 on stress-related neuroendocrine neurons. *J. Neurosci.* 1994; 14: 897-913.
47. Kovacs, K.J. and Elenkov, I.J. Differential dependence of ACTH secretion induced by various cytokines on the integrity of the paraventricular nucleus. *J. Neuroendocrinol.* 1995; 7: 15-23.
48. Bernardini, R., Kamilaris, T.C., Calogero, A.E., et al. Interactions between tumor necrosis factor-alpha, hypothalamic corticotropin-releasing hormone, and adrenocorticotropin secretion in the rat. *Endocrinol.* 1990; 126: 2876-2881.
49. Kageyama, K., Watanobe, H. and Takebe, K. In vivo evidence that arginine vasopressin is involved in the adrenocorticotropin response induced by interleukin-6 but not by tumor necrosis factor-alpha in the rat. *Neuroimmunomod.* 1995; 2: 137-140.
50. Naitoh, Y,, Fukata, J., Tominaga, T., et al. Interleukin-6 stimulates the secretion of adrenocorticotropic hormone in conscious, freely-moving rats *Biochem. Biophys.Res. Comm.* 1988; 155: 1459-1463.
51. van der Meer, M.J., Sweep, C.G., Rijnkels, C.E., et al. Acute stimulation of the hypothalamic-pituitary-adrenal axis by IL-1 beta, TNF alpha and IL-6: a dose response study. *J. Endocrinol. Invest.* 1996; 19: 175-182.
52. Navarra, P., Tsagarakis, S., Faria, M.S., et al. Interleukins-1 and -6 stimulate the release of corticotropin-releasing hormone-41 from rat hypothalamus in vitro via the eicosanoid cyclooxygenase pathway. *Endocrinol.* 1991; 128: 37-44.
53. Spinedi, E., Hadid, R., Daneva, T., et al. Cytokines stimulate the CRH but not the vasopressin neuronal system: evidence for a median eminence site of interleukin-6 action. *Neuroendocrinol.* 1992; 56: 46-53.
54. Matta, S.G., Weatherbee, J. and Sharp, B.M. A central mechanism is involved in the secretion of ACTH in response to IL-6 in rats: comparison to and interaction with IL-1 beta. *Neuroendocrinol.* 1992; 56: 516-525.
55. McCoy, J.G., Matta, S.G. and Sharp, B.M. Prostaglandins mediate the ACTH response to interleukin-1-beta instilled into the hypothalamic median eminence. *Neuroendocrinol.* 1994; 60: 426-435.
56. Sharp, B.M., Matta, S.G., Peterson, P.K., et al. Tumor necrosis factor-alpha is a potent ACTH secretagogue: comparison to interleukin-1 beta. *Endocrinol.* 1989; 124: 3131-3133.
57. Watanobe, H. and Takebe, K. Intravenous administration of tumor necrosis factor-alpha stimulates corticotropin releasing hormone secretion in the push-pull cannulated median eminence of freely moving rats. *Neuropep.* 1992; 22: 81-84.
58. Callahan, T.A. and Piekut, D.T. Differential Fos expression induced by IL-1beta and IL-6 in rat hypothalamus and pituitary gland. *J. Neuroimmunol.* 1997; 73: 207-211.
59. Harbuz, M.S., Stephanou, A., Sarlis, N., et al. The effects of recombinant human interleukin (IL)-1 alpha, IL-1 beta or IL-6 on hypothalamo-pituitary-adrenal axis activation. *J. Endocrinol.* 1992; 133: 349-355.
60. Vallieres, L., Lacroix, S. and Rivest, S. Influence of interleukin-6 on neural activity and transcription of the gene encoding corticotrophin-releasing factor in the rat brain: an effect depending upon the route of administration. *Eur. J. Neurosci.* 1997; 9: 1461-1472.

61. Vallieres, L. and Rivest, S. Interleukin-6 is a needed pro-inflamatory cytokine in the prolonged neural activity and transcriptional activation of corticotropin-releasing factor during endotoxemia. *Endocrinol.* 1999; 140: 3890-3903.
62. Niimi, M., Wada, Y., Sato, M., et al. Effect of continuous intravenous injection of interleukin-6 and pretreatment with cyclooxygenase inhibitor on brain c-fos expression in the rat. *Neuroendocrinol.* 1997; 66: 47-53.
63. Harbuz, M.S., Rees, R.G., Eckland, D., et al. Paradoxical responses of hypothalamic corticotropin-releasing factor (CRF) messenger ribonucleic acid (mRNA) and CRF-41 peptide and adenohypophysial proopiomelanocortin mRNA during chronic inflammatory stress. *Endocrinol.* 1992; 130: 1394-1400.
64. Patchev, V.K., Kalogeras, K.T., Zelazowski, P., et al. Increased plasma concentrations, hypothalamic content, and in vitro release of arginine vasopressin in inflammatory disease-prone, hypothalamic corticotropin-releasing hormone-deficient Lewis rats. *Endocrinol.* 1992; 131: 1453-1457.
65. Sakic, B., Laflamme, N., Crnic, L.S., et al. Reduced corticotropin-releasing factor and enhanced vasopressin gene expression in brains of mice with autoimmunity-induced behavioral dysfunction. *J.Neuroimmunol.* 1999; 96: 80-91.
66. Shanks, N., Moore, P.M., Perks, P., et al. Alterations in hypothalamic-pituitary-adrenal function correlated with the onset of murine SLE in MRL +/+ and lpr/lpr mice. *Brain Beh. Immunity* 1999; 13: 348-360.
67. Mulla, A. and Buckingham, J.C. Regulation of the hypothalamo-pituitary-adrenal axis by cytokines. *Best Pract. Res. Clin. Endocrinol. Metab.* 1999; 13: 503-521.
68. Schobitz, B., De Kloet, E.R. and Holsboer, F. Gene expression and function of interleukin 1, interleukin 6 and tumor necrosis factor in the brain. *Prog. Neurobiol.* 1994; 44: 397-432.
69. Laye, S., Parnet, P., Goujon, E., et al. Peripheral administration of lipopolysaccharide induces the expression of cytokine transcripts in the brain and pituitary of mice. *Brain Res.-Molec. Brain Res.* 1994; 27: 157-162.
70. Lechan, R.M., Toni, R., Clark, B.D., et al. Immunoreactive interleukin-1 beta localization in the rat forebrain. *Brain Res.* 1990; 514: 135-140.
71. Ban, E., Haour, F. and Lenstra, R. Brain interleukin 1 gene expression induced by peripheral lipopolysaccharide administration. *Cytokine* 1992; 4: 48-54.
72. Eriksson, C., Nobel, S., Winblad, B., et al. Expression of interleukin 1 alpha and beta, and interleukin 1 receptor antagonist mRNA in the rat central nervous system after peripheral administration of lipopolysaccharides. *Cytokine* 2000; 12: 423-431.
73. Van Dam, A.M., Bauer, J., Tilders, F.J., et al. Endotoxin-induced appearance of immunoreactive interleukin-1 beta in ramified microglia in rat brain: a light and electron microscopic study. *Neurosci.* 1995; 65: 815-826.
74. Pitossi, F., del Rey, A., Kabiersch, A., et al. Induction of cytokine transcripts in the central nervous system and pituitary following peripheral administration of endotoxin to mice. *J. Neurosci. Res.* 1997; 48: 287-298.
75. Wong, M.L., Bongiorno, P.B., Rettori, V., et al. Interleukin (IL) 1beta, IL-1 receptor antagonist, IL-10, and IL-13 gene expression in the central nervous system and anterior pituitary during systemic inflammation: pathophysiological implications. *Proc. Natl. Acad. Sci.* U.S.A. 1997; 94: 227-232.
76. Spangelo, B.L., Judd, A.M., MacLeod, R.M., et al. Endotoxin-induced release of interleukin-6 from rat medial basal hypothalami. *Endocrinol.* 1990; 127: 1779-1785.
77. Nadeau, S. and Rivest, S. Regulation of the gene encoding tumor necrosis factor alpha (TNF-alpha) in the rat brain and pituitary in response in different models of systemic immune challenge. *J. Neuropathol. Exp. Neurol.* 1999; 58: 61-77.

78. Breder, C.D., Tsujimoto, M., Terano, Y., et al. Distribution and characterization of tumor necrosis factor-alpha-like immunoreactivity in the murine central nervous system. *J. Comparative Neurol.* 1993; 337: 543-567.
79. Gatti, S. and Bartfai, T. Induction of tumor necrosis factor-alpha mRNA in the brain after peripheral endotoxin treatment: comparison with interleukin-1 family and interleukin-6. *Brain Res.* 1993; 624: 291-294.
80. Breder, C.D., Hazuka, C., Ghayur, T., et al. Regional induction of tumor necrosis factor alpha expression in the mouse brain after systemic lipopolysaccharide administration. *Proc. Natl. Acad. Sci.*U.S.A. 1994; 91: 11393-11397.
81. Bethin, K.E., Vogt, S.K. and Muglia, L.J. Interleukin-6 is an essential, corticotropin-releasing hormone-independent stimulator of the adrenal axis during immune system activation. *Proc. Natl. Acad. Sci.* U.S.A. 2000; 97: 9317-9322.
82. Silverman, M.N., Pearce, B.D., Biron, C.A., et al. CRH-dependency of ACTH and corticosterone responses during viral infection. *Soc. Neurosci.* 2001; 27: 843-18.
83. Turnbull, A.V., Smith, G.W., Lee, S., et al. CRF type I receptor-deficient mice exhibit a pronounced pituitary-adrenal response to local inflammation. *Endocrinol*1999; 140: 1013-1017.
84. Parnet, P., Brunke, D.L., Goujon, E., et al. Molecular identification of two types of interleukin-1 receptors in the murine pituitary gland. *J. Neuroendocrinol.*1993; 5: 213-219.
85. Whiteside, M.B., Quan, N. and Herkenharn, M. Induction of pituitary cytokine transcripts by peripheral lipopolysaccharide. *J. Neuroendocrinol.*1999; 11: 115-120.
86. French, R.A., Zachary, J.F., Dantzer, R., et al. Dual expression of p80 type I and p68 type II interleukin-I receptors on anterior pituitary cells synthesizing growth hormone. *Endocrinol.*1996; 137: 4027-4036.
87. Ohmichi, M., Hirota, K., Koike, K., et al. Binding sites for interleukin-6 in the anterior pituitary gland. *Neuroendocrinol.*1992; 55: 199-203.
88. Wolvers, D.A., Marquette, C., Berkenbosch, F., et al. Tumor necrosis factor-alpha: specific binding sites in rodent brain and pituitary gland. *Eu. Cytokine Netw.* 1993; 4: 377-381.
89. Kobayashi, H., Fukata, J., Murakami, N., et al. Tumor necrosis factor receptors in the pituitary cells. *Brain Res.*1997; 758: 45-50.
90. Bernton, E.W., Beach, J.E., Holaday, J.W., et al. Release of multiple hormones by a direct action of interleukin-1 on pituitary cells. *Science* 1987; 238: 519-521.
91. Brown, S.L., Smith, L.R. and Blalock, J.E. Interleukin 1 and interleukin 2 enhance proopiomelanocortin gene expression in pituitary cells. *J. Immunol.* 1987; 139: 3181-3183.
92. Kehrer, P., Turnill, D., Dayer, J.M., et al. Human recombinant interleukin-1 beta and -alpha, but not recombinant tumor necrosis factor alpha stimulate ACTH release from rat anterior pituitary cells in vitro in a prostaglandin E2 and cAMP independent manner. *Neuroendocrinol.* 1988; 48: 160-166.
93. Lyson, K. and McCann, S.M. The effect of interleukin-6 on pituitary hormone release in vivo and in vitro. *Neuroendocrinol.*1991; 54: 262-266.
94. Fukata, J., Usui, T., Naitoh, Y., et al. Effects of recombinant human interleukin-1 alpha, -1 beta, 2 and 6 on ACTH synthesis and release in the mouse pituitary tumour cell line AtT-20. *J. Endocrinol.* 1989; 122: 33-39.
95. Katahira, M., Iwasaki, Y., Aoki, Y., et al. Cytokine regulation of the rat proopiomelanocortin gene expression in AtT-20 cells. *Endocrinol.*1998; 139: 2414-2422.
96. Woloski, B.M., Smith, E.M., Meyer, W.J., 3rd, et al. Corticotropin-releasing activity of monokines. Science 1985; 230: 1035-1037.

97. Gaillard, R.C., Turnill, D., Sappino, P., et al. Tumor necrosis factor alpha inhibits the hormonal response of the pituitary gland to hypothalamic releasing factors. *Endocrinol.* 1990; 127: 101-106.
98. Sacerdote, P., Brini, A.T., Locatelli, L., et al. Tumor necrosis factor alpha differentially regulates beta-endorphin concentrations and proopiomelanocortin RNA in the anterior and neurointermediate pituitary in vivo. *Neuroimmunomod.* 1994; 1: 357-360.
99. Payne, L.C., Weigent, D.A. and Blalock, J.E. Induction of pituitary sensitivity to interleukin-1: a new function for corticotropin-releasing hormone. *Bioch. Biophys. Res. Comm.* 1994; 198: 480-484.
100. Webster, E.L., Tracey, D.E. and De Souza, E.B. Upregulation of interleukin-1 receptors in mouse AtT-20 pituitary tumor cells following treatment with corticotropin-releasing factor. *Endocrinol.* 1991; 129: 2796-2798.
101. Takao, T., Tojo, C., Nishioka, T., et al. Corticotropin-releasing factor treatment upregulates interleukin-1 receptors in the mouse pituitary: reversal by dexamethasone. *Brain Res.* 1995; 688: 219-222.
102. Elenkov, I.J., Kovacs, K., Kiss, J., et al. Lipopolysaccharide is able to bypass corticotrophin-releasing factor in affecting plasma ACTH and corticosterone levels: evidence from rats with lesions of the paraventricular nucleus. *J. Endocrinol.* 1992; 133: 231-236.
103. Koenig, J.I., Snow, K., Clark, B.D., et al. Intrinsic pituitary interleukin-1 beta is induced by bacterial lipopolysaccharide. [erratum appears in *Endocrinology* 1990; 127: 657]. *Endocrinol.* 1990; 126: 3053-3058.
104. Spangelo, B.L., MacLeod, R.M. and Isakson, P.C. Production of interleukin-6 by anterior pituitary cells in vitro. *Endocrinol.* 1990; 126: 582-586.
105. Spangelo, B.L., Jarvis, W.D., Judd, A.M., et al. Induction of interleukin-6 release by interleukin-1 in rat anterior pituitary cells in vitro: evidence for an eicosanoid-dependent mechanism. *Endocrinol.* 1991; 129: 2886-2894.
106. Spangelo, B.L., Judd, A.M., Isakson, P.C., et al. Interleukin-1 stimulates interleukin-6 release from rat anterior pituitary cells in vitro. *Endocrinol.* 1991; 128: 2685-2692.
107. Vankelecom, H., Carmeliet, P., Van Damme, J., et al. Production of interleukin-6 by folliculo-stellate cells of the anterior pituitary gland in a histiotypic cell aggregate culture system. *Neuroendocrinol.* 1989; 49: 102-106.
108. Baes, M., Allaerts, W. and Denef, C. Evidence for functional communication between folliculo-stellate cells and hormone-secreting cells in perifused anterior pituitary cell aggregates. *Endocrinol.* 1987; 120: 685-691.
109. Inoue, K., Couch, E.F., Takano, K., et al. The structure and function of folliculo-stellate cells in the anterior pituitary gland. *Arch. Histol. Cytol.* 1999; 62: 205-218.
110. Lohrer, P., Gloddek, J., Nagashima, A.C., et al. Lipopolysaccharide directly stimulates the intrapituitary interleukin-6 production by folliculostellate cells via specific receptors and the p38alpha mitogen-activated protein kinase/nuclear factor-kappaB pathway. *Endocrinol.* 2000; 141: 4457-4465.
111. Kurotani, R., Yasuda, M., Oyama, K., et al. Expression of interleukin-6, interleukin-6 receptor (gp80), and the receptor's signal-transducing subunit (gp130) in human normal pituitary glands and pituitary adenomas. *Modern Pathol.* 2001; 14: 791-797.
112. Muramami, N., Fukata, J., Tsukada, T., et al. Bacterial lipopolysaccharide-induced expression of interleukin-6 messenger ribonucleic acid in the rat hypothalamus, pituitary, adrenal gland, and spleen. *Endocrinol.* 1993; 133: 2574-1578.
113. Ehrhart-Bornstein, M., Haidan, A., Alesci, S., et al. Neurotransmitters and neuropeptides in the differential regulation of steroidogenesis in adrenocortical-chromaffin co-cultures. *Endoc. Res.* 2000; 26: 833-842.

114. Bornstein, S.R., Ehrhart-Bornstein, M., Scherbaum, W.A., et al. Effects of splanchnic nerve stimulation on the adrenal cortex may be mediated by chromaffin cells in a paracrine manner. *Endocrinol.*1990; 127: 900-906.
115. Bornstein, S.R., Gonzalez-Hernandez, J.A., Ehrhart-Bornstein, M., et al. Intimate contact of chromaffin and cortical cells within the human adrenal gland forms the cellular basis for important intraadrenal interactions. *J. Clin. Endocrinol. Metabol.* 1994; 78: 225-232.
116 Willenberg, H.S., Bornstein, S.R., Hiroi,N., et al. Effects of a novel corticotropin-releasing-hormone receptor type I antagonist on human adrenal function. Molec. Psychia. 2000; 5: 137-141.
117. Bornstein, S.R. and Chrousos, G.P. Clinical review 104: Adrenocorticotropin (ACTH)- and non-ACTH-mediated regulation of the adrenal cortex: neural and immune inputs. *J. Clin. Endocrinol. Metabol* 1999; 84: 1729-1736.
118. Ehrhart-Bornstein, M., Hinson, J.P., Bornstein, S.R., et al. Intraadrenal interactions in the regulation of adrenocortical steroidogenesis. [erratum appears in *Endocr Rev* 1998; 19: 301]. *Endocr. Rev.* 1998; 19: 101-143.
119. Path, G., Scherbaum, W.A. and Bornstein, S.R. The role of interleukin-6 in the human adrenal gland. *Eur. J. Clin. Investi.*2000; 30 Suppl 3: 91-95.
120. Barney, M., Call, G.B., McIlmoil, C.J., et al. Stimulation by interleukin-6 and inhibition by tumor necrosis factor of cortisol release from bovine adrenal zona fasciculata cells through their receptors. *Endocr. J.Uk* 2000; 13: 369-377.
121. Gadient, R.A., Lachmund, A., Unsicker, K., et al. Expression of interleukin-6 (IL-6) and IL-6 receptor mRNAs in rat adrenal medulla. *Neurosci.Lett.*1995; 194: 17-20.
122. Judd, A.M., Call, G.B., Barney, M., et al. Possible function of IL-6 and TNF as intraadrenal factors in the regulation of adrenal steroid secretion. *Ann. N.Y. Acad. Sci.* 2000; 917: 628-637.
123. Andreis, P.G., Neri, G., Belloni, A.S., et al. Interleukin-1 beta enhances corticosterone secretion by acting directly on the rat adrenal gland. *Endocrinol.*1991; 129: 53-57.
124. O'Connell, N.A., Kumar, A., Chatzipanteli, K., et al. Interleukin-1 regulates corticosterone secretion from the rat adrenal gland through a catecholamine-dependent and prostaglandin E2-independent mechanism. *Endocrinol.* 1994; 135: 460-467.
125. Tominaga, T., Fukata, J., Naito, Y., et al. Prostaglandin-dependent in vitro stimulation of adrenocortical steroidogenesis by interleukins. *Endocrinol.*1991; 128: 526-531.
126. Winter, J.S., Gow, K.W., Perry, Y.S., et al. A stimulatory effect of interleukin-1 on adrenocortical cortisol secretion mediated by prostaglandins. *Endocrinol.* 1990; 127: 1904-1909.
127. Mazzocchi, G., Malendowicz, L.K., Markowska, A., et al. Effect of hypophysectomy on corticotropin-releasing hormone and adrenocorticotropin immunoreactivities in the rat adrenal gland. *Molec. Cell. Neurosci.*1994; 5: 345-349.
128. Gwosdow, A.R., O'Connell, N.A., Spencer, J.A., et al. Interleukin-1-induced corticosterone release occurs by an adrenergic mechanism from rat adrenal gland. *Am. J. Physiol.*1992; 263: E461-E466.
129. Roh, M.S., Drazenovich, K.A., Barbose, J.J., et al. Direct stimulation of the adrenal cortex by interleukin-1. *Surgery* 1987; 102: 140-146.
130. Salas, M.A., Evans, S.W., Levell, M.J., et al. Interleukin-6 and ACTH act synergistically to stimulate the release of corticosterone from adrenal gland cells. *Clin. Exp. Immunol.* 1990; 79: 470-473.
131. Weber, M.M., Michl, P., Auernhammer, C.J., et al. Interleukin-3 and interleukin-6 stimulate cortisol secretion from adult human adrenocortical cells. *Endocrinol.*1997; 138: 2207-2210.

132. Nijsten, M.W., de Groot, E.R., ten Duis, H.J., et al. Serum levels of interleukin-6 and acute phase responses. *Lancet* 1987; 2:921.
133. Spath-Schwalbe, E., Born, J., Schrezenmeier, H., et al. Interleukin-6 stimulates the hypothalamus-pituitary-adrenocortical axis in man. *J Clin. Endocrinol. Metab.* 1994; 79: 1212-1214.
134. Turnbull, A.V., Dow, R.C., Hopkins, S.J., et al. Mechanisms of activation of the pituitary-adrenal axis by tissue injury in the rat. *Psychoneuroendocrinol.* 1994; 19: 165-178.
135. Jaattela, M., Ilvesmaki, V., Voutilainen, R., et al. Tumor necrosis factor as a potent inhibitor of adrenocorticotropin-induced cortisol production and steroidogenic P450 enzyme gene expression in cultured human fetal adrenal cells. *Endocrinol.* 1991; 128: 623-629.
136. Olsen, N.J., Nicholson, W.E., DeBold, C.R., et al. Lymphocyte-derived adreno-corticotropin is insufficient to stimulate adrenal steroidogenesis in hypophysectomized rats. *Endocrinol.* 1992; 130: 2113-2119.
137. Silverman, M.N., Pearce, B.D., Karalis, K., et al. Virus-induced corticosterone responses in CRH-deficient and hypophysectomozed mice. *Brain Behav. Immunity* 2001; 15: 183.
138. Gwosdow, A.R., Kumar, M.S., Bode, H.H. Interleukin 1 stimulation of the hypothalamic-pituitary-adrenal axis. *Am.J. Physiol.* 1990; 258: E65-E70.
139. Karalis, K., Muglia, L.J., Bae, D., et al. CRH and the immune system. *J. Neuro-immunol.*1997; 72: 131-136.
140. Muglia, L., Jacobson, L., Dikkes, P., et al. Corticotropin-releasing hormone deficiency reveals major fetal but not adult glucocorticoid need. *Nature* 1995; 373: 427-432.
141. Smith, G.W., Aubry, J.M., Dellu, F., et al. Corticotropin releasing factor receptor 1-deficient mice display decreased anxiety, impaired stress response, and aberrant neuroendocrine development. *Neuron* 1998; 20: 1093-1102.
142. Timpl, P., Spanagel, R., Sillaber, I., et al. Impaired stress response and reduced anxiety in mice lacking a functional corticotropin-releasing hormone receptor 1110ee comments]. *Nature* Genetics 1998; 19: 162-166.
143. Bartfai, T., Andersson, C., Bristulf, J., et al. Interleukin-1 in the noradrenergic chromaffin cells in the rat adrenal medulla. *Ann. N.Y.Acad. Sci.* 1990; 594: 207-213.
144. Gonzalez-Hernandez, J.A., Bornstein, S.R., Ehrhart-Bornstein, M., et al. IL-1 is expressed in human adrenal gland in vivo. Possible role in a local immune-adrenal axis. *Clin. Exp. Immunol.* 1995; 99: 137-141.
145. Gonzalez-Hernandez, J.A., Bornstein, S.R., Ehrhart-Bornstein, M., et al. Interleukin-6 messenger ribonucleic acid expression in human adrenal gland in vivo: new clue to a paracrine or autocrine regulation of adrenal function. *J. Clin. Endocrinol. Metab.* 1994; 79: 1492-1497.
146. Gonzalez-Hernandez, J.A., Ehrhart-Bornstein, M., Spath-Schwalbe, E., et al. Human adrenal cells express tumor necrosis factor-alpha messenger ribonucleic acid: evidence for paracrine control of adrenal function. *J. Clin. Endocrinol. Metab.* 1996; 81:807-813.
147. Gonzalez-Hernandez, J.A., Bornstein, S.R., Ehrhart-Bornstein, M., et al. Macrophages within the human adrenal gland. *Cell Tissue Res.*1994; 278: 201-205.
148. Currie, K.P., Zhou, Z., Fox, A.P. Evidence for paracrine signaling between macrophages and bovine adrenal chromaffin cell Ca(2+) channels. *J. Neurophysiol.*2000; 83: 280-287.
149. Nobel, C.S. and Schultzberg, M. Induction of interleukin-1 beta mRNA and enkephalin mRNA in the rat adrenal gland by lipopolysaccharides studied by in situ hybridization histochemistry. *Neuroimmunomod.* 1995; 2: 61-73.

150. Call, G.B., Husein, O.F., McIlmoil, C.J., et al. Bovine adrenal cells secrete interleukin-6 and tumor necrosis factor in vitro. *Gen. Comp. Endocrinol.* 2000; 118: 249-261.
151. Judd, A.M. and MacLeod, R.M. Adrenocorticotropin increases interleukin-6 release from rat adrenal zona glomerulosa cells. *Endocrinol.* 1992; 130: 1245-1254.
152. Judd. A.M. and MacLeod. R.M. Differential release of tumor necrosis factor and IL-6 from adrenal zona glomerulosa cells in vitro. *Am. J. Physiol.* 1995; 268: E114-E120.
153. Auernhammer, C.J. and Melmed, S.. Leukemia-inhibitory factor-neuroimmune modulator of endocrine function. *Endoc. Rev.* 2000; 21: 313-345.
154. Chesnokova, V. and Melmed, S. Leukemia inhibitory factor mediates the hypothalamic pituitary adrenal axis response to inflammation. *Endocrinol.* 2000; 141: 4032-4040.
155. Petrovsky, N. and Bucala, R. Macrophage migration inhibitory factor (MIF). A critical neurohumoral mediator. *Ann. N.Y. Acad. Sci.* 2000; 917: 665-671.
156. Rook, G.A. Glucocorticoids and immune function. *Best Pract. Res. Clin. Endocrinol. Metab.* 1999; 13: 567-581.
157. Smith, E.M., Cadet, P., Stefano, G.B., et al. IL-10 as a mediator in the HPA axis and brain. *J. Neuroimmunol.* 1999; 100: 140-148.
158. Conti, B., Jahng, J.W., Tinti, C., et al. Induction of interferon-gamma inducing factor in the adrenal cortex. *J. Biol. Chem.* 1997; 272: 2035-2037.
159. Conti, B., Park, L.C., Calingasan, N.Y., et al. Cultures of astrocytes and microglia express interleukin 18. *Brain Res.- Molec. Brain Res.* 1999; 67: 46-52.
160. Sugama, S., Kim, Y., Baker, H., et al. Tissue-specific expression of rat IL-18 gene and response to adrenocorticotropic hormone treatment. *J. Immunol.* 2000; 165: 6287-6292.
161. Rosner, W. Plasma steroid-binding proteins. *Endocrinol. Metab. Clin. N. Am.* 1991; 20: 697-720.
162. Spencer, R.L., Miller, A.H., Moday, H., et al. Chronic social stress produces reductions in available splenic type II corticosteroid receptor binding and plasma corticosteroid binding globulin levels. *Psychoneuroendocrinol.*1996; 21: 95-109.
163. Savu, L., Lombart, C. and Nunez, E.A. Corticosterone binding globulin: an acute phase "negative" protein in the rat. *FEBS Lett.*1980; 113: 102-106.
164. Shanks, N., Windle, R.J., Perks, P., et al. The hypothalamic-pituitary-adrenal axis response to endotoxin is attenuated during lactation. [erratum appears in *J Neuroendocrinol* 2000; 12: 471]. *J. Neuroendocrinol.*1999; 11: 857-865.
165. Savu, L., Zouaghi, H., Carli, A., et al. Serum depletion of corticosteroid binding activities, an early marker of human septic shock. *Biochem. Biophys. Res.Comm.* 1981; 102: 411-419.
166. Lambert, M. Thyroid dysfunction in HIV infection. Baillieres *Clin. Endocrinol. Metab.* 1994; 8: 825-835.
167. Miller, A.H., Spencer, R.L., Pearce, B.D., et al. 1996 Curt P. Richter Award. Effects of viral infection on corticosterone secretion and glucocorticoid receptor binding in immune tissues. *Psychoneuroendocrinol.* 1997; 22: 455-474.
168. Hammond, G.L., Smith, C.L., Paterson, N.A., et al. A role for corticosteroid-binding globulin in delivery of cortisol to activated neutrophils. *J. Clin.Endocrinol. Metab.*1990; 71: 34-39.
169. Seckl, J.R. and Walker, B.R. Minireview: 11beta-hydroxysteroid dehydrogenase type 1-a tissue-specific amplifier of glucocorticoid action. *Endocrinol.* 2001; 142: 1371-1376.
170. Hennebold, J.D., Ryu, S.Y., Mu, H.H., et al. 11 beta-hydroxysteroid dehydrogenase modulation of glucocorticoid activities in lymphoid organs. *Am.J. Physiol.* 1996; 270: R1296-R306.

171. Escher, G., Galli, I., Vishwanath, B.S., et al. Tumor necrosis factor alpha and interleukin 1beta enhance the cortisone/cortisol shuttle. *J. Exp. Med.* 1997; 186: 189-198.
172. Chin, K.V., Chauhan, S.S., Pastan, I., et al. Regulation of mdr RNA levels in response to cytotoxic drugs in rodent cells. *Cell Growth Different.*1990; 1: 361-365.
173. Karssen, A.M., Meijer, O.C., van der Sandt, I.C., et al. Multidrug resistance P-glycoprotein hampers the access of cortisol but not of corticosterone to mouse and human brain. *Endocrinol.*2001; 142: 2686-2694.
174. Hartmann, G., Kim, H. and Piquette-Miller, M. Regulation of the hepatic multidrug resistance gene expression by endotoxin and inflammatory cytokines in mice. *Internat.Immunopharmacol.*2001; 1: 189-199.
175. Sukhai, M., Yong, A., Pak, A., et al. Decreased expression of P-glycoprotein in interleukin-1beta and interleukin-6 treated rat hepatocytes. *Inflammation Res.*2001; 50: 362-370.
176. Stein, U., Walther, W. and Shoemaker, R.H. Modulation of mdr1 expression by cytokines in human colon carcinoma cells: an approach for reversal of multidrug resistance. *Brit.J. Cancer* 1996; 74: 1384-1391.
177. Mandi, Y., Ocsovszki, I., Szabo, D., et al. Nitric oxide production and MDR expression by human brain endothelial cells. *Anticancer Res.*1998; 18: 3049-3052.
178. Llorente, L., Richaud-Patin, Y., Diaz-Borjon, A., et al. Multidrug resistance-1 (MDR-1) in rheumatic autoimmune disorders. Part I: Increased P-glycoprotein activity in lymphocytes from rheumatoid arthritis patients might influence disease outcome. *Joint, Bone, Spine: Rev. du Rhumat.* 2000; 67: 30-39.
179. Farrell, R.J., Murphy, A., Long, A., et al. High multidrug resistance (P-glycoprotein 170) expression in inflammatory bowel disease patients who fail medical therapy. *Gastroenterol.*2000; 118: 279-288.
180. Diaz-Borjon, A., Richaud-Patin, Y., Alvarado de la Barrera, C., et al. Multidrug resistance-1 (MDR-1) in rheumatic autoimmune disorders. Part II: Increased P-glycoprotein activity in lymphocytes from systemic lupus erythematosus patients might affect steroid requirements for disease control. *Joint, Bone, Spine: Rev. du Rhumat.* 2000; 67: 40-48.
181. Miller, A.H., Pearce, B.D., Ruzek, M.C., et al. "Interactions between the hypothalamic-pituitary-adrenal axis and immune system during viral infection: pathways for environmental effects on disease expression." In *Handbook of Physiology*, McEwen, B S (ed): Oxford University Press, New York, 2001, pp 425-450.
182. Kam, J.C., Szefler, S.J., Surs, W., et al. Combination IL-2 and IL-4 reduces glucocorticoid receptor-binding affinity and T cell response to glucocorticoids. *J. Immunol.* 1993; 151: 3460-3466.
183. Adcock IM, Brown CR, Barnes PJ. Tumour necrosis factor alpha causes retention of activated glucocorticoid receptor within the cytoplasm of A549 cells. *Biochem. Biophys. Res. Comm.*1996; 225: 545-550.
184. Pariante, C.M., Pearce, B.D., Pisell, T.L., et al. The pro-inflamatory cytokine, interleukin-1alpha, reduces glucocorticoid receptor translocation and function. *Endocrinol.* 1999; 140: 4359-4366.
185. De Bosscher, K., Vanden Berghe, W., Haegeman, G. Mechanisms of anti-inflammatory action and of immunosuppression by glucocorticoids: negative interference of activated glucocorticoid receptor with transcription factors. *J. Neuroimmunol.* 2000; 109: 16-22.
186. Kovalovsky, D., Refojo, D., Holsboer, F., et al. Molecular mechanisms and Th1/Th2 pathways in corticosteroid regulation of cytokine production. *J. Neuroimmunol.* 2000; 109: 23-29.

187. Webster, J.C. and Cidlowski, J.A. Mechanisms of glucocorticoid-receptor-mediated repression of gene expression. *Trends in Endocrinol. Metab.*1999; 10: 396-402.
188. Adcock, I.M. and Caramori, G. Cross-talk between pro-inflammatory transcription factors and glucocorticoids. *Immunol. Cell Biol.* 2001; 79: 376-384.
189. Hollenberg, S.M., Weinberger, C., Ong, E.S., et al. Primary structure and expression of a functional human glucocorticoid receptor cDNA. *Nature* 1985; 318: 635-641.
190. Oakley, R.H., Sar, M. and Cidlowski, J.A. The human glucocorticoid receptor beta isoform. Expression, biochemical properties, and putative function. *J. Biol. Chem.*1996; 271: 9550-9559.
191. Webster, J.C., Oakley, R.H., Jewell, C.M., et al. Pro-inflamatory cytokines regulate human glucocorticoid receptor gene expression and lead to the accumulation of the dominant negative beta isoform: a mechanism for the generation of glucocorticoid resistance. *Proc. Natl. Acad. Sci.* U.S.A. 2001; 98: 6865-6870.
192. Sousa, A.R., Lane, S.J., Cidlowski, J.A., et al. Glucocorticoid resistance in asthma is associated with elevated in vivo expression of the glucocorticoid receptor beta-isoform. *J. Allergy Clin. Immunol.* 2000; 105: 943-950.
193. Honda, M., Orii, F., Ayabe, T., et al. Expression of glucocorticoid receptor beta in lymphocytes of patients with glucocorticoid-resistant ulcerative colitis. [see comments]. *Gastroenterol.* 2000; 118: 859-866.
194. Yehuda, R. Biology of posttraumatic stress disorder. *J. Clin. Psych.*2001; 62 Suppl 17: 41-46.
195. Pariante, C.M. and Miller, A.H. Glucocorticoid receptors in major depression: relevance to pathophysiology and treatment. *Biol.Psych.* 2001; 49: 391-404.
196. Maes, M., Lin, A.H., Delmeire, L., et al. Elevated serum interleukin-6 (IL-6) and IL-6 receptor concentrations in posttraumatic stress disorder following accidental man-made traumatic events. *Biol. Psych.*1999; 45: 833-839.
197. Sluzewska, A. Indicators of immune activation in depressed patients. *Adv. Exp. Med. Biol.* 1999; 461: 59-73.
198. Spivak, B., Shohat, B., Mester, R., et al. Elevated levels of serum interleukin-1 beta in combat-related posttraumatic stress disorder. *Biol.Psych.*1997; 42: 345-348.
199. Manolagas, S.C. and Jilka, R.L. Bone marrow, cytokines, and bone remodeling. Emerging insights into the pathophysiology of osteoporosis. *N E J M* 1995; 332: 305-311.
200. Fernandez-Real, J.M., Vayreda, M., Richart, C., et al. Circulating interleukin 6 levels, blood pressure, and insulin sensitivity in apparently healthy men and women. *J. Clin. Endocrinol. Metab.* 2001; 86: 1154-1159.
201. Kent, S., Bluthe, R.M., Kelley, K.W., et al. Sickness behavior as a new target for drug development. *Trends in Pharmacol. Sci.*1992; 13: 24-28.
202. Cannon, J.G., Angel, J.B., Ball, R.W., et al. Acute phase responses and cytokine secretion in chronic fatigue syndrome. *J. Clin. Immunol.* 1999; 19: 414-421.
203. Maes, M., Libbrecht, I., Van Hunsel, F., et al. The immune-inflammatory pathophysiology of fibromyalgia: increased serum soluble gp130, the common signal transducer protein of various neurotrophic cytokines. *Psychoneuroendocrinol.* 1999; 24: 371-383.
204. Crofford, L.J., Pillemer, S.R., Kalogeras, K.T., et al. Hypothalamic-pituitary-adrenal axis perturbations in patients with fibromyalgia. *Arthrit. Rheumat.* 1994; 37: 1583-1592.
205. Demitrack, M.A., Dale, J.K., Straus, S.E., et al. Evidence for impaired activation of the hypothalamic-pituitary-adrenal axis in patients with chronic fatigue syndrome. *J. Clin. Endocrinol. Metab.*1991; 73: 1224-1234.

Chapter 6

CYTOKINES AND NEUROPEPTIDES

Nicholas P. Plotnikoff[1], Robert E. Faith[2], and Anthony J. Murgo[3*]
[1]College of Pharmacy, University of Illinois, Chicago, Illinois; [2]Center for Comprehensive Medicine, Baylor College of Medicine, Houston, Texas; [3]National Cancer Institute, Rockville, Maryland

1. INTRODUCTION

Clinical experience shows that cytokines have side effects of a central nervous system origin. Flu-like systems (fever, chills, headache, fatigue) develop rapidly with cytokine therapy. The latter symptoms are followed by neuroasthenia or fatigue syndrome (asthenia, malaise, lethargy, somnolence, headache, low-grade fevers, and Anorexia). Psychomotor, cognitive, and psychiatric changes are also observed (1-4). Numerous groups have considered the effects of cytokines on the central nervous system (5).

One of the most common effects produced in clinical depression (6). Although the blood brain barrier may impede direct entrance of cytokines into the CNS for access from the choroid plexus and circum ventricular organs (7). Benveniste proposes that cytokines are synthesized by glial cells and neurons which may explain neurological diseases such as multiple sclerosis and stroke (8). Anorectic effects have been attributed to Interleukin 1 alpha, 1 beta IL6, and TNF (9). Low serum levels of IL-2 and TGF-beta 2 were found in anorexic women (10). McCarthy suggested using agents to block cytokine activity to improve food intake (11). Leonard reviews the impact of pro-inflamatory cytokines in major depression and Alzheimers disease (IL1, IL6, and TNF-alpha) (12). Antidepressant treatment reduces the effects of cytokines presumably via biogenic amine systems in clinically depressed patients (13). In major depression there is hypersecretion of adrenal glucocorticoid leading to a desensitization of central glucocorticoid receptors. In addition, malfunctioning of noradrenergic and serotonergic neurotransmission in the brain is attributed to elevated gluocorticoids and pro-inflammatory cytokines (13). Thus clinical depression has been

*This work was co-authored by Dr. A.J. Murgo in his private capacity. No official support or endorsement by the NCI is intended or should be inferred.

described as a "sickness behavior" via macrophage activation (IL1, IL6, and TNF) (6). Finally antidepressants reduce release of pro-inflammatory cytokines, increase release of IL-1 receptor antagonist and IL10 and inhibit cyclooxygenase (13). Sheng et al discusses the numerous activities of opioids, neurotransmitters and their interactions with cytokines, betachemokines, and free radicals (ROI reactive oxygen intermediates and NO) produced by microglia and astrocytes (14). In rats alpha interferon produced analgesia (which could be reversed with naloxone) supporting the role of opioid peptides and their receptors (15). IL-2 was found to exert analgesic actions by binding to opioid receptors [via leucine enkephalin changes in paraventricular hypothalamic nucleus and locus ceruleus (16).]

The AIDS epidemic has focused the effects of endogenous and erogenous cytokines on brain function and behaviour, especially clinical depression and suicide attempts (17). Cerebral levels of transforming growth factor-beta 1 and interferon alpha appear to correlate with HIV RNA load and development of AIDS-dementia complex (18). Tumor necrosis factor released by glial cells and infiltrating macrophages contribute to HIV-induced neurological disease (19). High levels of IL-6 in (brain tissue and CSF) were found in AIDS dementia complex and other diseases (20). Traumatic and metabolic injury resulted in release of IL-1, IL-6, TNF alpha, interferon gamma from astrocytes (21).

2. OPIOID PEPTIDES

The mechanisms of action of the interferons in producing clinical depression may be explained via actions on opioid receptors (22). Alpha interferon binds to opioid receptors resulting in analgesia, reduced motor activity and catatonia. These actions may account for the antogonism of dependence on endogenous brain opioids. Using single-unit recording IFN alpha stimulated or depressed various neurons in the ventro-medial hypothalamus which effects could be reversed with naloxone (mu, delta, kappa receptors) (23). Since there are interactions between alpha interferon and opioid receptors it is probably that the resultant depression, fatigue as well as changes of appetite and sleep involve other neuropeptides, serotonin, acetylcholine, nor-epinephrine and nitric oxide (24). However, many of the central effects of alpha-interferon can be reversed with naloxone (25). Alpha-interferon injected into the lateral ventricle of rats increased the pain threshold which could be reversed by a specific mu receptor antagonist (26). Alpha interferon administered intracerebrally causes fever, anorexia, and analgesia which are inhibited by the opioid receptor antagonist naloxone (27). Alpha interferon injected intracerebrally in mice induced transient analgesia that was reversible with naloxone (28). Interleukin 2 effects of analgesia have been related to similarities in structure and binding to opioid receptors in arcuate hypothalamic nucleus, paraventricular hypothalamic nucleus and locus ceruleus (29). IL2 introduced into the third ventricle changed neuronal firing resulting in a decrease in the ventromedical nucleus of the hypothalamus and an increase in the supraoptic and paraventricular nuclei (30). IL-1 beta injected into the ventromedial hypotholamus induced analgesia as measured by the paw-withdrawal model (31). Small doses of LPS and IL-1 beta produced hyperalgesia and analgesia at high doses as measured in the preoptic area and ventromedial hypothalamus (32). After central administration both TNF and IL-1 increase nociceptive thresholds on the hot plate test in rats (33) (34). In contrast microinjection of IL-1 beta into the medial part of the preoptic area reduced the paw-withdrawal latency (hyperalgesia) whereas injection into the ventromedial hypothalamus produces analgesia (35).

3. BIOGENIC AMINES

Cytokines (IL-1, IL-2, IL-6) have been shown to influence the release of biogenic amines in the nucleus accumbens (36). IL-1 activates the hypothalama-pituitary-adreno-cortical (HPA) axis and stimulates cerebral non-adrenaline and serotonin metabolism (37). Alpha interferon given by a single intracerebro-ventricular injection reduced serotomin and noradrena-line in the mid-brain, striatum, and frontal cortex (38). Systemic IL-2 administered repeatedly increased NE utilization within the medium eminence and in the hippocampus. Serotonin levels were also altered in the hippo-campus and prefrontal cortex (39). When administered for 7 days, IL-2 increased NE utilization within median eminence, arcuate nucleus, hippocampus, central amygdala and medial prefrontal cortex. IL-2 immuno-therapy in patients is associated with cognitive/memory impairment.

Direct interactions of opioid peptides with catecholamine cell groups, in ‘the ventral medial medulla as well as well dopaminergic neurons in rat fetal mesencephalic cells and basal ganglia further support the opioid catecholamine hypothesis of depression induced by IL2 and alpha interferon (40). However, other neuropeptides such as vaso-active intestinal polypeptide, neurotensin and substance P may also be involved in the central effects of cytokines (41).

4. CONCLUSION

In conclusion we are impressed by the findings that clinical depression induced by cytokines can be treated with antidepressants. These clinical observations further support the basic findings that the cytokines bind to opioid receptors in the hypothalamus and septum. Interaction of the opioid receptors with catecholamines may help explain the profound depression seen in patients treated with cytokines. Further studies of the cytokines and other neuropeptides are in order (42). Thus, systemic injection of interleukin-1 beta into male rats resulted in an increase in c-fox mRNA expression in the bed nucleus of the stria terminalis and the central nucleus of the amygdala. All IL-1 responsive cells were GABAergic and the majority of the cells expressed enkephalin mRNA and some expressed neurotensin mRNA. IL-1 activates corticotrophin-releasing hormone in the hypothalamus. Central administration of IL-1 increased the beta-endorphin in splenocytes lymph node cells and peripheral blood mononuclear cells. Blocking the catecholaminergic system with 6-hydroxydopamine prevented the increase in beta-endorphin. Thus catecholamines as well as CRH are involved (43). Clinical depression associated with use of interferons may be associated with their anti-serotonergic effects (44). Parkinsonism and clinical depression associated with repeated alpha-interferon treatment have been attributable to a decrease in dompamine and 3-4-dihydroxyphenylacetic acid levels (45).

REFERENCES

1. Turowski, R.C. and Triozzi,, P.L Central nervous system toxicities of cytokine therapy. In *Cytokines, Stress and Immunity*, NP Plotnikoff, RE Faith, AJ Murgo, RA Good, eds. CRC Press, Boca Raton, FL 1999, pp 93-114.
2. Yirmiya, R., Pollak, Y., Morag, M. et al. Illness, cytokines, and depression. *Ann. N.Y. Acad. Sci.* 2000; 917:478-487, .
3. Kronfol, Z. Depression and immunity. In *Cytokines, Stress and Immunity*, NP Plotnikoff, RE Faith, AJ Murgo, RA Good, eds. CRC Press, Boca Raton, FL 1999, pp 51-61.
4. Dunn, A.J. Role of cytokines in infection-induced stress. *Ann. N.Y. Acad. Sci.* 1993; 697:189-202.
5. Szelenyi, J. Cytokines and the central nervous system. *Brain Res. Bull.* 2001; 54: 329-38.
6. Dantzer, R. Cytokine-induced sickness behavior: where do we stand? *Brain Behav. Immun.* 2001 Mar; 15: 7-24.
7. Banks, W.A. Anorectic effect of circulating cytokines: role of the vascular blood-brain barrier. *Nutrition*. 2001; 17: 434-7.
8. Benveniste, E.N. Cytokine actions in the central nervous system. *Cytokine Growth Factor Rev.* 1998; 9: 259-75.

9. Plata-Salaman, C.R. Cytokines and feeding. *News Physiol. Sci.* 1998; 13:298-304.
10. Corcos, M., Guilbaud, O., Chaovat, G., et al. Cytokines and anorexia. *Nervous Psychosom .Med.* 2001; 63: 502-4.
11. McCarthy, D.O. Cytokines and the anorexia of infection: potential mechanisms and treatments. *Biol. Res. Nurs.* 2000; 1: 287-98.
12. Leonard, B.E. Changes in the immune system in depression and dementia: causal or co-incidental effects? *Int. J. Dev. Neurosci.* 2001; 19: 305-12.
13. Leonard, B.E. The immune system, depression and the action of antidepressants. *Prog. Neuropsychopharmacol. Biol. Psychiatry* 2001; 25: 767-80.
14. Sheng, W.S., Hus, S., Gekker, G., et al. Immunomodulatory role of opioids in the central nervous system. *Arch. Immunol. Ther. Exp.* 1997; 45: 359-66.
15. Jiang, C.L., Son, L.X., Lu, C.L., et al. Analgesic effect of intereron-alpha via mu opioid receptor in the rat. *Neurochem. Int.* 2000; 36: 193-6.
16. Jiang, C.L., You, Z.D., Lu, C.L., et al. Lev-enkephalin induced by IL-2 administration mediates analgesic effect of IL-2. *Neuroreport* 2000; 11: 1483-5.
17. Kelly, B., Raphael, B., Judd, F., et al. Suicidal ideation, suicide attempts, and HIV infection *Psychosomatics.* 1998; 39: 405-15.
18. Perrella, O., Carreiri, P.B., Perrella, A., et al. Transforming growth factor beta-1 and interferon-alpha in the AIDS dementia complex (ADC): possible relationship with cerebral viral load? *Eur. Cytokine Netw.* 2001; 12: 51-5.
19. Mayne, M., Bratanich, A.C., Chen, P., et al. HIV-1 tat molecular diversity and induction of TNF-alpha: impications for HIV-induced neurological disease. *Neuroimmuinomodulation* 1998; 5: 184-92.
20. Terrenil De Simoni, M.G.. Role of the brain in interleukin-6 modulation. Neuroimmunomodulation 1998; 5: 214-9.
21. Lau, L.T. and Yu, A.C. Astrocytes produce and release interleukin-1, interleukin-6, tumor necrosis factor alpha and interferon-gamma following traumatic and metabolic injury. *J. Neurotrauma* 2001; 18: 351-9.
22. Dafny, N. Interferon and the Central Nervous System. In *Cytokines Stress and Immunity* eds. N.P. Plotnikoff, R.E. Faith, A.J. Murgo, R.A. Good, CRC Press 1999, pp 221-233.
23. Holden, J.E. and Proudfit, H.K. Enkephalin neurons that project to the A7 catecholamine cell group are located in nuclei that modulate nociception ventromedial medulla. *Neuroscience* 1998; 83: 929-47.
24. Borlongan, C.V., Su, P.T. and Wang, Y. Treatment with delta opioid peptide enhances in vitro and in vivo survival of rat dopaminergic neurons, *Neuroreport.* 2000: 11: 923-6.
25. You, Z.B., Herrera-Marschitz, M. and Terenius, L. Modulation of neurotransmitter release in the basal ganglia of the rat brain by dynorphin peptides. *J. Pharmacol. Exp. Ther.* 1999; 290: 1307-15.
26. Jiang, C.L., Son, L.X., Lu, C.L., et al. Analgesic effect of interferon-alpha via mu opioid receptor in the rat. *Neurochem. Int.* 2000; 36: 193-6.
27. Hori, T., Katafuchi, T., Take, S., et al. Neuroimmuno-modulatory actions of hypothalamic interferon-alpha. *Neuroimmunomodulation* 1998; 5: 172-7.
28. Smith, E.M., Diom, L.D. and Blalock, I.E. Opioid receptor mediated effects of INF-alpha and lymphocyte derived endorphin-like peptides. *Prog. Clin. Biol. Res.* 1985; 192: 259-64.
29. Jiang, C.L., Xu, D., You, Z.D., et al. Possible mechanism of the analgesic effect of interleukin-2 in central nervous system. *Shen LI Xue Bao* 1996; 48: 243-8.
30. Jiang, C.L., You, Z.D., Lu, C.L., et al. Leu-enkephalin induced by IL-2 administration mediates analgesic effect of IL-2. *Neuroreport* 2000; 11(7): 1483-5.

31. Hori, T., Oka, T., Hosoi, M., et al. Pain modulatory actions of cytokines and prostaglandin E2 in the brain. *Ann. N.Y. Acad. Sci.* 1998; 840: 269-81.
32. Hori, T., Oka, T., Hosoi , M., et al. Hpothalamic mechanisms of pain modulatory actions of cytokines and prostaglandin E2. *Ann. N.Y. Acad. Sci.* 2000; 917:106-20.
33. Bianchi, M., Sacerdote, P., Ricciardi-Castagnoli, P., et al. Central effects of tumor necrosis factor alpha and interleukin-1 alpha on nociceptive thresholds and spontaneous locomotor activity. *Neurosci. Lett.* 1992; 148: 76-80.
34. Bianchi, M., Sacerdote, P., Locatelli, L., et al. Corticotropin releasing hormone, interleukin-1 alpha, and tumor necrosis factor-alpha share characteristics of stress mediators. *Brain Res.* 1991; 546: 139-42.
35. Oka, T., Oka, K., Mosoi, M., et al. The opposing effects of interleukin-1 beta microinjected into the preoptic hypothalamus and the ventromedial hypothalamus on nociceptive behavior in rats. *Brain Res.* 1995; 700: 271-8.
36. Song, C., Merali, Z. and Anisman, H. Variations of nucleus accumbens dopamine and serotonin following systemic interleukin-1, interleukin-2, or interleukin-6 treatment. *Neuroscience* 1999; 88: 823-36.
37. Dunn, A.J., Wang, J. and Ando, T. Effects of cytokines on cerebral neurotransmission. *Adv. Exp. Med. Bio.* 1999; 461:117-27.
38. Kamata, M., Higuchi, H., Yoshimoto, M., et al. Effect of single intracerebroventricular injection of alpha interferon on monoamine concentrations in the rat brain. *Eur. Neurophyschopharmacol.* 2000; 10: 129-32.
39. Lacosta, S., Merali, Z. and Anisman, H. Central monoamine activity following acute and repeated systemic interleukin-2 administration. *Neuroimmunomodulation* 2000; 8: 83-90.
40. Merrill, J.E. Interleukin-2 effects in the central nervous system. *Ann. N.Y. Acad. Sci.* 1990; 594: 188-199.
41. Eskay, R.L. and Eiden, L.E. Interleukin-1 alpha and tumor necrosis factor-alpha differentially regulate enkephalin, vasoactive intestinal polypeptide, neurotensin, and substance P biosynthesis in chromaffin cells. *Endocrinology* 1992; 130: 2252-2258.
42. Day, H.E., Curran, E.J., Watson, S.J. Jr., et al. Distinct neurochemical populations in the rat central nucleus of the amygdala and bed nucleus of the stria terminalis: evidence for their selctive activation by interleukin-1 beta. *J. Comp. Neurol.* 1999; 413: 113-28.
43. Sacerdote, P., Bianchi, M., Manfredi, B., et al. Intracerebroventricular interleuki-1 alpha increases immonucyte beta-endorphin concentrations in the rat: involvement of conticotropin-releasing hormone, catecholamines, and serotonin. *Endocrinology* 1994; 135: 1346-1352.
44. Menkes, D.B., MacDonald, J.A. Inteerons, serotonin and neurotoxicity. *Psychol. Med.* 2000; 30: 259-268.
45. Shuto, H., Hataoka, Y., Morikawa, T., et al. Repeated interferon-alpha administration inhibits dopaminergic neural activity in the mouse brain. *Brain Res.*, 1997; 747: 348-351.

Chapter 7

CYTOKINES AND SICKNESS BEHAVIOR

Robert Dantzer
Inegrative Neurobiology, INRA-INSERM U394, Bordeaux, France

1. INTRODUCTION

Non specific symptoms of infection and inflammation include fever and profound physiological and behavioral changes. Sick individuals experience weakness, malaise, listlessness and inability to concentrate. They become depressed and lethargic, show little interest in their surroundings and stop eating and drinking. This constellation of non-specific symptoms is collectively referred to as "sickness behavior". Due to their commonality, sickness symptoms are frequently ignored by physicians. They are considered as an uncomfortable, but rather banal, component of the pathogen-induced debilitation process that affects an ill individual.

This view is, however, totally inadequate. The behavioral symptoms of sickness represent, together with the fever response and the associated neuroendocrine changes, a highly organized strategy that reorganizes the organism's priorities and enables it to fight infection (1). In physiological terms, fever corresponds to a new homeostatic state that is characterized by a raised set-point of body temperature regulation. A feverish individual feels cold at usual environmental temperatures. Therefore it not only seeks warmer temperatures but also enhances heat production (increased thermogenesis) and reduces heat loss (decreased thermolysis). The higher body temperature that is achieved during fever stimulates proliferation of immune cells and is unfavorable for the growth of many bacterial and viral pathogens. In addition, the reduction of zinc and iron plasma levels that occurs during fever decreases the availability of these vital elements for growth and multiplication of microorganisms. The adaptive nature of the fever response is supported by the observation that organisms infected with a bacteria or virus and unable to mount an appropriate fever response because they are kept in a cold environment or treated with an antipyretic drug have a lower survival rate than organisms which develop a normal fever (2).

The amount of energy that is required to increase body temperature during the febrile process is quite high since, in human beings, metabolic rate needs to increase by 13 % to achieve a rise of 1°C in body temperature. Because of the high metabolic cost of fever, there is little room for activities other than those favoring heat production (e.g., shivering) and minimizing thermal losses (e.g., rest, curl up posture, piloerection).

In recent years, evidence has rapidly accumulated to demonstrate that the necessary synchrony between metabolic, physiological and behavioral components of the systemic response to infection is dependent on the same molecular signals than those that are already responsible for the local inflammatory response. These signals are pro-inflamatory cytokines, such as interleukin-1 (IL-1), interleukin-6 (IL-6), tumor necrosis factor α (TNFα) and interferons (IFNs). These cytokines are released by activated monocytes and macrophages during the course of an infection. They are local rather than long distance communication signals and act on the brain in two successive waves. The first wave is triggered by activation of the primary afferent neurons that innervate the body site where the inflammatory reaction takes place. The second wave involves cytokines produced within the brain structures that interface with the periphery and slowly diffuse from there to brain targets such as the amygdaloid complex.

2. CYTOKINES INDUCE SICKNESS BEHAVIOR

Systemic or central infusion of recombinant cytokines induces the full blown repertoire of non specific symptoms of sickness in both experimental animals and human beings. The same effects are obtained in response to the administration of molecules that induce the synthesis of endogenous cytokines (e.g., lipopolysaccharide (LPS), the active fragment of endotoxin from Gram negative bacteria).

The behavior of animals injected with LPS or cytokines at the periphery or into the lateral ventricle of the brain has been studied extensively (3). Rats injected with LPS or IL-1β after drinking a new taste solution develop a conditioned aversion to the taste of that solution (4,5). Conditioned taste aversions have also been established to TNF-α. However, treatment with IFN-α fails to induce conditioned taste aversion (6). The reasons for these differences between cytokines are still unclear.

Decreased social exploration of juvenile conspecifics has been used as a convenient way to assess sickness behavior in laboratory rodents. It involves olfactory sampling of the partner and is normally used for social recognition. It offers the advantage of being reproducible and quantifiable. Because of the use of juvenile conspecifics, the emergence of other behavioral patterns such as sexual behavior or aggression that normally occurs with adult partners is usually kept minimal. Systemic and central administration of

LPS, IL-1β and recombinant human TNF-α to adult laboratory animals, rats or mice, consistently decreases the time they spend in exploration of juveniles in a time- and dose-dependent manner (7-9).

Another important component of sickness behavior is the decreased food intake that develops in sick individuals. Systemic administration of IL-1β and TNF-α consistently suppresses feeding. This effect has been observed using various measurements of food intake under *ad libitum* as well as deprived conditions (10,11). Food deprived animals are usually more resistant to the depressing effects of cytokines on food intake than non deprived animals. In contrast to the decrease in social exploration that takes about 2 hours to develop, the suppression of food intake occurs earlier, within one hour following treatment. Water intake can also be suppressed by the administration of pro-inflamatory cytokines, but the effects of cytokines on drinking are less consistent than on eating. Much of the reported attenuation of water intake could actually be due to the discomfort and even the pain experienced by sick animals which have to stand up on their hind feet in order to get access to the water spout that protrudes in the cage from the ceiling.

When injected into the lateral ventricle of the brain or directly into the brain parenchyma, IL-1 and LPS are as effective to induce sickness behavior as when administered at the periphery. However, much smaller amounts are needed, about a few hundreds times less. Surprisingly, the time interval for the appearance of effect is actually longer, whereas the duration of effect is shorter.

Pro-inflamatory cytokines act in a network fashion, each cytokine being able to induce its own synthesis and the synthesis of other cytokines that mimic its effects. At the periphery for instance TNFα induce the synthesis of IL-1β and then that of IL-6. This is sequentially followed by the production of other cytokines that oppose the production and action of pro-inflamatory cytokines and are therefore known as anti-inflammatory cytokines. Anti-inflammatory cytokines can either be specific, such as the specific antagonist of IL-1 receptors, IL-1ra, that opposes all the known biological effects of IL-1 including sickness behavior (12), or non specific, such as IL-10 that opposes the expression and action of most of the pro-inflamatory cytokines. At the behavioral level, IL-1β is the main active cytokine (13). Peripheral and central injections of other species-specific cytokines such as recombinant rat IL-6 (14) and TNFα have no effect on rat behavior, unless combined with IL-1. The anti-inflammatory cytokine IL-10 attenuates LPS-induced sickness behavior (15), whereas other anti-inflammatory cytokines such as IL-4 (16) and IL-13 (17) have mixed action, depending on when they are injected in relation to the administration of LPS.

The behavioral effects of cytokines are potently down regulated both at the periphery and in the brain by endogenous glucocorticoids (18). Bilateral adrenalectomy sensitizes mice to the behaviorally depressing effects of LPS and IL-1. This effect is attenuated but not abrogated by a subcutaneous implant of corticosterone, indicating that the phasic hypothalamic-pituitary-adrenal response to cytokines is certainly important in limiting the development of cytokine-induced sickness behavior.

Other peptides able to oppose the behavioral effects of LPS include vasopressin (19) and IGF-1 (20). The involvement of vasopressin is androgen-dependent since the protective effect of this peptide is observed only in intact male rats and in castrated male rats supplemented with testosterone.

3. NEUROANATOMICAL BASIS OF SICKNESS BEHAVIOR

Our understanding of the way sickness behavior is organized in the brain is based on several methodological approaches. Molecular biology and pharmacology studies have demonstrated that in order to promote the development of sickness behavior, peripheral immune stimuli have to induce the synthesis and release of cytokines in the brain. LPS for instance induces the expression of brain IL-1α, IL-1β and TNFα, followed by that of IL-6 (21, 22). The main cellular sources of IL-1 in the brain are represented by microglial cells and perivascular and meningeal macrophages (23). Central administration of cytokine antagonists or cytokine receptor antagonists attenuates the behavioral effects of cytokines. For instance, administration of the IL-1 receptor antagonist (IL-1ra) into the lateral ventricle of the brain to block brain IL-1 receptors abrogated the depressing effect of peripherally administered IL-1 on social exploration in rats exposed to a juvenile (12). However, central IL-1ra did not block the effect of peripheral LPS on social exploration (7) whereas it attenuated the depressing effects of LPS on food intake (24). These findings indicate that other brain cytokines than IL-1 can compensate for the lack of IL-1 in the brain, their relative importance depending on the behavior under consideration.

Using expression of the early gene *c-fos* as a marker of neuronal activation in those brain areas that are activated by stressors, neuroanatomists have identified the brain targets of peripheral immune stimuli. Wan et al. (25) were the first authors to show that intraperitoneal injection of LPS activates the primary projection area of the vagus nerve in the brain, that is represented by the nucleus tractus solitarius, and the secondary projections of this nerve, including the parabrachial nucleus, the hypothalamic paraventricular and supraoptic nuclei, the central nucleus of the amygdala, and the bed nucleus of the stria terminalis. A subdiaphragmatic section of the vagus nerves abrogated the expression of Fos in these brain areas. The key role of the vagus nerves in the transmission of periperal immune signals to the brain was further confirmed by the demonstration that vagotomy attenuates the behavioral actions of peripheral cytokines (26, 27), and abrogates the induction of IL-1β in the brain in response to peripheral LPS or IL-1β (28,29). Furthermore, electrical stimulation of the vagus nerves was shown to induce the expression of brain IL-1 at the mRNA and protein levels in the hypothalamus and hippocampus of rats (30).

The abdominal vagus nerve has the important peculiarity of being intimately associated with immune cells that express not only IL-1 receptors (31) but also receptors specialized in the innate recognition of microbial products associated patterns and known as toll-like receptors (TLR). TLR-4 recognizes membrane products of Gram-negative bacteria, including lipopolysaccharide, whereas TLR-2 recognizes membrane products of Gram-positive bacteria such as staphylococcus enterotoxin B. Activation of TLRs results in the production of IL-1β and other pro-inflamatory cytokines. These communication signals are geared to act locally on adjacent cells, in a paracrine and autocrine manner, rather than at distance. In the case of immune cells associated with the vagus nerves, the locally produced IL-1β binds to vagal fibers to increase vagal discharge activity. Glutamate is released at the level of the nucleus tractus solitarius where vagal fibers terminate (32). Glutamate acts on catecholaminergic neurons of the NTS that project to the paraventricular and preoptic nuclei (33,34). The central nucleus of the amygdala can be reached via this pathway or, more probably, via the parabrachial nuclei (35). These pathways appear to be responsible for the activating effects of inflammatory stimuli on the hypothalamic-pituitary-adrenal axis and their depressing effects on behavior. However, the pyrogenic activity of inflammatory stimuli may also involve still another pathway represented by prostaglandin synthesis by cyclooxygenase-2 in endothelial cells of the brain capillaries (36). Whether these endothelial cells respond only to the high levels of cytokines that have been injected intravenously in the experimental studies of fever or can also detect low circulating levels of endogenously produced cytokines is still controversial.

The brain production of IL-1β in response to peripheral inflammatory stimuli is first restricted to the choroid plexus and circumventricular organs (37). This production is certainly mediated by the activation of TLRs that are located on the membrane of macrophage-like cells in these structures. The locally produced IL-1β then slowly diffuses to the brain side of the blood-brain barrier by volume transmission (38). This IL-1β can activate neurons originating from the area postrema and projecting via the NTS and parabrachial area to the amygdala and/or act directly on neurons in the basolateral amygdala after slowly diffusing from its source (39). Projections from the basolateral amygdala to the periacqueducal gray via the central amygdala and the bed nucleus of the stria terminalis are likely to mediate the depressing effects of IL-1β on behavior. The observation that a late (4 hours post LPS) infusion of IL-1ra into the lateral ventricle of the brain attenuates the depressing effects of systemic LPS on social exploration concomitantly with an abrogation of Fos expression in the central amygdala and bed nucleus of the stria terminalis points to the role of these two structures in cytokine-induced behavioral depression.

In summary, activation of afferent nerve fibers by peripherally released cytokines represents a fast pathway of transmission of immune signals from the periphery to the brain. This pathway certainly sensitizes the brain target areas to the action of brain produced cytokines that relay and amplify the action of peripheral cytokines (40).

4. RECEPTOR MECHANISMS OF THE EFFECT OF IL-1 IN THE BRAIN

The recent elucidation of the signal transduction pathways that are activated by IL-1 receptors has prompted a surge of interest for the receptor mechanisms of IL-1 actions in the brain. Binding of IL-1 to the type I IL-1 receptor (IL-1RI) results in the formation of an heterodimeric complex between this receptor and the IL-1 receptor accessory protein (IL-1RacP). IL-1RI and IL-1RacP knock-out mice do not respond any longer to intraperitoneal or intracerebroventricular IL-1β (41, 42). In the same manner, blockade of the type I IL-1 receptor by a neutralising antibody injected into the lateral ventricle of the brain abrogates the behavioural effects of intracerebroventricular IL-1β (43). In contrast, blockade of the type II IL-1 receptor that is known to function as a decoy, has no effect when the inhibition of social behaviour is the endpoint but potentiates the anorexic effect of IL-1β (44).

The signalling pathways that mediate the behavioural effects of IL-1 have not yet been elucidated. Depending on the cell target, activation of IL-1 receptors signals via either the nuclear transcription factor NF-κB or several MAP-kinase pathways. Since IκBα is strongly expressed in the CVOs and choroid plexus in conditions of peripheral immune stimulation (45), the effects of IL-1 at this level are likely to be NF-κB dependent. In accordance with this hypothesis, central administration of a NF-κB inhibitor peptide was found to block the sleep-inducing and pyrogenic effects of peripheral IL-1β in rabbits (46). The same is not necessarily true in the brain parenchyma, since activation of mitogen activated protein (MAP) kinases pathways appear to be predominantly involved in the neural effects of IL-1 that have been investigated so far. Those include IL-1β-induced inhibition of long term potentiation in perforant path-granule cell synapses (47) and IL-1β-induced elevation in intracellular free calcium levels in rat cortical synaptosomes (48).

5. MOTIVATIONAL ASPECTS OF SICKNESS BEHAVIOR

Sickness behavior is usually viewed by physicians as the result of debilitation and physical weakness which inevitably occur in an organism whose all resources are engaged in the defensive process against pathogens. An alternative hypothesis is that sickness behavior is the expression of a highly organized strategy that is critical to the survival of the organism. If it is the case, then it follows that sick individuals should be able to reorganize their behavior depending on its consequences and the internal and external constraints they are exposed to. This flexibility is characteristic of what psychologists call a motivation. Psychologists define motivation as a central state that reorganizes perception and action (49). Fear is a good example of how a motivational state organizes perception and action. In order to escape a potential threat, a fearful individual must be attentive to all the events that are likely to occur in his or her environment. At the same time, the fearful person must be ready to engage in the most appropriate defensive behavioral pattern that is available in his/her behavioral repertoire. In other terms, a motivational state does not trigger an inflexible behavioral pattern. It enables to uncouple perception from action and therefore to select the appropriate strategy depending on the eliciting situation (49).

The first evidence that sickness behavior is the expression of a motivational state rather than the result of weakness was provided by Miller long before cytokines were discovered (50). Miller was then looking for possible motivational stimuli responsible for the cessation of thirst. Following the advice of a microbiologist colleague, René Dubos, he injected rats with endotoxin and showed that they stopped bar pressing for water. However, endotoxin-treated rats were still able to drink when water was made freely available, although they did so to a lesser extent than normally. This effect was not specific to thirst since the endotoxin treatment also reduced bar pressing for food and even blocked responding in rats trained to press a bar for the rewarding effects of electrical stimulation in the lateral hypothalamus. Interestingly enough, when rats were trained to turn off an aversive electrical stimulation in this brain area, endotoxin also reduced the rate of responding, but to a lesser extent than bar pressing for a rewarding brain stimulation. However, when rats were placed in a rotating drum they could stop for brief periods by pressing a lever, endotoxin treatment resulted in an increase rather than a decrease in their response rate. The mere fact that endotoxin treatment could decrease or increase behavioral output depending on its consequences gave strong support to the motivational interpretation of the behavioral effects of such a treatment. However, these findings were only reported in the form of a review paper, and the original data were never published.

An example of behavioral reorganization in response to sickness comes from the effects of LPS on macronutrient intake (51). When rats are given the opportunity to select components of their diet, the selection pattern they display reflects their nutritional and energetic requirements. To determine whether this selection pattern is altered during sickness, rats were submitted to a dietary self selection protocol in which they had free access to carbohydrate, protein and fat diets for 4 hours a day. After a 10-day habituation to this regimen, they were injected with LPS or IL-1β. Under the effect of this treatment, they decreased their total food intake but reorganized their self-selection pattern so as to ingest relatively more carbohydrate and less protein whereas fat intake remained unchanged. This change in macronutrient intake contrasts with the increased fat intake that occurs in rats exposed to cold. It is important to note that although eating fat would be a better way for feverish animals to cope with their increased energy requirements, it would not be of much use since cytokines have adverse metabolic effects resulting in increased lipolysis and hypertriglyceridemia. An increased intake of fat would actually be counterproductive in cytokine-treated rats since it would further enhance hyperlipidemia without positively contributing to lipid metabolism.

An important characteristic of a motivational state is that it necessarily competes with other motivational states for behavioral output. The normal expression of behavior requires a hierarchical organization of motivational states that is continuously updated according to circumstances. When an infection occurs, the sick person is at a life or death juncture and his/her physiology and behavior must be altered so as to overcome the disease. However, this is a relatively long-term process that needs to make room to more urgent needs when necessary. It is easy to imagine that if a sick person lying in his/her bed hears a fire alarm ringing in his/her house and sees flames and smoke coming out of the basement, he/she should be able to momentarily overcome his/her sickness behavior to escape danger. In motivational terms, fear competes with sickness, and fear-motivated behavior takes precedence over sickness behavior. An example of this competition between fear and sickness is provided by the observation that the depressing effects of IL-1β on behavior of mice are more pronounced when experimental animals are tested in the safe surroundings of their home cage than when they are placed into a new environment.

Another example of the motivational aspects of sickness behavior comes from a study of the effects of cytokines on maternal behavior (52). If fitness is the key issue, it is evident that dams should care for their infants despite sickness. In motivational terms, the components of maternal behavior that are crucial for the survival of the progeny should be more resilient, i.e., less sensitive to the depressing effects of pyrogens, than those behavioral patterns that are less important. In accordance with this prediction, administration of LPS to lactating mice did not disrupt pup retrieval but impaired nest-building. However, LPS treatment was less effective to depress nest building when the dams and their litters were exposed to 6°C, to increase the fitness value of nest building than when dams were tested at 22°C.

To show that sickness does not interfere with the animal's ability to adjust its behavioral strategies with regard to its needs and capacities, Aubert et al. (53) assessed the effects of LPS on food hoarding and food consumption in rats receiving or not a food supplement in addition to the amount of food they obtained in the situation. Rats were trained to get food for 30 min in an apparatus consisting of a cage connected to an alley with free food at its end. In this apparatus, rats normally bring back to their home cage the food that is available at the end of the alley, and the amount of food they hoard is lower when they receive a food supplement than when they have no supplement. In response to LPS, food intake was decreased to the same extent whether rats received a food supplement or not. However, food hoarding was less affected in rats that did not receive a food supplement compared to rats provided with the food supplement. These results indicate that the internal state of sickness induced by LPS is more effective in suppressing the immediate response to food than the anticipatory response to future needs.

6. PSYCHOPATHOLOGICAL IMPLICATIONS OF THE EXISTENCE OF A MOTIVATIONAL STATE OF SICKNESS

The demonstration that the immune system is able to influence behavior and mental states has important implications for our understanding of the relationships between psychological factors and disease. In the case of cancer for example, such psychological features as the feelings of hopelessness and helplessness that are commonly associated with the onset and progression of the disease might be secondary to the effects on the central nervous system of soluble factors released from inflammatory cells surrounding the tumoral cells during the early stage of the neoplastic process. The same possibility applies to the relationship between psychological factors and auto-immune diseases. The possible causal role of cytokines in the mental and behavioral symptoms that occur in various pathological conditions, has hardly been investigated, except in a few cases, such as infection and fever, cachexia, AIDS dementia complex, chronic fatigue syndrome and depression.

There is already evidence demonstrating that pro-inflamatory cytokines are responsible for the development of subjective and behavioral symptoms of sickness during infection with a bacterial or viral pathogen. For instance, patients treated with IFN-α show fever, anorexia, fatigue, headache, myalgia and arthralgia. These symptoms culminate in lethargy and withdrawal from the surroundings. The same symptoms are observed in volunteers injected with low doses of LPS. The possibility that the release of cytokines accounts for more subtle changes in cognition and performance has been assessed by Smith et al (54). On the basis of earlier work showing that infection with upper respiratory viruses decreased the efficiency with which psychomotor tasks were performed, volunteers of both sexes were injected with IFN-α. Volunteers injected with the larger dose were significantly slower at responding in a reaction time task when they were uncertain when the target stimulus would appear. Simultaneously, they displayed hyperthermia and experienced feelings of illness. However, they were not impaired on a pursuit tracking task or syntactic reasoning task. These effects were similar to the alterations in performance observed in patients with influenza.

The possibility that pro-inflamatory cytokines have relatively specific effects on cognitive processes has been further investigated in animal models. IL-1β, but not IL-6, impaired spatial navigation learning in rats (55). A similar deficit in spatial learning was observed in mice injected with IL-1β or infected with the pathogenic agent *Legionella pneumophilia* (56). Interference of cytokines with formation of new memories has also been demonstrated in an autoshaping task in which rats learned to press a lever which was introduced into the cage before food delivery (57). These effects of cytokines appear to be independent of their pyrogenic activity since they were observed whether body temperature increased or decreased in response to the treatment under study.

There has been much speculation on the possible pathogenic role of cytokines in the chronic fatigue syndrome (CFS). Always feeling tired is a common complaint of patients afflicted with a viral infection and represents the core symptom of the so-called post-viral fatigue syndrome. CFS patients feel the same but in the absence of any persistent viral infection (58). Their symptoms are real, pervasive and often incapacitating. The fact that a substantial proportion of these patients fulfills criteria for major depression and other psychiatric illness does not facilitate the classification of this disorder. Whatever the case, and in view of the similarities between the subjective effects of cytokines and the symptoms reported by CFS patients, many researchers have looked for possible overproduction of cytokines in this condition. Elevated plasma levels of pro-inflamatory cytokines have been reported in a number of studies of CFS patients (59,60), but these result have not been found in other studies (61-63). Such inconsistent results can be easily explained by heterogeneity of the clinical population under study, technical problems associated with the detection of cytokines in biological fluids, and the poor correlation between plasma levels of cytokines and local activity of these mediators. A better way of assessing peripheral cytokine function is to study the ability of peripheral blood mononuclear cells (PBMC) to produce cytokines when put in culture and stimulated with adequate stimuli. Using such a strategy, a few studies have found evidence of association of chronic fatigue with low level activation of the immune system (64-66). However, it is important to note that abnormalities, if any, of the cytokine network, are not necessarily present at the periphery but might instead affect preferably cytokines that are expressed in the central nervous system.

Besides their commonality in CFS patients, lack of energy and loss of interest are very frequent in depressed patients. These symptoms are actually incorporated in the basic description of depressive episodes. The 10th revision of the International Classification of Disease begins with the statement that "the subject suffers from a lowering of mood, reduction of energy, and decrease in activity. Capacity for enjoyment, interest and concentration are impaired, and marked tiredness after even minimum effort is common." The possibility that activation of peripheral blood monocytes and T lymphocytes play a role in the pathophysiology of major depression has been proposed by Maes et al (67). In addition to the evidence pointing out the profound effects of cytokines on behavior and the hypothalamic-pituitary-adrenal axis, this hypothesis is based on the observation of an increased production of cytokines by monocytes and T-lymphocytes of depressed patients. For example, elevated levels of acute phase proteins and increased concentrations of IL-6 and its soluble receptor have been found in the plasma of subjects with major depression and there was a close relationship between IL-6 levels and acute phase proteins. However, in view of the contradictory results that have been found by other authors, and the still unresolved question of the exact symptoms with which elevated cytokine levels are associated, there is clearly a need for more research at both the clinical and experimental levels (68,69).

If the evidence in favor of an association between depression and an acute phase response appears still contradictory, it is clear that administration of cytokines to non-psychiatric patients can induce true depressive episodes. Cytokines are commonly administered in the medical treatment of metastatic kidney cancer, melanoma, chronic myeloma, and chronic viral infection. The flu-like symptoms that develop very early in all of the treated patients are followed more or less rapidly by depressed mood and alterations in cognition. Severe depression occurs in about 30 to 45 percent of the patients. Patients treated with IL-2 display depressed mood after a few days of treatment whereas patients treated with IFN-α become depressed after a few weeks (70). The risk of developing a depressive episode is positively correlated to the score of depressed mood at the initiation of treatment despite the fact that the range of variation of the initial scores stays within the normal range (71). The occurrence of depression in response to IFN-α immunotherapy can be prevented by pretreatment with antidepressants such as paroxetine (72). However, this treatment is more effective on the psychic symptoms of cytokine-induced depression than on the neurovegetative core symptoms (73). In particular, fatigue remains unchanged.

The pathophysiology of cytokine-induced depression has not yet been elucidated (16). At the symptomatic level, it is probably mediated by cytokine-induced alterations of neurotransmitter function. In particular, there is indirect clinical evidence that cytokines alter the metabolism of tryptophan, certainly as a result of the induction of the enzyme indoleamine 2,3-dioxygenase. This enzyme that is present in monocytes, macrophages and microglia is responsible for the degradation of tryptophan into kynurenine and quinolinic acid, making less tryptophan available for the synthesis of serotonin. Another enzyme that contributes to the degradation of tryptophan is the hepatic tryptophan pyrrolase that is induced by glucocorticoids. In view of these data and the close relationship between variations in tryptophan availability and depressed mood in vulnerable patients, it is therefore not surprising to find a positive correlation between the fall in tryptophan circulating levels and the increase in depressed mood in cancer patients treated with cytokines (74).

The possibility of a role of cytokines in depression has also been studied in animal models of depression. In the absence of any knowledge on the causal factors of depression, most animal models of depression are based on behavioral and pharmacological analogies. At the behavioral level, the two main symptoms that are usually considered include the deficit in escape/avoidance learning and the anhedonia, or more precisely the diminished capacity to experience pleasure, which are typically displayed by experimental animals exposed to uncontrollable electric shocks. At the pharmacological level, chronic but not acute treatment with antidepressant drugs blocks the development of these symptoms. A number of studies provide some evidence for a role of cytokines in animal models of depression. Intracerebroventricular administration of the interleukin-1 receptor antagonist (IL-1ra) which blocks the access of endogenous IL-1 to its receptors, attenuated the escape-avoidance deficit induced by inescapable electric shock in rats (75). In a different series of experiments, cytokine treatment was found to induce anhedonia in rats, as evidenced by decreased responding for rewarding lateral hypothalamic self-stimulation in response to IL-2 and LPS but not to IL-1β and IL-6 (76), and attenuated preference for a saccharin solution in response to LPS (77). This last effect was antagonized by chronic but not acute treatment with the anti-depressant drug imipramine (77). The atypical antidepressant drug tianeptine had the same action. It attenuated the effects of LPS and IL-1β injected peripherally on behavior (78) and hypothalamic-pituitary-adrenal activity (79), but it failed to alter the behavioral effects of LPS and IL-1β when these molecules were injected into the lateral ventricle of the brain. The anti-cytokine effect of tianeptine is associated to its ability to shift the pattern of brain cytokine production in response to peripheral immune stimuli from a predominance of pro-inflamatory cytokines over anti-inflammatory cytokines to a reverse pattern.

7. CONCLUSION

Sufficient evidence is now available to accept the concept that cytokines are interpreted by the brain as molecular signals of sickness. Sickness can actually be considered as a motivation, that is, a central state that organizes perception and action in face of this particular threat that is represented by infectious pathogens. A sick individual does not have the same priorities as a well one, and this reorganization of priorities is mediated by the effects of cytokines on a number of peripheral and central targets. The elucidation of the mechanisms that are involved in these effects should give new insight on the way sickness and recovery processes are organized in the brain.

ACKNOWLEDGEMENTS

Supported by INSERM, INRA, Université de Bordeaux II, Pôl e Médicament Aquitaine, European Community (Biomed CT97-2492, TMR CT97-0149), and NIH (MH51569-02 and DK49311).

REFERENCES

1. Hart, B.L. Biological basis of the behavior of sick animals. *Neurosci. Biobehav. Rev.* 1988; 12: 123-137.
2. Kluger, M.J. *Fever: its biology, evolution and function.* Princeton University Press, Princeton, 1979.
3. Kent, S., Bluthé, R.M., Kelley, K.W. et al. Sickness behavior as a new target for drug development. *Trends Pharmacol. Sci.*1992; 13: 24-28.
4. Tazi, A., Dantzer, R., Crestani, F. et al. Interleukin-1 induces conditioned taste aversion in rats: A possible explanation for its pituitary-adrenal stimulating activity. *Brain Res.*1988; 473: 369-371.
5. Tazi, A., Crestani, F. and Dantzer, R. Aversive effects of centrally injected interleukin-1 are independent of its pyrogenic activity. *Neurosci. Res. Comm.*1990; 7: 159-165.
6. Segall, M.A. and Crnic, L.S. A test of conditioned taste aversion with mouse interferon-α. *Brain Behav. Immun.*1990; 4: 223-231.
7. Bluthé, R.M., Dantzer, R. and Kelley, K.W. Effects of interleukin-1 receptor antagonist on the behavioral effects of lipopolysaccharide in rat. *Brain Res.*1992; 573: 318-320.
8. Bluthé, R.M., Crestani, F., Kelley, K.W. et al. Mechanisms of the behavioral effects of interleukin-1. Role of prostaglandins and CRF. *Ann. N.Y. Acad. Sci.*1992; 650: 268-275.
9. Bluthé, R.M., Pawlowski, M., Suarez. S. et al. Synergy between tumor necrosis factor α and interleukin-1 in the induction of sickness behavior in mice. *Psychoneuroendocrinology* 1994; 19: 197-207.
10. Kent, S., Bret-Dibat, J.L., Kelley, K.W. et al. Mechanisms of sickness-induced decreases in food-motivated behavior. *Neurosci. Biobehav. Rev.* 1995; 20: 171-175.
11. Plata-Salaman, C.R. Cytokines and ingestive behavior: methods and overview. *In Methods in Neuroscience.* E. De Souza, Ed. Academic Press. Orlando, 1993; pp 152-168.
12. Kent, S., Bluthé, R.M., Dantzer, R. et al. Different receptor mechanisms mediate the pyrogenic and behavioral effects of interleukin-1. *Proc. Natl. Acad. Sci.* USA 1992; 89: 9117-9120.
13. Anforth, H.R., Bluthé, R.M., Bristow, A. et al. Biological activity and brain actions of recombinant rat interleukin-1alpha and interleukin-1beta. *Eur. Cytokine Network* 1998; 9: 279-288.
14. Lenczowski, M.J.P., Bluthe, R., Roth, J. et al. Central administration of rat IL-6 induces HPA activation and fever but not sickness behavior in rats. *Am. J. Physiol.* 1999; 276: R652-R658.
15. Bluthé, R.M., Castanon, N., Pousset, F. et al. Central injection of IL-10 antagonizes the behavioural effects of lipopolysaccharide in rats. *Psychoneuroendocrinology* 1999; 24: 301-311.
16. Bluthé, R.M., Lestage, J., Rees, G. et al. Dual effect of central injection of recombinant interleukin-4 on lipopolysaccharide-induced sickness behavior in rats, *Neuropsychopharmacology* 2002; 26: 86-93.

17. Bluthé, R.M., Bristow, A., Lestage, J. et al. Central injection of interleukin-13 potentiates LPS-induced sickness behavior in rats. *NeuroReport* 2001; 12: 3979-3983.
18. Goujon, E., Layé, S., Parnet, P. et al. Regulation of cytokine gene expression in the central nervous system by glucocorticoids: mechanisms and functional consequences. *Psychoneuroendocrinology* 1997; 22: Suppl. 1, 875-880.
19. Dantzer, R. and Bluthé, R.M. Vasopressin involvement in antipyresis, social communication and social recognition: a synthesis. *Crit. Rev. Neurobiol.* 1992; 16: 243-255.
20. Dantzer, R., Gheusi, G., Johnson, R. et al. Central administration of insulin-like growth factor I inhibits lipopolysaccharide-induced sickness behavior in mice. *NeuroReport* 1999; 10: 289-292.
21. Gatti, S. and Bartfai, T. Induction of tumor necrosis factor-α mRNA in the brain after peripheral endotoxin treatment: comparison with interleukin-1 family and interleukin-6. *Brain Res.*1993; 624: 291-295.
22. Layé, S., Parnet, P., Goujon, E. et al. Peripheral administration of lipopolysaccharide induces the expression of cytokine transcripts in the brain and pituitary of mice. *Mol. Brain Res.*1994; 27: 157-162.
23. Van Dam, A.M., Brouns, M., Louisse, S. et al. Appearance of interleukin-1 in macrophages and ramified microglia in the brain of endotoxin-treated rats: a pathway for the induction of non specific symptoms of sickness. *Brain Res.*1992; 588: 291-296.
24. Layé, S., Gheusi, G., Cremona, S., et al. Endogenous brain IL-1 mediates LPS-induced anorexia and hypothalamic cytokine expression. *Am. J. Physiol.*2000; 279: R93-R98.
25. Wan, W., Wetmore, L., Sorensen, C.M. et al. Neural and biochemical mediators of endotoxin and stress-induced *c-fos* expression in the rat brain. *Brain Res. Bull.*1994; 34: 7-14.
26. Bluthé, R.M., Walter, V., Parnet, P. et al. Lipopolysaccharide induces sickness behavior in rats by a vagal mediated mechanism. *C.R. Acad. Sci.* Paris, Sciences de la Vie 1994; 317: 499-503.
27. Bluthé, R.M., Michaud, B., Kelley, K.W. et al. Vagotomy attenuates behavioral effects of interleukin-1 injected peripherally but not centrally. *Neuroreport* 1996; 7: 1485-1488.
28. Layé, S., Bluthé, R.M., Kent, S. et al. Subdiaphragmatic vagotomy blocks induction of IL-1β mRNA in mice brain in response to peripheral LPS. *Am. J. Physiol.*1995; 268: R1327-R1331.
29. Hansen, M.K., Taishi, P., Chen, Z. et al. Vagotomy blocks the induction of interleukin-1beta (IL-1beta) mRNA in the brain of rats in response to systemic IL-1beta. *J. Neurosci.*1998; 18: 2247-2253.
30. Hosoi, T., Okuma, Y., Nomura, Y. Electrical stimulation of afferent vagus nerve induces IL-1beta expression in the brain and activates HPA axis, *Am. J. Physiol.*2000; 279, R141-R147.
31. Goehler, L.E., Gaykema, R.P., Nguyen, K.T. et al. Interleukin-1 beta in immune cells of the abdominal vagus nerve: a link between the nervous and immune systems? *J. Neurosci.*1999; 19: 2799-2806.
32. Mascarucci, P., Perego, C., Terrazzino, S. et al. Glutamate release in the nucleus tractus solitarius induced by peripheral lipopolysaccharide and interleukin-1 beta. *Neuroscience*1998; 86: 1285-1290.
33. Ericsson, A., Kovacs, K.J. and Sawchenko, P.E. A functional anatomical analysis of central pathways subserving the effects of interleukin-1 on stress-related neuroendocrine neurons. *J. Neurosci.*1994; 14: 897-913.

34. Sawchenko, P.E. and Swanson, L.W. The organization of noradrenergic pathways from the brainstem to the paraventricular and supraoptic nuclei in the rat.*Brain Res.*1982; 257: 275-325.
35. Tkacs, N. and Li, J. Immune stimulation induces Fos expression in brainstem amygdala afferents. *Brain Res. Bull.*1999; 48: 223-231.
36. Cao, C., Matsumura, K., Yamagata, K. et al. Involvement of cyclooxygenase-2 in LPS-induced fever and regulation of its mRNA by LPS in the rat brain. *Am. J. Physiol.*1997; 272: R1712-R1725.
37. Konsman, J.P., Kelley, K.W. and Dantzer, R. Temporal and spatial relationships between lipopolysaccharide-induced expression of Fos, interleukin-1β and inducible nitric oxide synthase in rat brain. *Neuroscience* 1999; 89: 535-548.
38. Konsman, J.P., Tridon, V. and Dantzer, R. Diffusion and action of intracerebroventricularly injected interleukin-1 in the CNS, *Neuroscience* 2000; 101, 957-967.
39. Konsman, J.P., Parnet, P., Dantzer, R. Cytokine-induced sickness behaviour : Mechanisms and implications, *Trends Neurosci.*2002; 25: 154-159.
40. Dantzer, R., Konsman, J.P., Bluthé, R.M. et al. Neural and humoral pathways of communication from the immune system to the brain: parallel or convergent? *Auton. Neurosci.*2000; 85: 60-65.
41. Bluthé, R.M., Layé, S., Michaud, B. et al. Role of interleukin-1beta and tumour necrosis factor-alpha in lipopolysaccharide-induced sickness behaviour: a study with interleukin-1 type I receptor-deficient mice. *Eur. J. Neurosci.*2000; 12: 4447-4456.
42. Liege, S., Layé, S., Li, K.S. et al. Interleukin 1 receptor accessory protein (IL-1RAcP) is necessary for centrally mediated neuroendocrine and immune responses to IL-1beta. *J. Neuroimmunol.*2000; 110: 134-139.
43. Cremona, S., Goujon, E., Kelley, K.W. et al. Brain type I but not type II IL-1 receptors mediate the effects of IL-1 beta on behavior in mice. *Am. J. Physiol.*1998; 274: R735-740.
44. Cremona, S., Layé, S., Dantzer, R. et al. Blockade of brain type II interleukin-1 receptors potentiates IL1beta- induced anorexia in mice. *Neurosci. Lett.*1998; 246: 101-104.
45. Quan, N., Whiteside, M., Kim, L. et al. Induction of inhibitory factor kappaBalpha mRNA in the central nervous system after peripheral lipopolysaccharide administration: an in situ hybridization histochemistry study in the rat. *Proc. Natl. Acad. Sci.* U.S.A.1997; 94: 10985-10990.
46. Kubota, T., Kushikata, T., Fang, J. et al. Nuclear factor-kappaB inhibitor peptide inhibits spontaneous and interleukin-1beta-induced sleep. Am. J. Physiol. Regul. Integr. *Comp. Physiol.*2000; 279: R404-413.
47. Vereker, E., O'Donnell, E. and Lynch, M.A. The inhibitory effect of interleukin-1beta on long-term potentiation is coupled with increased activity of stress-activated protein kinases. *J. Neurosci.*2000; 20: 6811-6819.
48. Campbell, V. and Lynch, M.A. The role of ceramide in the modulation of intracellular Ca2+ levels by interleukin 1 beta in rat cortical synaptosomes. *Cytokine* 2000; 12: 487-490.
49. Bolles, R.C. Species-specific defense reactions and avoidance learning. *Psychol. Rev.*1970; 77: 32-48.
50. Miller, N.E. Some psychophysiological studies of motivation and of the behavioral effects of illness. *Bull. Brit. Psychol. Soc.*1964; 17: 1-20.
51. Aubert, A., Goodall, G. and Dantzer, R. Compared effects of cold ambient temperature and cytokines on macronutrient intake in rats. *Physiol. Behav.*1995; 57: 869-873.

52. Aubert, A., Goodall, G., Dantzer, R. et al. Differential effects of lipopolysaccharide on pup retrieving and nest building in lactating mice. *Brain Behav. Immun.*1997; 11: 107-118.
53. Aubert, A., Kelley, K.W. and Dantzer, R. Differential effects of lipopolysaccharide on food hoarding behavior and food consumption in rats. *Brain Behav. Immun.*1997; 11: 229-238.
54. Smith, A., Tyrrell, D., Coyle, K. et al. Effects of interferon alpha in man: a preliminary report. *Psychopharmacology* 1988; 96: 414-416.
55. Oitzl, M.S., van Oers, H., Schöbitz, B. et al. Interleukin-1β, but not interleukin-6, impairs spatial navigation learning. *Brain Res.*1993; 613: 160-163.
56. Gibertini, M., Newton, C., Friedman, H. et al. Spatial learning impairment in mice infected with Legionella pneumophila or administered exogenous interleukin-1β. *Brain Behav. Immun.*1995; 9: 113-128.
57. Aubert, A., Vega, C., Dantzer, R. et al. Pyrogens specifically disrupt the acquisition of a task involving cognitive processing in the rat. *Brain Behav. Immun.*1995; 9: 129-148.
58. Kendell, R.E. Chronic fatigue, viruses and depression. *Lancet* 1991; 337: 160-162.
59. Cannon, J.G., Angel, J.B., Ball, R.W. et al. Acute phase response and cytokine secretion in chronic fatigue syndrome. *J. Clin. Immunol.*1999; 19: 414-421.
60. Moss, R.B., Mercandetti, A. and Vojdani, A. TNF-alpha and chronic fatigue syndrome. *J. Clin. Immunol.*1999; 19: 314-316.
61. LaManca, J.J., Sisto, S.A., Zhou, X.D. et al. Immunological responses in chronic fatigue syndrome following a graded exercise test to exhaustion. *J. Clin. Immunol.*1999; 19: 135-142.
62. Zhang, Q. Zhou, X.D., Denny, T. et al. Changes in immune parameters seen in Gulf War veterans but not in civilians with chronic fatigue syndrome. *Clin. Diagn. Lab. Immunol.*1999; 6: 6-13.
63. Buchwald, D., Wener, M.H., Pearlman, T. et al. Markers of inflammation and immune activation in chronic fatigue and chronic fatigue syndrome. *J. Rheumatol.*1997; 24: 372-376.
64. Gupta, S., Aggarwal, S., See, D. et al. Cytokine production by adherent and non-adherent mononuclear cells in chronic fatigue syndrome. *J. Psychiatr. Res.*1997; 31: 149-156.
65. Cannon, J.G., Angel, J.B., Abad, L.W. et al. Interleukin-1 beta, interleukin-1 receptor antagonist, and soluble interleulin-1 receptor type II secretion in chronic fatigue syndrome. *J. Clin. Immunol.*1997; 17: 253-261.
66. Chao, C.C., Janoff, E.N., Hu, S. et al. Altered cytokine release in peripheral blood mononuclear cell cultures from patients with the chronic fatigue syndrome. *Cytokine* 1991; 3: 292-295.
67. Maes, M, Smith, R. and Scharpe, S. The monocyte-T-lymphocyte hypothesis of major depression. *Psychoneuroendocrinology* 1995; 20: 111-116.
68. Dantzer, R., Vitkovic, L., Wollman E.E. et al. Cytokines and depression: fortuitous or causative association? *Mol. Psychiat.*1999; 4: 328-332.
69. Dantzer, R., Wollman, E.E. and Yirmiya, R., eds. *Cytokines, Stress, and Depression.* Kluwer Academic/Plenum, New York, 1999.
70. Capuron, L., Ravaud, A. and Dantzer, R. Early depressive symptoms in cancer patients receiving interleukin-2 and/or interferon alfa-2b therapy. *J. Clin. Oncol.*2000; 18: 2143-2151.
71. Capuron, L. and Ravaud, A. Prediction of the depressive effects of interferon alfa therapy by the patient's initial affective state. *N. Engl. J. Med.*1999; 340: 1370.
72. Musselman, D.L., Lawson, D.H., Gumnick, J.F. et al. Paroxetine for the prevention of depression induced by high dose interferon-alfa. *N. Engl. J. Med.* 2001; 344: 961-966.

73. Capuron, L., Gumnick, J.F., Musselman, D.L. et al. Neurobehavioral effects of interferon-alpha: phenomenology and antidepressant responsivenss of symptom dimensions. *Neuropsychopharmacology* 2002, 26; 643-652.
74. Capuron, L., Ravaud, A., Neveu, P.J. et al. Association between decreased serum tryptophan concentrations and depressive symptoms in cancer patients undergoing cytokine therapy. *Mol. Psychiat.*2002, 7: 468-473.
75. Maier, S.F. and Watkins, L. Intracerebroventricular interleukin-1 receptor antagonist blocks the enhancement of fear conditioning and interference with escape produced by inescapable shock. *Brain Res.*1995; 695: 279-282.
76. Anisman, H. and Merali Z. Anhedonic and anxiogenic effects of cytokine exposure. *Adv. Exp. Biol. Med.*1999; 46: 199-234.
77. Yirmiya, R. Endotoxin produces a depressive-like episode in rats. *Brain Res.*1996; 711: 163-174.
78. Castanon, N., Bluthé, R.M. and Dantzer, R. Chronic treatment with the atypical antidepressant tianeptine attenuates sickness behavior induced by peripheral but not central lipopolysaccharide and interleukin-1β in the rat. *Psychopharmacology* 2001; 154: 50-60.
79. Castanon, N., Konsman, J.P., Medina, C. et al. Chronic treatment with the antidepressant tianeptine attenuates lipopolysaccharide-induced Fos expression in the rat paraventricular nucleus and HPA axis activation. *Psychoneuroendocrinology* 2002, in press.

Chapter 8

CYTOKINES AND SLEEP REGULATION

James M. Krueger and Lynn Churchill
Washington State University, Pullman, Washington

1. INTRODUCTION

There is substantial evidence implicating the brain cytokine network in sleep regulation; this network seems to be intimately involved in the regulation of daily physiological sleep as well as sleep responses to perturbations such as sleep loss and infectious challenge. Two cytokines are extensively studied for their involvement in sleep regulation; they are tumor necrosis factor α (TNFα) and interleukin-1β (IL1β) and we will focus on their involvement in non-rapid eye movement sleep (NREMS) regulation. In addition, many other cytokines are expressed in brain and affect NREMS and these will be summarized. We will mention some likely downstream biochemical mechanisms by which pro-somnogenic cytokines affect sleep; these include nitric oxide (NO), adenosine and growth hormone releasing hormone (GHRH). We will also provide an overview of the brain circuitry known to be involved in NREMS regulation and summarize the relatively limited data germane to cytokine influences on these circuits. Finally, we will provide a short summary of the implications of these findings for sleep function, which remains one of the most important enigmas in all of neurobiology.

2. SLEEP AND SLEEP REGULATION

Sleep is defined using several physiological parameters such as the electroencephalogram (EEG), the electromyogram (EMG), electrooculogram, brain temperature (T_{br}), and movement. Sleep is usually divided into two major states, NREMS and rapid eye movement sleep (REMS), though subclassifications of each vigilance state are often used within the sleep literature. REMS, also called paradoxical sleep, is associated with an EEG composed of relatively fast frequency low-voltage waves, muscle relaxation (a flat EMG) interrupted by phasic twitches, and a relatively rapid rise in T_{br}.

In contrast, NREMS, also called slow-wave sleep in animals, is characterized by an EEG of high-voltage slow waves (1/2-4 Hz) and a regulated drop in T_{br} (1,2). There is a continuous structured cycling between waking, NREMS, and REMS. There is also a strong influence of circadian rhythms on sleep/wake cycle organization (reviewed 2). Historically, there has been some debate as to whether sleep is regulated by humoral (chemical) or neural mechanisms. However, today, in view of our knowledge that neurons produce and respond to a large number of biochemicals, these previously seemingly dichotomous theories of sleep regulation are viewed as complementary to each other. For example, IL1β (3) and TNFα (4) alter the firing rates of anterior hypothalamic (AH) neurons; this brain region is involved in sleep regulation (see below).

The accumulation of somnogenic substances in cerebrospinal fluid (CSF) or brain tissue during prolonged wakefulness provides very strong support for the hypothesis that sleep is regulated, in part, by humoral agents (reviewed 2,5,6,7). Humoral agents must operate within a biochemical milieu; there are precursors, enzymes, receptors, inducers, inhibitors, synergists, cofactors, second messengers, etc. related to the sleep regulatory substance, all of which are essential for normal function. Many substances can affect sleep (reviewed 2,5,7). However, only a handful of humoral agents are strongly implicated in sleep regulation. It is important to recognize that those agents implicated in NREMS and REMS regulation affect each others' production and seem to act in concert with each other to affect sleep (5). For instance, TNFα induces production of IL1β, nerve growth factor (NGF), prostaglandin, NO, adenosine, and growth hormone.

3. TNFα AND IL1β IN SLEEP REGULATION

Administration of exogenous IL1β or TNFα to mice, rats, rabbits, and cats enhances NREMS (reviewed 5). These effects can be large. For instance, mice given 3 μg TNF intraperitoneally have about 80 minutes of extra NREMS during the first 9 hours postinjection (8). The excess NREMS induced by either IL1β or TNFα appears physiological in the sense that after low somnogenic doses, sleep architecture remains normal. Thus, animals continue to cycle through stages of sleep and wakefulness, although within each cycle there may be more NREMS. After IL1β or TNFα treatment, sleep remains reversible, e.g., the entry of an individual into the recording room quickly awakes IL1β- or TNFα-treated animals. After such a disturbance, the return to NREMS is generally more rapid in IL1β- or TNFα-treated animals than in control animals. After somnogenic doses of IL1β or TNFα, there are no gross behavioral abnormalities, sleep postures remain normal and no motor dysfunction is evident. Changes in sleep-coupled autonomic functions also remain intact, e.g., changes in T_{br} associated with sleep states persist in TNFα- or IL1β-treated animals. Further, after central administration of IL1β or TNFα the amplitudes of EEG slow waves (½ - 4 Hz) during NREMS increase. Such "supranormal" EEG slow waves also occur during the deep sleep after sleep deprivation (6) and are thus thought to be indicative of the intensity of NREMS. High doses of either IL1β or TNFα disrupt sleep.

Substances that enhance the production of IL1β or TNFα, e.g., bacterial and viral products, enhance sleep after low doses (high doses of these microbial products disrupt sleep) (9,10). Substances which inhibit production of IL1β or TNFα inhibit spontaneous sleep. For instance, corticotrophin releasing hormone, the soluble IL1 receptor and the soluble TNF receptor all inhibit spontaneous sleep and are found in normal brain. In addition, the expected sleep rebound, which normally occurs after sleep loss, is attenuated if animals are pretreated with IL1 or TNF inhibitors. Inhibition of TNF (but not IL1) also attenuates the increases in NREMS induced by mild increases in ambient temperature. Mice that lack the TNF 55 kD receptor (8) or mice that lack the IL1-Type I receptor (11) sleep less than strain controls. Interestingly, the sleep deficits in these two mutant strains occur at different times of the day. Finally, inhibition of either IL1 or TNF inhibits muramyl dipeptide-(a bacterial cell wall product) induced sleep thereby implicating these cytokines in the sleep responses associated with bacterial infection (reviewed 5,9,10).

Some clinical conditions associated with excessive sleepiness or fatigue also seem to involve TNF. Thus, sleep apneic patients have elevated plasma TNF levels (12,13). HIV-AIDS patients have disrupted TNF rhythms (14). Treatment of rheumatoid arthritis patients with the TNF 75 kD soluble TNF receptor (15) alleviates the fatigue associated with this disease (even though it is the 55 kD TNF receptor that is involved in sleep, both soluble TNF receptors should bind TNF and thereby inhibit its actions). TNFα is elevated in chronic fatigue patients (16). Elevated TNF also occurs in preeclampsia patients accompanied by excess sleep (17). Postdialysis fatigue is also associated with elevated TNF levels (18,19) and cancer patients receiving TNF report fatigue (e.g., 20).

IL1β and TNFα proteins, their mRNAs and their receptors and other associated molecules are found in many areas of brain including the AH, hippocampus, and frontal cortex (21,22,23). These cytokines seem to be in neurons, glia and brain endothelial cells. In rats, their levels vary with sleep propensity, e.g., highest levels occur at the onset of daylight hours, a time at which sleep is maximum in rats. Further, IL1β and TNFα mRNA levels in brain increase during sleep deprivation (reviewed 5). TNF plasma levels also correlate with EEG slow wave activity (14). Further, in humans, plasma levels of TNFα and IL1 increase after sleep loss and the ability of circulating white blood cells to produce them increases (24,25).

Collectively, these data strongly implicate IL1β and TNFα in physiological sleep regulation. However, it seems likely that they operate within the context of the large cytokine network expressed in brain (5,21) (Fig. 1).

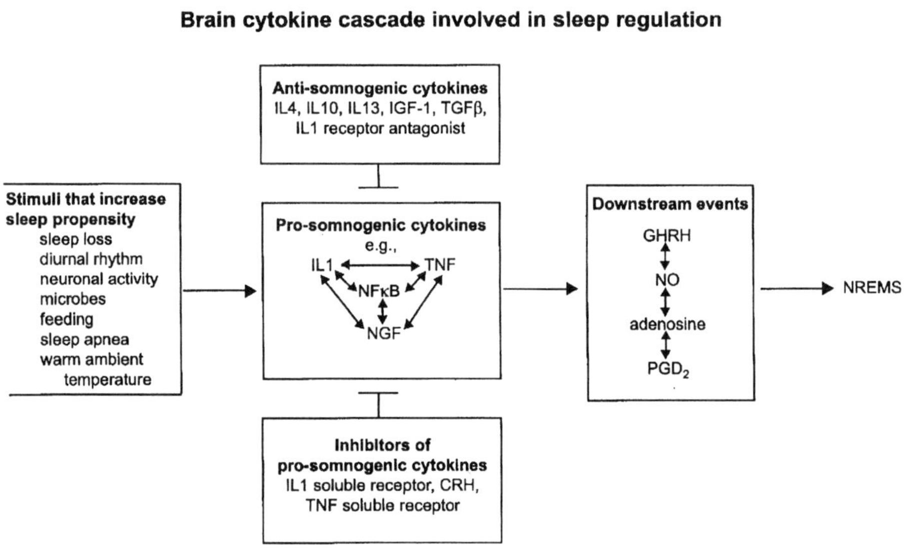

Figure 1. Sleep is regulated in part by a series of biochemical events including the cytokine cascade shown. Many cytokines likely play a role in sleep regulation although only a few are illustrated. A major challenge for sleep researchers is to determine the cellular localization of these various events. As indicated in the text and Figure 2, we are beginning to decipher some of the cytokine cellular localization of somnogenic activity. Abbreviations: IL, interleukin; IGF, insulin-like growth factor; TGF, transforming growth factor; TNF, tumor necrosis factor; NFκB, nuclear factor kappa B; NGF, nerve growth factor; CRH, corticotropin releasing hormone; GHRH, growth hormone releasing hormone; NO, nitric oxide; PGD_2, prostaglandin D_2.

4. ADDITIONAL CYTOKINES IN SLEEP REGULATION

Administration of NGF or brain-derived neurotrophic factor (BDNF) enhances both NREMS and REMS (26,27). In addition, both these cytokines upregulate in brain after sleep loss (28,29). The localization of NGF mRNA in the olfactory bulb, hippocampus, hypothalamus, and frontal lobe is similar to that of IL1β mRNA (22). Finally, destruction of NGF-receptive neurons is associated with changes in sleep (30).

A variety of additional cytokines, if injected, have the capacity to enhance NREMS. The list includes IL1α, TNFβ, IL2, IL15, IL18, acidic fibroblast growth factor, glial-derived neurotrophic factor, neurotrophin-3, neurotrophin-4, interferon α, interferon γ, and epidermal growth factor (reviewed 5,7). In contrast, other cytokines usually considered as anti-inflammatory cytokines, inhibit spontaneous sleep. These substances include the IL1-receptor antagonist, IL4, IL10, IL13, transforming growth factor β and insulin-like growth factor (reviewed 5,7). Finally, some cytokines do not appear, at least on first testing, to greatly affect sleep; these include IL6, interferon β, and basic fibroblast growth factor. More extensive studies of the sleep-related properties of these cytokines are needed before informed judgment can be made as to their roles in sleep regulation. However, that they affect sleep and that they and their receptors are expressed in normal brain suggests that a cytokine cascade operates within brain to regulate sleep.

5. SLEEP-RELATED MOLECULAR STEPS DOWNSTREAM FROM IL1β AND TNFα

IL1β and TNFα elicit GH secretion (5). The NREMS and GH responses to IL1 require, in part, GHRH because immunoneutralization of GHRH inhibits IL1-induced NREMS and GH release (31). There is an independent literature documenting the involvement of GHRH in physiological sleep regulation (reviewed 32). For instance, administration of GHRH via nasal spray, systemic injection, intraventricular injection, or by microinjection into the anterior hypothalamus enhances NREMS. Inhibition of GHRH reduces spontaneous sleep and sleep deprivation-induced sleep rebound. Hypothalamic levels of GHRH and GHRH-mRNA vary with sleep propensity. Finally, IL1β induces an upregulation of GHRH mRNA and GHRH-receptor mRNA (Taishi, et al, unpublished).

GHRH and IL1 induce increases in intracellular Ca^{++}. Ca^{++} was one of the first substances studied for its somnogenic activity. Many reports in the

1920s and 1930s indicated that injection of Ca^{++}, but not other divalent cations, into the anterior hypothalamus or cerebral ventricles induces sleep after short latencies (reviewed 33). Within the past few years, Ca^{++} has again become important in sleep research. For instance, low threshold Ca^{++} channels are involved in producing neuronal network oscillations (34), such as those thalamo-cortical currents involved in sleep regulation (reviewed 35-37). Further, L-type and T-type Ca^{++} channels are involved in the actions of some hypnotics (e.g., 38).

Nuclear Ca^{++} signaling controls, in part, the regulation of gene transcription by synaptic activity (39). The transient dendritic Ca^{++} increases induced by synaptic activity are transformed into a longer-lasting nuclear Ca^{++} transient allowing the firing patterns to be integrated into gene expression. Recently we (De et al, unpublished) have developed preliminary data showing that some cultured fetal hypothalamic neurons increase their intracellular Ca^{++} when perfused with either IL1 or GHRH. These GHRH-IL1 receptive cells comprise a small fraction of the cultured cells. There are additional cells that are responsive to only GHRH or only to IL1; but these also are only a small fraction of the total number of cells . The cells that are responsive to both IL1 and GHRH, are also GABAergic. These results are exciting for several reasons. First, they are consistent with the results from pituitary cells indicating that somatotrophs also contain IL1 receptors (40). Second, they suggest there are at least two populations of IL1-responsive hypothalamic cells, which is consistent with the finding that IL1 is excitatory to sleep active cells and inhibitory to wake-active cells (3). Third, that these GHRH-IL1 receptive cells are GABAergic suggests that they could be the hypothalamic GABAergic cells mentioned below involved in sleep regulation.

IL1β and TNFα, as well as other somnogenic cytokines such as NGF and TNFβ activate the transcription factor nuclear factor kappa B (NFκB). Activation of NFκB occurs when a regulatory subunit, IκB dissociates from NFκB after IκB is phosphorylated by a Ca^{++}-dependent kinase. NFκB is then translocated to the nucleus where it binds to DNA. NFκB usually acts as an enhancer element. In mice, during sleep deprivation and during daylight hours when sleep propensity is high, NFκB is activated in brain. Further, an NFκB inhibitory protein inhibits spontaneous sleep (41). IL1β, TNFα, and NGF genes also have a NFκB binding site and thus, for each of these, NFκB forms part of a positive feedback loop since each also stimulates NFκB and each other's production. Several substances are capable of dampening this positive feedback loop, including IL4 and IL10 which also inhibit sleep.

NFκB also seems to be involved in additional downstream events involved in sleep regulation. NFκB promotes production of NO synthase;

the enzyme responsible for the production of NO which is a retrograde neurotransmitter. NO is implicated in sleep regulation (reviewed 5,7); for instance inhibition of NO production inhibits spontaneous sleep and IL1β-induced sleep. NFκB also promotes cyclooxygenase-2 production; this enzyme is involved in the production of prostaglandin D_2 and PGD_2 is also involved in sleep regulation (reviewed 42). IL2 production, another promoter of NREMS, is also increased by NFκB activation. NFκB mediates other substances implicated in sleep regulation such as c-Fos (AP-1) and the adenosine A1 receptor. Finally, several anti-somnogenic cytokines that inhibit NFκB, e.g., IL4, IL10, IL13 and transforming growth factor β, inhibit sleep (reviewed 5,7).

In conclusion, the hypothesis that cytokines are involved in physiological processes is gaining acceptance (e.g., 21). The central idea invoked by this hypothesis is that, under normal conditions, low basal levels of cytokine production or actions vary in subtle ways with one or more physiological processes. Specificity of response for any one physiological function, e.g., sleep, would arise from a network of interactions of several cytokines/hormones with multiple neurons, each having a different array of sensitivities to different cytokines/hormones. After pathological disturbances, the production of one or more cytokines would be greatly amplified via microbial (or other pathological) stimuli in a site-specific manner. Such amplified cytokine production would then induce pathology in a manner analogous to the pathologies produced by excessive production of hormones.

6. NEURONAL CIRCUITRY INVOLVED IN NREMS REGULATION

6.1 The anterior hypothalamus in NREMS regulation

The localization of cytokine influences on the coordination of NREMS might facilitate our understanding of the role of growth factors such as cytokines in sleep function. Overall coordination of NREMS along with other behavioral states involves an integrated network of cells through the basal region of the hindbrain and forebrain (43). Historically, both the anterior and posterior hypothalamus are implicated within this integrated network. More recently, the data implicating the interconnections between the hypothalamus and the basal forebrain and cortex in NREMS regulation have been reviewed (44,45).

Humans with viral-induced lesions localized near the third ventricle within the AH (46) or rats that received electrolytic lesions in this region (47,48) have demonstrated insomnia. Conversely, electrical or chemical stimulation of this region, induces sleep (49). In particular, electrical stimulation or microinjection of sleep-promoting substances including cytokines into the AH/preoptic area (POA) increases NREMS (44,50,51).

IL1β excites sleep-active neurons and inhibits wake-active neurons (as measured by extracellular recording) within the AH/POA (3). The electrophysiological modulation of wake-active and sleep-active neurons in the AH/POA by sleep promoting substances, such as the cytokines, IL1 β or TNFα, is also consistent with the idea that this area is involved in the homeostatic regulation of sleep (3,4).

Recently, sleep-active neurons within small restricted zones within the hypothalamus, such as the ventrolateral preoptic area (VLPO) and the median preoptic nucleus (MnPN) (49) have been identified by c-fos activation during sleep. An excitotoxic lesions of the VLPO and surrounding neurons reduces NREMS time and EEG delta power (52). These c-fos sleep-active neurons are GABAergic and project to the posterior lateral hypothalamus as well as the basal forebrain and other components of the ascending monoaminergic arousal system (53). However, these neurons are not activated during sleep deprivation suggesting that they are not involved in the homeostatic regulation of sleep although they are likely involved in the maintenance of sleep.

6.2 The posterior lateral hypothalamus in wake state regulation

The posterior hypothalamus has been implicated in coordinating the wake state rather than the NREMS state as described above. However, the coordination between sleep and wake states may involve a GABAergic projection from the anterior to the posterior hypothalamus (54,55). In von Economo's studies, humans that had viral-induced lesions in the posterior hypothalamus developed excessive sleepiness (46).

The loss of orexin (also called hypocretin)-containing neurons within the posterior lateral hypothalamus, which project to many arousal systems, are implicated in the excessive daytime sleepiness and cataplexy of narcolepsy (56,57). Successful treatment of daytime sleepiness in narcoleptic patients with amphetamine or modafinal suggest that posterior hypothalamic orexin/hypocretin neurons project to the noradrenergic cells in the locus coeruleus (58) and are important for the maintenance of wakefulness. Also an increase in fos expression within these neurons induced by sleep deprivation suggest that these cells may play a role in sleep homeostasis (59).

The posterior lateral hypothalamus is not only an integrated part of the neuronal network involved in behavioral state control (43) but also a part of the central autonomic nervous system (60), which extends to the brain stem nucleus tractus solitarius (NTS). The NTS region receives vagal and glosso-pharyngeal afferents regulating gastrointestinal, cardiovascular and respiratory control (60). Stimulation of vagal afferents as well as the NTS mediates peripheral influences on NREMS (61,62,63). Vagotomy blocks the induction of NREMS by peripheral administration of several sleep-promoting substances, including IL1β and TNFα (64,65,66,67). Systemically administered IL1β induces brain production of IL1β and this action is also blocked by vagotomy.

In conclusion, there is a consensus in the sleep research field that the AH/POA and its interconnections with the posterior hypothalamus are taking part in the coordination of physiological NREMS, although there is controversy as to the specific subregions involved.

6.3 Integration of the hypothalamic network with the basal forebrain-cortical networks involved in vigilance, arousal and sleep regulation

A large-scale network involved in attention and vigilance is the "reticular activating system" which includes a core of cells through the brainstem reticular formation and mesencephalon into the posterior lateral hypothalamus and the magnocellular basal forebrain which extends from the subpallidal region through the diagonal band to the medial septum (reviewed 44,45). Although the basal forebrain often includes the POA/AH (68), we discussed this region separately. Efferents from the posterior lateral hypothalamus reach the cortex directly as well as indirectly through the paraventricular thalamus and the basal forebrain (69). Specific subregions of the lateral hypothalamus interface with different subregions within the cholinergic magnocellular basal forebrain (69). The paraventricular thalamus and the basal forebrain also receive direct inputs from the POA/AH (45,60).

The coordination of behavioral state control, such as forebrain EEG activity, by the hypothalamic network may be, in part, due to the localization of this network of cells in juxtaposition to the medial forebrain bundle that contains the fibers from the adrenergic cells from the locus coeruleus, dopaminergic cells from ventral tegmentum, and serotonergic cells from the dorsal raphe (70,71). The magnocellular cholinergic and GABAergic projection neurons within the magnocellular basal forebrain are often located adjacent to the medial forebrain bundle. The monoaminergic arousal systems along with the magnocellular basal forebrain provide enhanced attentional and vigilance states within the wake state in response to motivationally salient stimuli (45). The discharge patterns of neurons within the basal forebrain are elevated during desynchronized EEG such as in awake and REMS states, especially when there are head or limb movements (49). Tonic inhibition of these cells mediated by adenosine appears during both waking and sleep states, while the increased local GABAergic inhibition of cholinergic neurons contributes to NREMS (45). The output pathways for the basal forebrain also include nitric oxide synthase (NOS)-containing cells (72). The NOS-cells in the basal forebrain are involved in innervating microvessels as well as NOS neurons within the cortex (73).

The cytokine/growth factor, NGF, is elevated in the cortex during forced wakefulness (28). The role of this growth factor in the network between the basal forebrain and the cortex has been implicated by the presence of a low affinity receptor for NGF (p75) in the cholinergic projection neurons of the basal forebrain (74). Since NGF is retrogradely transported from the cortex to the basal forebrain by the cholinergic axons with the p75 NGF receptor (75), then this process may be involved in the reentrant integration between the basal forebrain and cortex. Removal of the NGF-receptive cholinergic neurons in the basal forebrain results in both NREMS and REMS sleep loss (30). The role of these magnocellular basal forebrain neurons in attention has been reviewed extensively because their loss has such a dramatic impact on learning and memory in Alzheimer's disease (76,77). These cholinergic inputs have also been proposed to be important in sleep regulation (44,45,49).

7. BRAIN ORGANIZATION OF SLEEP: CHALLENGE TO AN OLD PARADIGM

The circuits discussed above are well-documented for their roles in sleep regulation. Further, many of the NREMS regulatory substances such as IL1β, TNFα, GHRH, adenosine, and NO can induce NREMS if applied directly to those circuits. Such findings form the basis of the dominant paradigm in sleep research: The concept that sleep is imposed upon the brain by certain neural networks. However, this paradigm is beginning to be challenged. Thus, sleep loss results in the excess production of IL1, TNF, BDNF, adenosine, and possibly other molecules involved in sleep regulation in areas of brain normally not considered to be part of the behavioral control column for sleep regulation, e.g., the cerebral cortex. Other evidence also challenges this paradigm by suggesting that sleep is not imposed upon the brain by a small group of neurons, nor is sleep a phenomenon of the whole brain (79). For instance, there has been a failure to identify a specific neuronal network necessary for sleep. Many stimulation, lesion, and post stroke investigations lead to the conclusion: If an individual or animal survives the injury, they sleep. Marine mammals such as dolphins, have EEG slow waves characteristic of deep NREMS, only in one half of the brain at a time (80).

Fig. 2. A model of hypothesized networks involved in coordinating sleep and the influence of cytokines/growth factors. The behavioral state control column receives inputs from the reticular activating system through the brainstem, hypothalamus, and basal forebrain.

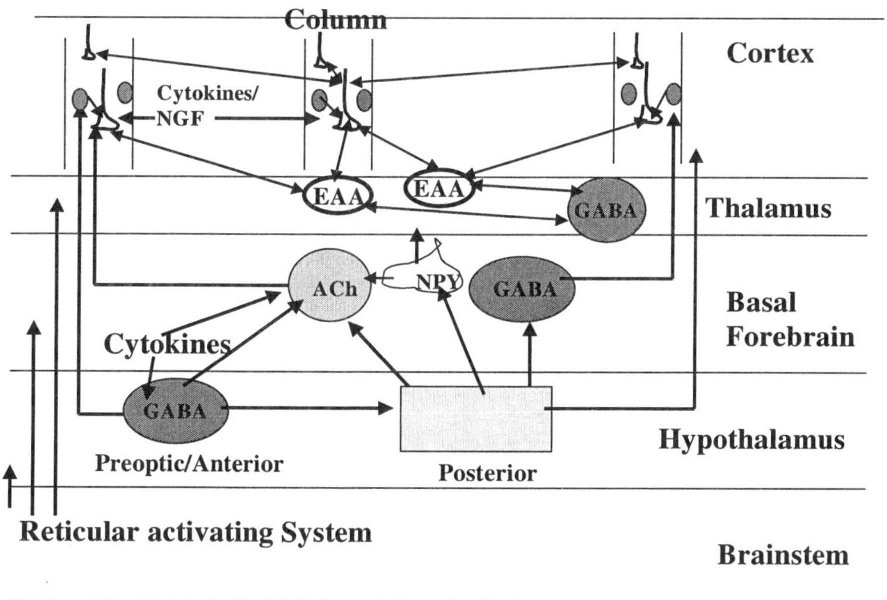

The columns of smaller neuronal networks within the cortex coordinate their activity with each other through the excitatory neurons (EAA) in the cortex and thalamus and the GABAergic neurons in the reticular thalamus (36,78). The interactions of the wake-active, cholinergic and GABAergic projection neurons with the sleep-active, NPY neurons in the basal forebrain may also influence the thalamocortical loops (45). GABAergic neurons in the preoptic/anterior hypothalamus inhibit the wake-active neurons within the posterior lateral hypothalamus. The cytokines influence the hypothalamic network through GHRH-responsive cells in the preoptic/anterior hypothalamus as well as other cells within the basal forebrain and cortex (3). The wake-active neurons in the posterior hypothalamus interface with cholinergic neurons in the basal forebrain as well as other monoaminergic arousal systems within the reticular activating system (69). Some of these posterior hypothalamic cells directly communicate with the cortex as well. Perhaps the adaptations in NGF in cortical pyramidal neurons during the wake state feedback to the basal forebrain through retrograde transport in the cholinergic axons or influence the calcium levels involved in integration within the cortical-thalamic-cortical loop.

Unihemispheric deprivation of NREMS results in sleep rebound on the half deprived of sleep, but not on the other half (81). Birds are capable of detecting predators during unihemispheric sleep and the occurrence of unihemispheric sleep varies as a function of predation risk (82). In monkeys, electrophysiological signs of sleep occur in the visual association cortex before they occur in the primary visual cortex (83). In humans, there is a gradient, from anterior to posterior, in the power of the EEG delta waves during the first sleep cycle (84). Further, sleep loss-induced increases in delta wave activity are larger in the frontal cortex than in the parietal cortex (85). Similar results occur in rats after sleep deprivation (86). The distribution of cerebral activation during sleep, whether determined by EEG techniques or functional magnetic resonance imaging (fMRI), is dependent upon the distribution of neuronal activity during wakefulness. For instance, unilateral stimulation of the somatosensory cortex during wakefulness induces enhanced delta wave activity in the stimulated area in subsequent NREMS (87). Similarly, in another human study using fMRI, Drummond *et al* (88,89) reached the conclusion that the localized effects of sleep loss are dependent, in part, on the specific cognitive task performed during prior wakefulness. These data indicate that sleep seems to be differentially stimulated in various areas of brain depending upon prior neural activity. This hypothesis is a substantial departure from the long-held view that sleep is dependent upon duration (rather than prior neuronal activity) of prior wakefulness. The newer concept is also consistent with observations such as sleep is increased by mild increases in ambient temperature (90), excessive food intake (91), certain stressors (92) without prior sleep loss. Several investigators, including our laboratory, have concluded that sleep is a distributed process in brain with small highly interconnected groups of neurons forming the fundamental units of sleep organization. In this concept, neural networks known to be involved in sleep regulation, perform a coordination role to ensure that sleep is adapted to each species' specific ecological niche (reviewed 79).

8. INVOLVEMENT OF SLEEP IN EXPRESSION OF A CYTOKINE WITHIN A SPECIFIC NETWORK: IMPLICATIONS FOR SLEEP FUNCTION

A general feature of the cerebral cortex is that the loss or overuse of sensory inputs modifies the synaptic connections and representational maps involved (93,94). One of the specific networks that is modified by loss of sensory input is the somatosensory system for the mystacial whiskers in rats (95,96,97,98). The mystacial whiskers in a rat are very important in spatial exploration; there is a one-to-one correspondence between individual whiskers and individual barrels within the somatosensory cortex (99) as well as a dynamic reorganization of the barrel cortex after a whisker trim. There is a network of neural connections that includes, in part, the somatosensory innervation from the whiskers to the ventroposterior thalamus, which relays the information to the cerebral cortex, in particular, the barrel field. Thalamocortical information terminates in the barrels (100). Finally, cortical pyramidal cells exit the cortex to the brainstem and thalamus as well as projecting back into the cortex in a feedback loop.

It is posited that repetitive neuronal activity enhances the expression, secretion and/or actions of the neurotrophins at the synapse to modify synaptic transmission and connectivity (101). NGF is one of these cytokine-neurotrophins whose production is activity-dependent (102,103). Topical application of NGF to the cerebral cortex increases the functional representation of the whisker in the barrel cortex (104) as well as increases the immunoreactivity of NGF localized within layer V pyramidal neurons (105). The plasticity within the somatosensory cortex may also involve cholinergic NGF-receptive cells from the basal forebrain (106,107).

Sleep loss impacts this model system of synaptic plasticity. The number of NGF-immunoreactive pyramidal neurons in Layer V increases after 6 hours of sleep deprivation. In contrast, sleep deprivation in combination with a whisker trim produces an increase in NGF-immunoreactivity only on the side that received whisker input while the side contralateral to the whisker cut has decreased NGF-immunoreactivity (28). These results suggest that NGF gene expression is influenced by the interactions between sleep, afferent inputs and the nature of the ongoing synaptic reorganization. A similar conclusion was reached after determining GAD mRNA levels in the barrel cortex. The number of GAD mRNA containing neurons increased after sleep deprivation and a whisker trim only in the barrel cortex that received input from the remaining lip hairs (108).

These results are also consistent with the general hypothesis that sleep serves a synaptic plasticity function. We have proposed that with use, neurons, and perhaps nearby glia, produce cytokine growth factors. For instance, the NGF-related data discussed above are consistent with this proposal. Second, these growth factors then influence the electrical properties of neurons within the immediate vicinity of where they are produced and thereby, alter the input-output relationships of the network within which they operate. The altered input-output relationship for a neuronal group can be viewed as being a different state; we believe that this represents sleep at the local neuronal group (network) level. Finally, the cytokine growth factors also affect the production of molecules involved in synaptic formation and efficiency and thereby alter synapses. This can be viewed as a function of sleep. It is inherent in this hypothesis that sleep mechanisms (use-induced production of growth factors) and sleep function (growth factor-altered synapses) cannot be separated and that the dissociation of environmentally-driven input from output (the unconsciousness associated with sleep) is a result of the growth factor-induced altered input-output relationship at the local level. At the organism level, sleep (and consciousness) is a statistical property of a population of neuronal groups; the sleep regulatory networks serve to coordinate the sleep of large numbers of these neuronal groups. For instance, cortical neurons have a large-scale coherence in multiple sites during sleep which is enhanced by the actions of acetylcholine (109). Regardless, it seems likely that cytokines play a central role in sleep regulation and impact synaptic plasticity.

ACKNOWLEDGEMENTS

This work was supported in part by the National Institutes of Health (Grant nos. NS25378, NS27250, NS31453, HD36520, and MH60308).

REFERENCES

1. Heller H.C., Glotzbach S.F. Thermoregulation during sleep and hibernation. *Int. Rev. Physiol.* 1977; 15:147-187.
2. Borbély A.A., Tobler I. Endogenous sleep-promoting substances and sleep regulation. *Physiol Rev* 1989; 69:605-670.
3. Alam N., McGinty D., Imeri L., et al. Efects of interleukin-1 beta on sleep- and wake-related preoptic/anterior hypothalamic neurons in unrestrained rats. *Sleep* 2001; 24: A59.
4. Shibata M., Blatteis C.M. Human recombinant tumor necrosis factor and interferon affect the activity of neurons in the organum vasculosum laminae terminalis. *Brain Res.* 1991; 562:323-326.
5. Krueger J.M., Obal F. Jr., Fang J., et al. The role of cytokines in sleep regulation. *Ann. N.Y. Acad. Sci.* 2001; 933:211-221.
6. Pappenheimer J.R., Koski G., Fencl V., et al. Extractions of sleep-promoting factor S from cerebrospinal fluid and from brains of sleep-deprived animals. *J. Neurophysiol.* 1975; 38:1299.
7. Krueger J.M., Fang J. Sleep and host defense. In *Principles and Practice of Sleep Medicine*, 3rd edition. M.H. Kryger, T. Roth, W . Dement, eds. W.B. Saunders Co., Philadelphia, 2000, pp 255-265.
8. Fang J., Wang Y., Krueger J.M. Mice lacking the TNF 55-kD receptor fail to sleep more after TNFα treatment. *J. Neurosci.* 1997; 17:5949-5955.
9. Krueger J.M., Majde J.A. Microbial products and cytokines in sleep and fever regulation. *Crit. Rev. Immunol.* 1994; 14:355-379.

10. Krueger J.M., Fang J., Majde J.A. Sleep in health and disease. In *Psychoneuroimmunology*, 3rd edition. B. Ader, D. Felten, N. Cohen eds. Academic Press, San Diego. 2001, pp. 667-685.
11. Fang J., Wang Y., Krueger J.M. The effects of interleukin-1β on sleep are mediated by the type I receptor. *Am. J. Physiol.* 1998; 274:R655-R660.
12. Entzian P., Linnemann K., Schlaak M., et al. Obstructive sleep apnea syndrome and circadian rhythms of hormones and cytokines. *Am. J. Respir. Crit. Care Med.* 1996; 153:1080-1086.
13. Vgontzas A.N., Papanicolaou D.A., Bixler E.O., et al. Elevation of plasma cytokines in disorders of excessive daytime sleepiness: Role of sleep disturbance and obesity. *J. Clin. Endocrinol. Metab.* 1997; 82:1313-1316.
14. Darko D.F., Miller J.C., Gallen C., et al. Sleep electroencephalogram delta-frequency amplitude, night plasma levels of tumor necrosis factor alpha, and human immunodeficiency virus infection. *Proc. Natl. Acad. Sci. USA* 1995; 92:12080-12084.
15. Franklin C.M. Clinical experience with soluble TNF p75 receptor in rheumatoid arthritis. *Semin. Arthritis Rheum.* 1999; 29:172-181.
16. Moss R.B., Mercandetti A., Vojdani A. TNF-alpha and chronic fatigue syndrome. *J. Clin. Immunol.* 1999; 19:314-316.
17. Edwards N., Blyton D.M., Kesby G.J., et al. Pre-eclampsia is associated with marked alterations in sleep architecture. *Sleep* 2000; 23:619-625.
18. Dreisbach A.W., Hendrickson T., Beezhold D., et al. Elevated levels of tumor necrosis factor alpha in postdialysis fatigue. *Int. J. Artif. Organs* 1998; 21:83-86.
19. Sklar A.H., Beezhold D.H., Newman N., et al. Postdialysis fatigue: Lack of effect of a biocompatible membrane. *Am. J. Kidney Dis.* 1998; 31:1007-1010.
20. Eskander E.D., Harvey H.A., Givant E., et al. Phase I study combining tumor necrosis factor with interferon-alpha and Interleukin-2. *Am. J. Clin. Oncol.* 1997; 20:511-514.
21. Vitkovic L., Bockaert J., Jacque C. "Inflammatory" cytokines: Neuromodulators in normal brain? *J. Neurochem.* 2000; 74:457-471.
22. Bandtlow C.E., Meyer M., Lindholm D., et al. Regional and cellular codistribution of interleukin 1β and nerve growth factor mRNA in the adult rat brain: possible relationship to the regulation of nerve growth factor synthesis. *J. Cell Biol.* 1990; 111: 1701-1711.
23. Taishi P., Chen Z., Obál F. Jr., et al. Sleep-associated changes in interleukin-1β mRNA in the brain. *J. Interferon Cytokine Res.* 1998; 18:793-798.
24. Yamasu K., Shimada Y., Sakaizumi M., et al. Activation of the systemic production of tumor necrosis factor after exposure to acute stress. *Eur. Cytokine Netw.* 1992; 3:391-398.
25. Uthgenannt D., Schoolmann D., Pietrowsky R., et al. Effects of sleep on the production of cytokines in humans. *Psychosom. Med.* 1995; 57:97-104.
26. Takahashi S., Krueger J.M. 1999. Nerve growth factor enhances sleep in rabbits. *Neurosci. Lett.* 1999; 264:149-152.
27. Kushikata T., Fang J., Krueger J.M. Brain-derived neurotrophic factor enhances spontaneous sleep in rats and rabbits. *Am. J. Physiol.* 1999; 276: R1334-1338.
28. Brandt J.A., Churchill L., Guan Z., et al. Sleep deprivation but not a whisker trim increases nerve growth factor within barrel cortical neurons. *Brain Res.* 2001; 898: 105-112.
29. Taishi P., Sanchez C., Wang Y., et al. Sleep alters the expression of molecules associated with synaptic plasticity. *Am. J. Physiol.* 2001; 281:R839-R845.
30. Kapás L., Obal F. Jr., Book A.A., et al. The effects of immunolesions of nerve growth factor-receptive neurons by 192 IgG-saporin on sleep. *Brain Res.* 1996; 712:53-59.

31. Obál F. Jr., Fang J., Payne L.C., et al. Growth hormone-releasing hormone mediates the sleep-promoting activity of interleukin-1 in rats. *Neuroendocrinology* 1995; 61:559-565.
32. Obál F. Jr., Krueger J.M. Hormones, cytokines and sleep. In *Coping with the Environment Neural and Endocrine Mechanisms: Handbook of Physiology.* B.S. McEwen ed. Oxford Univ. Press, New York, 2000, pp 331-339.
33. Kleitman N. Sleep and Wakefulness, University of Chicago Press, Midway Reprint, Chicago, 1963.
34. Lee J.H., Daud A.N., Cribbs L.L., et al. Cloning and expression of a novel member of the low-voltage activated T-type calcium channel family. *J. Neurosci.* 1999; 19:1912-1921.
35. Sejnowski T.J. Computational neurobiology of sleep rhythms. In *The Regulation of Sleep,* A.A. Borbély, O. Hayaishi, T.J. Sejnowski, J.S. Altman eds. Human Frontier Science Program. Strasbourg, 2000, pp 57-65.
36. Steriade M. Neuronal substates for short-term plasticity during sleep oscillations in corticothalamic networks. In *The Regulation of Sleep*, A.A. Borbély, O Hayaishi, T.J. Sejnowski, J.S. Altman eds. Human Frontier Science Program. Strasbourg, 2000, pp 30-44.
37. McCormick D.A. Dynamic properties of thalamocortical activity. In *The Regulation of Sleep*, A.A. Borbély, O Hayaishi, T.J. Sejnowski, J.S. Altman eds. Human Frontier Science Program. Strasbourg, 2000, pp 45-57.
38. Takahashi H., Yoshimoto M., Higuchi H., et al. Different effects of L-type and T-type calcium channel blockers on the hypnotic potency of triazolam and zolpidem in rats. *Eur Neuropsychopharmacol.* 1999; 9:317-321.
39. Hardingham G.E., Arnold F.J.L., Bading H. Nuclear calcium signaling controls CREB-mediated gene expression triggered by synaptic activity. *Nature Neurosci.* 2001; 4: 261-267.
40. French R.A., Zachary J.F., Dantzer R., et al. Dual expression of p80 type I and p68 type II interleukin-1 receptors on anterior pituitary cells synthesizing growth hormone. *Endocrinology*; 1966; 137: 4027-4036.
41. Kubota T., Kushikata T., Fang J., Krueger J.M. A nuclear factor kappa B (NFkB) inhibitor peptide inhibits spontaneous and interleukin-1β-induced sleep. *Am J Physiol.* 2000; 279:R404-R413.
42. Hayaishi O. Molecular mechanisms of sleep-wake regulation: a role of prostaglandin D2. *Philos. Trans. R Soc. Lond. B Bio.l Sci.* 2000; 355: 275-280.
43. Swanson L.W. Cerebral hemisphere regulation of motivated behavior. *Brain Res.* 2000; 886:113-164.
44. Szymusiak R. Magnocellular nuclei of the basal forebrain: substrates of sleep and arousal regulation. *Sleep* 1995; 18:478-500.
45. Zaborszky L., Duque A. Local synaptic connections of basal forebrain neurons. *Behav. Brain Res.* 2000; 115:143-158.
46. von Economo C. Sleep as a problem of localization. *J. Nerv. Ment. Dis.* 1930; 71:249-259.
47. Nauta W.J.H. Hypothalamic regulation of sleep in rats. An experimental study. *J. Neurophysiol.* 1946; 9:285-316.
48. McGinty D.J. Physiological equilibrium and the control of sleep stages. In *Brain Mechanisms of Sleep.* D.J. McGinty, R. Drucker-Colin, A. Morrison, P.L. Parmeggiani eds. Raven Press, New York, 1985, pp 301-384.
49. Szymusiak R., Steininger T., Alam N., et al. Preoptic area sleep-regulating mechanisms. *Arch. Ital. Biol.* 2001; 139:77-92.

50. Zhang J., Obal F. Jr., Zheng T., et al. Intrapreoptic microinjection of GHRH or its antagonist alters sleep in rats. *J. Neurosci.* 1999; 19:2187-2194.
51. Terao A., Matsumura H., Saito M. Interleukin-1 induces slow-wave sleep at the prostaglandin D_2-sensitive sleep-promoting zone in the rat brain. *J. Neurosci.* 1998; 18:6599-6607.
52. Lu J., Zhang Y.H., Chou T.C., et al. Contrasting effects of ibotenate lesions of the paraventricular nucleus and subparaventricular zone on sleep-wake cycle and temperature regulation. *J. Neurosci.* 2001; 21:4864-4874.
53. Sherin J.E., Elmquist J.K., Torrealba F., et al. Innervation of histaminergic tuberomammillary neurons by GABAergic and galaninergic neurons in the ventrolateral preoptic nucleus of the rat. *J. Neurosci.* 1998; 18:4705-4721.
54. Lin J.-S., Sakai K,. Vanni-Mercier G., et al. A critical role of the posterior hypothalamus in the mechanisms of wakefulness determined by microinjection of muscimol in freely moving cats. *Brain Res.* 1989; 479:225-240.
55. Sallanon M., Denoyer M., Kitahama K., et al. Long-lasting insomnia induced by preoptic neuron lesions and its transient reversal by muscimol injection into the posterior hypothalamus in the cat. *Neuroscience* 1989; 32:669-683.
56. Hara J., Beuckmann C.T., Nambu T., et al. Genetic ablation of orexin neurons in mice results in narcolepsy, hypophagia, and obesity. *Neuron* 2001; 30: 345-354.
57. Hungs M., Mignot E. Hypocretin/orexin, sleep and narcolepsy. *Bioessays* 2001; 23:397-408.
58. Bourgin P., Huitron-Resendiz S., Speir A.D., et al. Hypocretin-1 modulates rapid eye movement sleep through activation of locus coeruleus neurons. *J. Neurosci.* 2000; 20: 7760-7765.
59. Estabrooke I.V., McCarthy M.T., Ko E., et al. Fos expression in orexin neurons varies with behavioral state. *J. Neurosci.* 2001; 21: 1656-1662.
60. Saper C.B. The central autonomic nervous system In *The Rat Nervous System*, G. Paxinos ed, Academic Press, San Diego, CA, 1995, pp 107-135.
61. Chase M.H., Nakamura Y., Clemente C.D., et al. Afferent vagal stimulation: neurographic correlates of induced EEG synchronization and desynchronization. *Brain Res.* 1967; 5:236-247.
62. Steriade M., McCarley R.W. *Brainstem Control of Wakefulness and Sleep*, Plenum Press, New York, NY 1990.
63. Vertes R.P. Brainstem mechanisms of slow-wave sleep and REM sleep. In *Brainstem Mechanisms of Behavior*, W.R. Klemm, R.P. Vertes eds. R.P. John Wiley & Sons, New York, NY, 1990, pp 535-583.
64. Kubota T., Fang J., Guan Z., et al. Vagotomy attenuates tumor necrosis factor-alpha-induced sleep and EEG delta-activity in rats. *Am. J. Physiol.* 2001; 280: R1213-R1220.
65. Hansen M.K., Krueger J.M. Subdiaphragmatic vagotomy blocks the sleep- and fever-promoting effects of interleukin-1ß. *Am. J. Physiol.* 1997; 273:R1246-R1253.
66. Fleshner M., Goehler L.E., Schwartz B.A., et al. Thermogenic and corticosterone responses to intravenous cytokines (IL-1 beta and TNF-alpha) are attenuated by subdiaphragmatic vagotomy. *J. Neuroimmunol.* 1998; 86:134-141.
67. Kapás L., Hansen M.K., Chang H.Y., et al. Vagotomy attenuates but does not prevent the somnogenic and febrile effects of lipopolysaccharide in rats. *Am. J. Physiol.* 1998; 274:R406-R411.
68. Heimer L., Harlan R.E., Alheid G.F., et al. Substantia innominata: a notion which impedes clinical-anatomical correlations in neuropsychiatric disorders. *Neuroscience* 1997; 76:957-1006.

69. Cullinan W.E., Zaborszky L. Organization of ascending hypothalamic projections to the rostral forebrain with special reference to the innervation of cholinergic projection neurons. *J. Comp. Neurol.*1991; 306:631-667.
70. Vertes R.P. A lectin horseradish peroxidase study of the origin of ascending fibers in the medial forebrain bundle of the rat. The lower brainstem. *Neuroscience* 1984; 11: 651-668.
71. Vertes R.P. A lectin horseradish peroxidase study of the origin of ascending fibers in the medial forebrain bundle of the rat. The upper brainstem. *Neuroscience* 1984; 11: 669-690.
72. Semba K. Multiple output pathways of the basal forebrain: organization, chemical heterogeneity, and roles in vigilance. *Behav. Brain Res.* 2000; 115:117-141.
73. Tong X.K., Hamel E. Basal forebrain nitric oxide synthase (NOS)-containing neurons project to microvessels and NOS neurons in the rat neocortex: cellular basis for cortical blood flow regulation. *Eur. J. Neurosci.* 2000; 12:2769-2780.
74. Heckers S., Ohtake T., Wiley R.G., et al. Complete and selective cholinergic denervation of rat neocortex and hippocampus but not amygdala by an immunotoxin against the p75 NGF receptor. *J. Neurosci.*1994; 14:1271-1289.
75. Kramer, B.M.R., Van Der Zee C.E.E.M., Hagg T. p75 nerve growth factor receptor is important for retrograde transport of neurotrophins in adult cholinergic basal forebrain neurons. *Neuroscience* 1999; 94:1163-1172.
76. Buzsaki G., Bickford R.G., Ponomareff G., et al. Nucleus basalis and thalamic control of neocortical activity in the freely moving rat. *J. Neurosci.* 1988; 8:4007-4026.
77. Price D.L., Whitehouse P.J., Struble R.G. Cellular pathology in Alzheimer's and Parkinson's diseases. *Trends Neurosci.* 1986; 9:29-33.
78. Krimer L.S., Goldman-Rakic P.S. (2001) Prefrontal microcircuits: membrane properties and excitatory input of local, medium, and wide arbor interneurons. *J. Neurosci.* 2001; 21:3788-3796.
79. Krueger J.M., Obál F. A neuronal group theory of sleep function. *J. Sleep Res.* 1993; 2:63-69.
80. Mukhametov L.M., Supin A.V., Polyakova I.G., et al. Interhemispheric asymmetry of the electroencephalographic sleep patterns in dolphins. *Brain Res.* 1977; 134:581-584.
81. Oleksenko A.I., Mukhametov L.M., Polyakova I.G., et al. Unihemispheric sleep deprivation in bottlenose dolphins. *J. Sleep Res.* 1992; 1:40-44.
82. Rattenborg N. Half-awake to the risk of predation. *Nature* 1999; 397:397-398.
83. Pigarev I.N., Nothdurft H.C., Kastner S. Evidence for asynchronous development of sleep in cortical areas. *Neuroreport* 1997; 8:2557-2560.
84. Werth E., Achermann P., Borbély A.A. Fronto-occipital EEG power gradients in human sleep. *J. Sleep Res.* 1997; 6:102-112.
85. Cajochen C., Foy R., Dijk D.J. Frontal predominance of a relative increase in sleep delta and theta EEG activity after sleep loss in humans. *Sleep Res.* Online 1999; 2:65-69.
86. Schwierin B., Achermann P., Deboer T., et al. Regional differences in the dynamics of the cortical EEG in the rat after sleep deprivation. *Clin. Neurophysiol.* 1999; 110:869-875.
87. Kattler H., Dijk D.J., Borbély A.A. Effect of unilateral somatosensory stimulation prior to sleep on the sleep EEG in humans. *J. Sleep Res.* 1994; 3:159-164.
88. Drummond A., Brown G.G., Gillin J.C., et al. Altered brain response to verbal learning following sleep deprivation. *Nature* 2000; 403:655-657.
89. Drummond S.P., Brown G.G., Stricker J.L., et al. Sleep deprivation-induced reduction in cortical functional response to serial subtraction. *Neuroreport* 1999; 10:3745-3748.

90. Obál F. Jr. Thermoregulation and sleep. *Exp. Brain Res. Suppl.* 1984; 8:157-172.
91. Hansen M.K., Kapás L., Fang J., et al. Cafeteria diet-induced sleep is blocked by subdiaphragmatic vagotomy in rats. *Am. J. Physiol.* 1998; 274:R168-R174.
92. Bodosi B., Obál F. Jr., Gardi J., et al. An ether stressor increases REM sleep in rats: Possible role of prolactin. *Am. J. Physiol.* 2000; 279:R1590-R1598.
93. Purves D. *Neural activity and the growth of the brain,* Cambridge: Cambridge University Press, 1994.
94. Buonomano D., Merzenich M. Cortical plasticity: from synapses to maps. *Ann. Rev. Neurosci.* 1998; 21:149-186.
95. Diamond M., Armstrong-James M., Ebner F. Experience-dependent plasticity in adult rat barrel cortex. *Proc. Natl. Acad. Sci. USA*, 1993; 90(5) 2082-2086.
96. Armstrong-James M., Diamond M.E., Ebner F.F. An innocuous bias in whisker use in adult rats modifies receptive fields of barrel cortex neurons. *J. Neurosci.* 1994; 14:6978-6991.
97. Steiner H., Gerfen C.R. Tactile sensory input regulates basal and apomorphine-induced immediate-early gene expression in rat barrel cortex. *J. Comp. Neurol.* 1994; 344:297-304.
98. Dietrich W.D., Ginsberg M.D., Busto R., et al. Metabolic alterations in rat somatosensory cortex following unilateral vibrissal removal. *J. Neurosci.* 1985; 5:874-880.
99. Woolsey T.A., Van der Loos H. The structural organization of layer IV in the somatosensory region (SI) of mouse cerebral cortex. The description of a cortical field composed of discrete cytoarchitectonic units. *Brain Res.* 1970; 17:205-242.
100. Huang W., Armstrong-James M., Rema V., et al. Contribution of supragranular layers to sensory processing and plasticity in adult rat barrel cortex. *J. Neurophysiol.* 1998; 80:3261-3271.
101. Schinder A.F., Poo M. The neurotrophin hypothesis for synaptic plasticity. *Trends Neurosci.* 2000; 23:639-645.
102. Lindholm D., Castren E., Berzaghi M., et al. Activity-dependent and hormonal regulation of neurotrophin mRNA levels in the brain--implications for neuronal plasticity. *J. Neurobiol.* 1994; 25:1362-1372.
103. Schmidt-Kastner R., Humpel C., Wetmore C., et al. Cellular hybridization for BDNF, trkB, and NGF mRNAs and BDNF-immunoreactivity in rat forebrain after pilocarpine-induced status epilepticus. *Exp. Brain Res.* 1996; 107:331-347.
104. Prakash N., Cohen-Cory S., Frostig R.D. RAPID and opposite effects of BDNF and NGF on the functional organization of the adult cortex in vivo. *Nature* 1996; 381:702-706.
105. Kolb B., Gorny G., Cote S., et al. Nerve growth factor stimulates growth of cortical pyramidal neurons in young adult rats. *Brain Res.* 1997; 751:289-294.
106. Sachdev R.N.S., Lu S.M., Wiley R.G., et al. Role of the basal forebrain cholinergic projection in somatosensory cortical plasticity. *J. Neurophysiol.* 1998; 79:3216-3228.
107. Zhu X.O., Waite P.M. Cholinergic depletion reduces plasticity of barrel field cortex. *Cereb. Cortex* 1998; 8:63-72.
108. Churchill L.P. Taishi Z., Guan L., et al. Sleep modifies glutamate decarboxylase mRNA within the barrel cortex of rats after a mystacial whisker trim. *Sleep* 2001; 24: 261-266.
109. Destexhe A., Contreras D., Steriade M. Cortically-induced coherence of a thalamic-generated oscillation. *Neuroscience* 1999; 92:427-443.

Chapter 9

CYTOKINES AND APPETITE

Wolfgang Langhans and Brian J. Hrupka
Institute of Animal Sciences, Swiss Federal Institute of Technology (ETH), 8092 Zurich, Switzerland

1. INTRODUCTION

Cytokines are known to orchestrate non-specific and specific immune reactions and are broadly categorized as being pro-inflammatory or anti-inflammatory. In addition to their pleiotropic effects in the immune system, cytokines affect other physiologic functions and cause CNS-mediated effects such as fever, sleep, and an activation of the hypothalamic-pituitary-adrenal axis. Moreover, several pro-inflammatory cytokines, such as interleukin-1 (IL-1), IL-2, IL-6, IL-8, tumor necrosis factor-α (TNFα) and interferon-γ (IFNγ) suppress appetite. The cytokine-induced suppression of appetite, i.e. the resulting anorexia and hypermetabolism are major contributors to the cachexia of chronic diseases, and part of the organism's acute phase response to various immune stimuli. Acute, short-term anorexia during an immune challenge is considered to be beneficial for the host. For example, force-feeding experimentally infected mice to the control levels reduced survival time and increased mortality (1). Sustained cytokine-mediated anorexia, however, which occurs during chronic infections, cancer and other chronic diseases can compromise the host's ability to fight disease. Cytokines also presumably contribute to the anorexia and cachexia that promote fragility in aging people. Thus, the appetite suppressing effect of cytokines is a practically relevant and scientifically interesting research problem with various facets. This chapter addresses several critical questions in this context, such as: What are the effects of cytokines alone and in combination on eating, when are endogenous cytokines produced in sufficient quantities to suppress eating, and how do cytokines affect a behavior as complex as eating, i.e. what are the mechanisms of the suppressive effect of pro-inflammatory cytokines on eating?

2. EFFECTS OF EXOGENOUS CYTOKINES ON EATING

Pro-inflammatory cytokines inhibit eating after peripheral and central administration (2-4), suggesting that their enhanced production contributes to the anorexia that accompanies various pathological states. The type of the cytokine, the administered dose, and the route of administration determine the magnitude of the food intake suppression (4). The anorexia induced by cytokines is independent of any suppressive effect on water intake or their fever and sleep promoting properties (5). Moreover, the anorectic effect appears to be specific because it is blocked by cytokine receptor antagonists or specific antibodies (5-7)

In a macronutrient choice situation, IL-1β specifically reduced protein intake (8), whereas percent fat intake was unaltered and carbohydrate intake slightly increased. This is in line with empirical evidence suggesting that illness induces a preference for high carbohydrate-high fat/low protein diets (9). It is unknown whether other cytokines also exert nutrient-specific effects. Changes in nutrient selection in response to cytokines could be related to chemosensory changes and/or taste aversions. Aversions against an unfamiliar taste can be produced by either peripheral or central administration of high doses of pro-inflammatory cytokines (10-12). This may be relevant for some diseases and become apparent in nutrient choice situations because proteins are presumably a more salient target for taste aversions than carbohydrates or fats. Yet, taste aversions do not appear to play a major role in IL-1β and TNFα-induced anorexia when laboratory animals are continuously fed normal chow (10-12).

Finally, some studies suggest that cytokine-induced anorexia is blunted by prior body weight reduction (13,14). Thus, IL-1 or experimental inflammation failed to reduce food intake in rats after prior body weight reduction, and all rats achieved the same level of body weight independent of body weight at test onset. Based on such findings it was proposed that cytokines might cause a temporary reduction in body weight set point that is spontaneously maintained by reduced feeding (14). This may be based on interactions between pro-inflammatory cytokines and long-term signals of adiposity (13), but the exact mechanisms of these interactions have not been identified yet.

Tolerance to the anorectic effects of IL-1 and TNFα often develops with chronic peripheral administration of these cytokines (14-16). Cytokine tolerance is not the result of altered absorption and clearance of cytokines or due to a humoral immune reaction (16,17). Some evidence indicates that, at least for TNFα tolerance, a blockade of receptor mechanisms is involved (18). Tolerance is not observed with continuous intracerebroventricular

(ICV) infusion of IL-1β (19), with repetitive IP injections of IFNα (20) or IFNγ (21) in mice and with repetitive IP injections of IL-1β in rats, given when the anorectic effect of the previous injection has subsided (16). In the latter situation sensitization occurs, i.e. the anorectic effect of IL-1β is enhanced. This is interesting because cytokines may be released in a cyclic fashion during several diseases. The reasons for these differential tolerance responses are unknown.

ICV administration of cytokines at doses in the picogram–nanogram range reduces food intake (19,22). These doses yield cytokine concentrations in the CSF which correspond to the cytokine concentrations found during different diseases or injuries of the brain. Much higher cytokine doses have to be administered peripherally to reduce food intake. This indicates that centrally administered cytokines inhibit feeding by a direct action in the brain. Interestingly, however, exogenous cytokines appear to have different effects on eating after peripheral and central administration: Whereas ICV administered cytokines reduced food intake by primarily reducing meal size (23), peripherally administered IL-1β and TNFα have been shown to affect meal size and meal frequency (24-26). Moreover, whereas centrally administered cytokines inhibit water intake in addition to eating, peripherally administered anorectic doses of IL-1β and TNFα do not (26).

Substantial evidence indicates that sex steroids influence various immune functions (27). Females often show enhanced immunoreactivity and cytokine production compared to males (28,29). In premenopausal women, the LPS-induced cytokine production by peripheral blood monocytes was lower during the luteal phase than during the follicular phase (30), suggesting that estrogen increases this response. Gender differences have also been observed in effects of IL-1β on locomotor activity, sexual responsivity, and fever (see (31)). Estrogen administration in rats has been shown to enhance the fever response to IL-1β (32) and to sensitize liver Kupffer cells to LPS (33). Finally, estradiol appears to enhance the anorectic effect of IL-1β (34). Recent data from Nori Geary's laboratory show that cyclic 17β-estradiol replacement in ovariectomized rats enhances the anorectic effect of LPS (31). Thus, women may actually experience a more severe anorexia of infection during the follicular stage of the menstrual cycle, but to our knowledge, this possibility has not been critically examined yet.

Pro-inflammatory cytokines can act in sequence, i.e. that one cytokine stimulates the synthesis and release of another cytokine, or in parallel, and the final response depends on their synergistic and antagonistic interaction pattern. A synergistic feeding suppressive effect has so far mainly been described for IL-1β and TNFα (22,35,36), sometimes also for these two cytokines and IL-8 (22). This synergism is presumably related to the cytokines overlapping effects and to the fact that they act through converging intracellular signaling pathways. Activation of the transcription factor NFκB is a common feature of the postreceptor mechanisms of several cytokines (37). Given these complex interactions, it is clear that administration of single cytokines yields an incomplete and at times misleading picture and does not really model the complex interactions that occur during natural diseases. To understand the mechanisms and relevance of the feeding suppressive effects of cytokines, it is better to investigate the effects of appropriate cytokine combinations or to look at situations in which the production and release of endogenous cytokines is enhanced.

3. INCREASED ENDOGENOUS CYTOKINE PRODUCTION

3.1 General Aspects

Pro-inflammatory cytokines are produced by immune and other cells in response to different infections and in a variety of other pathologic situations. During such diseases, a network of pro-inflammatory and anti-inflammatory cytokines is activated and plasma levels of cytokines are elevated. Pro-inflammatory cytokines are often involved in both protection against, and pathogenesis of the disease. Changes in plasma concentrations of some cytokines have at times been considered to be a predictor of disease severity and prognosis. Typically, however, increases in plasma levels of cytokines depend on the characteristics of the pathogen, are individually variable, and can not be used as a reliable means of prognosis.

Several immune functions are altered in mice made genetically deficient of a single cytokine or cytokine receptor (knockout = KO mice) (38,39), reflecting the various immune functions of the pertinent cytokines. The situation is more complex regarding the cachexia and anorexia observed in various disease models. In many studies, treatments designed to model infection and/or inflammation still reduce food intake in various KO preparations (see below), although the anorexia is sometimes attenuated. This is presumably related to the redundancy and synergistic interactions of cytokines.

3.2 Infections and Models of Infection or Local Inflammation

3.2.1 Toxoplasma gondii infection in mice

A useful tool to study the role of cytokines in chronic infections is the mouse model of chronic infection with Toxoplasma gondii (40). In Swiss Webster mice, Toxoplasma gondii produces a progressive and substantial loss in body weight within 2-3 weeks. Intitially (during days 1-7 after infection), food intake is typically unaltered, but energy expenditure is increased. This has been reported to be associated with an elevated expression of several pro-inflammatory cytokines, including TNFα, IL-1β, and IFNγ, in spleen and brain. The hypermetabolic state typically persists during the subsequent anorexia, and at the end of the acute phase of cachexia, the dual anorexic and hypermetabolic states were associated with an increased expression of all cytokines examined (IL-1β, IL-2, IL-5, IL-6, IL-10, IFNγ). In the chronic phase of the infection, the mice showed either partial weight recovery (gainers) or no weight regain (nongainers). The infected gainers, though still hypophagic, were no longer hypermetabolic, and their cytokine mRNAs were no longer elevated, except for TNFα and IL-10. In contrast, the infected nongainers remained anorectic and hypermetabolic; they displayed elevations in all cytokines examined and particularly those of the TH2 profile (IL-4 and IL-5) and IL-6. Interestingly, mice lacking uncoupling protein-2 (UCP-2) generated more reactive oxygen species than wild type control mice in response to Toxoplasma gondii inoculation and were resistant to the infection, suggesting a role for UCP-2 in the limitation of ROS production and macrophage-mediated immunity (41). Thus, the Toxoplasma infection model reveals a distinct pattern of cytokine mRNA expression underlying 1) hypermetabolism vs. anorexia, 2) acute vs. chronic cachexia, and 3) stable weight loss vs. partial weight recovery (40). This two-way dissociation of cytokine-mediated hypermetabolism and anorexia makes the Toxoplasma model very interesting for future studies into the mechanisms of both effects.

3.2.2 Sterile inflammation

Sterile inflammation is often produced by subcutaneous (SC) injection of turpentine, which results in fever, lethargy, body weight loss, muscle breakdown and anorexia in laboratory animals (42). SC turpentine injection increased plasma IL-6 (43), and a complete elimination of anorexia and other sickness symptoms in response to SC turpentine injection was observed in IL-6 KO mice (44,45). These results indicate that IL-6 is essential for the anorexia in this model of sterile local inflammation, and are consistent with the presumed general role of IL-6 as an essential mediator of the responses to sterile localized inflammation. In the same study (45), sickness symptoms were only slightly attenuated during influenza virus infection in IL-6 KO mice, suggesting that IL-6 independent mechanisms mediate the anorexia (and other symptoms) during viral infection. The feeding suppressive effect of SC turpentine administration was also absent in IL-1 receptor-1 KO mice (46), suggesting that, in addition to IL-6, IL-1-mediated pathways are involved in the feeding response to local sterile inflammation. Results from an older study in which antagonism of either IL-6 or IL-1 receptor-1 by monoclonal antibodies equally spared body weight and food intake responses to SC turpentine in mice are consistent with this interpretation (47). Finally, the hypermetabolic but not the feeding response to SC turpentine was attenuated by pretreatment with TNFα antiserum (43), suggesting an involvement of TNFα in the metabolic reaction to local inflammation. Thus, IL-1β, IL-6, and, to a lesser extent, TNFα, all appear to be involved in mediating the feeding suppressive effect of sterile inflammation.

3.2.3 Models of inflammatory bowel disease

Inflammatory bowel disease, such as ulcerative colitis and Crohn's disease are often associated with marked weight loss due to increased energy expenditure, malabsorption and reduced food intake. Prolonged anorexia has been identified as the most important etiological factor (48). Inflammatory bowel disease has been modeled in laboratory animals by intrarectal administration of irritants such as trinitrobenzenesulfonic acid (TNB). TNB-treated animals typically eat less and loose body weight (49). TNB produces locally high levels of pro-inflammatory cytokines (50). TNB-induced anorexia was attenuated by human recombinant IL-1 receptor antagonist (rhIL-1ra) administered chronically via osmotic mini-pumps either peripherally or centrally after TNB treatment. With peripheral administration, a higher dose of rhIL-1ra was necessary to attenuate TNB-induced anorexia than with central administration (51). These data implicate an activation of the IL-1 system in the feeding suppressive effect of intrarectal TNB treatment. The roles of other pro-inflammatory cytokines in the responses to this model of inflammatory bowel disease are unknown.

3.2.4 Peripheral LPS administration as a model of systemic bacterial infection

Systemic microbial infections usually trigger a profound pro-inflammatory cytokine response, with the increases in cytokine production and serum levels mainly depending on microbial characteristics and severity of infection (52-54). Parenteral administration of bacterial lipopolysaccharide (LPS) is widely used to model bacterial infections. LPS constitutes a major part of the outer layer of Gram-negative bacterial cell walls. It is released after bacteriolysis or during periods of rapid proliferation and triggers many of the host's reactions to severe Gram-negative bacterial infections and generalized inflammation (55). LPS has no structural homolog among multicellular organisms and is probably the most powerful stimulus of innate immune responses. It is now well established that LPS acts through the Toll-like receptor 4 (TLR-4)(56,57). TLRs are a family of transmembrane proteins that are grouped into the same gene family (56,58). The postreceptor signaling pathways of TLRs and the IL-1 receptor-1 are similar, and result in activation of the transcription factor NFκB. The activation of macrophages by LPS requires the cell surface glycoprotein CD14. In cells devoid of CD14, such as endothelial cells, the soluble form of CD14 present in serum can replace membrane-bound CD14 (56). Through these mechanisms, LPS induces marked increases in circulating pro-inflammatory cytokine levels, and they potently reduces food intake after parenteral administration, a response that is mediated by several pro-inflammatory cytokines (see below).

Genetic ablation of IL-1β and TNFα or these cytokines' receptors usually does not alter the anorectic effect of peripheral LPS or influenza virus infection (46,59-62). Thus, IL-1β and TNFα appear to be unnecessary for the anorectic response under these conditions and may actually be able to replace each other in the cascade of pro-inflammatory cytokines. This appears reasonable given their overlapping actions. The potential for a reciprocal substitution of IL-1β and TNFα in the mediation of LPS anorexia is supported by the finding that IL-1 receptor KO mice reduce food intake as much as wild type control mice in response to IP LPS, but that concomitant administration of a fragment of soluble TNF receptor eliminates the effect (60). This indicates that neither cytokine receptor is necessary for the feeding suppressive and other sickness-related effects of LPS, but that the IL-1β and TNFα pathways can replace each other in mediating LPS-induced anorexia because blockade of both eliminates the anorectic effect (60).

An attenuation of LPS-induced reduction in body weight and social exploration was seen in IL-6 KO mice (63). This was seen when LPS was injected either IP (2.5 μg/mouse) or into the lateral ventricle (5 ng/mouse).

These results suggest that peripheral and/or central IL-6 plays a role in some behavioral and body weight responses to LPS (63). In another study (64) the change in body weight observed during sepsis induced by cecal ligation and puncture was similar in wild type and IL-6 KO mice, and the suppression of food intake was prolonged in the KO preparation. These discrepant results with the different models in genetically IL-6 deficient mice presumably reflect the complexity and the multiple overlap and redundancies of the cytokine systems. Nevertheless, the results are compatible with a role of IL-6 in the feeding response to bacterial infections. IL-6 is a pleiotropic cytokine produced by many different cell types (65). The main cellular sources for IL-6 are macrophages, fibroblasts, and endothelial cells (65). The effects of IL-6 include a stimulation of antibody production, hematopoiesis, and acute phase protein synthesis in the liver, as well as T cell activation (65).

A clear abolition of the anorectic and hypermetabolic effects of peripherally administered LPS, even at high doses, has so far only been observed in IFNγ-receptor KO mice (59), suggesting that IFNγ action is necessary for the suppression of feeding and the stimulation of metabolism by LPS. Exogenous IFNγ potently reduces food intake after parenteral administration (3,21). Endogenous IFNγ is known to be produced primarily by T cells and natural killer (NK) cells (66). Pretreatment with Anti-Asialo GM1 antibodies, which are known to deplete NK cells, also attenuated the anorectic effect of LPS to an extent comparable to that observed in IFNγ-receptor KO mice (59). This implicates NK cells in the IFNγ production that is essential for LPS-induced anorexia. LPS has been shown to stimulate IFNγ production, but this effect is presumably indirect because TLR4 has not been identified on T cells or NK cells (67). IFNγ has receptors on virtually all cell types, and its main function as a pro-inflammatory cytokine is to activate macrophages and endothelial cells, partially in synergism with macrophage-derived cytokines (66). In turn, the presence of macrophages is required for the stimulation of IFNγ production by gram negative bacteria or LPS, which is probably related to the stimulation of T and NK cells by macrophage-derived IL-12 and IL-18 as well as TNFα (66,68). Thus, it is tempting to speculate that, without this positive reinforcing effect of IFNγ, the LPS-induced immune reaction does not proceed to a level that causes anorexia. It is also interesting in this context that anti IFNγ antibodies decreased the mortality from TNFα and LPS, whereas pretreatment with IFNγ increased LPS lethality as well as plasma levels of TNFα and IL-6, indicating that IFNγ is needed for the full expression of the detrimental effects of LPS and TNFα (68,69). From these data, a picture emerges for the mechanism of LPS-induced anorexia which comprises the subsequent sequence of events: LPS stimulates macrophages and other antigen presenting cells to secrete an array of pro-inflammatory cytokines including IL-1β, IL-6, TNFα and the NK and T cell activation factors IL-12 and IL-18. As a result, IFNγ production increases dramatically and provides the necessary positive feedback for the production of these and other macrophage-derived cytokines, which jointly affect eating (Figure 1).

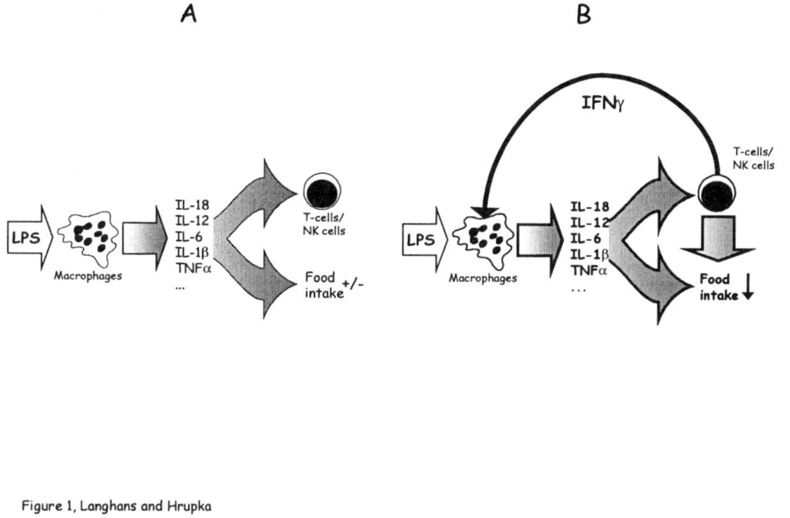

Figure 1. *Possible mediation of the anorectic effect of bacterial lipopolysaccharide (LPS) by pro-inflammatory cytokines. A: Initially, LPS triggers the release of cytokines from macrophages. B: Interferon-γ (IFNγ) released from T cells and/or NK cells enhances the release of cytokines from macrophages, thus providing a positive feedback that is necessary for LPS to reduce food intake. See Text for further details.*

Compared to cytokine or cytokine receptor KO models, a somewhat different picture emerges when particular cytokines and/or their receptors are acutely antagonized by receptor antagonists, soluble receptors, or other pharmacologically active substances. Consistent with the model presented in Figure 1, such studies basically implicate all of the pro-inflammatory cytokines discussed above (IL-1β, IL-6, TNF-α, and IFN-γ) in peripheral LPS- or microbial infection-induced anorexia (6,59,70,71). In our hands, rats made tolerant to exogenous TNFα did not reduce food intake in response to IP LPS (15), and complete blockade of TNF production together with an attenuation of IL-1β production as measured by the plasma levels of these two cytokines blocked the feeding suppressive effect of IP LPS (70). The latter result, again, is in line with the assumption that interference with both cytokine pathways is necessary to block LPS anorexia. Acute SC administration of IL-1ra also attenuated the feeding suppressive effects of LPS and influenza virus infection in mice (71). All in all, it is safe to say that various pro-inflammatory cytokines play a prominent role in the anorexia of peripheral infection, although developmental compensation, based on the redundancy and synergistic interactions of cytokines, may sometimes mask the role of a particular cytokine.

3.3 Cancer cachexia

Several cytokines, including IL-1β, IL-6, TNFα, and IFNγ, have been implicated in cancer cachexia and anorexia. This assumption is based on a) reports demonstrating that circulating levels of particular cytokines or cytokine expression in various organs is higher in cancer bearing individuals than in healthy controls (72-78), b) some reports that cytokine or cytokine receptor KO mice display differences in the magnitude of anorexia and cachexia induced by experimental tumors (79,80), and c) on results showing that pharmacologic antagonism of cytokines can attenuate cancer-induced anorexia and/or cachexia (81-83).

Assessment of circulating pro-inflammatory cytokine levels in cancer patients and their correlation with weight loss has also led to some controversial results (84,85). Often, an increase in circulating cytokine levels is detectable only in some patients and/or is discontinuous. A high correlation between serum IL-6 levels and disease extent as well as body weight loss was observed in pancreatic cancer patients (76). Serum IL-6 concentration was also positively related to clinical and pathologic parameters in breast cancer patients, whereas such a relationship was not observed for IL-1β, IL-2 and TNFα. Based on such findings, serum IL-6 was considered to be a useful prognostic factor for these types of cancer (76,78).

In addition to changing serum levels of pro-inflammatory cytokines directly, cancer may also change the sensitivity of cytokine producing cells to other stimuli: Thus, LPS-induced TNFα and IL-1β production of peripheral blood monocytes was significantly increased in patients with advanced renal cell carcinoma (74). Overexpression of these cytokines may therefore contribute to cancer cachexia and the accompanying malnutrition.

Reverse transcriptase-polymerase chain reaction (RT-PCR) analysis revealed significantly higher levels of IL-1β in the spleens of tumor-bearing C3H/He mice compared to those of freely fed control animals (86). IL-1β mRNA and IL-1 receptor antagonist (IL-1Ra) mRNA were also significantly upregulated in the spleen of Lobund-Wistar rats bearing prostate adenocarcinoma tumor cells (77). In this study, early anorexia in tumor-bearing Lobund-Wistar rats was also associated with an upregulation of IL-1β mRNA in the brain regions examined (cerebellum, cortex, and hypothalamus). IL-1Ra mRNA and IL-1 receptor type I mRNA levels were also increased in the cortex and hypothalamus of tumor bearing rats. Other cytokine components were not significantly different between tumor-bearing and pair-fed (control) rats. Together, these data are consistent with the idea that overproduction of IL-1β is relevant for the clinical manifestations of some cancer-burden states. In particular, the IL-1β mRNA upregulation in the brain may be relevant to the anorexia exhibited (77).

A causal relationship between cancer anorexia and certain cytokines is suggested by the results of studies in various KO preparations: In one of these studies, rcMCG 101 tumors were implanted subcutaneously in wild-type C57 B1 and KO mice to evaluate the role of host-produced cytokines [IL-6, IL-12, IFNγ, TNF receptor 1, and TNF receptor 2] in local tumor growth, anorexia, and carcass weight loss in a well-defined model with experimental cachexia. Wild-type tumor-bearing mice developed cachexia in relation to rapid tumor growth, whereas IL-6 KO mice displayed attenuated cachexia and tumor growth. Likewise, when anti-IL-6 was administered to wild type tumor-bearing mice, alterations in food intake were not directly related to systemic IL-6 but rather secondarily to IL-6-dependent tumor growth. In the other KO preparations, the absence of host-derived IL-12, IFNγ or the TNF receptor 1 or receptor 2 did not attenuate tumor growth or improve subsequent cachexia (79). These results demonstrate that host- and tumor-derived cytokines promote anorexia and cachexia in a more complex fashion than usually presented based on previous information in studies on either anti-cytokine experiments in vivo or, on gene knockouts with respect to a "single cytokine model" (79). IL-6 blockade (genetic and pharmacologic) has also been shown to attenuate the cachectic effects of a colon-adenocarcinoma model (87). When wild type and IL-6 KO mice were inoculated with Ehrlich ascites tumor cells, IL-6 KO mice gained weight due to tumor growth more rapidly than did the wild-type mice. Moreover, the daily food intake of wild-type mice declined on day 2 after tumor inoculation and was only 37% on day 8, whereas the daily food intake of IL-6 KO mice was constant for the first 7 days after tumor inoculation. Interestingly, however, although wild-type mice suffered from cachexia, their survival time was significantly longer than that of IL-6 KO mice. This indicates that IL-6 secretion and the cancer cachexia syndrome are involved in the defense mechanism against tumor progression (80).

Subcutaneous injection of a dimeric, pegylated 55-kDa TNF receptor construct significantly improved food intake and cachexia in Fischer 344 male rats injected with 10^6 viable methylcholanthrene cells. Anti TNFα antibodies also enhanced food intake in another study in the same tumor model (82). These data suggest that TNFα, in addition to other cytokines, contributes to the anorexia of tumor growth in this model. Finally, passive immunization against IFNγ increased food intake and body weight in cachectic, tumor bearing rats (21).

All in all, the currently available evidence suggests that several pro-inflammatory cytokines play a role in the anorexia/cachexia syndrome that accompanies progressive tumor growth in man and animals. It seems likely that any such role of a particular cytokine is exerted in concert with other cytokines. The type of the tumor as well as its location and size presumably determine the cytokine response of the organism and the final result of the disease.

3.4 Anorexia of aging

Aging involves several changes in cellular and humoral immune functions. The main alterations include: a) a thymic involution resulting in a decreased number of lymphoid precursor T- and B cells, b) a reduced proliferative capacity of T-cells and loss of lymphocyte subgroups as a consequence of the shortening of telomeres, c) a qualitative deficiency of B-lymphocytes with a reduced response to exogenous antigen, d) a compromised activity of the accessory cells, both directly by depressing the chemotactic and phagocytic responses, and indirectly by increasing the prostaglandin production which inhibit the proliferation of T-cells, e) alterations in the production and secretion of various cytokines, f) other factors like the general physiological conditions, the nutritional state, psychological habit and various hormone levels (88). Some studies report an increase in plasma levels of pro-inflammatory cytokines or their production in response to various stimulations in older subjects (89-92). Thus, the plasma TNFα concentration appears to increase significantly with advancing age and might predict the age-related progressive impairment in insulin action (93). A reduction in plasma concentrations of pro-inflammatory cytokines in geriatric cachexia has been reported to be associated with weight gain (92). In general, however, results on age-related differences in the production and plasma concentrations of pro-inflammatory cytokines are inconsistent. Given the complexity of the cytokine network, however, circulating pro-inflammatory cytokine levels do not necessarily have to increase to trigger a certain response; a shift in the balance between pro-and anti-inflammatory cytokines may be sufficient to enhance the detrimental effects of inflammatory processes. Often, pro-inflammatory cytokines increase whereas anti-inflammatory cytokines decrease in response to a given stimulus. Such a prevalence of pro-inflammatory cytokines, has been reported for old age (91). A summation of minor ailments and originally trivial infections with advancing age may therefore increase pro-inflammatory cytokine production sufficiently to shift the delicate balance between pro- and anti-inflammatory cytokines. All in all, the presently available evidence indicates that immune processes which depend on or are modulated by pro-inflammatory cytokines can differ between young and aged subjects as a consequence of the availability of regulatory soluble receptors and related agonists or antagonists. This imbalance of pro- and anti-inflammatory cytokines may well contribute to the anorexia and cachexia that is often observed with advancing age.

3.5 High intensity exercise

It has been known for several years that the acute phase response to exercise is similar to the immune reactions in general medical and surgical conditions. Although there is considerable variation with respect to the type of the pro-inflammatory cytokines involved as well as to the kind and duration of exercise, it is clear that strenuous exercise increases circulating levels of several pro- and anti-inflammatory cytokines, naturally occurring cytokine inhibitors and chemokines. For example, increased serum levels of TNFα, IL-1β, IL-6, IL-1 receptor antagonist (IL-Ira), TNF-receptors (TNF-R), IL-10, IL-8 and macrophage inflammatory protein (MIP)-1 were found after strenuous exercise (90). In general, pro-inflammatory cytokines increase first and are then balanced by the secretion of anti-inflammatory cytokines (90).

In particular IL-6 is produced in response to strenuous exercise. The serum concentration of IL-6 increases up to 100-fold after a marathon race (94). This increase is tightly related to the duration and the intensity of the exercise (95), and there appears to be a logarithmic relationship (96). Circulating monocytes are probably not the source of the increase in serum IL-6 because ex-vivo isolated leukocytes actually produced less IL-6 and other pro-inflammatory cytokines after exhaustive exercise than under control conditions (97,98). As IL-6 is also produced locally in the skeletal muscle in response to exercise (99) and is known to have growth factor abilities, it is likely that IL-6 plays a beneficial role and may be involved in mediating exercise-related metabolic changes (96). The mild endotoxemia that presumably results from gut ischemia-related leakage of LPS into the circulation (100) may be another trigger of the pro-inflammatory cytokine response to strenuous exercise.

As food intake is usually reduced immediately after an exhaustive exercise, it is feasible that the markedly increased plasma levels of pro-inflammatory cytokines somehow contribute to this early anorexia in response to exercise. Clearly, however, further studies are necessary to critically examine such a causal relationship.

3.6 Obesity

The discovery of leptin has increased the recognition of the adipocyte as an endocrine organ. It is now clear that adipocytes produce pro-inflammatory cytokines and other immune-active substances which may well contribute to some of the metabolic and cardiovascular complications associated with obesity, and which might also influence food intake and energy balance regulation. Conversely, the development of obesity may be accompanied by and related to an insensitivity to the actions of these substances.

Adipose tissue expresses TNFα, and adipose tissue TNFα production is increased in obese humans (101,102) and rodents (103). A polymorphism in the 5-flanking region of the TNFα gene has been considered to represent a genetic marker for increased susceptibility for obesity in Caucasians (104). Adipose tissue TNFα activity, measured on secreted protein collected from medium of minced, incubated fat pads of mature animals was directly related to adipocyte size (105) and increased with a high fat diet (106). TNFα inhibits lipoprotein lipase activity, increases lipolysis in adipose tissue, decreases insulin receptor tyrosine kinase activity, inhibits glucose transporter gene expression, and decreases PPARγ expression, which enhances most of the other effects (107-109). Together, these effects efficiently counteract further nutrient uptake and, hence, lipid accumulation in adipocytes. Thus, it is tempting to speculate that adipose tissue-derived TNFα is part of a local/short feedback loop that controls adipocyte size.

Adiposity is also associated with increased basal plasma levels of TNFα (110). Kern et al. reported a 7.5-fold increase in TNFα secretion from adipose tissue of obese (BMI 30-40) compared to lean (BMI<25) subjects. The TNFα secretion was inversely related to insulin sensitivity (108). Whereas higher circulating TNFα levels were found in abdominal obesity compared with peripheral obesity (111), others did not find TNFα release from subcutaneous adipose tissue (112) and, in general, the evidence for the secretion of TNFα from adipose tissue is inconclusive. Therefore, it is also questionable whether adipose tissue-derived TNFα can have a direct effect on CNS mechanisms of energy balance regulation. Nevertheless, the possible local adipostat function of TNFα (see above) and its stimulating effect on adipose tissue leptin secretion (113) fit a role of TNFα in the maintenance of energy balance because leptin is known to inhibit eating and to stimulate energy expenditure. A synergistic contribution of TNFα and leptin to the control of adipocyte size and energy balance appears also likely considering the increasing evidence that one important role of leptin is to dissipate excess fat by stimulating the β-oxidation pathway of fatty acids in non-adipose tissue cells (114).

The local expression of TNFα is higher in subjects with obesity-related insulin resistance (108), and reduced insulin receptor tyrosine kinase activity by TNFα may contribute to insulin resistance and type II diabetes in obesity (107). In fact, genetically TNFα deficient mice had improved insulin sensitivity in both diet-induced obesity and in the *ob/ob* model of obesity (115). We recently observed that rats genetically deficient in TNFα and TNFβ or in any one of the two TNF receptors ate consistently more than wild type controls. Moreover, on a high fat diet the TNFα and the TNF receptor 2-deficient mice gained substantially more weight than on normal lab chow, whereas the wild type controls did not (Hrupka et al., unpublished). Although still preliminary, these results might reflect a role of endogenous TNFα in the long-term regulation of food intake and body weight. Thus, several lines of evidence suggest that TNFα is involved in obesity-related disturbances of energy balance which might also comprise changes in food intake, but the exact role and mechanisms of these phenomena need to be further explored.

IL-6 is another pleiotropic pro-inflammatory cytokine expressed in adipose tissue, and arterial plasma concentrations of IL-6 correlated significantly with BMI and percent body fat (116). Ingestion of a very low calorie diet decreased adipose tissue and plasma IL-6 (116), suggesting that circulating IL-6 at least in part reflects adipose tissue IL-6 production. Thus, in contrast to the situation with TNFα, there is no doubt that IL-6 release is also increased in hypertrophied adipocytes (112,116). IL-6 acts on hepatocytes to stimulate the formation of acute-phase reactants that increase blood viscosity and promote thrombus formation. This may explain why plasma fibrinogen is increased in visceral obesity and poorly controlled diabetes (117). Elevated IL-6 production by peripheral blood mononuclear cells in obese individuals (118) provides another potential link between obesity and various cytokine functions. Yet, the relevance and the exact mechanisms of these changes remain to be identified. It is also unclear whether adipose tissue derived IL-6 has any centrally mediated effect on food intake and/or energy balance.

Recently, a marked stimulation of IL-8 production by IL-1β (3 nM) and TNFα (0.6 nM) in human subcutaneous adipose tissue has been reported in vitro (119). This raises the possibility that this cytokine, too, may be involved in obesity-related phenomena that are mediated by cytokines.

3.7 Normal and disordered eating

In addition to orchestrating the immune response during diseases and causing the ensuing anorexia, cytokines might also play a role in the control of normal eating. Substantial amounts of cytokines are produced in the stomach and intestine (120,121). Lamina propria mononuclear cells appear to be a natural source for these cytokines, but their exact cellular origin still needs to be identified. Food intake often results in a high intestinal antigen exposure requiring host defense. Besides local immune activation, this defense may include a coordinated systemic immune response. Moreover, dietary factors such as malnutrition or fiber content affect bacterial translocation from the gastrointestinal tract to the blood and extraintestinal organs (122). A high protein meal did not affect the in vitro production or plasma levels of IL-1β and TNFα, but it decreased the plasma level of interferon-γ (123); it also increased the numbers of neutrophils in the blood and decreased monocyte and lymphocyte counts, presumably reflecting emigration of these cells into extravascular lymphoid tissues. Very likely this results in an enhanced local production of pro-inflamatory cytokines, which could well be involved in limiting the antigenic "impact", i.e. in inhibiting further ingestion, as part of their defense function. A role of IL-1β in some aspects of normal food intake is also suggested by recent findings of an increase in IL-1β mRNA expression in rat liver and hypothalamus in response to cafeteria diet feeding (124). Cafeteria diet feeding also decreased the IL-1R AcP levels in liver and brain stem, suggesting that eating has profound modulating effects on the IL-1 system. Pro-inflammatory cytokines released by normal eating may mediate the sleep-promoting effect of large meals (124), and this may be related to the well-known stimulating effects of IL-1β on vagal afferents (125,126), because the sleep-promoting effect of peripheral IL-1β was critically dependent on vagal afferents (127). On the other hand, neither daily IP injection nor continuous ICV infusion of an IL-receptor antagonist at doses that were sufficient to block the feeding suppressive effect of exogenous IL-1 enhanced normal eating in rats (128). This suggests that IL-1 is not essential for the physiologic control of normal eating. Interestingly, rat hypothalamus, hippocampus, and cerebral cortex have been found to display marked circadian variations in bioactive TNFα with the highest levels at light onset, when feeding is usually low (129). It is unknown, however, whether these changes bear any relevance for the control of normal eating and/or energy balance.

Occasionally, cytokines have also been related to eating disorders such as anorexia and bulimia nervosa (130), but the results appear to be inconsistent (118,131,132), and the pathophysiological relevance of cytokines in disordered eating is presently unclear.

4. MECHANISMS OF CYTOKINE-INDUCED ANOREXIA

4.1 Central cytokines

To inhibit eating, cytokines must act on the neural circuitry that controls food intake. Cytokines produced in the brain can easily act on neurons within the blood brain barrier (BBB). There is evidence indicating that some pro-inflammatory cytokines and there accessory proteins are expressed by neurons and glia cells in normal brain (133). Moreover, central nervous system (CNS) cytokine production is clearly promoted during acute and chronic CNS diseases such as meningitis, encephalitis, multiple sclerosis, Alzheimer's disease and brain tumors (37). Local ischemia (=stroke) also leads to massive production of pro-inflammatory cytokines. In general, IL-1β appears to be the main IL-1 agonist induced in the brain in response to local (e.g. injury, stroke) insults (134). In rodents, IL-1β mRNA expression is increased within 15-30 minutes and protein within 1h of experimental cerebral ischemia, brain injury, or infection (134). Neurons and astrocytes also produce TNFα in response to various pathologic stimuli (135). The feeding suppression observed in response to ICV administration of low doses of pro-inflammatory cytokines (22) is certainly relevant for the anorexia in clinical situations in which marked CNS cytokine synthesis occurs. Yet, it is currently unclear where centrally produced (or ICV administered) cytokines act to reduce food intake. ICV administered IL-1β rapidly penetrates into perivascular tissue and spreads along white matter fiber bundles and blood vessels (136). Thus, IL-1β may act as a "volume transmission signal" and activate neurons that express the type 1 IL-1 receptor. Neuronal expression of type 1 IL-1 receptor is found in the basolateral amygdala. The strong projections from there to the central amygdala might therefore mediate the central amygdala activation by ICV IL-1β. Such a pathway may also be involved in behavioral effects of centrally administered or produced IL-1β. Interestingly, the IL-1 type 1 receptor mRNA appears to be absent in the supraoptic and paraventricular nuclei of the rat hypothalamus (137). These areas appear to be activated by ICV IL-1β through an indirect, prostaglandin-dependent pathway (138).

De novo pro-inflammatory cytokine synthesis in the brain has also been shown in response to some types of peripheral immune stimulation. In-situ hybridization, immunohistochemistry, RNase protection and reverse transcriptase PCR studies show that systemic immune stimulation induces various cytokines and their receptors in several brain regions, including the hypothalamus (133,139-143). Cytokines synthesized in the brain appear to mediate some centrally mediated systemic effects of peripheral immune stimuli (144,145) and have also been implicated in the concomitant suppression of food intake. The situation is complicated, however, because peripheral immune stimulation can affect brain cytokine and cytokine receptor expression differentially, i.e. while local IL-1 synthesis is increased, IL-1 receptor expression may be decreased by peripheral LPS (141). Moreover, very high doses of peripheral LPS were often used to show an increase in brain cytokine expression (143,146), whereas IP injection of subseptic doses of LPS induced IL-1β and TNFα mRNA expression only in the choroid plexus, the circumventricular organs, and the meninges (147). Even after IP injection of high LPS doses, the initial increase in IL-1 activity is mainly derived from cells of the BBB and the circumventricular organs (CVOs)(146,148), and thereafter, the more sustained IL-1 activity is derived from glia (145). Perivascular and meningal macrophages and microglia represent the main cellular sources of IL-1β in the brain (149), whereas TNFα appears to be expressed in response to peripheral LPS in microglial cells and neurones (150). Finally, some data indicate that the majority of IL-1β appearing in areas of the brain in response to intravenous IL-1β administration is due to passage across the BBB and not to release from sites within the brain (151). In a study that we did in collaboration with Dr. Plata-Salamàn and his group, IP injection of LPS (100μg/kg) induced marked increases in the mRNA levels of the components of the IL-1 system (IL-1β, IL-1 receptor antagonist, IL-1 receptor 1, IL-1 receptor accessory proteins I and II) as well as TNFα in several brain areas (mostly cerebellum and parieto-frontal cortex) as measured by RNase protection assay (152). The changes induced in the hypothalamus were comparatively small, and whether or not they are causally related to the behavioral effects of this peripheral LPS dose is unclear.

In relation to LPS-triggered CNS-mediated effects, it must also be considered that IFNγ, which is important for the effects of peripheral LPS on food intake and metabolism, is only expressed in the periphery in response to LPS and not in the brain (59). Recently, we employed the very sensitive Taqman real-time quantitative rtPCR technique to search for IFNγ mRNA expression in hypothalamic and cortex samples and could also not detect any increase in brain IFNγ mRNA in response to peripheral LPS (Lugarini et al., unpublished results). These data suggest that there is at least one cytokine-related peripheral step in the cascade of events that triggers CNS responses to peripheral LPS. Together, these results raise questions about a major role of de novo synthesis of pro-inflammatory cytokines within the BBB in the anorectic effect of peripheral immune stimulation. Consequently, although central administration of cytokines may model the clinical features of brain injuries and CNS infections, it does not necessarily reflect the pathophysiologic mechanisms relevant for systemic infections.

Central cytokines may play a role in the anorexia in response to acute colitis induced by rectal administration of TNB in the rat, where ICV administration of IL-1Ra attenuated the accompanying inhibition of feeding (51). In another study (81), intra-VMH injection of IL-1Ra improved food intake in anorectic tumor bearing rats, suggesting that centrally acting IL-1 contributes also to cancer anorexia. In contrast, the majority of findings from genetic or pharmacologic antagonism studies in fact argues against a crucial role of central cytokines in the anorectic effect of peripheral LPS: Thus, IL-1β-converting enzyme-deficient mice resisted the anorectic effect of ICV but not IP administered LPS (153,154), suggesting that IL-1β is crucial for the anorexia in response to central, but not peripheral LPS administration. In line with these findings, ICV administration of IL-1 receptor antagonist inhibited the effect of ICV injected LPS in rats, but failed to attenuate the food intake suppressive effects of IP LPS administration under the same conditions (6,155). A more recent study from the same laboratory reported an attenuation of IP LPS-induced anorexia by ICV IL-1 receptor antagonist administration in mice (156). Species differences (e.g. mouse vs. rat) may account for this discrepancy. There is evidence for more species differences which may also account for other discrepancies with respect to brain cytokine responses to peripheral immune stimulation: Thus, whereas no increase in pro-inflammatory cytokine mRNA expression was detected in Trypanosoma brucei infected mice (157), a chronic overexpression of pro-inflammtory cytokines was evident in brains of rats infected with the same Trypanosoma species (158).

4.2 Peripheral cytokines

4.2.1 General considerations

In response to most of the situations described above, cytokines may act on the brain but are mainly produced in the periphery, or at least outside the BBB. *C-fos* studies have shown that peripheral immune stimulation activates several CNS regions that are involved in the control of eating (159,160). The mechanisms of this activation and how this leads to anorexia are unclear, but neural and humoral pathways are probably involved.

4.2.2 Mediation by afferent nerves

Immune stimulation induces IL-1β immunoreactivity in dendritic cells and macrophages in connective tissue associated with the abdominal vagus (161). Locally released cytokines (162) might therefore act on the brain and inhibit feeding by activating peripheral sensory fibers directly or indirectly, and without a concomitant increase in circulating cytokines. Subdiaphragmatic vagal afferents in rats appear to be responsive to peripheral IL-1β because intravenous (IV) administration of IL-1β increased multiunit activity in gastric and hepatic branch vagal afferent fibers (125,126).

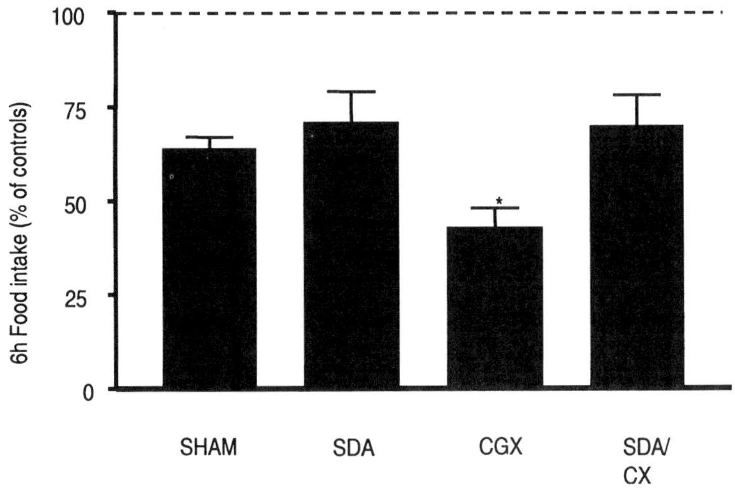

Figure 2: *Subdiaphragamtic vagal deafferentation (SDA), celiac superior mesenteric ganglionectomy (CGX) and a combination of both surgical procedures (SDA/CGX) do not attenuate the anorectic effect of intraperitoneally injected IL-1β (2μg/kg) in rats. SHAM = sham-operated control rats. Data are mean ± SEM percent food intakes of 9 IL-1β injected rats in each surgical group (corresponding vehicle = 100%). Compared to vehicle injection, IL-1β reduces food intake significantly in all surgical groups. *IL-1β-induced anorexia is more pronounced (p>0.05, post-hoc test after significant ANOVA) than in all other groups.*

Subdiaphragmatic vagotomy has been used to investigate the role of the vagus in cytokine-induced hyperalgesia, fever, neuroendocrine activation, depression of social exploration, decrease in general activity, and reduction in food intake (see (163)). Whereas vagotomy consistently eliminated cytokine effects on some of these phenomena, the results are less clear concerning the other responses (163). Vagal afferents have been implicated in the feeding suppressive effects of cytokines because subdiaphragmatic vagotomy attenuated IP LPS and IL-1β-induced suppression of instrumental responses to obtain food in mice (164). Systemic capsaicin pretreatment, however, failed to block the effects of IP administered IL-1β and LPS on food-motivated behavior in rats (165). Finally, we observed that subdiaphragmatic vagal deafferentation, alone and in combination with celiac-superior mesenteric ganglionectomy, did also not alter the anorexia after IP injection of LPS and IL-1β in rats (166) (Figure 2). These findings indicate that sensory visceral afferents are not necessary for the anorectic effects of IP injected LPS and IL-1β, at least in rats.

Cytokines dose-dependently increase leptin expression in adipose tissue and plasma levels of leptin (167-169,113), and these increases are closely related to the food intake suppression by the pertinent cytokines (169,170). The anorectic effect of LPS was enhanced in *ob/ob* (leptin-deficient) mice and attenuated in *db/db* (leptin receptor-deficient) mice (171), suggesting that leptin somehow contributes to LPS-induced anorexia in this species. We found that a single IP LPS injection (100μg/kg) reduced food intake similarly in lean (*Fa/?*) and obese (*fa/fa*) Zucker rats, which have only some functional leptin receptors. Likewise, high dose (500μg or 1.0mg/kg) IP LPS administration caused a similar initial (day 1) feeding suppression in lean and obese Zucker rats. During days 3 and 4 post-LPS, lean rats' food intake recovered similarly regardless of dose and returned to control levels by day 5. In contrast, the food intake in obese rats was more depressed after the highest LPS dose on days 3 and 4, and took one day longer to return to control levels, when compared to rats adminstered the 500 μg/kg dose.

4.2.3 Humoral mediation by leptin

After chronic IP administration of LPS (2.45 or 9.8μg/h) through osmotic minipumps, postimplantation recovery of food intake occurred at the same rate in obese rats regardless of dose and returned to control levels by day 5. In contrast, lean rats took longer to recover and recovery time was dose related (Lugarini et al., unpublished results). Clearly functional leptin receptors are not necessary for the expression of the feeding suppressive effect of LPS, especially the early response. Nonetheless, after acute supraphysiolocial, or chronic pathophysiological doses of LPS, leptin may be involved in the adaptive response to LPS.

It is also worth to mention that low dose parenteral administration of LPS in healthy humans does not affect plasma leptin concentration (172,173). We could also not detect a differential effect of IP injected Il-1β on food intake in lean and obese Zucker rats. In contrast, the anorectic effect of systemic TNFα was enhanced in obese Zucker rats in a previous study by others (174). In a recent study, concomitant ICV leptin administration prolonged the anorectic effect of an IP LPS injection: 24 hour food intake in normal Sprague Dawley rats administered with IP LPS alone (100μg/kg) or with IP LPS and ICV leptin (0.8μg/rat) was 93 and 70% ($P<0.05$) of control animals (175). The leptin dose alone did not reduce food intake under the conditions tested. This result may be somehow related to the fact that IL-1 and LPS lost much of their potency to suppress food intake after prior body weight reduction (13). All in all, the available data suggest that leptin is not necessary for the feeding suppressive effect of pro-inflammatory cytokines, but it may contribute in several ways, and further studies are necessary to critically examine these interactions.

4.2.4 Central action of peripheral cytokines

Cytokines produced outside the BBB can reach CNS cytokine receptors through active or passive transport mechanisms (176) or through CVOs. The anteroventral region of the third ventricle (AV3V) that includes the organum vasculosum laminae terminalis is one site of IL-1 action on the brain, because electrolytic lesions of the AV3V in various animal species decrease the febrile and plasma ACTH responses to systemic administration of IL-1β (177). Neurons in the AV3V region that send axons to hypothalamic nuclei are responsive to systemically administered IL-1β (178), but it is unknown whether the AV3V and its hypothalamic projections play a role in the feeding suppressive effect of IL-1β. Another potential site of cytokine action on the brain, the area postrema (AP), does not seem to be involved in the anorexia during infection. AP lesion(s) did not alter hypothalamic *c-fos* expression in response to IV IL-1β (179), and lesions of the AP and the adjacent part of the nucleus of the solitary tract (AP/NST) did also not change the anorectic effects of IP administered IL-1β (10) and TNFα (11). Together with the failure of subdiaphragmatic vagal deafferentation to attenuate IP IL-1β and LPS-induced anorexia, these results are consistent with the assumption that the neural afferent pathway is not essential for the feeding suppressive effect of peripheral pro-inflammatory cytokines.

The surface area of the BBB vastly exceeds that of the CVOs (180), and efficient BBB transport systems regulate the passage of many substances, including cytokines, between the circulation and the CNS (176). Pro-inflammatory cytokines and LPS may promote cytokine uptake into the brain by increasing BBB permeability (59,181). Therefore, BBB passage may play a role for the actions of circulating cytokines on the brain. In addition, circulating cytokines may affect CNS function without crossing the BBB, by acting at the level of the brain vasculature endothelial cells (182).

Peripheral administration of IL-1β or LPS leads to a rapid induction of *c-fos* mRNA in non-neural cells of the BBB including blood vessels, outer meninges, and astrocytes (183,184). This is consistent with the view that

BBB endothelial cells play a major role in the effects of circulating cytokines on CNS neurons. Cytokine receptors are present on the surface of endothelial cells of the brain capillaries and on perivascular cells (microglia, macrophages)(185-187). The binding of pro-inflammatory cytokines to those receptors leads to activation of the transcription factor NF-κB and to the subsequent release of messengers, such as nitric oxide or prostanoids (185,187-192). Small capillary endothelial cells are in fact considered to be the major source of prostaglandins (PGs) in the brain (191). Mediators are presumably also released from perivascular microglia (143). Brain endothelial cells express cyclooxygenase-2 (COX-2), the inducible form of cyclooxygenase, and cytokines increase the expression of COX-2 mRNA in blood vessels (190,193-196) and potently stimulate the production of eicosanoids by endothelial cells (193). Endothelial cells not only react to pro-inflammatory cytokines, they also produce cytokines in response to various stimuli (182,185), which leads to signal amplification. Further signal amplification is based on the functional synergism between IL-1β, TNFα, and INFγ (197). LPS also increases COX-2 mRNA in endothelial cells and stimulates PGE_2 production (189,194,196,198). IV LPS induced COX-2 like immunoreactivity in perivascular microglia and in meningeal macrophages throughout the brain (198). LPS-induced activation of endothelial cells and stimulation of prostanoid production depends on soluble CD14 (185) and is independent of IL-1β (199). The latter result might explain some of the failures of IL-1 antagonism to block peripheral LPS effects on CNS functions (e.g. (153,155)).

PGE_2 appears to be the major metabolite formed from cerebral blood vessel endothelial cells in response to stimulation by LPS, IL-1β and IL-6 (200). Interestingly, the feeding suppressive effects of IL-1β and LPS have been shown to be attenuated by non-specific inhibitors of eicosanoid synthesis (25,201,202). Indomethacin also attenuated the LPS-induced increase in paraventricular nucleus (PVN) *c-fos* mRNA expression (203). Moreover, we recently showed that the anorectic effect of IP LPS is blocked by the COX-2 inhibitor NS-398, but unaffected by the COX-1 inhibitor resveratrol (Figure 3). Acetaminophen also attenuated IL-1β-induced anorexia (25), and more recent studies indicate that the attenuating effect of acetaminophen on LPS-induced phenomena is due to inhibition of COX-2 (204). The febrile response to peripheral LPS is absent in COX-2 KO, but not in COX-1 KO mice (205), suggesting that COX-2 is also crucial in LPS-induced fever. Further, under conditions where food intake was markedly reduced, peripheral LPS markedly increased CSF PGE_2 concentration, but not plasma PGE_2 (Lugarini et al., unpublished results). Similar to the anorectic effect of LPS, the LPS-induced increase in CSF PGE_2 was blocked by the COX-2 inhibitorNS-398, but not by the COX-1 inhibitor resveratrol (Figure 4). These data are consistent with the idea that CSF

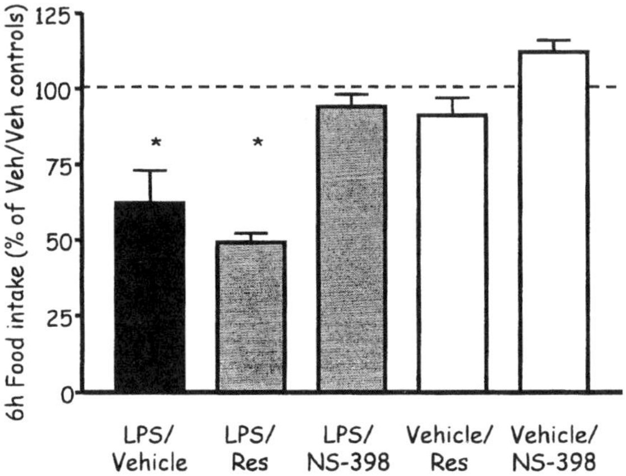

Figure 3.The cyclooxygenase-2 (COX-2) inhibitor NS-398 (10 mg/kg, intraperitoneally = IP), but not the COX-1 inhibitor resveratrol (Res, 10 mg/kg IP) blocks the anorectic effect of IP lipopolysaccharide (LPS, 100 µg/kg) in rats. Data are compiled from two separate experiments with NS-398 and resveratrol. Therefore, data are mean ± SEM percent food intakes of 14 (LPS/vehicle) or 7 (all other treatments) rats. Corresponding vehicle/vehicle injected rats' intake =100%. *Significantly ($p<0.05$ in post-hoc tests after significant ANOVA) lower than 100%.

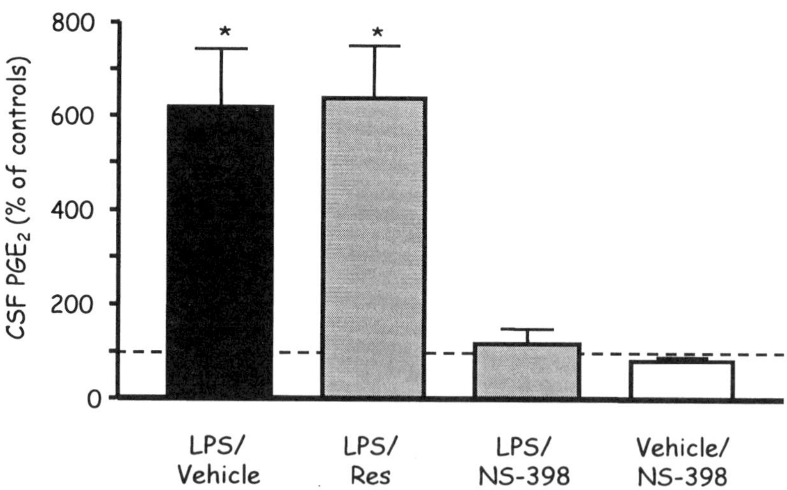

Figure 4: Intraperitoneally (IP) injected lipopolysaccharide (LPS, 100 µg/kg) increases cerebrospinal fluid (CSF) prostaglandin-E_2 (PGE_2) concentration, and this increase is blocked by the cyclooxygenase-2 (COX-2) inhbitor NS-398 (10 mg/kg, IP) but not by the COX-1 inhibitor resveratrol (Res, 10 mg/kg, IP). Data are mean ± SEM percent changes in CSF PGE_2. Corresponding vehicle/vehicle treated controls = 100%. N = 7 for each group. *Significantly ($p<0.05$ in post-hoc test after significant ANOVA) higher than vehicle/vehicle treated controls.

PGE_2 is somehow involved in the feeding suppressive effect of LPS. Interestingly, in another study, the increase in CSF PGE_2 after LPS challenge was not attenuated by subdiaphragamtic vagotomy (206), indicating that vagal connections between periphery and brain are not necessary for this response. As mentioned above, surgical elimination of the vagal afferent and all sympathetic connections between periphery and brain also did not eliminate the anorectic response to IP LPS and IL-1β in our hands (166). Together, all these studies point to central PGE_2 as a major mediator of the feeding suppressive effects of pro-inflammatory cytokines derived from circulating leukocytes and their interaction with endothelial and perivascular cells. Thus, concerning the anorectic effect of pro-inflammatory cytokines, the following picture emerges for signal processing at the blood brain barrier (Figure 5): 1) Pro-inflammatory cytokines act on brain capillary endothelial

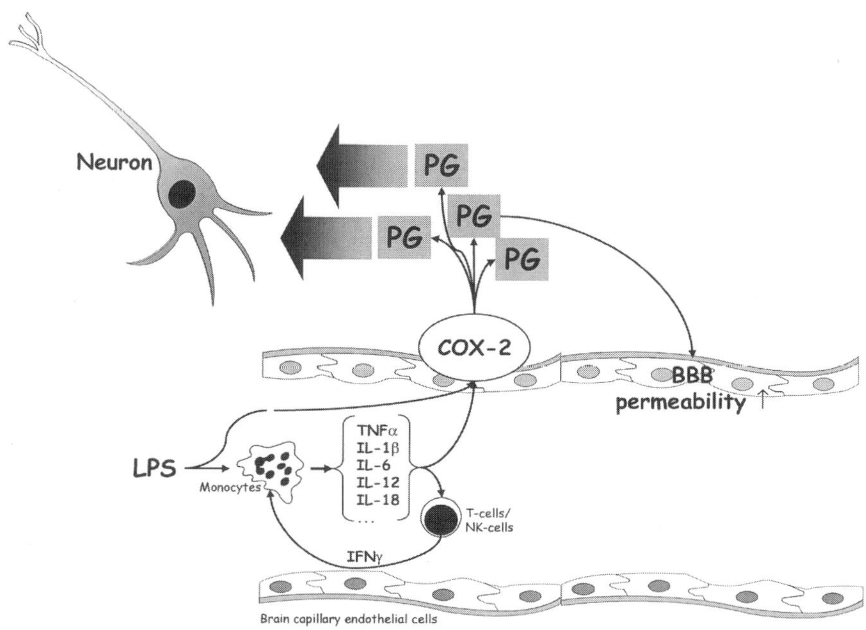

Figure 5: Possible role of brain capillary endothelial cells and prostaglandins (PG) in the activation of neurons by pro-inflammatory cytokines. BBB = blood brain barrier, COX-2 = cyclooxygenase-2, LPS = lipopolysaccharide.

and perivascular cells. This effect is enhanced by the synergism between various cytokines. The action of cytokines on the BBB endothelial and perivascular cells triggers the production and release of mediators such as prostanoids, in particular of PGE_2. PGE_2 (and/or other prostanoids) act on prostanoid receptors on neurons that project to brain sites involved in food intake control, in particular to the paraventricular nucleus of the hypothalamus.

A similar mechanism might also play a role for the anorexia in response to sterile peripheral inflammatory reactions such as the turpentine administration model. LPS and cytokine-induced activation of endothelial cells depends on activation of the transcription factor NFκB (185). Il-1β is essential for activation of NFκB and PG synthesis in endothelial cells by turpentine, but not LPS (199). This corresponds to the finding that turpentine, but not LPS-induced anorexia is crucially dependent on IL-1β (see above). Finally, activation of endothelial cells of the brain vasculature might even contribute to some of the responses seen after ICV administration of LPS and cytokines, because these treatments also induce COX-2 in endothelial cells (193,207,208).

4.3 Neurochemical mediation

4.3.1 Central propagation of the cytokine-derived signal

Cytokines and LPS strongly activate medullary and hypothalamic paraventricular neurons (179,203,209). Indomethacin dose-dependently blocked this activation by LPS and IL-1β (179,203). IP LPS and IL-1β evoked a dose-dependent potentiation of the ex vivo release of PGE_2 from brainstem slices (210). Microinjection of PGE_2 into the rostral ventrolateral medulla provoked cellular activation in the PVN (179). This is interesting because PGE_2 receptors have been located on medullary but not on PVN neurons see (179). Finally, the stimulatory effect of intravenous IL-1β infusion on hypothalamic paraventricular nucleus (PVN) neurosecretory neurons depended on the integrity of catecholaminergic projections originating in the caudal medulla (211). All these data suggest that prostanoids mediate the cytokine effect on hindbrain catecholaminergic neurons which project to the hypothalamus.

In addition to catecholaminergic neurons, serotonergic neurons might carry the cytokine-derived signal to the hypothalamus. Serotonin (5-HT) is abundant in neurons originating from the midbrain dorsal raphe nucleus and projecting to the hypothalamus, including the PVN and ventromedial nucleus (VMH). Increases in serotonergic activity or stimulation of postsynaptic serotonergic receptors decreases, and antagonism of 5-HT increases food intake (see (212) for review). 5-HT is thought to affect eating through the 5-HT_{1b} and/or 5-HT_{2c} receptors (212). 5-HT might contribute to the anorectic effect of pro-inflammatory cytokines because the dorsal raphe nucleus contains IL-1 receptors type 1 (213) and because central and peripheral administration of IL-1β and TNFα increase serotonergic activity in this area (214). Central IL-1β increases 5-HT metabolism also in other brain areas (215), and peripheral IL-1β stimulates 5-HT turnover in the hypothalamus (216). The effect of central IL-1β on serotonergic activity can be blocked by prior administration of an IL-1 receptor antagonist (217). This is consistent with the idea that 5-HT mediates the feeding suppressive effect of central IL-1β because this effect is also blocked by ICV administration of IL-1 receptor antagonist (7). Further, it is worth mentioning that central 5-HT appears to mediate the feeding suppressive effect of rectal TNB administration in the rat (218), which is presumably mediated by central IL-1β (51). Interestingly, IFNγ stimulated serotonergic activity only after intracerebroventricular injection (214), suggesting that this cytokine plays a different role/uses different mechanisms for the communication with the CNS. Similar to pro-inflammatory cytokines, peripheral LPS treatment produced marked increases in the concentrations of dopamine (DA), 5-HT and its metabolite 5-hydroxyindoleacetic acid (5-HIAA) in the hypothalamic PVN, and these effects were completely blocked by treatment with IL-Ira in combination with LPS (219). This indicates that the highly specific changes in monamine metabolism produced by LPS in the hypothalamus are mediated at least in part by IL-1β. Interestingly, subdiaphragmatic vagotomy attenuated the effect of IP LPS on body temperature but not on PVN monoamines (220), suggesting that this effect, as the feeding suppressive effect of LPS, is not dependent on intact vagal afferents. In our hands, 8-OH-DPAT (a 5-HT_{1A} agonist which reduces 5-HT synthesis and transmission) and ritanserin (a $5HT_{2A/C}$ antagonist), were able to attenuate the anorectic effect of LPS and IL-1β after peripheral and central administration (221)(Hrupka et al., unpublished results). 8-OH-DPAT injected directly into the raphe nucleus also blocked the feeding suppressive effect of IP injected IL-1β (Figure 6). On the other hand, Swiergiel and Dunn recently reported that numerous 5-HT antagonists did not attenuate IL-1β-induced suppression of short-term milk intake in mice (222). Whether this discrepancy is due to a species difference or to methodological differences is not clear. CNS 5-HT antagonism also yielded conflicting results concerning cancer anorexia. Whereas the 5-HT receptor antagonist mianserin administered into the VMH attenuated the food intake reduction in tumor-bearing rats (223), intracisternal injection of 5,7-dihydroxytryptamine or the systemic injection of para-chloramphetamine did not affect the onset or severity of the anorectic response to cancer (224), although both compounds reduced brain 5-HT and 5-HIAA by approximately 50%. The effect of central 5-HT antagonism on the feeding suppressive effects of peripheral cytokines also needs to be tested. In general, the mediation of cytokine-induced suppression of feeding by central serotonergic systems appears to be situationally variable and may be related to changes in serotoninergic systems in several brain areas. Clearly, further studies are necessary to characterize the exact role of 5-HT in the anorexia induced by pro-inflammatory cytokines.

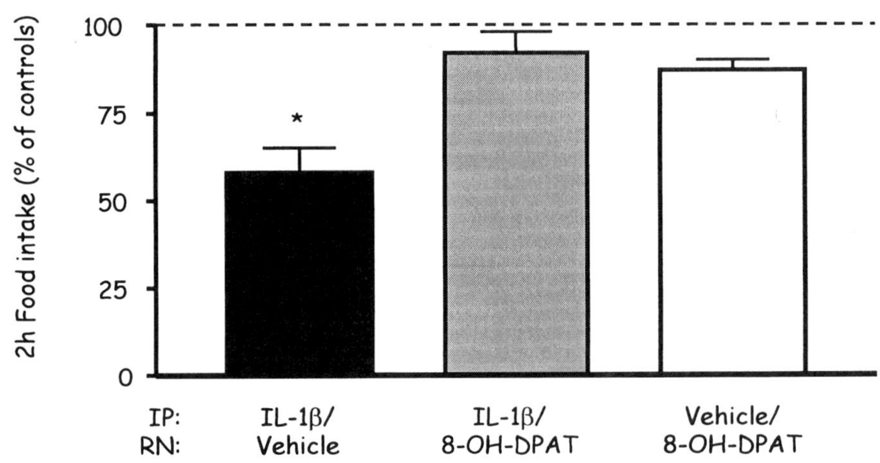

Figure 6: 8-OH-DPAT (4 nmol) injected directly into the raphe nucleus (RN) blocks the anorectic effect of intraperitoneally injected IL-1β (2 μg/kg). Data are mean ± SEM percent food intakes of 7 rats/group. Corresponding vehicle/vehicle injected rats' intake =100%. *Significantly ($p<0.05$ in post-hoc tests after significant ANOVA) lower than 100%

Finally, peptidergic neurons appear to link the hindbrain and the hypothalamic PVN, and to mediate signals related to eating. Rinaman and colleagues have shown that LPS activates GLP-1 neurons (225), and that GLP-1 neurons project to the PVN (225). More recently, the same group could also show that ICV administration of a GLP-1 receptor antagonist blocked the fever inducing and feeding suppressive effects of IP LPS (226,227), suggesting that GLP-1 is in fact involved in mediation of these LPS-induced phenomena. Whether the same holds true for cytokine-induced phenomena awaits further clarification.

4.3.2 Hypothalamic integration

Cytokines can directly change the activity of hypothalamic neurons that are implicated in the control of food intake (228,229) and this might be involved in the feeding suppressive effects of cytokines in some situations. Nevertheless, the weight of evidence currently favors the aminergic and peptidergic transmission of the cytokine signal to the hypothalamus outlined above. Downstream of these pathways, several other neurochemicals may be involved in the hypothalamic integration of the cytokine signal.

Peripheral injection of IL-1β increases hypothalamic CRF mRNA (230) and IL-1β-induced anorexia is attenuated by ICV administration of a CRF antagonist (231), suggesting that CRF is involved in the neurochemical mediation of IL-1β-induced anorexia. Interestingly, prostanoids appear to mediate the effect of IL-1β on hypothalamic CRF release (232), which might provide a link between the presumed roles of PGE_2 and CRF in the feeding suppressive effect of IL-1β.

Neuropeptide Y is a 36 amino acid peptide that is widely distributed throughout the brain. It potently stimulates food intake, and NPYergic neurons of the arcuate nucleus (ARC) which project to the PVN are presumably important for this action. Intracerebroventricularly infused IL-1β has been shown to cause a small (18%), but significant, reduction in hypothalamic NPY mRNA levels (140), and ICV injections of NPY can prevent or reverse ICV IL-1β-induced anorexia (233). Anorectic tumor-bearing rats have decreased PVN NPY levels, increased ARC NPY levels, decreased NPY release, and eat less than controls after exogenous ICV NPY administration (234,235). These findings suggest an antagonistic interaction between IL-1β and NPY, and that the feeding suppressive effects of IL-1β and tumors are in part due to a modulation of NPYergic mechanisms. Thus, it is possible that a cytokine-induced decrease in NPY attenuates feeding that would normally occur in response to an energy deficit (236).

Intraperitoneal IL-1β injection increased hypothalamic histamine turnover and the activity of histidine decarboxylase and histamine-N-methyltransferase. In addition, depletion of neuronal histamine induced by IP injection of alpha-fluoromethylhistidine attenuated the suppressive effect of IL-1β on food intake in one study in rats (237), implicating a histaminergic mechanism in the feeding suppressive effect of IL-1β. Yet, the milk intake suppression by IP IL-1β and LPS in mice was not affected by several histaminergic antagonists in another study (238). We recently observed that the anorexia in response to IP IL-1β was unaffected in genetically histamine receptor 1-deficient mice. These discrepancies may reflect species differences or developmental compensation (the effectiveness of IL-1β to reduce food intake in histamine receptor KO mice), but in any case, the role of central histamine in the anorectic effect of pro-inflammatory cytokines is open to question.

Finally, alpha-melanocyte-stimulating hormone (αMSH) has long been recognized as a key component in neuroimmunomodulation (239). LPS stimulates the release of αMSH (240). Alpha-MSH appears to antagonize inflammatory and acute phase reactions at various levels (cytokine production, cytokine action) in the periphery and the brain (241). Alpha-MSH binds to central melanocortin receptors (MC3-R and MC4-R). Central administration of MC4-R agonists inhibits food intake, increases energy expenditure, and reduces body weight (see (242)). In contrast, deletion of the MC4-R increases food intake and body weight (243). Huang et al. (244) reported that αMSH enhanced LPS-induced anorexia in rats, whereas administration of the melanocortin (MC3/MC4) receptor antagonist SHU9119 attenuated it. More recently, the anorexia and cachexia induced by IP LPS or by a syngenic carcinoma were attenuated in MC4-R KO mice and also by the MC3-R/MC4-R antagonist Agouti-related peptide (AgrP)(245). These data strongly implicate the central melanocortin system in mediation of cytokine-induced anorexia.

5. SUMMARY

The production of pro-inflammatory cytokines is enhanced during microbial infections, cancer, and a variety of other pathologic situations. Given the potent feeding suppressive effect of several pro-inflammatory cytokines such as such as IL-1, IL-2, IL-6, IL-8, TNFα and IFNγ, it is reasonable to assume that these cytokines also contribute to the anorexia that accompanies many pathologic situations. When pro-inflammatory cytoklines are produced in the brain, i.e. witin the BBB, they can reduce food intakae by acting directly on neural circuitries involved in control of food intake and energy balance. On the other hand, cytokines produced outside the BBB presumably act at the level of the brain capillary endothelial and perivascular cells to trigger the release of other messengers such as prostanoids. These messengers activate aminergic and peptidergic pathways projecting to the hypothalamus, and to modulate the hypothalamic neuropeptide systems of food intake control.

REFERENCES

1 Murray, M.J., Murray, A.B. Anorexia of infection as a mechanism of host defense. *Am J Clin Nutr* 1979; 32:593-596.

2 Langhans, W., Hrupka, B. Interleukins and tumor necrosis factor as inhibitors of food intake. *Neuropeptides* 1999; 33:415-424.

3 Plata-Salamàn, .CR. Interferons and central regulation of feeding. *Am J Physiol* 1992; 263:R1222-R1227.

4 Plata-Salamàn, C.R. Cytokines and feeding suppression: An integrative view from neurologic to molecular levels. *Nutrition* 1995; 11:674-677.

5 Plata-Salamàn, C.R. Anorexia during acute and chronic disease. *Nutrition* 1996; 12:69-78.

6 Kent, S., BretDibat, .L., Kelley, K.W., et al. Mechanisms of sickness-induced decreases in food-motivated behavior. *Neurosci Biobehav Rev* 1996; 20:171-175.

7 Plata-Salamàn, C.R., FfrenchMullen, J.M.H. Intracerebroventricular administration of a specific IL-1 receptor antagonist blocks food and water intake suppression induced by interleukin-1 beta. *Physiol Behav* 1992; 51:1277-1279.

8 Aubert, A., Goodall, G., Dantzer, R. Compared effects of cold ambient temperature and cytokines on macronutrient intake in rats. *Physiol Behav* 1995; 57:869-873.

9 Winograd, C.H., Brown, E.M. Aggressive oral refeeding in hospitalized-patients. *Am J Clin Nutr* 1990; 52:967-968.

10 Bauer, C., Weingarten, S., Senn, M., Langhans W. Limited importance of a learned aversion in the hypophagic effect of interleukin-1 beta. *Physiol Behav* 1995; 57:1145-1153.

11 Bernstein, I.L. Neural mediation of food aversions and anorexia induced by tumor necrosis factor and tumors. *Neurosci Biobehav Rev* 1996; 20:177-181.

12 Goehler, L.E., Busch, C.R., Tartaglia, N., et al. Blockade of cytokine induced conditioned taste aversion by subdiaphragmatic vagotomy: Further evidence for vagal mediation of immune-brain communication. *Neurosci Lett* 1995; 185:163-166.

13 Lennie, T.A. Relationship of body energy status to inflammation-induced anorexia and weight loss. *Physiol Behav* 1998; 64:475-481.

14 Mrosovsky, N., Molony, L.A., Conn, C.A., et al. Anorexic effects of interleukin 1 in the rat. *Am J Physiol* 1989; 257:R1315-R1321.

15 Porter, M.H., Arnold, M., Langhans, W. TNF-alpha tolerance blocks LPS-induced hypophagia but LPS tolerance fails to prevent TNF-alpha-induced hypophagia. *Am J Physiol* 1998; 274:R741-R745.

16 Weingarten, S., Savoldelli, D., Langhans, W. Enhancement or loss of the hypophagic effect of interleukin-1 upon chronic administration. *Physiol Behav* 1992; 52:831-837.

17 Fraker, D.L., Stovroff, M.C., Merino, M.J., et al. Tolerance to tumor necrosis factor in rats and the relationship to endotoxin tolerance and toxicity. *J Exp Med* 1988; 168:95-105.

18 Takahashi, N., Brouckaert, P., Fiers, W. Mechanism of tolerance to tumor necrosis factor: Receptor-specific pathway and selectivity. *Am J Physiol* 1995; 269:R398-R405.

19 Plata-Salamàn, C.R., Sonti, G., Borkoski, J.P., et al. Anorexia induced by chronic central administration of cytokines at estimated pathophysiological concentrations. *Physiol Behav* 1996; 60:867-875.

20 Segall, M.A., Crnic, L.S. An animal model for the behavioral effects of interferon. *Behav Neurosci* 1990; 104:612-618.

21 Laŋgstein, H.N., Doherty, G.M., Fraker, D.L., et al. The roles of gamm-interferon and tumor-necrosis-factor-alpha in an experimental rat model of cancer cachexia. *Cancer Res* 1991; 51:2302-2306.

22 Sonti, G., Ilyin, S.E., Plata-Salamàn, C.R. Anorexia induced by cytokine interactions at pathophysiological concentrations. *Am J Physiol* 1996; 270:R1394-R1402.

23 Plata-Salamàn, C.R. Meal patterns in response to the intracerebroventricular administration of interleukin-1 beta in rats. *Physiol Behav* 1994; 55:727-733.

24 Debonis, D., Meguid, M.M., Laviano, A., et al. Temporal changes in meal number and meal size relationship in response to rHu IL-1 alpha. *Neuroreport* 1995; 6:1752-1756.

25 Langhans, W., Savoldelli, D., Weingarten, S. Comparison of the feeding responses to bacterial lipopolysaccharide and interleukin-1 beta. *Physiol Behav* 1993; 53:643-649.

26 Langhans, W., Balkowski, G., Savoldelli, D. Further characterization of the feeding responses to interleukin and tumor necrosis factor. In: Murison, R., (ed.) *Endocrine and Nutritional Control of Basic Biological Functions*. Hogrefe & Huber Publishers, Toronto (1992) pp. 137-144.

27 Angele, M.K., Knoferl, M.W., Schwacha, M.G., et al. Sex steroids regulate pro- and anti-inflammatory cytokine release by macrophages after trauma-hemorrhage. *Am J Physiol* 1999; 277:C35-C42.

28 Li, Z.G., Danis, V.A., Brooks, P.M. Effect of gonadal steroids on the production of IL-1 and IL-6 by blood mononuclear cells in vitro. *Clin Exp Rheumatol* 1993; 11:157-162.

29 Lynch, E.A., Dinarello, C.A., Cannon, J.G. Gender differences in IL-1 alpha, IL-1 beta, and IL-1 receptor antagonist secretion from mononuclear cells and urinary excretion. *J Immunol* 1994; 153:300-306.

30 Schwarz, E., Schafer, C., Bode, J.C., et al. Influence of the menstrual cycle on the LPS-induced cytokine response of monocytes. *Cytokine* 2000; 12:413-416.

31 Geary, N. Sex differences in disease anorexia. *Nutrition* 2001; 17:499-507.

32 Mouihate, A., Chen, X., Pittman, Q.J. Interleukin-1beta fever in rats: gender difference and estrous cycle influence. *Am J Physiol* 1998; 275:R1450-R1454.

33 Ikejima, K., Enomoto, N., Iimuro, Y., et al. Estrogen increases sensitivity of Kupffer cells to endotoxin. *Alcoholism - Clin Exp Res* 1998; 22:768-769.

34 Butera, P.C., Doerflinger, A.L., Roberto, F. Cyclic estradiol treatmet enhances the effects of interluekin-1beta on food intake in female rats. *Brain Behav Immun* 2002; 16: 275-281.

35 VanderMeer, M.J.M., Sweep, C.G.J.F., Pesman, G.J., et al. Synergism between IL-1 beta and TNF-alpha on the activity of the pituitary-adrenal axis and on food intake of rats. *Am J Physiol* 1995; 268:E551-E557.

36 Yang, Z.J., Koseki, M., Meguid, M.M., et al. Synergistic effect of rhTNF-alpha and rhIL-1 alpha in inducing anorexia in rats. *Am J Physiol* 1994; 267:R1056-R1064.

37 Plata-Salamàn, C.R., Turrin, N.P. Cytokine interactions and cytokine balance in the brain: relevance to neurology and psychiatry. *Mol Psychiatry* 1999; 4:303-306.

38 Alheim, K., Chai, Z., Fantuzzi, G., et al. Hyperresponsive febrile reactions to interleukin (IL) 1 alpha and IL-1 beta, and altered brain cytokine mRNA and serum cytokine levels, in IL-1 beta-deficient mice. *Proc Natl Acad Sci USA* 1997; 94:2681-2686.

39 Chai, Z., Gatti, S., Toniatti, C., et al. Interleukin (IL)-6 gene expression in the central nervous system is necessary for fever response to lipolysaccharide or IL-1 beta: A study on IL-6-deficient mice. *J Exp Med* 1996; 183:311-316.

40 Arsenijevic, D., Girardier, L., Seydoux, J., et al. Altered energy balance and cytokine gene expression in a murine model of chronic infection with Toxoplasma gondii. *Am J Physiol* 1997; 272:E908-E917.
41 Arsenijevic, D., Onuma, H., Pecqueur, C., et al. Disruption of the uncoupling protein-2 gene in mice reveals a role in immunity and reactive oxygen species production. *Nature Genetics* 2000; 26:435-439.
42 Wusteman, M., Wight, D.G.D., Elia, M. Protein-metabolism after injury within turpentine - A rat model for clinical trauma. *Am J Physiol* 1990; 259:E763-E769.
43 Cooper, A.L., Brouwer, S., Turnbull, A.V., et al. Tumor necrosis factor-alpha and fever after peripheral inflammation in the rat. *Am J Physiol* 1994; 36:R1431-R1436.
44 Kopf, M., Baumann, H., Freer, G., et al. Impaired immune and acute-phase responses in interleukin-6-deficient mice. *Nature* 1994; 368:339-342.
45 Kozak, W., Poli, V., Soszynski, D.,et al. Sickness behavior in mice deficient in interleukin-6 during turpentine abscess and influenza pneumonitis. *Am J Physiol* 1997; 272:R621-R630.
46 Leon, L.R., Conn, C.A., Glaccum, M., et al. IL-1 type I receptor mediates acute phase response to turpentine, but not lipopolysaccharide, in mice. *Am. J. Physiol.* 1996; 271:R1668-R1675.
47 Oldenburg, H.S.A., Rogy, M.A., Lazarus, D.D., et al. Cachexia and the acute-phase protein response in inflammation are regulated by interleukin-6. *Eur J Immunol* 1993; 23:1889-1894.
48 Gee, M.I., Grace, M.G.A., Wensel, R.H., et al. Protein-energy malnutrition in gastroeneterology outpatients – Increased risk in Crohns-disease. *J Am Diet Assoc* 1985; 85:1466-1474.
49 Mchugh, K., Castonguay, T.W., Collins, S.M., et al. Characterization of suppression of food intake following acute colon inflammation in the rat. *Am J Physiol* 1993; 265:R1001-R1005.
50 Rachmilewitz, D., Simon, P.L., Schwartz, L.W., et al. Inflammatory mediators of experimental colitis in rats. *Gastroenterol* 1989; 97:326-337.
51 Mchugh, K.J., Collins, S.M., Weingarten, H.P. Central interleukin-1 receptors contribute to suppression of feeding after acute colitis in the rat. *Am J Physiol* 1994; 266:R1659-R1663.
52 Abram, M., Vuckovic, D., Wraber, B., et al. Plasma cytokine response in mice with bacterial infection. *Med Inflamm* 2000; 9:229-234.
53 Imanishi, J. Expression of cytokines in bacterial and viral infections and their biochemical aspects. *J Biochem* 2000; 127:525-530.
54 Poveda, F., Camacho, J., Arnalich, F.,et al. Circulating cytokine concentrations in tuberculosis and other chronic bacterial infections. *Infection* 1999; 27:272-274.
55 Rietschel, E.T., Schletter, J., Weidemann, B., El S, V, et al. Lipopolysaccharide and peptidoglycan: CD14-dependent bacterial inducers of inflammation. *Microb Drug Resist* 1998; 4:37-44.
56 Akira, S. Toll-like receptors: lessons from knockout mice. *Biochem Soc Trans* 2000; 28 Part 5:551-556.
57 Beutler, B. Endotoxin, toll-like receptor 4, and the afferent limb of innate immunity. *Curr Opin Microbiol* 2000; 3:23-28.
58 Muzio, M., Mantovani, A. Toll-like receptors. *Microbes Inf* 2000; 2:251-255.
59 Arsenijevic, D., Garcia, I., Vesin, C., et al. Differential roles of tumor necrosis factor-alpha and interferon-gamma in mouse hypermetabolic and anorectic responses induced by LPS. *Eur Cytok Netw* 2000; 11:662-668.

60 Bluthe, R.M., Laye, S., Michaud, B., et al. Role of interleukin-1 beta and tumour necrosis factor-alpha in lipopolysaccharide-induced sickness behaviour: a study with interleukin-1 type I receptor-deficient mice. *Eur J Neurosci* 2000; 12:4447-4456.
61 Kozak, W., Zheng, H., Conn, C.A., et al. Thermal and behavioral effects of lipopolysaccharide and influenza in interleukin-1 beta-deficient mice. *Am J Physiol* 1995; 269:R969-R977.
62 Leon, L.R., Kozak, W., Peschon, J., et al. Exacerbated febrile responses to LPS, but not turpentine, in TNF double receptor-knockout mice. *Am J Physiol* 1997; 272:R563-R569.
63 Bluthe, R.M., Michaud, B., Poli, V., et al. Role of IL-6 in cytokine-induced sickness behavior: a study with IL-6 deficient mice. *Physiol Behav* 2000; 70:367-373.
64 Leon, L.R., White, A.A., Kluger, M.J. Role of IL-6 and TNF in thermoregulation and survival during sepsis in mice. *Am J Physiol* 1998; 275:R269-R277.
65 Matsuda, T., Hirano, T. IL-6. In: Oppenheim JJ, Feldmann M, editors. *Cytokine Reference, Volume 1: Ligands*. San Diego: Academic Press, 2001: 537-563.
66 Billiau, A., Vendenbroeck, K. INF-gamma. In: Oppenheim JJ, Feldmann M, editors. *Cytokine Reference, Volume 1: Ligands.* San Diego: Academic Press, 2001: 641-688.
67 Beutler, B. TLR4: central component of the sole mammalian LPS sensor. *Curr Opin Immunol* 2000; 12:20-26.
68 Doherty, G.M., Lange, J.R., Langstein, H.N., et al. Evidence for IFN-gamma as a mediator of the lethality of endotoxin and tumor necrosis factor-alpha. *J Immunol* 1992; 149:1666-1670.
69 Chang, H.R., Bistrian, B. The role of cytokines in the catabolic consequences of infection and injury. *J Parent Enteral Nutr* 1998; 22:156-166.
70 Porter, M.H., Hrupka, B.J., Altreuther, G.,et al. Inhibition of TNF-? production contributes to the attenuation of LPS-induced hypophagia by pentoxifylline. *Am J Physiol* 2000; 279:R2113-R2120.
71 Swiergiel, A.H., Dunn, A.J. The roles of IL-1, IL-6, and TNF alpha in the feeding responses to endotoxin and influenza virus infection in mice. *Brain Behav Immun* 1999; 13:252-265.
72 Barton, B.E. IL-6-like cytokines and cancer cachexia - Consequences of chronic inflammation. *Immunol Res* 2001; 23:41-58.
73 Bossola, M., Muscaritoli, M., Bellantone, R., et al. Serum tumour necrosis factor-alpha levels in cancer patients are discontinuous and correlate with weight loss. *Eur J Clin Invest* 2000; 30:1107-1112.
74 Ikemoto, S., Sugimura, K., Yoshida, N., et al. TNF alpha, IL-1 beta and IL-6 production by peripheral blood monocytes in patients with renal cell carcinoma. *Anticancer Res* 2000; 20:317-321.
75 Mantovani, G., Maccio, A., Mura, L., et al. Serum levels of leptin and pro-inflamatory cytokines in patients with advanced-stage cancer at different sites. *J Mol Med-Jmm* 2000; 78:554-561.
76 Okada, S., Okusaka, T., Ishii, H., et al. Elevated serum interleukin-6 levels in patients with pancreatic cancer. *Jap J Clin Oncol* 1998; 28:12-15.
77 Plata-Salamàn, C.R., Ilyin, S.E., Gayle, D. Brain cytokine mRNAs in anorectic rats bearing prostate adenocarcinoma tumor cells. *Am J Physiol* 1998; 275:R566-R573.
78 Zhang, G.J., Adachi, I. Serum interleukin-6 levels correlate to tumor progression and prognosis in metastatic breast carcinoma. *Anticancer Res* 1999; 19:1427-1432.
79 Cahlin, C., Korner, A., Axelsson, H.,et al. Experimental cancer cachexia: The role of host-derived cytokines interleukin (IL)-6, IL-12, interferon-gamma, and tumor

necrosis factor alpha evaluated in gene knockout, tumor-bearing mice on C57 Bl background and eicosanoid-dependent cachexia. *Cancer Res* 2000; 60:5488-5493.
80 Molotkov, A., Satoh, M., Tohyama, C. Tumor growth and food intake in interleukin-6 gene knock-out mice. *Cancer Lett* 1998; 132:187-192.
81 Laviano, A., Renvyle, T., Meguid, M.M., et al. Relationship between interleukin-1 and cancer anorexia. *Nutrition* 1995; 11:680-683.
82 Smith, B.K., Kluger, M.J. Anti-TNF-alpha antibodies normalized body temperature and enhanced food intake in tumor-bearing rats. Am J Physiol 1993; 265:R615-R619.
83 Torelli, G.F., Meguid, M.M., Moldawer, L.L., et al. Use of recombinant human soluble TNF receptor in anorectic tumor-bearing rats. *Am J Physiol* 1999; 277:R850-R855.
84 Moldawer, L.L., Rogy, M.A., Lowry, S.F. The role of cytokines in cancer cachexia. *J Parent Ent Nutr* 1992; 16 Suppl:43S-49S.
85 Noguchi, Y., Yoshikawa, T., Matsumoto, A., et al. Are cytokines possible mediators of cancer cachexia? *Surg Today* 1996; 26:467-475.
86 Nakatani S, Iwagaki H, Okabayashi T, Isozaki H, Takakura N, Tanaka N. Is increased IL-1 beta mRNA expression in spleen of tumor- bearing mice relevant to cancer cachexia? *Res Comm Mol Pathol Pharmacol* 1998; 102:241-249.
87 Strassmann, G., Fong, M., Kenney, J.S., et al. Evidence for involvement of interleukin-6 in experimental cancer cachexia. *J Clin Invest* 1992; 89:1681-1684.
88 Malaguarnera, L., Ferlito, L., Imbesi, R.M., et al. Immunosenescence: a review. *Arch Gerontol Geriat* 2001; 32:1-14.
89 Bruunsgaard, H., Pedersen, A.N., Schroll, M., et al. TNF-alpha, leptin, and lymphocyte function in human aging. *Life Sci* 2000; 67:2721-2731.
90 Pedersen, B.K., Bruunsgaard, H., Ostrowski, K., et al. Cytokines in aging and exercise. *Int J Sports Med* 2000; 21:S4-S9.
91 Saurwein, T.M., Blasko, I., Zisterer, K., et al. An imbalance between pro- and anti-inflammatory cytokines, a characteristic feature of old age. *Cytokine* 2000; 12:1160-1161.
92 Yeh, S.S., Schuster, M.W. Geriatric cachexia: the role of cytokines. *Am J Clin Nutr* 1999; 70:183-197.
93 Paolisso, G., Rizzo, M.R., Mazziotti, G., et al. Advancing age and insulin resistance: role of plasma tumor necrosis factor-alpha. *Am J Physiol* 1998; 275:E294-E299.
94 Ostrowski, K., Rohde, T., Asp, S., et al. Pro- and anti-inflammatory cytokine balance in strenuous exercise in humans. *J Physiol London* 1999; 515:287-291.
95 Ostrowski, K., Schjerling, P., Pedersen, B.K. Physical activity and plasma interleukin-6 in humans - effect of intensity of exercise. *Eur J Appl Physiol* 2000; 83:512-515.
96 Pedersen, B.K. Exercise and cytokines. *Immunol Cell Biol* 2000; 78:532-535.
97 Baum, M., MullerSteinhardt, M., Liesen, H., et al. Moderate and exhaustive endurance exercise influences the interferon-gamma levels in whole-blood culture supernatants. *Eur J Appl Physiol* 1997; 76:165-169.
98 Starkie, R.L., Angus, D.J., Rolland, J., et al. Effect of prolonged, submaximal exercise and carbohydrate ingestion on monocyte intracellular cytokine production in humans. *J Physiol London* 2000; 528:647-655.
99 Ostrowski, K., Rohde, T., Zacho, M.,et al. Evidence that interleukin-6 is produced in human skeletal muscle during prolonged running. *J Physiol London* 1998; 508:949-953.

100 Jeukendrup, A.E., Vet, J.K., Sturk, A., et al. Relationship between gastro-intestinal complaints and endotoxaemia, cytokine release and the acute-phase reaction during and after a long-distance triathlon in highly trained men. *Clin Sci* 2000; 98:47-55.

101 Hotamisligil, G.S., Arner, P., Caro, J.F., et al. Increased adipose tissue expression of tumor necrosis factor-alpha in human obesity and insulin resistance. *J Clin Invest* 1995; 95:2409-2415.

102 Kern, P.A., Saghizadeh, M., Ong, J.M., et al. The expression of tumor necrosis factor in human adipose tissue - Regulation by obesity, weight loss, and relationship to lipoprotein lipase. *J Clin Invest* 1995; 95:2111-2119.

103 Hotamisligil, G.S., Shargill, N.S., Spiegelman, B.M. Adipose expression of tumor necrosis factor-alpha: direct role in obesity-linked insulin resistance. *Science* 1993; 259:87-91.

104 Brand, E., Schorr, U., Kunz, I., et al. Tumor necrosis factor-alpha-308 G/A polymorphism in obese Caucasians. *Int J Obes* 2001; 25:581-585.

105 Morin, C.L., Pagliassotti, M.J., Windmiller, D., et al. Adipose tissue-derived tumor necrosis factor-alpha activity is elevated in older rats. *J Gerontol Series A Biol Sci Med Sci* 1997; 52:B190-B195.

106 Morin, C.L., Eckel, R.H., Marcel, T., et al. High fat diets elevate adipose tissue-derived tumor necrosis factor-alpha activity. *Endocrinol* 1997; 138:4665-4671.

107 Hotamisligil, G.S., Budavari, A., Murray, D., et al. Reduced tyrosine kinase activity of the insulin receptor in obesity-diabetes. Central role of tumor necrosis factor-alpha . *J Clin Invest* 1994; 94:1543-1549.

108 Kern, P.A. Potential role of TNF alpha and lipoprotein lipase as candidate genes for obesity. *J Nutr* 1997; 127:S1917-S1922.

109 Tanaka, T., Itoh, H., Doi, K., et al. Down regulation of peroxisome proliferator-activated receptor gamma expression by inflammatory cytokines and its reversal by thiazolidinediones. *Diabetologia* 1999; 42:702-710.8

110 Dandona, P., Weinstock, R., Thusu, K., et al. Tumor necrosis factor-alpha in sera of obese patients: fall with weight loss. *J Clin Endocrinol Metab* 1998; 83:2907-2910.

111 Tsigos, C., Kyrou, L., Chala, E., et al. Circulating tumor necrosis factor alpha concentrations are higher in abdominal versus peripheral obesity. *Metabolism* 1999; 48:1332-1335.

112 Mohamed-Ali, V., Goodrick, S., Rawesh, A., et al. Subcutaneous adipose tissue releases interleukin-6, but not tumor necrosis factor-alpha, in vivo. *J Clin Endocrinol Metabol* 1997; 82:4196-4200.

113 Kirchgessner, T.G., Uysal, K., Wiesbrock, S.M., et al. Tumor necrosis factor-alpha contributes to obesity-related hyperleptinemia by regulating leptin release from adipocytes. *J Clin Invest* 1997; 100:2777-2782.

114 Unger, R.H. Leptin physiology: a second look. *Regul Peptides* 2000; 92:87-95.

115 Uysal, K.T., Wiesbrock, S.M., Marino, M.W., et al. Protection from obesity-induced insulin resistance in mice lacking TNF-alpha function. *Nature* 1997; 389:610-614.

116 Bastard, J.P., Jardel, C., Bruckert, E., et al. Elevated levels of interleukin 6 are reduced in serum and subcutaneous adipose tissue of obese women after weight loss. *J Clin Endocrinol Metabol* 2000; 85:3338-3342.

117 McCarty, M.F. Interleukin-6 as a central mediator of cardiovascular risk associated with chronic inflammation, smoking, diabetes, and visceral obesity: down-regulation with essential fatty acids, ethanol and pentoxifylline. *Med Hypoth* 1999; 52:465-477.

118 Raymond, N.C., Dysken, M., Bettin, K., et al. Cytokine production in patients with anorexia nervosa, bulimia nervosa, and obesity. *Int J Eating Disord* 2000; 28:293-302.

119 Bruun, J.M., Pedersen, S.B., Richelsen, B. Regulation of interleukin 8 production and gene expression in human adipose tissue in vitro. *J Clin Endocrinol Metabol* 2001; 86:1267-1273.
120 Montuschi, P., Tringali, G., Mirtella, A., et al. Interleukin-1beta release from rat gastric fundus. *Am J Physiol* 1996; 271:G275-G281.
121 Youngman, K.R., Simon, P.L., West, G.A., et al. Localization of intestinal interleukin 1 activity and protein and gene expression to lamina propria cells. *Gastroenterol* 1993; 104:749-758.
122 Deitch, E.A. Bacterial translocation – The influence of dietary variables. *Gut* 1994; 35:S23-S27.
123 Hansen, K., Sickelmann, F., Pietrowsky, R., et al. Systemic immune changes following meal intake in humans. *Am J Physiol* 1997; 273:R548-R553.
124 Hansen, M.K., Taishi, P., Chen, Z.T., et al. Cafeteria feeding induces interleukin-1 beta mRNA expression in rat liver and brain. *Am J Physiol* 1998; 274:R1734-R1739.
125 Kurosawa, M., UvnasMoberg, K., Miyasaka, K., et al. Interleukin-1 increases activity of the gastric vagal afferent nerve partly via stimulation of type A CCK receptor in anesthetized rats. *J Auton Nerv Syst* 1997; 62:72-78.
126 Niijima, A. The afferent discharges from sensors for interleukin-1beta in the hepatoportal system in the anesthetized rat. *J Auton Nerv Syst* 1996; 61:287-291.
127 Hansen, M.K., Krueger, J.M. Subdiaphragmatic vagotomy blocks the sleep- and fever-promoting effects of interleukin-1 beta. *Am J Physiol* 1997; 273:R1246-R1253.
128 Pu, S.Y., Anisman, H., Merali, Z. Central infusion of interleukin-I receptor antagonist fails to alter feeding and weight gain. *Neuroreport* 2000; 11:1699-1702.
129 Floyd, R.A., Krueger, J.M. Diurnal variation of TNF alpha in the rat brain. *Neuroreport* 1997; 8:915-918.
130 Holden, R.J., Pakula, I,S. The role of tumor necrosis factor-alpha in the pathogenesis of anorexia and bulimia nervosa, cancer cachexia and obesity. *Med Hypoth* 1996; 47:423-438.
131 Brambilla, F., Bellodi, L., Brunetta, M., et al. Plasma concentrations of interleukin-1 beta, interleukin-6 and tumor necrosis factor-alpha in Anorexia and Bulimia Nervosa. *Psychoneuroendocrinol* 1998; 23:439-447.
132 Nakai, Y., Hamagaki, S., Takagi, R., et al. Plasma concentrations of tumor necrosis factor-alpha (TNF- alpha) and soluble TNF receptors in patients with bulimia nervosa. *Clin Endocrinol* 2000; 53:383-388.
133 Vitkovic, L., Bockaert, J., Jacque, C. Inflammatory cytokines: Neuromodulators in normal brain? *J Neurochem* 2000; 74:457-471.
134 Rothwell, N.J. Cytokines - killers in the brain? *J Physiol London* 1999; 514:3-17.
135 Pan, W., Zadina, J.E., Harlan, R.E., et al. Tumor necrosis factor-alpha: a neuromodulator in the CNS. *Neurosci Biobehav Rev* 1997; 21:603-613.
136 Konsman, J.P., Tridon, V., Dantzer, R. Diffusion and action of intracerebroventricularly injected interleukin-1 in the CNS. *Neuroscience* 2000; 101:957-967.
137 Ericsson, A., Liu, C., Hart, R.P., et al. Type-1 interleukin receptor in the rat brain – Distribution, regulation, and relationship to sites IL-1-induced cellular activation. *J Comp Neurol* 1995; 361:681-698.
138 Zhang, J., Rivest, S. A functional analysis of EP4 receptor-expressing neurons in mediating the action of prostaglandin E-2 within specific nuclei of the brain in response to circulating interleukin-1 beta. *J Neurochem* 2000; 74:2134-2145.
139 Gabellec, M.M., Griffais, R., Fillion, G., et al. Expression of interleukin 1 alpha, interleukin 1 beta and interleukin 1 receptor antagonist mRNA in mouse brain:

Regulation by bacterial lipopolysaccharide (LPS) treatment. *Mol Brain Res* 1995; 31:122-130.

140 Gayle, D., Ilyin, S.E., Plata, S.C. Central nervous system IL-1 beta system and neuropeptide Y mRNAs during IL-1 beta-induced anorexia in rats. *Brain Res Bull* 1997; 44:311-317.

141 Haour, F., Marquette, C., Ban, E., et al. Receptors for interleukin-1 in the central nervous and neuroendocrine systems - Role in infection and stress. *Ann D Endocrinol* 1995; 56:173-179.

142 Marquette, C., VanDam, A.M., Ban, E., et al. Rat interleukin-1 beta binding sites in rat hypothalamus and pituitary gland. *Neuroendocrinol* 1995; 62:362-369.

143 Wong, M.L., Bongiorno, P.B., Rettori, V., et al. Interleukin (IL) 1 beta, IL-1 receptor antagonist, IL-10, and IL-13 gene expression in the central nervous system and anterior pituitary during systemic inflammation: Pathophysiological implications. *Proc Natl Acad Sci USA 1*997; 94:227-232.

144 Cartmell, T., Luheshi, G.N., Rothwell, N.J. Brain sites of action of endogenous interleukin-1 in the febrile response to localized inflammation in the rat. *J Physiol London* 1999; 518:585-594.

145 Kakizaki, Y., Watanobe, H., Kohsaka, A., et al. Temporal profiles of interleukin-1 beta, interleukin-6, and tumor necrosis factor-alpha in the plasma and hypothalamic paraventricular nucleus after intravenous or intraperitoneal administration of lipopolysaccharide in the rat: Estimation by push-pull perfusion. *Endocrine J* 1999; 46:487-496.

146 Eriksson, C., Nobel, S., Winblad, B., et al. Expression of interleukin 1 alpha and beta, and interleukin 1 receptor antagonist mRNA in the rat central nervous system after peripheral administration of lipopolysaccharides. *Cytokine* 2000; 12:423-431.

147 Quan, N., Stern, E.L., Whiteside, M.B., et al. Induction of pro-inflammatory cytokine mRNAs in the brain after peripheral injection of subseptic doses of lipopolysaccharide in the rat. *J Neuroimmunol* 1999; 93:72-80.

148 Konsman, J.P., Kelley, K., Dantzer, R. Temporal and spatial relationships between lipopolysaccharide-induced expression of fos, interleukin-1 beta and inducible nitric oxide synthase in rat brain. *Neuroscience* 1999; 89:535-548.

149 VanDam, A.M., Brouns, M., Louisse, S., et al. Appearance of interleukin-1 in macrophages and in ramified microglia in the brain of endotoxin-treated rats - A pathway for the induction of non-specific symptoms of sickness. *Brain Res* 1992; 588:291-296.

150 Breder, C.D., Hazuka, C., Ghayur, T., et al. Regional induction of tumor necrosis factor alpha expression in the mouse brain after systemic lipopolysaccharide administration. *Proc Natl Acad Sci USA* 1994; 91:11393-11397.

151 Banks, W.A., Kastin, A.J., Gutierrez, E.G. Interleukin-1 alpha in blood has direct access to cortical brain cells. *Neurosci Lett* 1993; 163:41-44.

152 Turrin, N.P., Gayle, D., Ilyin, S.E., et al. Pro-inflammatory and anti-inflammatory cytokine mRNA induction in the periphery and brain following intraperitoneal administration of bacterial lipopolysaccharide. *Brain Res Bull* 2001; 54:443-453.

153 Burgess, W., Gheusi, G., Yao, J.H., et al. Interleukin-1 beta-converting enzyme-deficient mice resist central but not systemic endotoxin-induced anorexia. *Am J Physiol* 1998; 274:R1829-R1833.

154 Yao, J.H., Ye, S.M., Burgess, W., et al. Mice deficient in interleukin-1 beta converting enzyme resist anorexia induced by central lipopolysaccharide. *Am J Physiol* 1999; 277:R1435-R1443.

155 Bluthe, R.M., Dantzer, R., Kelley, K.W. Effects of interleukin-1 receptor antagonist on the behavioral effects of lipopolysaccharide in rat. *Brain Res* 1992; 573:318-320.

156 Laye, S., Gheusi, G., Cremona, S., et al. Endogenous brain IL-1 mediates LPS-induced anorexia and hypothalamic cytokine expression. *Am J Physiol* 2000; 279:R93-R98.

157 Brochu, S., Olivier, M., Rivest, S.. Neuronal activity and transcription of pro-inflamatory cytokines, I kappa B alpha, and iNOS in the mouse brain during acute endotoxemia and chronic infection with Trypanosoma brucei brucei. *J Neurosci Res* 1999; 57:801-816.

158 Quan, N., Mhlanga, J.D.M., Whiteside, M.B., et al. Chronic overexpression of pro-inflamatory cytokines and histopathology in the brains of rats infected with Trypanosoma brucei. *J Comp Neurol* 1999; 414:114-130.

159 Callahan, T.A., Piekut, D.T. Differential Fos expression induced by IL-1 beta and IL-6 in rat hypothalamus and pituitary gland. *J Neuroimmunol* 1997; 73:207-211.

160 Wan, W.H., Wetmore, L., Sorensen, C.M., et al. Neural and biochemical mediators of endotoxin and stress-induced c-fos expression in the rat brain. *Brain Res Bull* 1994; 34:7-14.

161 Goehler, L.E., Gaykema, R.P.A., Nguyen, K.T., et al. Interleukin-1 beta in immune cells of the abdominal vagus nerve: a link between the immune and nervous systems? *J Neurosci* 1999; 19:2799-2806.

162 Miller, A.J., Hopkins, S.J., Luheshi, G.N. Sites of action of IL-1 in the development of fever and cytokine responses to tissue inflammation in the rat. *Br J Pharmacol* 1997; 120:1274-1279.

163 Dantzer, R., Konsman, J.P., Bluthe, R.M., et al. Neural and humoral pathways of communication from the immune system to the brain: parallel or convergent? *Auton Neurosci Basic and Clin* 2000; 85:60-65.

164 BretDibat, J.L., Bluthe, R.M., Kent, S., et al. Lipopolysaccharide and interleukin-1 depress food-motivated behavior in mice by a vagal-mediated mechanism. *Brain Behav Immun* 1995; 9:242-246.

165 BretDibat, J.L., Creminon, C., Couraud ,J.Y., et al. Systemic capsaicin pretreatment fails to block the decrease in food-motivated behavior induced by lipopolysaccharide and interleukin-1 beta. *Brain Res Bull* 1997; 42:443-449.

166 Porter, M.H., Hrupka, B.J., Langhans, W., et al. Vagal and splanchnic afferents are not necessary for the anorexia produced by peripheral IL-1 beta, LPS, and MDP. *Am J Physiol* 1998; 275:R384-R389.

167 Faggioni, R., Fantuzzi, G., Fuller, J., et al. IL-1 beta mediates leptin induction during inflammation. *Am J Physiol* 1998; 274:R204-R208.

168 Finck, B.N., Kelley, K.W., Dantzer, R., et ak. In vivo and in vitro evidence for the involvement of tumor necrosis factor-alpha in the induction of leptin by lipopolysaccharide. *Endocrinology* 1998; 139:2278-2283.

169 Grunfeld, C., Zhao, C., Fuller, J., et al. Endotoxin and cytokines induce expression of leptin, the ob gene product, in hamsters - A role for leptin in the anorexia of infection. *J Clin Invest* 1996; 97:2152-2157.

170 Sarraf, P., Frederich, R.C., Turner, E.M., et al. Multiple cytokines and acute inflammation raise mouse leptin levels: Potential role in inflammatory anorexia. *J Exp Med* 1997; 185:171-175.

171 Faggioni, R., Fuller, J., Moser, A., et al. LPS-induced anorexia in leptin-deficient (ob/ob) and leptin receptor-deficient (db/db) mice. *Am J Physiol* 1997; 273:R181-R186.

172 Bornstein, S.R., Preas, H.L., Chrousos, G.P., et al. Circulating leptin levels during acute experimental endotoxemia andanti-inflammatory therapy in humans. *J Infect Dis* 1998; 178:887-890.
173 Granowitz, E.V., Porat, R., Dinarello, C.A. Circulating leptin during experimental endotoxemia in humans. *J Infect Dis* 1999; 179:1313-1314.
174 Vaselli, J.R., Casey, D. Increased responseiveness of Zucker obese rats to the feeding-inhibitory effect of systemic TNFalpha. *FASEB J* 1996; 10:A823.
175 Langhans, W., Lugarini, F., Hrupka, B.J., et al. Chronic lipopolysaccharide (LPS) anorexia is unaltered in obese Zucker rats, but leptin enhances acute LPS anorexia in normal rats. *Soc Neurosci Abstr* 2000; 26:1071.
176 Banks, W.A., Kastin, A.J. Passage of peptides across the blood-brain barrier: Pathophysiological perspectives. *Life Sci* 1996; 59:1923-1943.
177 Stitt, J.T. Evidence for the involvement of the organum vasculosum laminae terminalis in the febrile response of rabbits and rats. *J Physiol Lond* 1985; 368:501-511.
178 Ota, K., Katafuchi, T., Takaki, A., et al. AV3V neurons that send axons to hypothalamic nuclei respond to the systemic injection of IL-1beta. *Am J Physiol* 1997; 272:R532-R540.
179 Ericsson, A., Arias, C., Sawchenko, P.E. Evidence for an intramedullary prostaglandin-dependent mechanism in the activation of stress-related neuroendocrine circuitry by intravenous interleukin-1. *J Neurosci* 1997; 17:7166-7179.
180 Kastin, A.J., Pan, W., Maness, L.M., et al. Peptides crossing the blood-brain barrier: some unusual observations. *Brain Res* 1999; 848:96-100.
181 DeVries, H.E., BlomRoosemalen, M.C.M., vanOosten, M., et al. The influence of cytokines on the integrity of the blood-brain barrier in vitro. *J Neuroimmunol* 1996; 64:37-43.
182 Licinio, J., Wong, M.L. Pathways and mechanisms for cytokine signaling of the central nervous system. *J Clin Invest* 1997; 100:2941-2947.
183 Herkenham, M., Lee, H.Y., Baker ,RA. Temporal and spatial patterns of c-fos mRNA induced by intravenous interleukin-1: A cascade of non-neuronal cellular activation at the blood-brain barrier. *J Comp Neurol* 1998; 400:175-196.
184 Quan, N., Whiteside, M., Herkenham, M. Time course and localization patterns of interleukin-1 beta messenger RNA expression in brain and pituitary after peripheral administration of lipopolysaccharide. *Neurosci* 1998; 83:281-293.
185 Bierhaus, A., Chen, J., Liliensiek, B., et al. LPS and cytokine-activated endothelium. *Sem Thromb Hemost* 2000; 26:571-587.
186 Deckert-Schluter, M., Bluethmann, H., Kaefer, N., et al. Interferon-gamma receptor-mediated but not tumor necrosis factor receptor type 1- or type 2-mediated signaling is crucial for the activation of cerebral blood vessel endothelial cells and microglia in murine Toxoplasma encephalitis. *Am J Pathol* 1999; 154:1549-1561.
187 Nadeau, S., Rivest, S. Effects of circulating tumor necrosis factor on the neuronal activity and expression of the genes encoding the tumor necrosis factor receptors (p55 and p75) in the rat brain: A view from the blood-brain barrier. *Neurosci* 1999; 93:1449-1464.
188 VanDam, A.M., DeVries, H.E., Kuiper, J., et al. Interleukin-1 receptors on rat brain endothelial cells: A role in neuroimmune interaction? *FASEB J* 1996; 10:351-356.
189 Cao, C., Matsumura, K., Yamagata, K., et al. Induction by lipopolysaccharide of cyclooxygenase-2 mRNA in rat brain; its possible role in the febrile response. *Brain Res* 1995; 697:187-196.

190 Cao, C.Y., Matsumura, K., Yamagata, K., et al. Endothelial cells of the rat brain vasculature express cyclooxygenase-2 mRNA in response to systemic interleukin-1 beta: A possible site of prostaglandin synthesis responsible for fever. *Brain Res* 1996; 733:263-272.

191 Rivest, S. What is the cellular source of prostaglandins in the brain in response to systemic inflammation? Facts and controversies. *Mol Psychiatry* 1999; 4:501-507.

192 VanDam, A.M., Brouns, M., Manahing, W., et al. Immunocytochemical detection of prostaglandin-E(2) in microvasculature and in neurons of rat brain after administration of bacterial endotoxin. *Brain Res* 1993; 613:331-336.

193 Cao, C.Y., Matsumura, K., Yamagata, K., et al. Cyclooxygenase-2 is induced in brain blood vessels during fever evoked by peripheral or central administration of tumor necrosis factor. *Mol Brain Res* 1998; 56:45-56.

194 Lacroix, S., Rivest, S. Effect of acute systemic inflammatory response and cytokines on the transcription of the genes encoding cyclooxygenase enzymes (COX-1 and COX-2) in the rat brain. *J Neurochem* 1998; 70:452-466.

195 Perkins, D.J., Kniss, D.A. Tumor necrosis factor-alpha promotes sustained cyclooxygenase-2 expression: Attenuation by dexamethasone and NSAIDs. *Prostaglandins* 1997; 54:727-743.

196 Quan, N., Whiteside, M., Herkenham, M. Cyclooxygenase 2 mRNA expression in rat brain after peripheral injection of lipopolysaccharide. *Brain Res* 1998; 802:189-197.

197 Kalaria RN. Cerebral endothelial activation and signal transduction mechanisms during inflammation and infectious disease. *Am J Pathol* 1999; 154:1311-1314.

198 Elmquist, J.K., Breder, C.D., Sherin, J.E., et al. Intravenous lipopolysaccharide induces cyclooxygenase 2-like immunoreactivity in rat brain perivascular microglia and meningeal macrophages. *J Comp Neurol* 1997; 381:119-129.

199 Laflamme, N., Lacroix, S., Rivest, S. An essential role of interleukin-1 beta in mediating NF-kappa B activity and COX-2 transcription in cells of the blood-brain barrier in response to a systemic and localized inflammation but not during endotoxemia. *J Neurosci* 1999; 19:10923-10930.

200 de Vries, H.E., Hoogendoorn, K.H., van Dijk, J., et al. Eicosanoid production by rat cerebral endothelial cells: stimulation by lipopolysaccharide, interleukin-1 and interleukin-6 . *J Neuroimmunol* 1995; 59:1-8.

201 Langhans, W., Harlacher, R., Scharrer, E. Verapamil and indomethacin attenuate endotoxin-induced anorexia. *Physiol Behav* 1989; 46:535-539.

202 Swiergiel, A.H., Smagin, G.N., Dunn, A.J. Influenza virus infection of mice induces anorexia: Comparison with endotoxin and interleukin-1 and the effects of indomethacin. *Pharmacol Biochem Behav* 1997; 57:389-396.

203 Lacroix, S., Rivest, S. Functional circuitry in the brain of immune-challenged rats: Partial involvement of prostaglandins. *J Comp Neurol* 1997; 387:307-324.

204 Simmons, D.L., Wagner, D., Westover, K. Nonsteroidal anti-inflammatory drugs, acetaminophen, cyclooxygenase 2, and fever. *Clin Inf Dis* 2000; 31 Suppl 5:S211-S218.

205 Li, S., Wang, Y., Matsumura, K., et al. The febrile response to lipopolysaccharide is blocked in cyclooxygenase-2(-/-), but not in cyclooxygenase-1(-/-) mice. *Brain Res* 1999; 825:86-94.

206 Matsumura, K., Kaihatsu, S., Imai, H., et al. Cyclooxygenase in the vagal afferents: is it involved in the brain prostaglandin response evoked by lipopolysaccharide? *Auton Neurosci* 2000; 85:88-92.

207 Cao, C.Y., Matsumura, K., Ozaki, M., et al. Lipopolysaccharide injected into the cerebral ventricle evokes fever through induction of cyclooxygenase-2 in brain endothelial cells. *J Neurosci* 1999; 19:716-725.

208 Cao, C.Y., Matsumura, K., Shirakawa, N., et al. Pyrogenic cytokines injected into the rat cerebral ventricle induce cyclooxygenase-2 in brain endothelial cells and also upregulate their receptors. *Eur J Neurosci* 2001; 13:1781-1790.

209 Elmquist, J.K., Saper, C.B. Activation of neurons projecting to the paraventricular hypothalamic nucleus by intravenous lipopolysaccharide. *J Comp Neurol* 1996; 374:315-331.

210 MolinaHolgado, F., Borrell, J., Guaza, C. Effect of endotoxin and interleukin-1 beta on corticotropin-releasing factor and prostaglandin release by rat brainstem slices. *J Neuroendocrinol* 1998; 10:429-436.

211 Ericsson, A., Kovacs, K.J., Sawchenko, P.E. A functional anatomical analysis of central pathways subserving the effects of interleukin-1 on stress-related neuroendocrine neurons. *J Neurosci* 1994; 14:897-913.

212 Simansky, K.J. Serotonergic control of the organization of feeding and satiety. *Behav Brain Res* 1996; 73:37-42.

213 Cunningham, E.T., Wada, E., Carter, D.B., et al. Insitu histochemical-localization of type-I interleukin-1 receptor messenger mRNA in the central-nervous-system, pituitary, and adrenal-gland of the mouse. *J Neurosci* 1992; 12:1101-1114.

214 Clement, H.W., Buschmann, J., Rex, S., et al. Effects of interferon-gamma, interleukin-1 beta, and tumor necrosis factor-alpha on the serotonin metabolism in the nucleus raphe dorsalis of the rat. *J Neur Transmission* 1997; 104:981-991.

215 Gemma, C., Imeri, L., DeSimoni, M.G., et al. Interleukin-1 induces changes in sleep, brain temperature, and serotonergic metabolism. *Am J Physiol* 1997; 41:R601-R606.

216 Mohankumar, P.S., Thyagarajan, S., Quadri, S.K. Interleukin-1-beta Increases 5-hydroxyindoleacetic acid release in the hypothalamus in vivo. *Brain Res Bull* 1993; 31:745-748.

217 Gemma, C., Ghezzi, P., De-Simoni, M.G. Activation of the hypothalamic serotoninergic system by central interleukin-1. *Eur J Pharmacol* 1991; 209:139-140.

218 Ballinger, A., El Haj, T., Perrett, D., et al. The role of medial hypothalamic serotonin in the suppression of feeding in a rat model of colitis. *Gastroenterol* 2000; 118:544-553.

219 MohanKumar, S.M.J., Mohankumar, P.S., Quadri, S.K. Lipopolysaccharide-induced changes in monoamines in specific areas of the brain: blockade by interleukin-1 receptor antagonist. *Brain Res* 1999; 824:232-237.

220 MohanKumar, S.M.J., Mohankumar, P.S., Quadri, S.K. Effects of bacterial lipopolysaccharide on central monoamines and fever in the rat: involvement of the vagus. *Neurosci Lett* 2000; 284:159-162.

221 Hrupka, B.J., Langhans, W. A role for serotonin in lipopolysaccharide-induced anorexia in rats. *Pharmacol Biochem Behav* 2001; 68:355-362.

222 Swiergiel, A.H., Dunn, A.J. Lack of evidence for a role of serotonin in interleukin-1-induced hypophagia. *Pharmacol Biochem Behav* 2000; 65:531-537.

223 Laviano, A., Cangiano, C., Preziosa, I., et al. Serotoninergic block in the ventromedial nucleus of hypothalamus improves food intake in anorectic tumor bearing rats. *Adv Exp Med Biol* 1996; 398:551-553.

224 Chance, W.T., von Meyenfeldt, M., Fischer, J.E. Serotonin depletion by 5,7-dihydroxytryptamine or para-chloroamphetamine does not affect cancer anorexia. *Pharmacol Biochem Behav* 1983; 18:115-121.

225 Rinaman, L. Interoceptive stress activates glucagon-like peptide-1 neurons that project to the hypothalamus. *Am J Physiol* 1999; 277:R582-R590.

226 Comer, J., Rinaman, L. Role of central glucagon-like peptide-1 receptor signaling in lipopolysaccharide-induced fever and anorexia. *FASEB J* 2000; 14:A87.

227 Rinaman, L., Comer, J. Antagonism of central glucagon-like peptide-1 receptors enhances lipopolysaccharide-induced fever. *Auton Neurosci* 2000; 85:98-101.

228 Katafuchi, T., Motomura, K., Baba, S., et al. Differential effects of tumor necrosis actor-alpha and -beta on rat ventromedial hypothalamic neurons in vitro. *Am J Physiol* 1997; 272:R1966-R1971.

229 Plata-Salamàn, C.R., FfrenchMullen, J,M.H. Interleukin-1 beta inhibits Ca2+ channel currents in hippocampal neurons through protein kinase. *C. Eur J Pharmacol* 1994; 266:1-10.

230 Suda, T., Tozawa, F., Ushiyama, T., et al. Interleukin-1 stimulates corticotropin-releasing factor gene expression in rat hypothalamus. *Endocrinol* 1990; 126:1223-1228.

231 Uehara, A., Sekiya, C., Takasugi, Y., et al. Anorexia induced by interleukin 1: involvement of corticotropin-releasing factor. *Am J Physiol* 1989; 257:R613-R617.

232 Watanabe, T., Morimoto, A., Sakata, Y., et al. ACTH response induced by interleukin-1 is mediated by CRF secretion stimulated by hypothalamic PGE. *Experientia* 1990; 46:482-484.

233 Sonti, G., Ilyin, S.E., Plata-Salamàn, C.R. Neuropeptide Y blocks and reverses interleukin-1 beta-induced anorexia in rats. *Peptides* 1996; 17:517-520.

234 Chance, W.T., Balasubramaniam, A., Fischer, J.E. Neuropeptide Y and the development of cancer anorexia. *Ann Surg* 1995; 221:579-587.

235 Kang, M., Yoshimatsu, H., Chiba, S., et al. Hypothalamic neuronal histamine modulates physiological responses induced by interleukin-1 beta. Am J Physiol 1995; 38:R1308-R1313.

236 McCarthy, H.D., McKibbin, P.E., Perkins, A.V., et al. Alterations in hypothalamic NPY and CRF in anorexic tumor-bearing rats. *Am J Physiol* 1993; 264:E638-E643.

237 Inui, A. Cancer anorexia-cachexia syndrome: Are neuropeptides the key? *Cancer Res* 1999; 59:4493-4501.

238 Swiergiel, A.H., Burunda, T., Patterson, B., et al. Endotoxin- and interleukin-1-induced hypophagia are not affected by adrenergic, dopaminergic, histaminergic, or muscarinic antagonists. *Pharmacol Biochem Behav* 1999; 63:629-637.

239 Catania, A., Lipton, J.M. The neuropeptide alpha-melanocyte-stimulating hormone: a key component of neuroimmunomodulation. *Neuroimmunomodulation* 1994; 1:93-99.

240 Catania, A., Suffredini, A.F., Lipton, J.M. Endotoxin causes release of alpha-melanocyte-stimulating hormone in normal human subjects. *Neuroimmunomodulation* 1995; 2:258-262.

241 Lipton, J.M., Catania, A. Mechanisms ofanti-inflammatory action of the neuroimmunomodulatory peptide alpha-MSH. *Ann N Y Acad Sci* 1998; 840:373-380.

242 Tritos, N.A., Maratos-Flier, E. Two important systems in energy homeostasis: melanocortins and melanin-concentrating hormone. *Neuropeptides* 1999; 33:339-349.

243 Fan, W., Boston, B.A., Kesterson, R.A., et al. Role of melanocortinergic neurons in feeding and the agouti obesity syndrome. *Nature* 1997; 385:165-168.

244 Huang, Q.H., Hruby, V.J., Tatro, J.B. Role of central melanocortins in endotoxin-induced anorexia. *Am J Physiol* 1999; 276:R864-R871.

245 Marks, D.L., Ling, N., Cone, R.D. Role of the central melanocortin system in cachexia. *Cancer Res* 2001; 61:1432-1438.

Chapter 10

CYTOKINES EFFECTS ON LEARNING AND MEMORY

John M. Petitto, Martin J. Repetto, Zhi Huang, Ray D. Beck, Jr. and David A. Hartemink
McKnight Brain Institute, University of Florida, College of Medicine, Gainesville, Florida

1. INTRODUCTION

One of the earliest observations suggesting that cytokines could influence cognitive function in humans came from cancer studies where IL-2 immunotherapy was used to treat some types of neoplasias. Treatment with this pivotal immunoregulatory cytokine induced prominent, untoward neuropsychiatric side effects with patients exhibiting cognitive dysfunction (1,2). Seminal basic studies showed that peripheral immunization led to activation of hypothalamic neurons, and also demonstrated that lymphocytes had the capacity to produce neuropeptides (for reviews see 3,4). This sparked considerable growth in this area of research, and it is now well established that cytokines can communicate with endocrine and brain cells. Just as researchers were surprised several decades ago to discover that digestive peptides such as CCK and VIP were in the brain, cytokines and their receptors once believed to be derived solely from lymphoid cells have been identified in normal brain. They function as neuromodulators and neurotrophic factors, as well as mediators of immune-like responses involved in CNS pathology. As described in subsequent sections of the chapter, in the case of IL-2, a number of basic investigations have shown that this cytokine can modify various neurobiological processes associated with learning and memory. Thus, both clinical and basic studies have led to our current knowledge of the role that this pivotal immunoregulatory cytokine plays in regulating cognitive neural circuitry. Whereas the focus of clinical treatment trials has been primarily on the Th-1 cytokines, IL-2 and interferon-α, basic research in this area has been somewhat skewed towards the examination of IL-1 and other pro-inflamatory cytokines including TNF-α and IL-6. Such studies have laid the foundation for the growing body of research that has sought to dissect the potential mechanisms whereby cytokines can influence complex neurobiological processes involved in understanding learning and memory, and neuropsychiatric diseases affecting this domain of behavior.

The aim of this chapter is to examine some of the salient literature that addresses the role of cytokines in fundamental processes in learning and memory. We highlight some of the key findings, fundamental processes, and ideas guiding research in this area, and their potential relevance to the pathogenesis of certain neuropsychiatric syndromes and diseases. This short chapter is not designed to comprehensively catalogue all of the available research assessing cytokine effects on cognitive behavior, but rather seeks to convey some of the prinicipal, methodological issues and ideas guiding ongoing and future research. Selected studies are discussed that illustrate the essential phenomenology and mechanisms that are either known or theorized to link cytokine function to cognition. The cytokines IL-1 and IL-2 are the major focus of our discussion as they have been the most extensively studied cytokines in the CNS. Since the focus of research in our laboratory has been on the effect of IL-2 and brain IL-2 receptors, a number of examples from neuroimmunological studies of IL-2 are used to indicate how cytokines may influence learning and memory. In light of the early stage of our current understanding of the relationship between cytokines and cognition, and what is known about the redundant actions of subsets of cytokines outside of the brain, other cytokines will likely prove to possess equally significant neurobiological actions. The potential clinical application of preclinical information gleened from this area of research to normal and pathophysiological cognitive processes in humans and to neuropsychiatric disorders is also discussed.

2. CYTOKINES, THEIR RECEPTORS, AND THE HIPPOCAMPUS

Cytokines derived from peripheral immune cells (and in some cases perhaps other tissues), do not readily cross the blood-brain-barrier. The mechanisms by which this occurs varies for different cytokines (e.g., active vs. passive transport) and influences the degree to which they enter the CNS (5). More recently, work from the laboratories of Maier & Watkins (6) and others have shown that the afferent sensory fibers of the vagus can carry signals initiated by IL-1 to brainstem areas (e.g., nucleus tractus solitarius). It remains to be determined whether other cytokines can signal the CNS via the vagus or other afferent nerve in the periphery, and how such events might influence learning and memory.

There is evidence that some cytokines and cytokine receptors are synthesized by endogenous brain cells. In the brain, some classical immune cytokines (e.g., IL-1) have been found to have neuromodulatory and neurotrophic effects that are limited to very specific neural pathways, while at the same time exerting more general effects on the brain's endogenous immune cells (e.g., microglia, astrocytes). Cytokine receptors are generally more readily detectable in the brain than cytokines themselves. Although gene expression studies have been successful in detecting cytokine receptors, and under some conditions cytokines as well, it has been more challenging to unequivocally detect cytokines and cytokine receptors using immunohistocytochemistry (or to reliably detect cytokine receptors using radioligand receptor binding or autoradiography) (7). Receptors for IL-2 are expressed throughout CA1-CA4 of the hippocampus and dentate gyrus (7-10), and IL-1R (11), IL-1R antagonist (12) and IL-6R (13) are also detectable in the rodent dentate gyrus by in situ hybridization. Thus, receptors for these cytokines in the hippocampus place them in a position to influence learning and memory. Interestingly, IL-1 mRNA and protein have been found to be increased in the hippocampus following some forms of peripheral immune system activation such as occurs following LPS administration, suggesting a biological basis by which peripheral infection may influence cognition (14). As described below, several of these same cytokines that have been found to target receptors in the hippocampus have the capacity to modify learning and memory performance and to modulate neurobiological processes known to mediate these behaviors.

3. CYTOKINES EFFECTS ON LEARNING AND MEMORY

Several investigations in animals provide compelling evidence that IL-1 can modify learning and memory performance. Using contextual fear condition as a test of hippocampal-dependent learning, Pugh and colleagues (15) showed that I.C.V. adminstration of IL-1β impaired contextual fear conditioning. The same dosages of this cytokine did not significantly alter auditory-cue fear conditioning, a form of conditioning that is not dependent on the hippocampus. Gilbertini, et al (16) also demonstrated that systemically dosed LPS disrupts learning and memory performance as well, and this disruption was antagonized by antibody to the IL-1β. Along these lines, pyrogens which have reported to induce sickness behavior (e.g., fever, decreased activity and exploration, reduced social interaction, depressive-like symptoms), sometimes elicit reduced cognitive performance (17,18). Examining the effects of various cytokines, Bianchi et al. (19) found that IL-

1α augmented the effects of scopolamine-induced amnesia for passive avoidance (administration of the cytokine alone had no significant effects on this behavior, however).

As noted earlier, IL-2 immunotherapy has been found to elicit cognitive disturbances after repeated administration in treatment trials (1,2,20). Doses of the cytokine used to treat cancer are significantly above the physiological range, and repeated dosing appears to account for the untoward side effects. Chronic dosing of IL-2 in aging mice produces memory deficits and neuronal damage, which is selective to hippocampus (21). Whether older patients receiving IL-2 immunotherapy for cancer treatment are more vulnerable to such neurotoxic effects is unknown. Several complementary basic studies in animals have confirmed clinical observations that exogenous IL-2 administration can alter parameters of cognitive functioning. IL-2 has been found in several studies to enhance the effects of scopolamine-induced amnesia in a passive paradigm (19,22). This effect was antagonized completely by acetylcarnatine, a drug that enhances cholinergic activity (the actions of IL-2 on cholinergic neurotranmission in the hippocampus are described in the section that follows). Hanisch and colleagues (23) found that chronic I.C.V. infusion of human IL-2 in rats produced transient changes in behavior in the Morris water-maze test. This was accompanied by changes in hippocampal muscarinic cholinergic receptors. In a well-crafted set of experiments, Lacosta et al. (24) assessed the effect of exogenous IL-2 on learning and memory performance in mice in the Morris water-maze. Using standard submerged platform testing procedures where animals must locate a submerged platform that is always in the same position in the pool, they found that neither acute nor chronic IL-2 administration had an effect. The investigators then performed a modification of the standard submerged platform testing regimen where they changed the location of the submerged platform on each testing day (the platform remained in the same location however throughout each testing day). Using this modification they found that the between-trial improvement seen in saline treated mice was impaired in the IL-2 treated mice. This impairment was found in animals that received chronic dosing of IL-2, but not in animals receiving the acute dosing regimen. Thus, these data indicate that chronic IL-2 treatment interfered with the working component of spatial memory.

It is no doubt challenging to conduct methodologically sound experimental studies of this nature in humans and there are few data in this regard. Reichenberg and coworkers (25) recently performed such an experimental study in 20 healthy human volunteers. They examined the effects of a low doses of endotoxin using a randomized, double-blind, crossover design (endotoxin vs. saline) in order to test the effects of this challenge on several behavioral and biological endpoints. These endpoints included assessments of measures of cognitive function and levels of several inflammatory cytokines in peripheral blood. As expected, endotoxin increased circulating levels of IL-6, TNF-α and its soluble receptor, and IL-1 receptor antagonist. Neuropscyhological testing showed that the endotoxin induced significant decreases in verbal and nonverbal memory functions. Higher cytokine levels were associated with greater decreases in memory performance (the magnitude of these correlations were substantial for a clinical study, with r values ranging from 0.49 to 0.60). Although it was not possible to determine directly whether and which of these cytokines elicited the observed cognitive deficits, their findings are nonetheless consistent with data from the animal literature described earlier. This study therefore suggests that endotoxin-induced cytokine release likely elicited the observed alterations in cognitive status in these healthy volunteers.

4. EFFECTS OF CYTOKINES ON HIPPOCAMPAL LONG-TERM POTENTIATION AND OTHER RELEVANT MEASURES

Hippocampal long-term potentiation (LTP) is believed to be an important neurobiological mechanism involved in learning and memory storage at the cellular and molecular level. LTP has been found to correlate with performance in contextual fear conditioning (26), moreover blocking NMDA-dependent LTP results in alterations in spatial learning in the Morris water-maze (27). There is a substantial literature that has established a relationship between cytokines and LTP. This work suggests that some cytokines may play a role in synaptic modulation. Jankowsky and Patterson (28) have written an excellent comprehensive review of the literature on cytokine involvement in LTP. Several significant findings emerged from a study of IL-1β by Schneider and colleagues (29). This study showed that IL-1β mRNA increased over the course of LTP induction in the Schaffer collateral pathway in vitro, and that gene expression is also elevated in vivo following LTP induction. Jankowsky et al. (30) used RNase protection and RT-PCR assays to study the endogenous expression of 18 cytokines and growth factors and their receptors following LTP induction in the Schaffer collateral pathway. They found that gene expression levels of various cytokines were increased in vitro, however these increases were absent when electrodes were placed in vivo for an extended period of time (3 weeks) prior to testing. IL-6 gene expression, however, was found to be upregulated nearly 20-fold by LTP induction in vivo and was localized to the site of stimulation by in situ hybridization. Reduced LTP was found in the dentate gyrus of mice that overexpressed IL-6 (31). Exogenously administered cytokines have been found to modify LTP as well. Exogenous IL-6, TNF-α, and IL-2, for example, decrease LTP in vitro (32-34). IL-1 also inhibits LTP, and this inhibition is blocked by IL-1 antagonist (29,35,36). IL-1 antagonists have been found to decrease IL-1β as expected; however, surprisingly, the antagonist reduced IL-1ß levels and also decreased LTP (29).

Much evidence implicates cytokines as neuromodulators in areas of the mammalian brain involved in cognition. In the mature brain, cytokines have been found to modify release of various neurotransmitter, to modulate ion channels and intracellular pathways (e.g., MAP kinase, PKC) as well as mediate various neurobiological processes that serve as substrates of learning and memory (37,38). In the case of IL-2, for example, this cytokine can modulate calcium currents from dissociated adult hippocampal rat neurons in vitro (39). Septohippocampal cholinergic neurons play an integral role in many forms of learning and memory. IL-2 has been shown to modulate choline acetyltransferase (ChAT) activity in fetal septal cholinergic neurons and is the most potent modulator of acetylcholine release by cholinergic septohippocampal projection neurons. Quirion and colleagues at McGill University have demonstrated in a series of studies that IL-2 is among the most potent modulators of acetylcholine release (40). In hippocampal slices, IL-2 modulates acetylcholine in a dose-dependent biphasic manner, potentiating release at very low (fM) concentrations and inhibiting release at higher (nM) concentrations (41). In contrast, cholinergic interneurons in the striatum do not respond to IL-2. Because the hippocampus is critical in spatial learning and memory consolidation, it has been postulated that IL-2 may alter memory processing via interactions with septohippocampal cholinergic nerve terminals in hippocampus (42). Interestingly, IL-2 levels were found to be elevated in post-mortem hippocampi of Alzheimer's disease patients (43). Likewise, other cytokines have also been shown to have important actions in the hippocampus (38).

5. NEUROTROPHIC ACTIONS OF CYTOKINES IN THE HIPPOCAMPUS: IMPLICATIONS FOR NEURODEVELOPMENT

A number of cytokines are produced by glial cells and exert their effects on both neurons and glial cells. Cytokines also have trophic effects on glial cells and neurons, and under certain condition they can also exhibit toxicity. Although the majority of the literature of cytokines and the brain have attempted to understand how cytokines influence the brain's response to injury and disease, less attention has been given to the role of cytokines in neurodevelopmental processes.

Campbell et al. (44) found that transgenic mice with CNS expression of interferon-α developed significant loss of cholinergic neurons and changes in hippocampal excitability. Using IL-2 as an example again, there are data showing IL-2 may contribute to the normal growth and survival of hippocampal neurons (45-48). At very low concentration (picomolar), IL-2 produces receptor-mediated increases in ChAT activity in embryonic septal cholinergic neurons (49). We therefore sought to determine if the absence of IL-2 during development would influence hippocampal neurodevelopment, and learning and memory. Brain IL-2 receptors are enriched in the hippocampal formation, a region that is critical for the acquisition and consolidation of spatial learning and memory. Thus, we tested the hypothesis that mice lacking IL-2 would exhibit alterations in hippocampal-dependent learning and neurodevelopment. Compared with C57BL/6-IL-$2^{+/+}$ wild-type mice, we observed that C57BL/6-IL-$2^{-/-}$ gene knockout mice had markedly impaired spatial learning and memory in the Morris water maze (50). No significant deficits in parameters of learning and memory performance were found in severe combined immunodeficient (SCID) mice, however, suggesting that the impaired spatial learning and memory exhibited by IL-2 knockout mice is not attributable to generalized immunodeficiency resulting from the absence of endogenous IL-2. Examination of other domains of behavioral performance showed that the IL-2 knockout and wild-type mice did not differ in measures of fearfulness or locomotor activity in an elevated plus maze, or in reflexive startle responses to auditory stimuli, although prepulse inhibition of acoustic startle (PPI) was increased significantly in IL-2 knockout mice. The spatial learning and memory impairment in IL-2 knockout mice was accompanied by reductions in hippocampal infrapyramidal mossy neuronal fiber length, a factor shown previously to correlate positively with spatial learning ability. These findings indicate that, in addition to being a pivotal cytokine in immune regulation, IL-2 may play a role in the development and regulation of brain neurons involved in spatial learning and memory. Since IL-2 knockout mice develop active systemic autoimmunity, it cannot be ruled out that the observed impairments could be due to peripheral immune processes impacting the brain. This possibility is currently being examined in our lab.

6. CYTOKINE GENE ALLELES AND BEHAVIORAL QUANTITATIVE TRAIT LOCI (QTL): IS THERE A LINKAGE?

Research seeking to identify genetic mechanisms underlying brain-immune interactions has been limited to date and is an important area for future studies. The available literature suggests that subsets of genes involved in the expression of complex behavioral and immunological traits may be physically linked, interactive at the level of the genome, or shared by the brain and immune system. Only a few clinical studies on the topic have been published. Narcolepsy, for example, has been found to be associated with specific HLA alleles (51). Since it is a formidable challenge to address such questions in humans, animal models can provide valuable research paradigms to disentangle multifactorial behavioral-immune relationships. Enormous strides are being made in molecular genetics and there is a striking homology between the organization of the human and mouse (and more recently rat) genomes. Thus, genes of interest, specific chromosomal loci, and linkages identified in the rodent may be found in homologous regions of the human genome. Functional genetic relationships between behavior and immunity are not unprecedented. Several lines of investigation address the potential evolutionary significance of genetic mechanisms linking adaptive behavioral and immunological processes in vertebrate species.

We recently sought to assess whether known mouse candidate immunological genes and/or immune-related disease susceptibility loci (an interval on a chromosome where the gene(s) involved have not been identified or definitively determined) might be physically linked to quantitative trait loci (QTL) (52). QTL are individual loci involved in controlling complex traits where none of the individual genes involved is sufficient for the expression of the complex trait. To determine if specific candidate immunological genes and/or immune disease susceptibility loci might be physically linked to QTL for cognitive behavioral traits in animals, we recently examined the Mouse Genome Database for possible linkages. This assessment provided intriguing evidence that behavioral QTL involved in learning and memory may be linked to a subset of immune genes or susceptibility loci. Our assessment showed several potential genetic linkages between QTL for contextual fear conditioning and cytokine and cytokine receptor gene alleles. Contextual fear conditioning QTL on chromosome 1, for example, were within linkage range with alleles for IL-1 receptors type I and II and IL-10, a QTL on chromosome 10 with interferon receptor-γ, a QTL on chromosome 16 with IL-10 receptor-β and interferon receptors α, β, and γ type-2. The functional significance of such potential linkages remains to be determined, and could provide important new insights about how cytokines and their receptors may be related to cognitive behavioral traits.

7. CLINICAL RELEVANCE FOR NEUROPSYCHIATRIC CONDITIONS

There have been many reports of elevated levels of various cytokines associated with major psychiatric disorders. Unfortunately, there is a lack of compelling evidence that elevated levels of cytokines or serum soluble receptors (e.g., IL-2 receptor-α chain) are a valid reflection of prior or ongoing immunopathological or autoimmune-like processes affecting the brain in individuals with neuropsychiatric disorders. Nonetheless, it is noteworthy that lymphocytes expressing activation markers and certain adhesion molecules (molecules on the surface of lymphocytes that allow them to pass across endothelial cells into organ tissues including brain) are the cells that have the capacity to cross the blood-brain-barrier and enter the nondiseased brain (53). In those subsets of individuals with neuropsychiatric conditions that exhibit increased levels of activated lymphocytes and expression of certain adhesion molecules, it is therefore plausible that these lymphocytes have a greater opportunity to enter the brain. There, they may produce cytokines that could alter learning and memory performance, or even neurodevelopmental processes in the CNS (e.g., alter aspects of hippocampal neurodevelopment such as that seen in the brains of schizophrenic patients).

Support for the theory that elevated cytokine levels may be involved in the pathogenesis of various neuropsychiatric diseases (e.g., schizophrenia) comes largely from nonspecific correlative data. Frequently, mean values of neuropsychiatric patient groups for various laboratory measures of immune status including cytokine concentrations differ significantly from the group means of normal control subjects, however, the actual values themselves fall within the range of normal limits. Using studies of schizophrenia as an example, subsets of schizophrenic individuals including never medicated subjects have been found to have elevated levels of IL-2 in peripheral blood (54). Despite the strong corroborating evidence from the basic studies described earlier, it is unclear if elevated peripheral cytokine seen in schizophrenia, Alzheimer's disease or other neuropsychiatric disorders represent evidence of pathophysiological steps involved in these disorders or are epiphenomena.

Great interest has been given to the hypothesis that early neurodevelopmental alterations may account for key pathophysiological abnormalities seen in the mature brain of individuals with schizophrenia.

Epidemiological findings have contributed to a revisitation of the idea that prenatal viral infection or the maternal immune response to such a putative agent may serve as an environmental trigger for the expression of schizophrenia in individuals with genetic loading for the disorder (55). It is plausible, for example, that during early development an event like viral infection or birth trauma in a genetically susceptible individual could disrupt the normal timing at which IL-2 may stimulate neuronal growth and migration. If this occurred in a region like the hippocampus where IL-2 receptors are enriched, this could contribute to the alterations in this region (e.g., abnormal orientation of subsets of hippocampal neurons) that are seen in post-mortem brains of individuals with schizophrenia. Immunological disturbances in the peripheral immune system during development may also lead to abnormalities in neurodevelopment. There is considerable data, for example, indicating that thymic disturbance in vulnerable individuals during development can alter immune function including that of cytokines and induce autoimmune reactions that could impact on brain development (56). These and other processes that disrupt normal cytokine function could conceivably contribute to the pathogenesis of some neuropsychiatric disorders, and alter learning and memory processing.

8. CONCLUSION

Determining how cytokines can influence brain function may provide essential new insights into processes involved in the pathogenesis of some neuropsychiatrc conditions. Knowledge of basic immunological processes in the CNS has recently led to the discovery of immune-based drug therapies (e.g., new selective cyclooxygenase inhibitors) with the potential to slow the rate of cognitive decline in patients with Alzheimer's disease. Despite considerable efforts to relate cytokine changes to cognitive alterations in certain major neuropsychiatric disorders, it has been difficult to establish unequivocal connections between cytokine dysregulation and CNS pathology. We believe that the application of molecular genetic approaches to examine the relationship of cytokines to learning could prove to be a particularly fruitful new area of inquiry for clinical studies. In this regard, a new gene of the IL-1 receptor family that is highly expressed in the hippocampus has been found in X-linked mental retardation (57). Several cytokine gene polymorphisms have recently been associated with Alzheimer's disease and symptom progression (58). Thus, it is possible that future research could provide vital new clues about how cytokine and cytokine receptor gene alleles influence brain processes associated with cognition. These and other approaches could eventually lead to novel new forms of intervention.

ACKNOWLEDGEMENT

This work was supported by PHS grants RO1 NS34623 and RO1 NS38179 awarded to Dr. Petitto.

REFERENCES

1 Denicoff, K.D., Rubinow, D.R., Papa, M.Z., et al. The neuropsychiatric effects of treatment with interleukin-2 and lymphokine-activated killer cells. *Ann. Int. Med.* 1987; 107: 293-300.

2 West, W.H., Tauer, K.W., Yannelli, J.R., et al. Constant-infusion recombinant interleukin-2 in adoptive immunotherapy of advanced cancer. *NEJM* 1987; 316: 898-905.

3. Besedovsky, H.O., del Rey, A., Klusman, I., et al. Cytokines as modulators of the hypothalamus-pituitary-adrenal axis. *J. Ster. Biochem. Mole. Biol.* 1991; 40: 613-618.

4. Blalock, J.E. The syntax of immune-neuroendocrine communications. *Immunol. Today* 1994; 15: 504-511.

5. Pan, W. and Kastin, A.J. Penetration of neurotrophins and cytokines across the blood-brain/blood-spinal cord barrier . *Adv. Drug Deliv. Rev.* 1999; 36: 291-298

6. Maier, S.F. and Watkins, L.R. Cytokines for psychologists: Implications of bidirectional immune-to-brain communication for understanding behavior, mood and cognition. *Psychol. Rev.* 1998; 83-107

7. Petitto, J.M. and Huang, Z. Cloning the full-length IL-2/15 receptor-β cDNA sequence from mouse brain: evidence of enrichment in hippocampal formation neurons. *Reg. Pep.* 2001; 98: 77-87.

8. Lapchak, P.A., Araujo, D.M., Quirion, R., et al. Immunoautoradiographic localization of interleukin 2-like immunoreactivity and interleukin 2 receptors (Tac antigen-like immunoreactivity) in the rat brain. *Neurosci.* 1991; 44: 173-184..

9. Petitto, J.M. and Huang, Z. Molecular cloning of a partial cDNA of the interleukin-2 receptor-β in normal mouse brain: in situ localization in the hippocampus and expression by neuroblastoma cells. *Brain Res.* 1994; 650: 140-145.

10. Petitto, J.M., Huang, Z., Raizada, M., et al. Molecular cloning of the cDNA coding sequence of IL-2 receptor-γ (γ_c) from human and murine forebrain: expression in the hippocampus in situ and by brain cells in vitro. *Mole. Brain Res.* 1998; 53:152-62.

11. Cunningham, E.T.J., Wada, E., Carter, D.B., et al. In situ histochemical localization of type I interleukin-1 receptor messenger RNA in the central nervous system, pituitary, and adrenal gland of the mouse. *J. Neurosci.* 1992; 12: 1101-1114.

12. .Licinio, J., Wong, M.L., and Gold, P.W. Localization of interleukin-1 receptor antagonist mRNA in rat brain. *Endocrinology* 1991; 129: 562-564.

13. Schobitz, B., Voorhuis, D.A. and De Kloet, E.R. Localization of interleukin-6 mRNA and interleukin-6 receptor mRNA in rat brain. *Neurosci. Lett.* 1992; 136: 189-192.

14. Layę, S., Parnet, P. Goujon, E., et al. Peripheral administration of lipopolysaccharide induces the expression of cytokine transcripts in the brain and pituitary of mice. *Brain research. Mole. Brain Res* 1994; 27: 157-62.
15. Pugh, C.R., Nguyen, K.T., Gonyea, J.L., et al. Role of interleukin-1 beta in impairment of contextual fear conditioning caused by social isolation. *Behav. BrainRres.* 1999; 106: 109-18.
16. Gibertini, M., Newton, C., Klein, T.W., et al. Legionella pneumophila-induced visual learning impairment reversed by anti-interleukin-1 beta. *Proceed. Soc. Exp. Biol. Med.* 1995; 210: 7-11.
17. Aubert, A., Vega, C., Dantzer, R., et al. Pyrogens specifically disrupt the acquisition of a task involving cognitive processing in the rat. *Brain, Behav. Iimmun.* 1995; 9: 129-48.
18. Dantzer, R., Bluthe, R.M., Gheusi, G., et al. Molecular basis of sickness behavior. *Ann. NY Acad. Sci.* 1998; 856: 132-138.
19. Bianchi, M., Sacerdote, P., Panerai, A.E. Cytokines and cognitive functions in mice. *Biol. Signals. Recept.* 1998; 7: 45-54.
20. Capuron, L., Ravaud, A., Radat, F., et al. Effects of interleukin-2 and alpha interferon cytokine immunotherapy on the mood and cognitive performance of cancer patients. *Neuroimmunomodul.* 1998; 22: 9.
21. Nemni, R., Iannaccone, S., Quattrini, A., et al. Effect of chronic treatment with recombinant interleukin-2 on the central nervous system of adult and old mice. *Brain Res.* 1992; 591: 248-252.
22. Bianchi, M. and Panerai, A.E. Interleukin-2 enhances scopolamine-induced amnesia and hyperactivity in the mouse. *Neuroreport* 1993; 4: 1046-1048.
23. Hanisch, U., Neuhaus, J., Rowe, W., et al. Neurotoxic consequences of central long-term administration of interleukin-2 in rats. *Neurosci.* 1997; 79: 799-818.
24. Lacosta, S., Merali, Z., Animan, H. Influence of acute and repeated interleukin-2 administration on spatial learning, locomotor activity, exploratory behaviors and anxiety. *Behav. Neurosci.* 1999; 113: 1030-1041.
25. Reichemberg, A., Yirmiya, R., Schuld, A., et al. Cytokine-associated emotional and cognitive disturbances in humans. *Arch. Gen. Psychiat.* 2001; 58:445-452.
26. Maren, S., De Oca, B. and Fanselow, M.S. Sex differences in hippocampal long-term potentiation (LTP) and Pavlovian fear conditioning in rats: positive correlation between LTP and contextual learning. *Brain Research* 1994; 24: 66: 25-34.
27. Morris, R.G.M. Synaptic plasticity and learning: selective impairment of learning in rats and blockade of long–term potentiation in vivo by the N-methyl-D-aspartate receptor anatagonist AP5. *J. Neurosci.* 1989; 9: 3040-3057.
28. Jankowsky, J., Patterson, P. Cytokine and growth factor involvement in long-term potentiation. *Mole. Cell. Neurosci.* 1999; 14:273-286.
29. Schneider, H., Pitossi, F., Balschun, D., et al. A neuromodulatory role of interleukine-1b in the hippocampus. *Proceed. Natl. Acad. Sci. USA* 1998; 95:7778-7783.
30. Jankowsky, J.L., Derrick, B.E., Patterson, P.H. Cytokine responses to LTP induction in the rat hippocampus: a comparison of in vitro and in vivo techniques. *Learning & Memory* 2000; 7: 400-12.
31. Bellinger, F.P., Madamba, S.G., Campbell, I.L., et al. Reduced long-term potentiation in the dentate gyrus of transgenic mice with cerebral overexpression of interleukin-6. *Neurosci. Lett.* 1995; 198: 95-8.\
32. Li, A., Katafuchi, T., Oda, S., et al. Interleukin-6 inhibits long term potentiation in rat hippocampal slices. *Brain Res.* 1997; 748:30-38.

33. Tançredi, V., Zona, C., Velotti, F., et al. Interleukin-2 suppresses established long-term potentiation and inhibits its induction in the rat hippocampus. *Brain Res.* 1990; 525:149-151.
34. Tancredi, V., D'Archangelo, G., Grassi, F., et al. Tumor necrosis factor alters synaptic transmission in rat hippocampal slices. *Neurosci. Lett.* 1992; 146:176-178.
35. Cunningham, A.J., Murray, C.A., O'Neil, L.A., et al. Interleukin-1β (IL-1β) and tumour necrosis factor (TNF) inhibit long-term potentiation in the rat dentate gyrus in vitro. *Neurosci. Lett.* 1996; 203:17-20.
36. Katsuki, H., Nakai, S., Hirai, Y., et al. Interleukin-1β inhibits long-term potentiation in the CA3 region of mouse hippocampal slices. *Eur. J. Pharmacol.* 1990; 181:323-326.
37. Dantzer, R., Aubert, A., Bluthe, R.M., et al. Mechanisms of the behavioural effects of cytokines. *Adv. Exp. Med. Biol.* 1999; 461:83-105.
38. Ader, R., Felten, D., Cohen, N. (Eds.) *Psychoneuroimmunology, 3rd Edition*, Academic Press, San Diego, CA, 2001.
39. Plata-Salaman, C.R., Ffrench-Mullen, S.M. Interleukin-2 modulates calcium currents in dissociated hippocampal CA1 neurons. *Neuroreport.* 1993; 4: 579-581.
40. .Hanisch, U.K., Seto, D., and Quirion, R. Modulation of hippocampal acetylcholine release: a potent central action of interleukin-2. *J. Neurosci.* 1993; 13(8): 3368-3374.
41. Hanisch, U.K., Seto, D., Quirion, R. Modulation of hippocampal acetylcholine release: a potent central action of interleukin-2. *J. Neurosci.* 1993; 12: 3368-3374.
42. Seto, D., Satyabrata, K. and Quirion, R. Evidence for direct and indirect mechanisms in the potent modulatory action of interleukin-2 on the release of acetylcholine in rat hippocampal slices. *Brit. J. Pharmcol.y* 1997; 120: 1151-1157.
43. Hanisch, U.K. and Quirion R. Interleukin-2 as a neuroregulatory cytokine. *Brain Res. Rev.* 1996; 21: 246-284.
43. Araujo, D.M., Lapchak, P.A. Induction of immune system mediators in the hippocampal formation in Alzheimer's and Parkinson's diseases: Selective effects on specific interleukin receptors. *Neuroscience* 1994; 61: 745-754.
44. Campbell, I.L., Krucker, T., Steffensen, S., et al. Structural and functional neuropathology in transgenic mice with CNS expression of IFN-alpha. *Brain Res.* 1999; 835: 46-61.
45. Araujo, D.M., Lapchak, P.A., Collier, B., et al. Interleukin-2 interacts with opioid neuroregulatory systems in the rat hippocampus. *Prog. Clin. Biol. Res.* 1990; 328: 355-358.
46. Awatsuji, H., Furukawa, Y., Nakajima, M., et al. Interleukin-2 as a neurotrophic factor for supporting the survival of neurons cultured from various regions of fetal rat brain. *J. Neurosci. Res.* 1993; 35: 305-311.
47. Sarder, M., Saito, H. and Abe, K. Interleukin-2 promotes survival and neurite extension of cultured neurons from fetal rat brain. *Brain Res.* 1993; 625: 347-350.
48. Sarder, M., Abe, K., Saito, H., et al. Comparative effect of IL-2 and IL-6 on morphology of cultured hippocampal neurons from fetal rat brain. *Brain Res.* 1996; 715: 9-16.
49. Mennicken, F. and Quirion, R. Interleukin-2 increases choline acetyltransferase activity in septal-cell cultures. *Synapse* 1997; 26: 175-183.
50. Petitto, S.M., McNamara, R., Gendreau, P.L. et al. Impaired learning and memory and altered hippocampal neurodevelopment resulting from IL-2 gene deletion. *J. Neurosci Res.* 1999; 56: 441-446.
51. Kadotani, H., Faraco, J. and Mignot E. Genetic studies in the sleep disorder narcolepsy. *Genome Res.* 1998; 8:427-34.

52. Petitto, J.M. "Behavioral genetics and immunity". In *Psychoneuroimmunology,* Third Edition. Ader, Felten and Cohen eds. San Diego, CA: Academic Press, 2001, pp. 173-186.
53. Hickey, W.F., Hsu, B.L., and Kimura, H. T-lymphocyte entry into the central nervous system. *J. Neurosci. Res.* 1991; 28: 254-260.
54. McAllister, C.G., van-Kammen, D.P., Rehn, T.J., et al. Increases in CSF levels of interleukin-2 in schizophrenia: effects of recurrence of psychosis and medication status. *Am. J. Psychiat.* 1995; 152: 1291-1297.
55. Ganguli, R., Brar, J.S., Chengappa, K.N., et al. Autoimmunity in schizophrenia: a review of recent findings. *Ann. Med.* 1993; 25:489-496.
56. Nishiyama, N. Thymectomy-induced deterioration of learning and memory. *Cell. Mole. Biol.* 2001; 47: 161-165.
57. Carrie, A., Jun, L., Bienvenu, T., et al. A new member of the IL-1 receptor family highly expressed in hippocampus and involved in X-linked mental retardation. *Nature Genetics* 1999; 23: 25-31.
58. Kolsch, H., Ptok, U., Bagli, M., et al. Gene polymorphisms of interleukin-1alpha influence the course of Alzheimer's disease. *Ann.Neurol.* 2001; 49: 818-9

Chapter 11

CYTOKINE-ELICITED SENSITIZATION: *NEUROENDOCRINE, NEUROTRANSMITTER AND BEHAVIORAL RESPONSES*

Shawn Hayley[1], Zul Merali[2] and Hymie Anisman[1]
[1]Institute of Neuroscience, Carleton University and [2]School of Psychology and Department of Cellular and Molecular Medicine, University of Ottawa, Ottawa, Canada

1. INTRODUCTION

In addition to having short-term effects, stressors may proactively augment the response to subsequently encountered insults (sensitization) and may thus have protracted repercussions on health risk. It is known, for instance, that stressors alter central neurotransmitter activity and activate hypothalamic-pituitary-adrenal (HPA) axis activity (1). Upon later encounters with aversive stimuli, even of diminished severity, such neurochemical alterations may be exaggerated and more readily induced (2), and may come to impact on a variety of pathological states, including those related to mood disorders, cardiovascular illness, and immune processes (3,4).

In considering the impact of aversive events, it may be appropriate to broaden our definitions of what constitutes a stressor. Aversive stimuli are typically thought of as comprising "processive" stressors (i.e., those that involve appraisal of a situation or higher order sensory cortical processing, including events such as exposure to a novel environment, predators, or conditioned fear cues), which are either psychogenic or neurogenic (being purely psychological in nature or involving physcal or painful stimuli, respectively). It has been suggested, as well, that systemic (e.g. metabolic) insults, such as bacterial or viral infection (5), induce several neurochemical alterations reminiscent of (but not necessarily identical with) those induced by processive stressors. In this regard, stimulation of the immune system by various antigens, as well as viral and bacterial products, impacts on central neural circuits utilizing monoamine as transmitters (6-8) and elevates circulating levels of the stress-reactive hormones, including corticosterone and adrenocorticotropin (ACTH) (7-10). While processive and systemic stressors activate different neural circuits, they have some common end points, such as the activation of the HPA axis. Interestingly, systemic stressors, like their processive counterparts may have marked short-term effects, and may also provoke the sensitization of neurochemical systems, thereby leading to exaggerated stress responses upon susbequent environmental challenges.

Given that immune challenges may impact on autonomic, hormonal and central nervous systems (7,8,11), it was suggested that beyond its other functions, the immune system may act in a sensory capacity informing the brain of viral or bacterial insults (11). Moreover, in view of the similarities between the central neurochemical actions of stressors and those of immune activation, it was suggested that immune challenge (and particularly cytokine activation) may be interpreted by the brain as if it were a stressor (6,7,9). The present review indicates that cytokine activation will indeed induce neuroendocrine and central neurotransmitter changes akin to those elicited by traditional stressors, and proactively influences the neurochemical responses to subsequently encountered insults. In addition, we offer prelimary suggestions as to the mechanisms subserving the sensitization effects and discuss some of the behavioral ramifications of the sensitization.

2. MECHANISMS OF CENTRAL CYTOKINE EFFECTS

2.1 Direct and Indirect Central Actions of Cytokines

Cytokines are relatively large, hydrophilic polypeptides that do not readily cross the blood brain barrier (BBB). Nevertheless, when administered systemically, cytokines such as interleukin-β (IL-1β) stimulate *c-fos* expression in numerous brain regions important for autonomic and endocrine functions, including the paraventricular nucleus of the hypothalamus (PVN), dorsomedial hypothalmus, bed nucleus of the stria terminalis, central nucleus of the amygdala, lateral parabrachial nucleus and the ventrolateral medulla (12-14). In addition to potential direct effects, cytokines may also influence CNS functioning without actually passing through the blood brain barrier (BBB) into the brain parenchyma. In particular, afferent fibers of the vagus nerve may transmit information to the brain after stimulation by cytokines or in response to stressors (15,16). Indeed, receptors for IL-1 are present on the nodose ganglion which sends afferent projections to the brainstem nucleus tractus solitarous (NTS), and *c-fos* expression is elevated in these regions in response to systemic IL-1β or lipopolysaccharide (LPS) challenge (17). Among other effects, activation of vagal branches may stimulate the *de novo* synthesis of cytokines within various brainstem and hypothalamic nuclei (15,17), thus increasing local brain concentrations of these messengers.

Cytokine receptors are numerous at circumventricular and other vascular regions, and circulating cytokines may interact with such receptors to generate secondary messengers within the brain (18). In addition to these indirect routes, cytokines may also have direct effects on CNS functioning. In particular, IL-1β and tumor necrosis factor-α (TNF−α) have saturable carrier mediated transport mechanisms, which pump these cytokines into the brain. As well, cytokines may gain entry into the brain through circumventricular areas, which lack an efficient BBB (19). Indeed, the suggestion has been offered that cytokines may passively diffuse into the brain via the organum vasculosum laminae terminalis (OVLT) and other circumventricular organs, at which time they may reach various brain nuclei through the process of volume diffusion (18,20). Not only can cytokines penetrate these circumventricular regions, but it seems that systemic administration of IL-1β or TNF−α can lead to a disruption of the BBB (21) making it easier for these substances to gain entry to the brain. The proposition has also been entertained that cytokines may affect the BBB by the induction of adhesion molecules, such as ICAM-1 and VCAM-1 in the brain endothelium and astrocytes, which may guide inflammatory leukocytes into the brain parenchyma (22).

Once present within the brain, cytokines can provoke a functional response by binding to specific cytokine receptors at hypothalamic and extrahypothalamic brain regions (18,23,24). Indeed, IL-1 and TNF-α may have functions similar to classical neurotransmitters through their modulation of neuronal Ca^{2+} channels and activation of intracellular second messenger systems (25).

2.2 Central Cytokine Distribution

Although their levels are admittedly low under basal conditions, IL-1β, IL-6 and TNF-α, as well as their receptors are constitutively expressed within the brain, specifically within neuronal cell bodies, microglia and astrocytes in several brain regions (23,24,26). Moreover, mRNA for the various receptor subtypes appear to be differentially distributed within the brain; for instance, whereas the p55 TNF-α receptor was detected within the hypothalamus, amygdala, locus coeruleus, OVLT, subfornical organ as well around the lateral ventricles and other vascular associated sites (18), little basal expression of the p75 TNF-α receptor was detected at those sites (18).

In addition to being expressed in brain under basal conditions, central cytokine expression is subject to alterations in response to stressors as well as various immunological and neurological insults. For example, elevated IL-1β, IL-6 and/or TNF–α mRNA has been observed in the CNS following stressors, systemic or central LPS administration, viral infection, brain injury, cerebral ischemia and seizure (27-30).

3. NEUROCHEMICAL EFFECTS OF CYTOKINES

3.1 Cytokine-induced HPA Activation

Like stressors, various immunogenic and cytokine challenges (e.g., LPS and IL-1β) stimulate HPA activity (7,8,31). Intravenous and intracerbroventricular (icv) IL-1β administration increased *c-fos* expression within PVN neurons expressing corticotropin releasing hormone (CRH) (32) and icv IL-1 treatment elevated *c-fos* expression within the supraoptic nucleus and CRH mRNA within the PVN (33). Paralleling these findings, systemic and central IL-1β administration increased the expression and release of CRH and arginine vasopressin (AVP) secretion from PVN neurons of the hypothalamus (10,34). Likewise, direct infusion of IL-1β into the median eminence (site of CRH terminals from neurons originating within the PVN) also increased AVP and CRH secretion from these neurons and of ACTH from the pituitary corticotrophes (35); effects that were attenuated by CRH antagonists (36, 10). Together, these data suggest that IL-1β acts at the level of the hypothalamus to stimulate CRH and AVP, which synergistically provoke the secretion of ACTH and hence adrenal glucocorticoid secretion (10). The glucocorticoids stimulate hippocampal cells, to cause the feed-back inhibition of further CRH release from the PVN (1).

In addition to IL-1β, other cytokines, including IL-6 and TNF–α, influence HPA activity (31,37-39). TNF–α stimulates hypothalamic CRH release, suggesting that the cytokine may be stimulating HPA activity through activation of the PVN. Indeed, the ACTH and corticosterone secretion provoked by TNF–α administration were blocked by CRH antiserum (40). In contrast, neutralization of AVP (using antiserum) failed to affect ACTH release provoked by the cytokine (41), suggesting that CRH plays a more dominant role in the HPA activating effects of the TNF–α. However, in addition to the hypothalamic effects, TNF-α may also have direct actions on pituitary corticotroph cells (42) and adrenal gland (43).

While the acute effects of IL-1β and TNF–α on HPA activity have been relatively well documented, less attention has focused on the effects of chronic administration. Continuous systemic infusion of IL-1β over a 1 week period resulted in a persistent increase of plasma ACTH and corticosterone levels (44). It has also been shown that infusion of IL-1β into the cerebral ventricles for 3 days elicited substantial elevations of plasma ACTH and corticosterone, even at relatively low doses (0.5 μg/day) (45), and the co-administration of IL-1β and TNF-α synergistically increased ACTH and corticosterone (46). The HPA activation was inhibited by CRH immunoneutralization or systemic treatment with dexamethasone, supporting the notion of a hypothalamic site of action for IL-1β which is sensitive to immunoregulatory glucocorticoid feedback mechanisms (45).

3.2 Central Neurochemical Effects of Immune Activation or Cytokine Challenge

Cytokine treatment markedly influences monoamine activity within hypothalamic nuclei and at several extrahypothalamic sites, including limbic brain regions. Systemic IL-1β administration increased NE activity within the PVN, medial basal and lateral hypothalamic nuclei (7,47) and increased DA utilization within the hypothalamus and frontal cortex (48). Such effects, may be dependent upon the route of cytokine administration, as systemic but not intra-hypothalamic treatment with IL-1α elevated extracellular NE levels within the medial hypothalamus (49). In addition to the catecholaminergic changes, IL-1β increased serotonin (5-HT) activity within the hypothalamus, prefrontal cortex and hippocampus (8,31,50,51) and may have increased 5-HT synthesis, given that tryptophan levels were increased in the brainstem, hypothalamus and prefrontal cortex (8). As the effect of IL-1β on brain tryptophan was attenuated by the ganglionic antagonist, chlorisondamine, a role for sympathetic involvement is suggested in this respect (8)

As in postmortem studies, intracerbroventricular IL-1β administration increased *in vivo* hippocampal 5-HT release, as reflected by elevated interstitial amine concentrations (52). Moreover, direct application of IL-1β into the rat anterior hypothalamus markedly increased the release of NE, 5-HT and DA (53), although the time course for these effects differed from one another (53). Direct injection of IL-1β into the medial basal hypothalamus, like that observed following anterior hypothalamic injection, augmented 5-HT and dopamine (DA) release (54,55), and local injection of IL-1β increased NE release within the medial prefrontal cortex (56). Finally, the systemic administration of IL-1β increased the release of hypothalamic NE (8) and 5-HT at the nucleus accumbens and the hippocampus (57,58). These effects did not appear to be secondary to thermogenic actions, as subpyrogenic doses of IL-1β also elicited central amine alterations (8,31,47,59).

Like IL-1β, it appears that TNF–α and IL-6 also provoke alterations of CNS monoamine activity, albeit to a lesser extent. Systemic TNF–α increased hypothalamic NE activity (37,60), while turnover of the 5-HT precursor, tryptophan, was elevated within the hippocampus and the hypothalamus (37). Likewise, we observed that TNF-α increased NE activity within the PVN, central amygdala, dorsal hippocampus and locus coeruleus, while 5-HT utilization was increased within the PVN, medial prefrontal cortex and central amygdala (31,38,61), and once more the time course for the various aminergic changes varied appreciably (62). *In vivo* determinations of the effects of TNF-α administered icv revealed that this cytokine increased plasma corticosterone, but did not influence hippocampal 5-HT release (63). Thus it is possible that a peripheral route of action is necessary for the induction of some of the central effects of TNF-α. Indeed, we have observed that hippocampal amine variations are not readily induced by icv treatment, whereas 5-HT utilization was increased following systemic cytokine treatment (64).

3.3 Synergistic Effects of IL-1β, IL-6 and TNF–α

Different cytokines may have additive, synergistic or antagonistic effects (65). In this respect, IL-1β and TNF-α can induce each other's production as well as that of IL-6, IL-12 and IFN-γ, all of which may interact to potentiate or inhibit each other's actions. For instance, co-administration of IL-1β and TNF–α synergistically reduced sexual behavior, body weight and food consumption, intake of highly palatable substances, increased adrenal gland weight, and augmented plasma ACTH and/or corticosterone levels (31,46,66-68). In contrast to these findings, limited data are available concerning the possibility that cytokines synergistically influence central neurotransmitter activity. While co-administration of IL-1β and TNF-α was reported not to synergistcally affect brain monoamine activity (31), these studies involved postmortem tissues of mice taken at a single time point after exposure to single doses of cytokines. In fact, it has been reported that while IL-1β, even at high doses, does not affect DA activity within the nucleus accumbens (an effect ordinarilly elicited by stressors) (47), systemic LPS administration markedly influences accumbal DA and 5-HT functioning (69). It is possible that this qualitative difference relative to the effects of IL-1β reflects the synergistic actions of several cytokines released in response to LPS.

It is significant, that cytokines and stressors synergistically influence central processes, supporting the contention that these challenges activate common neural substrates. For instance, IL-1β acted synergistically with a mild stressor (air puff) to increase 5-HIAA, HVA and MHPG concentrations at several limbic brain regions (57,58). Thus, in evaluating the neurochemical consequences of cytokines, it may be important to consider the backdrop upon which these are administered.

4. CYTOKINES AND THE STRESS RESPONSE

As indicated earlier, stressor and cytokine treatments induce several comparable neuroendocrine and central neurotransmitter alterations. Indeed, augmented NE and/or 5-HT activity within the PVN or whole hypothalamus is evident in response to a variety of immune and non-immune associated stressors including restraint, footshock, predator exposure, viral infection, LPS, IL-1β and TNF-α (8,31,47,61,70,71). Likewise, IL-1β and various stressors increased *c-fos* expression in CRH-immunoreactive neurons within the PVN, and also increased PVN mRNA levels for the immediate early gene, NGFI-B, as well as that of CRH (72). Further, lesions of the PVN attenuated stressor and cytokine induced increases of plasma ACTH and corticosterone as well as the altered immune activation associated with such treatments (73,74). Interestingly, Shintani et al., (75) demonstrated that pretreatment with the IL-1 receptor antagonist, IL-1ra, blocked the hypothalamic NE alterations and the plasma ACTH increases induced by immobilization stress, suggesting that cytokines may play a role in the regulation of the stress response.

While a crucial role for the amygdala in processing stressful stimuli has long been proposed, only recently has evidence emerged suggesting amygdaloid involvement in the central response to cytokines and other immunogenic stimuli. Specifically, stressors and IL-1β both increase CRH mRNA expression at extrahypothalamic sites, including the amygdala and bed nucleus of the stria terminalis (76,77). Given that CRH mRNA is elevated in stressor reactive brain regions and anxiety is elicited by central CRH administration (78), this peptide may be important in subserving anxiety (79,80), and the possibility ought to be considered that IL-1, owing to its effects on CRH release, may influence anxiogenic processes.

In accordance with this proposition, the administration of IL-1β or TNF-α may elicit anxiogenic-like effects, although this has only been evaluated in a limited number of situations (81). Of course, these findings may be unrelated to CRH changes induced by the cytokine treatment. In fact the involvement of CRH in maintaining anxiety has been questioned, as CRH knockout mice were not distinguishable from wild type mice in several behavioral tests (82). Yet, there is the belief that CRH, while not necessary for the maintenance of an anxiety response, may be fundamental in the development of fear/anxiety (80). Further to this point, it is reasonable to suggest that distinct stressor-responsive brain circuits may mediate different phases of the anxiogenic response. While the basolateral nucleus of the amygdala may play a prominent role in the initial processing of fearful stimuli, the central nucleus may be more important for the generation of behavioral outputs to contend with the challenge (79).

Despite the similarity of outcomes associated with stressors and immune challenge, the neural pathways mediating the HPA effects of traditional stressors (e.g. footshock or restraint) may be dissociated from those subserving the neuroendocrine effects of systemic stressors (e.g. cytokines) (5). In this respect, Herman and Cullinan (1997) offered the view that while processive stressors influence HPA activity via the limbic forebrain, systemic stressors (e.g., respiratory, cardiovascular, bacterial and viral insults as well as cytokines) do so primarily through stimulation of brainstem afferents. In support of this contention, ablation of aminergic fibers originating from the medulla or removal of the area postrema (a brainstem, circumventricular organ) attenuated the HPA stimulating effects of IL-1β (32,83). However, contrary to the notion that limbic sites contribute minimally to the HPA variations elicited by cytokines, lesions of the central amygdala attenuated the plasma corticosterone and ACTH responses elicited by IL-1β, suggesting that amygdaloid-PVN communication may be important in regulating HPA responses to the cytokine (14).

5. SENSITIZATION

5.1 Cytokine and Stressor Sensitization Effects

Some degree of behavioral and neurochemical plasticity is necessary for an animal to cope effectively with environmental demands. In this respect, stressful events not only elicit acute responses, but may also invoke mechanisms that influence the organism's responses to later challenges. Indeed, a variety of stressors, including immunological stimuli, have the effect of priming (sensitization) biological systems so that an augmented response is elicited by later exposure to the same or somewhat different challenges (48,84).

Stressors were shown to provoke a phenotypic change in the co-localization of the hypothalamic neuropeptides, AVP and CRH, such that increased co-expression of AVP is evident within median eminence CRH terminals (85). Tilders and his associates (86,87) reported that following a chronic stressor regimen, characteristic changes were observed in hypothalamic CRH terminals in the external zone of the median eminence

(ZEME). Ordinarily, approximately 50% of the CRH terminals in the ZEME co-store AVP; however, following stressor or cytokine treatments there is a progressive increase of AVP stores in these CRH terminals (85,87,88). Interestingly, the shift in neuropeptide composition of CRH terminals in the ZEME was not stressor-specific and similar increases of AVP stores in CRH terminals were reported after immobilization or social defeat stressors (85,89) and after adrenalectomy (90,91). Furthermore, the changes in CRH terminals in the ZEME were paralleled by an increased proportion of CRH/AVP co-producing neurons within the mediodorsal PVN (88), as well as increased AVP mRNA expression in the parvocellular portion of the PVN (85). As AVP potentiates CRH-induced ACTH release from the pituitary corticotrophes, the increased co-expression of these neuropeptides would be expected to enhance ACTH and corticosterone secretion upon subsequent challenges (92).

In addition to the neuroendocrine processes, stressor-induced sensitization effects have been demonstrated within the hypothalamus, hippocampus and amygdala with respect to NE activity, and within the prefrontal cortex for DA utilization (9,93-95). Furthermore, exposure to a particular stimulus may enhance the response to a subsequently applied stimulus of a different form, including pharmacological challenges (cross-sensitization) (2,9,93,94,96-98).

At this juncture it should be underscored that the immediate and protracted effects of stressors are influenced by chronicity of such treatments. In this respect, the monoamine reductions ordinarily provoked by acute stressors may be absent following protracted or repeated insults (5,9, 99). Such an "adaptation" involves active processes, since monoamine levels in chronically stressed animals do not simply return to prestress levels, but exceed basal values (9). Evidently, chronic stressors promote a compensatory increase of amine synthesis, or moderation of excessive utilization, hence leading to the increased tissue concentrations.

Much like acute insults, chronic stressors also proactively influences responses to later stressors (94,100-102). Exposure to intermittent cold or repeated restraint augmented the effects of subsequent restraint, in terms of enhanced HPA activity coupled with elevated *c-fos* immunoreactivity within the PVN, amygdala, bed nucleus of the stria terminals, prefrontal cortex, nucleus accumbens and paraventricular nucleus of the thalamus (103,104). Chronic exposure to a series of different stressors also sensitized *in vivo* dopamine release from the frontal cortex (105), suggesting that although chronic predictable stressor exposure often results in behavioral/ neurochemical habituation, repeated unpredictable stressors may have the opposite effect (i.e. sensitization). The sensitization effects elicited by chronic stressors may stem, in part, from excessive wear and tear (allostatic load) on endogenous systems as attempts to adapt to environmental challenges proceed. It has been suggested that the repeated pertubations of stressor-sensitive systems may eventually result in adverse behavioral and physiological repercussions (106).

5.2 Cytokine-induced Neuroendocrine Sensitization

Administration of IL-1β and TNF−α may result in a time-dependent sensitization, such that responses to these cytokines are increased upon their reintroduction some time afterward (34,38,61,70,89,107). Specifically, as observed in response to stressors (86,92), several days after a single IL-1β treatment, a progressive increase of AVP stores was evident within CRH terminals of the ZEME (85,87,88). The increased co-localization of these neuropeptides only became apparent 4 days after acute administration of IL-1β, peaked at 7-11 days, and did not return to control levels until 40 days later (87). Commensurate with the notion that IL-1β results in a sensitization of HPA functioning, upon the second administration of IL-1β, plasma ACTH and corticosterone levels were increased (87). Thus, the augmented co-storage of AVP and CRH, and their subsequent release, was translated into an exaggerated corticosterone response upon reexposure to IL-1β.

One possible mechanism that may account for the augmented hypothalamic AVP and CRH activity is that the IL-1β reduced the degree of corticoid negative feedback, which typically keeps HPA activity in check. As already mentioned, hyper-responsive AVP/CRH and hormonal responses occur after adrenalectomy (108), and reduced glucocorticoid inhibitory actions on AVP production has been suggested to mediate some of the effects of stressors on HPA activity (90). Alternatively, the possibility also needs to be considered that cytokine-induced corticosterone and ACTH sensitization effects may stem from direct actions upon the pituitary or adrenal glands. As indicated earlier, TNF-α receptors have been demonstrated within the anterior pituitary and adrenal and the cytokine may directly provoke ACTH release from this region (42,43). Furthermore, the neuroendocrine consequences of various challenges, including pathogenic, chemical or other immunogenic insults, may be influenced by the state of the microenvironment. For example, hyperthermia sensitizes tumor cells to the well-documented cytotoxic effects of TNF-α (109). Thus, it is possible that treatment with a cytokine may affect subsequent thermogenic responses to this treatment, which may then influence later HPA responses induced by re-exposure to the cytokine or to stressors.

While greatest attention concerning neuro-immune communication has focused on IL-1β, other cytokines, including TNF-α, may also engender sensitization effects. Indeed, such an outcome has been demonstrated with respect to neuroendocrine functioning, central monoamine activity, as well as sickness behaviors (38). It will be recalled that when acutely administered, IL-1β and TNF-α dose-dependently elicit a constellation of symptoms (ptosis, piloerection, soporific effects, curled body posture, anorexia, anhedonia) generally referred to as sickness behaviors. The effects of TNF-α are considerably less pronounced than those elicited by IL-1β, although at higher doses the sickness profile becomes marked. Systemic administration of TNF-α was shown to induce sensitizating effects with respect to various indices of illness, and these effects varied with the passage of time following the initial treatment. Specifically, re-exposure to TNF-α 1-7 days following initial treatment hardly affected sickness behaviors, but upon re-exposure 2-4 weeks following initial cytokine treatment, marked signs of sickness were evident (e.g. reduced consumption, social exploration and locomotion) (38,61).

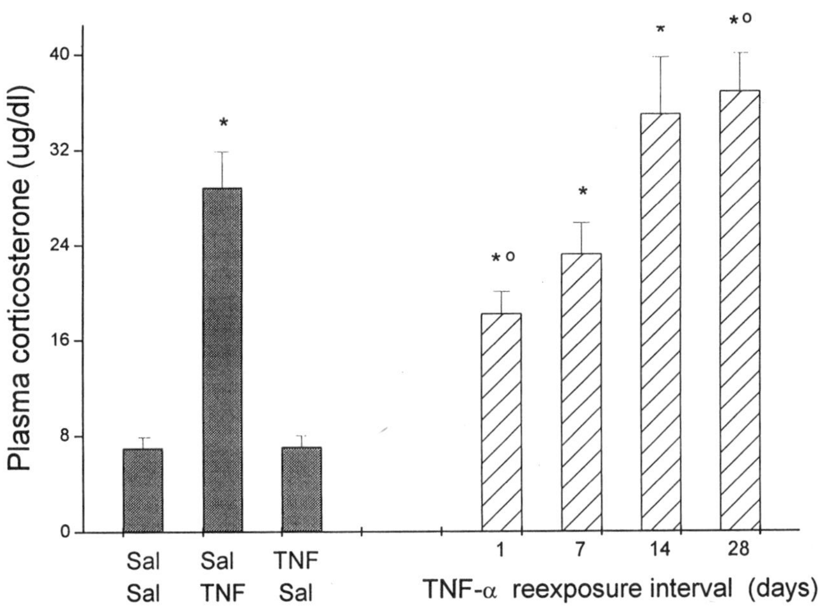

Figure 1. Plasma corticosterone levels (mean ± SEM) among mice exposed to various TNF-α regimens. All mice received two intraperitoneal injections; the three groups to the left (solid bars) received saline only, saline followed 2 weeks later by the low dose of TNF-α (1.0 μg), or a high dose of TNF-α (4.0 μg) followed 2 weeks later by saline treatment. The four groups on the right (hatched bars) received two injections of TNF-α: an initial 4.0 μg dose followed by a second 1.0 μg dose 1, 7, 14 or 28 days later. *p <0.05 relative to saline treated mice; °p < 0.05 relative to mice that received acute TNF-α injection. (38). Reprinted with permission from *J. Neurosci* 1999; 19: 5654-5665.

In concert with the sickness behaviors, a time-dependent sensitization of plasma corticosterone release was induced by TNF-α (Figure 1). Exaggerated corticoid responses were evident in mice re-exposed to the cytokine, 14 and 28 days following the initial TNF-α pretreatment, however, such a sensitization was absent upon reexposure at earlier times (1 and 7 days) (38). In fact re-exposure to human TNF-α 1 day after initial cytokine treatment resulted in a desensitization effect such that the response was less pronounced than that ordinarily elicited by acute treatment with this cytokine. Interestingly, the murine form of TNF-α (mTNF-α), like the human form, resulted in a time-dependent sensitization effect. However, the desensitization induced by human TNF-α (hTNF-α) in mice re-exposed 1 day afer initial treatment was not elicited by mTNF-α (64).

We cannot say with certainty that these differences were not related to the specific doses used or time at which blood was taken after the second injection. However, it needs to be considered that the differential desensitization may be related to characteristic differences between the two forms of TNF-α. A large degree of homology exists between the murine and human forms of TNF-α, and the human form of the cytokine has less potent anti-tumor and lethal effects in the mouse (110). The actions of TNF-α are mediated by binding to one of its two receptor subtypes (p55 and p75), which rapidly trimerize and activate various transcription factors and biochemical cascades (111). While mTNF-α acts upon both the p55 and p75 receptors, the hTNF-α selectively binds to the p55 receptor in the mouse. Given that both inhibitory and excitatory interactions have been proposed to occur between the p55 and p75 receptor subtypes (111), it remains possible that the differential desensitization effects associated with re-exposure to the two forms of TNF-α may be related to their effects on the two receptor subtypes.

As reported by Tilders and his associates regarding the effects of IL-1β, we observed that the systemic TNF-α administration resulted in time-dependent phenotypic changes within CRH neurons terminating in the external zone of the median eminence (64). In particular, a single dose of TNF-α provoked a shift towards increased AVP/CRH co-immunoreactive terminals within this site, largely attributable to increased AVP within CRH terminals (see Figure 2). While this effect increased over time, peaking 7-14 days following cytokine treatment, normalization of CRH/AVP co-expression was evident after 28 days. Given that TNF-α sensitized corticosterone release in mice re-exposed to the cytokine 28 days after pretreatment, but not after earlier intervals (38), it is unlikely that the altered median eminence AVP/CRH-ir plays a primary role with respect to the sensitization of the corticosterone response. It is tempting to suggest that since the cytokine induces sensitization of sickness behavior (38), the elevated corticosterone was secondary to the stress associated with illness or related factors (e.g. activation of acute phase proteins, histamine). However,

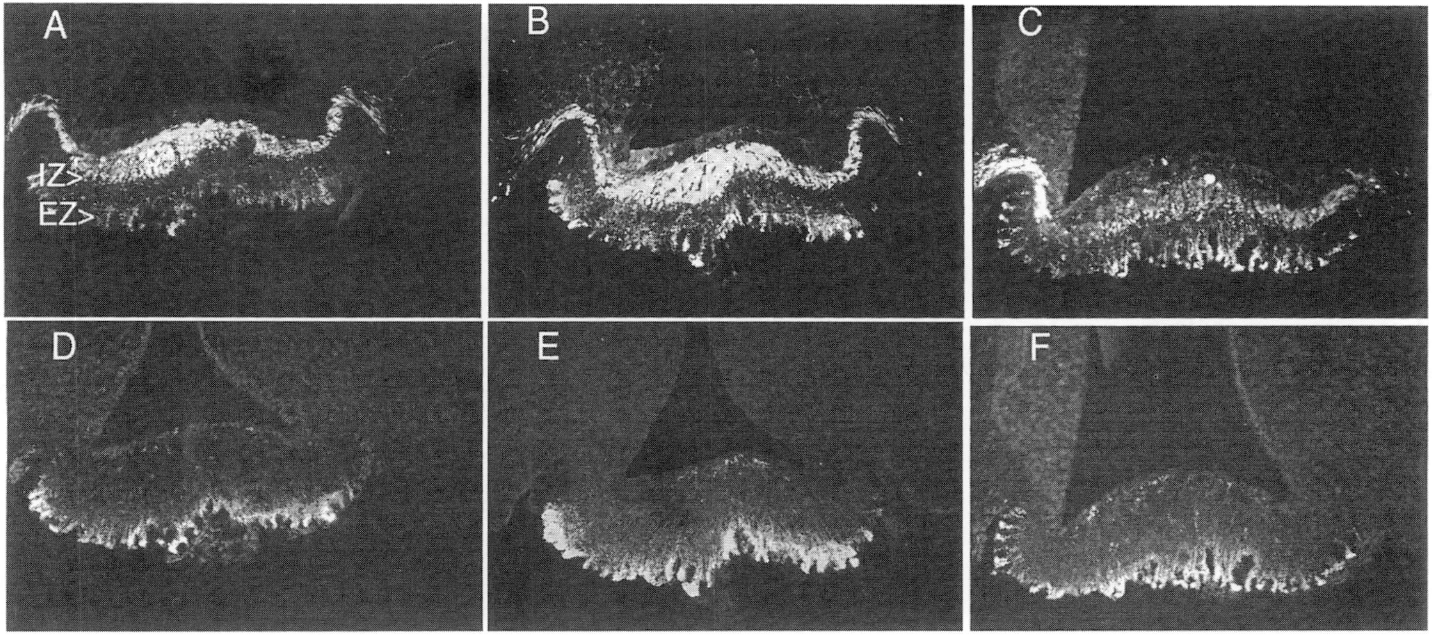

Figure 2. Immunoreactivity for AVP and CRH within the median eminence of mice treated with saline 7 days earlier (left), TNF-α (4.0 μg) 7 days earlier (middle) or TNF-α (4.0 μg) 14 days earlier (right). The top panels depicts AVP immunoreactivity, while the bottom panels show CRH positive terminals, within the same double labeled section. IZ denotes the internal zone of the median eminence. EZ denotes the external zone of the median eminence. Enhanced AVP and CRH immunoreactivity was evident within EZ 7 and 14 days after a single TNF-α injection, (B,C and E,F, respectively). However, AVP-ir was somewhat diminished within IZ after the 14 day interval (C). In mice treated with TNF-α either 1 or 28 days earlier, there was no effect on peptide immunoreactivity, and the data for these time points are not shown in the photomicrographs Magnification x 20. n = 5 / group. (120). Reprinted with permission from *Neuroscience* 2001; 106: 137-148.

it was observed that the development of the illness and the enhanced corticosterone levels were, in fact, independent of one another (e.g., corticosterone elevations could readily be induced at doses that did not elicit sickness).

The possibility exists that the TNF-α induced sensitization of corticosterone and sickness behavior involved peripheral processes (e.g. direct glandular actions). In fact, we observed that unlike the effects of systemically administered TNF-α, icv infusion of the cytokine did not provoke signs of illness, irrespective of dosage or re-exposure regimen employed. Moreover, the cytokine provoked only a modest increase of plasma corticosterone after acute icv infusion, an effect that was considerably less pronounced than that observed following ip treatment with the cytokine (64). In contrast to the pronounced corticosterone sensitization engendered by ip TNF-α, re-exposure to icv TNF-α at 1, 7 or 28 days following initial central infusion of the cytokine did not induce a sensitization with respect to circulating corticosterone. Taken together, these data suggest that the protracted effects of TNF-α on HPA activity and sickness behavior depends upon activation of peripheral targets, which may in turn, stimulate central processes.

5.3 Histamine Involvement in Sensitization

In addition to being secreted by macrophages, TNF-α is ordinarily produced by mast cells and secretion of the cytokine from these cells is increased in response to immune activation (112). As well, TNF-α receptors are found on mast cells and stimulation with the cytokine provokes histamine release (113). Mast cells that produce histamine are present at peripheral sites and within the brain (114), and are particularly abundant within the tuberomammillary nucleus (TM) located in the posterior portion of the hypothalamus (112). Like other monoamines, histamine acts within the brain and may influence behavioral and hormonal outputs (115,116). Both H_1 and H_2 receptors have been implicated in stimulation of the HPA axis, as agonists for these receptors increase PVN CRH mRNA coupled with elevated levels of plasma ACTH and corticosterone (117).

Although the H_1 and H_2 antagonists, diphenhydramine and cimetidine, when administered either singly or conjointly did not influence the acute effects of mTNF-α on plasma corticosterone or sickness, co-administration of these antihistamines prevented the sensitization ordinarily provoked by re-exposure to the cytokine after a 28 day interval (118). Given the integral role of histamine in the acute phase response, the fact that this transmitter appears to contribute to the sensitization and not the acute effects of mTNF-α, suggests that the HPA and sickness modulating effects of the amine are related to it's role in the shock-like syndrome induced by reexposure to TNF-α. It remains to be established whether histamine likewise contributes to the central neurochemical effects of the cytokine.

5.4 Cytokine-induced Sensitization of Central Processes: *c-fos* Activation

Paralleling the finding that intravenous TNF-α augmented central *c-fos* mRNA expression (119), intraperitoneal administration of this cytokine elicited the excitation of several stressor-reactive brain regions, including the PVN, supraoptic nucleus and central amygdala. In particular, although acute TNF-α had modest effects on immunoreactivity for the Fos protein, upon reexposure to the cytokine 7 or 14 days after initial TNF-α pretreatment, *c-fos*-ir was markedly increased (120). The increased immunoreactivity for the immediate early gene protein may represent altered intracellular calcium, cAMP or other cellular variations associated with transcriptional events (121). Thus, the elevation of *c-fos*-ir elicited by TNF-α re-exposure may reflect intracellular changes primed by the initial cytokine administration. Indeed, TNF-α altered synaptic transmission and modulated Ca^{2+} currents in hippocampal and sympathetic neurons (122).

It was reported that the administration of IL-1β induced the expression of Fos-ir within parvocellular (72) and magnocellular PVN neurons (123). Moreover, icv administered IL-1β also induced the transcription factor, nuclear factor kappa B, within the basolateral amygdala as well as several vascular associated sites, while *c-fos* immunoreactivity was elevated within the PVN, supraoptic nucleus and central amygdala (20). The pattern of *c-fos* upregulation induced by icv IL-1β in the Konsman et al (20) study was essentially indistinguishable from that observed after ip TNF-α treatment observed in our laboratory 120). It had been suggested (20) that IL-1β diffused through the brain parenchyma to activate nuclei through a process termed "volume transmission". It is likewise possible that peripherally administered TNF-α activates a central diffusable factor (including IL-1β itself) that ultimately comes to provoke the Fos changes.

In addition to the variations witnessed within the hypothalamus, it appeared that TNF-α also impacted on *c-fos*-ir within the central amygdala. Although the effects of acute TNF-α on amygdaloid *c-fos*-ir were modest, later reexposure to the cytokine resulted in a pronounced increase of the immediate early gene. Moreover, upon re-exposure to the cytokine CRH-ir within the central amygdala was appreciably enhanced regardless of whether the re-exposure occurred 1, 7, 14 or 28 days after initial treatment. Unlike the hypothalamic changes which evolved simply with the passage of time, the variations within the amygdala required that the cytokine be re-administered. The process by which TNF-α increased CRH-ir varicosities within the central amygdala is unclear. Yet, as *c-fos*-ir peaked 14 days after TNF-α, whereas CRH-ir peaked at 7 days, the *c-fos*-ir likely involves factors other than, or in addition to CRH.

The sensitizing effects of TNF-α on amygdaloid CRH-ir and *c-fos*-ir may reflect the engagement of biological cascades, involving numerous chemical messengers, essential for dealing with the insult over protracted intervals. Activation of CRH within the amygdala as well as the anatomically closely related bed nucleus of the stria terminalis, has been proposed to play an important role in behavioral responses to stressful or fear-related stimuli (124). These behavioral effects appear to be independent of HPA activation, but may involve interactions between CRH and noradrenergic projections between the amygdala and bed nucleus of the stria terminalis as well as the locus coeruleus (125). Indeed, tracing studies indicated that the central amygdala sends CRH projections to tyrosine hydroxylase immunoreactive dendrites within the locus coeruleus (125), which may affect behavioral arousal in response to stressors. There is reason to believe that TNF-α may influence central NE activity as well as that of CRH. Indeed, TNF-α altered pre-synaptic noradrenergic activity, and pharmacological manipulation of NE levels influenced central expression of TNF-α (126). Likewise, TNF-α was shown to provoke an increase of NE and MHPG levels within the locus coeruleus, while the cytokine sensitized MHPG accumulation within the central amygdala upon reexposure to TNF-α 1 day following it's pretreatment (38).

5.5 Cytokine-induced Sensitization of Central Processes: Monoamine Variations

In addition to the immediate effects of TNF-α on monoamine activity (38, 31), re-exposure to the cytokine altered levels of NE, DA and 5-HT in a region-specific, time-dependent fashion (38,61,64). Within the PVN, TNF-α induced a sensitization of NE activity, which followed a time-course similar to that of the corticosterone and sickness sensitization effects (38). In contrast, as seen in Figure 3, within the medial prefrontal cortex and central amygdala an earlier sensitization of MHPG accumulation was observed. Specifically, reexposure to TNF-α 1 day following pretreatment with the cytokine elicited a particularly marked augmentation of MHPG accumulation relative to acutely treated mice. At longer intervals this effect was not evident. Interestingly, however, a TNF-α provoked sensitization of 5-HT activity, as reflected by accumulation of 5-HIAA, was apparent within the central amygdala and prefrontal cortex upon re-exposure to the cytokine after an intermediate interval (7-14 days) following pretreatment (38).

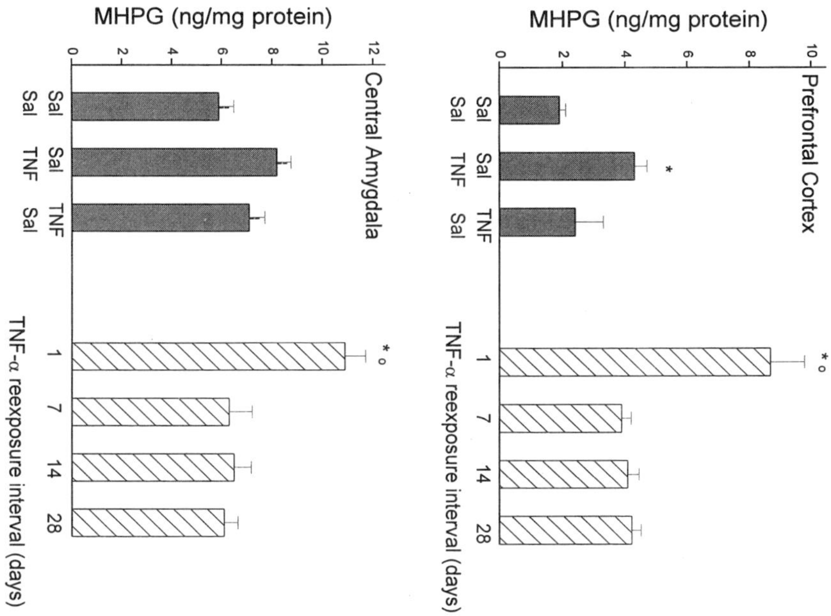

Figure 3. Mean ± SEM concentrations of MHPG within the prefrontal cortex (top) and central amygdala (bottom) as a function of the TNF-α treatments. The three groups on the left received intraperitoneal administration of saline on two occassions, saline followed two weeks later by the low dose of TNF-α (1.0 μg) or a high dose of TNF-α (4.0 μg) followed 2 weeks later by saline. The four groups on the right received an initial 4.0 μg dose of TNF-α followed by a scond 1.0 μg dose 1, 7, 14 or 28 days later. *$p < 0.05$ relative to saline treated mice; $^{o}p < 0.05$ relative to mice that received acute TNF-α injection. (38). Adapted from Hayley et al, *J. Neurosci* 1999; 19: 5654-5665.

Although icv administration of mTNF-α did not lead to sensitization of hormonal and behavioral outputs, central infusion of the cytokine was effective in promoting the sensitization of brain monoamine utilization. In particular, among mice pretreated with icv mTNF-α, later re-eexposure to the cytokine elicited augmented NE and DA utilization within the PVN and ME/ARC relative to acutely treated animals (see Figure 4). Likewise, mice that initially received ip mTNF-α and later challenged via icv administration with this cytokine displayed greatly increased 5-HT and DA activity within the ME/ARC, as well as augmented NE activity within the locus coeruleus.

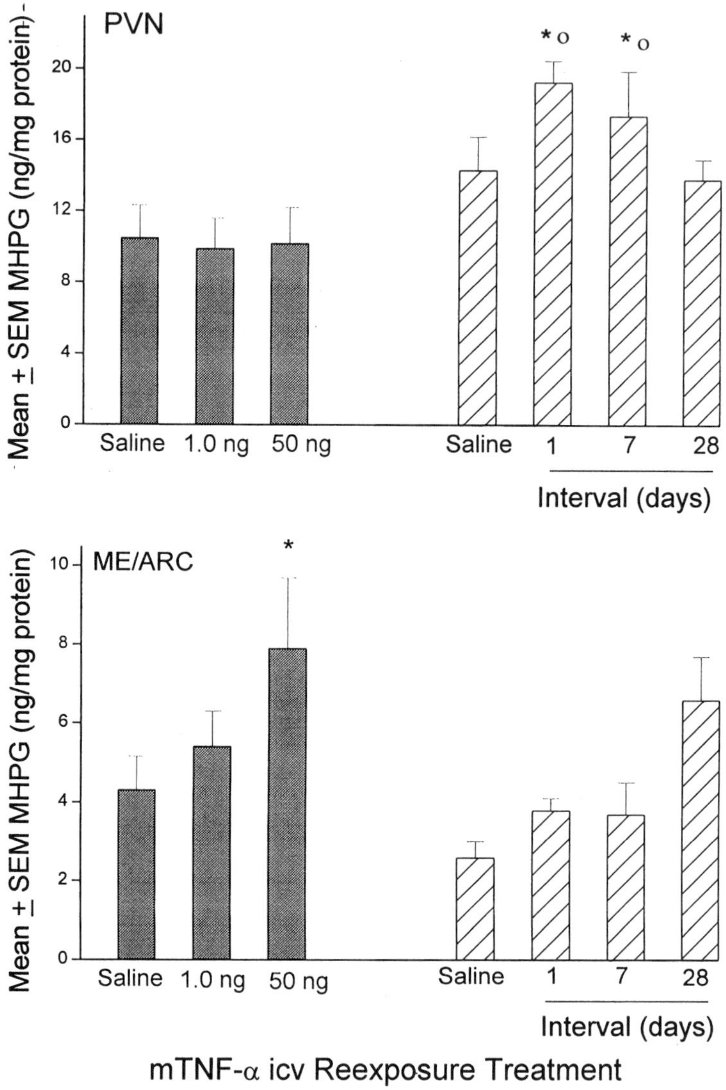

Figure 4. Mean ± SEM concentrations of MHPG within the paraventricular nucleus of the hypothalmus (top) and median eminence plus arcute nucleus of the hypothalmus (bottom) as a function of TNF-α infusions. Mice pretreated with saline (gray bars) received a second intracerbroventricular (icv) infusion of either saline or TNF-α (1.0 or 50 ng) 28 days following the initial injection. Animals initially administered icv TNF-α (hatched bars), were subsequently infused with saline or a second injection of the cytokine (1.0 ng) either 1, 7 or 28 days later. *$p < 0.05$ relative to saline treated mice; $^{o}p < 0.05$ relative to mice that received acute icv TNF-α infusion. Reprinted with permission from Hayley et al., *Eur. J. Neurosci* 2002; 15: 1061-1076.

Interestingly, with the exception of the locus coeruleus, these sensitization effects were restricted to hypothalamic brain regions (64). This contrasts sharply with the brain region sensitization profile induced when the cytokine was administered systemically on both occasions, wherein particularly pronounced effects were observed at extrahypothalamic sites, such as the PFC and amygdala (38,64). Thus, it appears that sensitization of monoamine activity may be related to the direct effects of TNF-α within specific brain sites, and clearly these effects are independent of sickness and corticosterone variations. However, it is equally clear that the effects of ip administration of the cytokine promotes immediate and protracted actions on central neurotransmission that are distinct from those seen after icv treatment. Whether this reflects differential patterns of diffusion or involvement of peripheral signaling processes (including activation of vagal afferents) is not known.

6. ACUTE PHASE RESPONSE

Thus far the present review focused on the influence of specific cytokines (IL-1β and TNF-α) on neuroendocrine and neurotransmitter processes. However, during the course of a typical immunological reaction, numerous cytokines are liberated, and their additive or synergistic effects could potentially impact on CNS processes. For instance, the acute phase response is a generalized host reaction, which involves alterations of metabolic, neurological and endocrine functioning that occurs in response to various environmental insults, including infection, inflammation as well as extreme physical or psychological stressors (127).

It appears that IL-1β, IL-6, and particularly TNF-α, are important mediators of the acute phase response and may act as mediators of the brain's response to such challenges (128, 129). For example, TNF−α modulates the firing of thermosensitive and glucose-responsive hypothalamic neurons, and has been shown to alter feeding and nutrient metabolism in a manner similar to that observed during systemic infection (130,131). It is thought that the acute phase response engenders profound central neurochemical changes, which may be of adaptive significance (16). However, these neurochemical alterations have also been proposed to act like stressors in eliciting behavioral pathologies such as depressive illness (132). If sufficiently severe and prolonged, the acute phase response reactions may result in shock and ultimately death.

Systemic administration of high doses of LPS readily provoke septic shock which typically involves a rapid loss of blood pressure and volume (hypovolemia) and multiple organ failure (133). Interestingly, pretreatment with a low dose of LPS may either reduce or increase (desensitization and sensitization, respectively) the shock inducing capacity of subsequently administered endotoxin (134), depending upon route of administration, timing of treatment and other variables. It has been suggested that TNF-α plays a primary role in the modulation of lethality resulting from LPS-induced shock (128). In fact, antagonists of the TNF-α receptor reduced the severity of symptoms resulting from bacterial-induced shock (129), and particularly high doses of systemically administered TNF-α were accompanied by elevated circulating IL-6, which correlated positively with the severity of symptomatology (110). Given that IL-6 has marked effects on liver acute phase proteins, and hepatic proteins have been reported to be essential in the sensitizing and desensitizing effects of cytokines on lethality (135), it is likely that the liver plays a major role in the shock provoking properties of TNF-α and bacterial endotoxins. As well, the elevation of circulating glucocorticoids that normally follows systemic IL-1β or TNF-α administration, act in an anti-inflammatory capacity thereby potentially reducing the shock-like symptoms provoked by these cytokines. In fact, systemic corticosterone reduces splenic IL-1β mRNA, fever and anorexia induced by LPS, but did not modulate the suppression of locomotion provoked by the endotoxin (136).

Some of the characteristic responses associated with a shock reaction can be elicited upon re-exposure to small doses of TNF-α. In particular, during an acute phase reaction blood volume and pressure decline, and the distribution of blood may be altered with a preferential pooling in certain visceral or limb regions (137). It was noted that upon re-exposure to TNF-α 28 days following an initial treatment with this cytokine, a sensitization effect was observed comprising pronounced reddening (cyanosis; indicating increased accumulation of blood cells and inflammation) of the tail, ears and nose, coupled with a marked reduction of blood volume and pressure, as well as signs of hypothermia (138). These effects were precluded by pretreatment with the combination of H_1 and H_2 histamine antagonists, implicating the involvement of this vasoactive amine. Other clinical manifestations of illness, including ptosis, piloerection, reduced locomotion and social interaction were also attenuated by this treatment (118).

Given the marked effects of cytokines on HPA functioning, it is not surprising that systemic LPS administration provokes particularly marked effects on central CRH, pituitary ACTH and adrenal glucocorticoid secretion (10). As well, LPS has been shown to elicit pronounced neurotransmitter alterations at both hypothalamic and extrahypothalamic sites (7,139,140). For instance, we observed that in mice, doses of LPS ranging from 0.5 to 10 μg increased NE and 5-HT utilization, as reflected by MHPG and 5-HIAA accumulation, within the PVN and median eminence arcuate, and hippocampus. As well, NE turnover was increased within the locus coeruleus, as was DA turnover in the nucleus accumbens (139). *In vivo*, DA release at the nucleus accumbens of rats was increased at approximately 90 min following systemic treatment with LPS and remained elevated for approximately 90 min. The levels of 5-HIAA were likewise elevated, but this outcome was apparent within 30 min and was sustained over the course of a 4 hr test session (69). Further, i*n vivo* studies indicated that ip LPS increased NE and DA concentrations within the medial prefrontal cortex and in the medial hypothalamus (8).

Like TNF-α, systemic LPS treatment provoked a sensitization effect with respect to neurochemical and behavioral functioning. However, the time-course for such effects were very different from those elicited by the cytokine (120). It was shown that repeated LPS administration promotes a progressive reduction in the effectiveness of the endotoxin. Such effects have not only been demonstrated with respect to behavior, but also with regard to circulating TNF-α and IL-1β levels (141). Likewise, low doses of systemically administered LPS may protect against the effects of subsequently applied lethal doses of the endotoxin (142,21). However, if LPS is initially injected into the footpad and then re-administered iv 24 h later, a sensitization of lethality is evident (generalized Shwartzman reaction) (143). Thus, the route of injection, chronicity of treatment, as well as the particular dose of LPS administered, interact to determine whether the endotoxin acts in a sensitizing or de-sensitizing fashion.

In our studies, a sensitization was provoked by ip LPS, such that plasma corticosterone levels and the sickness response were augmented in animals subsequently treated with LPS (i.e., sensitization) or TNF-α (i.e., cross-sensitization) As depicted in Figure 5, unlike the effects seen in mice initially treated with TNF-α and then re-exposed to the cytokine, sensitization of the behavioral changes and the plasma corticosterone response were appreciably increased among mice that had been pretreated with LPS 1 day earlier, while no such effect was apparent at a 28 day interval 71).

Given the divergent temporal profiles associated with LPS and TNF-α pretreatment, it is likely that different processes subserve these sensitizing effects. It is noteworthy, however, that LPS pretreatment effectively sensitized the sickness response to later IL-1β challenge, and the corticosterone response ordinarily provoked by a traditional stressor (restraint). In view of the broad sensitization effects elicited by LPS (i.e., cross-sensitization to various insults), it would appear that the endotoxin elicited a generalized priming effect involving several central (e.g. hypothalamus) and/or peripheral (e.g. liver, gut) mechanisms. However,

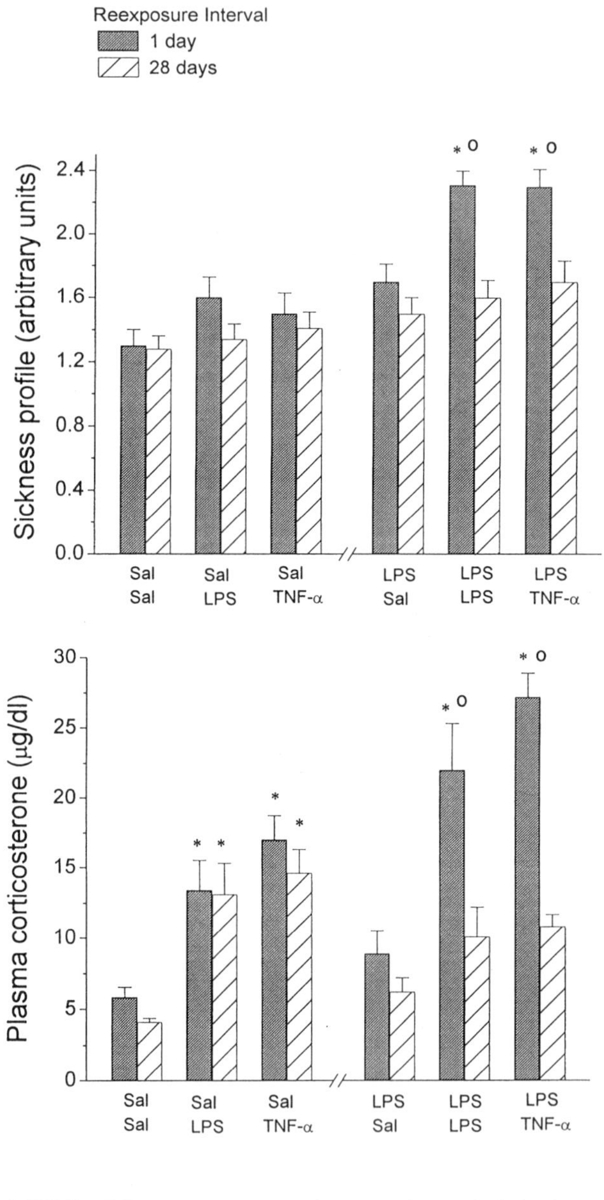

Figure 5. Mean (±SEM) sickness symptoms (e.g. ptosis, curled body posture, reduced locomotion and social interaction) determined using a 4-point scale (top) and plasma corticosterone levels (bottom) as a function of the LPS and TNF-α treatments. Mice were pretreated with saline (left bars) or LPS (5.0 μg; right bars) and either 1 day (gray bars) or 28 days (hatched bars) later exposed to saline, LPS (0.125 μg) or TNF- α (1.0 μg). * $P < 0.05$ relative to mice treated with saline on two occasions, ° $P < 0.05$ relative to animals receiving a single injection of LPS or TNF-α. (71). Reprinted from Hayley, et al. *Eur. J. Neurosci* 2001; 13: 1155-1165.

unlike the cross-sensitizatioon observed with respect to sickness, the endotoxin did not further modify plasma corticosterone responses to IL-1β administered 1 day later. The possibility, of course, should not be dismissed that TNF-α plays a more prominent role than IL-1β in the neuroendocrine sensitizing effects of the endotoxin, just as TNF-α may be the primary cytokine involved in endotoxic shock induced by LPS (128).

There is reason to believe that peripheral macrophage functioning may be fundamental in determining the plasma corticosterone alterations associated with cytokine treatment. In particular, selective macrophage depletion by a liposome-clodronate suspension blocked the corticosterone rise induced by acute TNF-α. However, this treatment did not attenuate the LPS-induced sensitization effect in that the impact of later TNF-α was unaffected by macrophage depletion (71). Thus, it is possible that the effects of macrophage depletion were unique to the immediate actions of LPS or TNF-α, and mechanisms independent of macrophages, were responsible for the sensitization. It is possible that the effectiveness of the liposome treatment was related to the magnitude of the corticosterone changes elicited by the challenge, and hence was absent in response to treatments that engendered particularly marked hormonal increases (as in the case of a sensitization effect). In fact, it was reported that liposome-clodronate pretreatment attenuated the rise of ACTH induced by subpyrogenic, but not by pyrogenic LPS dosages (144).

In addition to the neuroendocrine changes, the central monoamine alterations elicited by TNF-α were enhanced in mice that had previously received LPS. In this respect, it was observed that acute TNF-α increased MHPG accumulation in the PVN, PFC, hippocampus, locus coeruleus and central amygdala, as well as 5-HIAA within the PFC and hippocampus. Interestingly, in animals pretreated with a macrophage depleting agent the effects of acute TNF-α within the PVN were attenuated, while the amine variations in other regions were unaffected. It is unclear why the liposome treatment differentially influenced monoamine activity across brain regions, although it is certainly possible that macrophage activity may be fundamental to the cytokine-induced monoamine variations within hypothalamic sites, possibly owing to its greater accessibility from peripheral circulation.

Although TNF-α did not influence either 5-HIAA accumulation within the central amygdala or DOPAC within the PFC, in mice initially treated with LPS and then later exposed to TNF-α a sensitization effect was apparent. The macrophage depleting agent attenuated the 5-HIAA and DOPAC sensitization within the central amygdala and PFC, respectively. It is possible that the initial treatment with LPS elicited macrophage hyper-responsivity, so that subsequent challenges provoked more robust cellular responses. In fact, *in vitro*, macrophage reprogramming occurs wherein endotoxin application 1 day following pretreatment, increased IL-1β and nitric oxide release, but reduced that of TNF-α (134,145). Certainly, other factors could account for the sensitization effects of LPS, including variations of blood brain barrier permeability or alterations of neutrophil and macrophage infiltration into brain. While attractive, these explanations do not explain why the sensitization was not evident in PVN or other extrahypothalamic sites.

7. CONCLUSIONS

Numerous challenges, including stressful and immunological stimuli, may prime biological systems so that augmented responses are elicited upon later challenges with either the same or somewhat different stimuli. In addition to having immediate effects, cytokines appear to exert long lasting behavioral and neurochemical consequences. Indeed, like stressors, a single exposure to TNF-α primes CNS actvity, such that later reexposure to the cytokine provoked profound alterations of sickness, HPA activation and brain monoamine utilization. Such effects may stem from increased receptor sensitivity or the augmented release of various endogenous factors, such as other cytokines, prostaglandins or acute phase proteins. In this respect, enhanced co-activity of these factors would be expected to result in synergistic actions on central processes.

The processes responsible for the development of the sensitization remain to be established. However, it is clear that multiple sensitizing effects are elicited by various treatments. In the case of TNF-α, it seems that the altered sickness and corticosterone variations are independent of one another, and both of these are unrelated to the central monoamine changes. Indeed, icv treatment with TNF-α induces the sensitization of monoamine functioning, but not that of illness or corticosterone alterations. Moreover, the central sensitizing effects of TNF-α are modifiable by variations of macrophage functioning, and histamine appears to contribute to the peripheral sensitization associated with the cytokine. Finally, although LPS induces macrophage release of TNF-α, it is clear that these two treatments induce different types of effects. The LPS treatment engenders a sensitization that is marked at short intervals, while TNF-α induces a sensitization that progresses with the passage of time.

It appears that cytokines may play some role under conditions of impaired CNS functioning. Several studies have demonstrated the presence of IL-1, IL-6 and TNF-α and their receptors in discrete brain regions under normal conditions, and these cytokines may even play a role in brain development. Brain levels of these cytokines are markedly upregulated in a number of neurological conditions (e.g. multiple sclerosis and cerebral infections), as well as in response to traumatic events (e.g head injury or stroke). Furthermore, IL-1β and TNF-α provoke HPA and central monoamine variations reminesent of those observed in response to stressful stimuli. Moreover, increased circulating levels of IL-1β and TNF-α, as well as IL-2, are often associated with depressive or schizophrenic conditions, attesting to their potential involvement in psychiatric illnesses (4,132). It is unclear, however, whether the emergence of the depression is secondary to variations of cytokine functioning, or whether cytokine alterations simply represent an epiphenomenon (bystander effects) associated with the mood disorders. As some of the depressive effects attributable to LPS treatment are modifiable by antidepressant medication (146), as are the depressogenic actions associated with interferon-α immunotherapy (147), it is tempting to argue that cytokines play a pivotal role in the emergence of the affective condition. Yet, it has been demonstrated that the elevated cytokine levels associated with depressive illness do not necessarily normalize with the alleviation of symptoms following successful pharmacotherapy (4,132). Thus, the precise relationship between cytokines and depressive illness remains to be elucidated. This notwithstanding, the fact that cytokines may increase the neurochemical responses to later stressors, and stressors may act as a provocative agent for the induction of depression, the possibility needs to be considered that cytokine indirectly increase vulnerability to affective illness. Indeed, it has even been suggested that protracted alterations of HPA functioning stemming from early-life stressor or immunological challenges may contribute to the organism's subsequent ability to contend with environmental challenges, and hence may contribute to the development of stressor-related psychopathology (148,149,150).

ACKNOWLEDGEMENTS

Supported by grants from the Canadian Institutes of Mental Health Research. H.A. is an Ontario Mental Health Senior Research Fellow and holds a Canada Research Chair in Neuroscience.

REFERENCES

1. Sapolsky, R.M., Romero, L.M., Munck, A.U. How do glucocorticoids influence stress responses? Integrating permissive, suppressive, stimulatory, and preparative actions. *Endocr. Rev.* 2000; 21:55-89.

2. Kalivas, P.W. and Stewart, J. Dopamine transmission in the initiation and expression of drug- and stress-induced sensitization of motor activity. *Brain Res. Rev.* 1991; 16:223-244.
3. Anisman, H., Hayley, S., Staines, W., et al. Cytokines, stress, and neurochemical change: Immediate and proactive effects. In: *Towards a theory of Neuroplasticity* , CA Shaw, JC McEachern, eds. Psychology Press, Philadelphia. pp. 301-320, 2001.
4. Griffiths, J., Ravindran, A.V., Merali, Z., et al. Dysthymia: Neurochemical and behavioral perspectives. *Molecular Psychiatry* 2000; 5:242-261.
5. Herman, J.P. and Cullinan, W.E. Neurocircuitry of stress: Central control of hypothalamo-pituitary-adrenocortical axis. *Trends Neurosci* 1997; 20:78-84.
6. Anisman, H. and Merali Z. Anhedonic and anxiogenic effects of cytokine exposure. *Adv. Exp. Med. Biol.* 1999; 461:199-233.
7. Dunn, A.J. Interactions between the nervous system and the immune system. In *Psychopharmacology: The Fourth Generation of Progress,* F.E. Bloom, D.J. Kupfer, eds. Raven Press, New York. 719-731, 1995.
8. Dunn, A.J. Effects of cytokines and infections on brain neurochemistry. In *Psychoneuroimmunology, V. 2,* R. Ader, D.L Felten, N Cohen, eds. Academic Press, New York, 649-666, 2001.
9. Anisman, H., Zalcman, S. and Zacharko, R.M. The impact of stressors on immune and central neurotransmitter activity: Bidirectional communication. *Rev. Neurosci.* 1993; 4:147-180.
10. Rivier, C. The hypothalamo-pituitary-adrenal axis response to immune signals. In: *Psychoneuroimmunology, V. 2.*, R. Ader, D.L. Felten, N. Cohen (eds). Academic Press, New York, pp. 633-648, 2001.
11. Blalock, J.E. The syntax of immune-neuroendocrine communication. *Immunol Today* 1994; 15:504-511.
12. Ericsson, A., Arias, C. and Sawchenko, P.E. Evidence for an intramedullary prostaglandin-dependent mechanism in the activation of stress-related neuroendocrine circuitry by intravenous interleukin-1. *J. Neurosci.* 1997; 17:7166-7179.
13. Day, H.E., Curran, E.J., Watson, S.J. Jr., et al. Distinct neurochemical populations in the rat central nucleus of the amygdala and bed nucleus of the stria terminalis: evidence for their selective activation by interleukin-1beta. *J. Comp. Neurol.* 1999; 413: 113-128.
14. Xu, Y., Day, T.A. and Buller, K.M. The central amygdala modulates hypothalamic-pituitary-adrenal axis responses to systemic interleukin-1beta administration. *Neuroscience* 1999; 94:175-183.
15. Dantzer, R., Bluthe, R.M., Aubert, A., et al. Cytokine actions on behavior. In: *Cytokines and the Nervous System,* Rothwell N.J. (Ed.) Landes, London, pp 117-144, 1996.
16. Watkins, L.R., Maier, S.F. and Goehler, L.E. Cytokine-to-brain communication: A review & analysis of alternative mechanisms. *Life Sci.* 1995; 11:1011-1026.
17. Ek, M., Kurosawa, M., Lundeberg, T., et al. Activation of vagal afferents after intravenous injection of interleukin-1beta: role of endogenous prostaglandins. *J. Neurosci.* 1998; 18:9471-9479.
18. Laflamme, N. and Rivest, S. Effects of systemic immunogenic insults and circulating pro-inflamatory cytokines on the transcription of the inhibitory factor kappaB alpha within specific cellular populations of the rat brain. *J. Neurochem.* 1999; 73:309-321.
19. Banks, W.A. Cytokines, CVSs, and the blood-brain-barrier. In: *Psychoneuro-immunology, V2,* R. Ader, D.L. Felten, N. Cohen, (eds.). Academic Press, New York, pp. 483-498, 2001.

20. Konsman, J.P., Tridon, V. and Dantzer, R. Diffusion and action of intracerebroventricularly injected interleukin-1 in the CNS. *Neuroscience* 2000; 101:957-967.
21. Quagliarello, V.J., Wisplwey, B., Long, Jr. W.J., et al. Recombinant interleukin-1 induces meningitis and blood-brain barrier injury in the rat. *J. Clin. Invest.* 1991; 87:1360-1366.
22. Merrill, J.E. and Benveniste, E.N. Cytokines in inflammatory brain lesions: helpful and harmful. *Trends Neurosci.* 1996; 8:331-338.
23. Cunningham, E.T. Jr. and DeSouza, E.B. Interleukin-1 receptors in the brain and endocrine tissue. *Immunol. Today* 1993; 14:171-176.
24. Kinouchi, K., Brown, G., Pasternak, G., et al. Identification and characterization of receptors for tumor necrosis factor–α in the brain. *Biochem. Biophys. Res. Comm.* 1991; 181:1532-1538.
25. Tancredi, V., Darcangelo, G., Grassi, F., et al. Tumor necrosis factor alters synaptic transmission in rat hippocampal slices. *Neurosci.* Lett. 1992; 146:176-178.
26. Breder, C.D., Tsujimoto, M., Terano, Y., et al. Distribution and characterization of tumor necrosis factor–α-like immunoreactivity in the murine central nervous system. *J. Comp. Neurol.* 1993; 337:543-567.
27. Rothwell, N.J. Annual review prize lecture: Cytokines - killers in the brain? *J. Physiol.* 1999; 514:3-17.
28. Turrin, N.P., Gayle, D. and Ilyin, S.E. Pro-inflammatory and anti-inflammatory cytokine mRNA induction in the periphery and brain following intraperitoneal administration of bacterial lipopolysaccharide. *Brain Res. Bull.* 2001; 54:443-453.
29. Nguyen, K.T., Deak, T., Owens, S.M., et al. Exposure to acute stress induces brain interleukin-1b protein in the rat. *J. Neurosci.* 1998; 18:2239-2246.
30. Sato, S., Reiner, S.L., Jensen, M.A., et al. Central nervous system mRNA expression following Theilers murine encephalomyelitis virus infection. *J. Neuroimmunol.* 1997; 76:213-223.
31. Brebner, K., Hayley, S., Merali, Z., et al. Synergistic effects of interleukin-1beta, interleukin-6 and tumor necrosis factor-alpha: Central monoamine, corticosterone and behavioral variations. *Neuropsychopharmacol.* 1999; 22:566-580.
32. Ericsson, A, Kovacs, KJ and Sawchenko, PE. A functional anatomical analysis of central pathways subserving the effects of interleukin-1 on stress-related neuroendocrine neurons. *J. Neurosci.* 1994; 14:899-913.
33. Vellucci, SV, Parrott, R.F., da Costa, A.C., et al. Increased body temperature, cortisol secretion, and hypothalamic expression of c-fos, corticotrophin releasing hormone and interleukin-1 beta mRNAs, following central administration of interleukin-1 beta in the sheep. *Mol. Brain Res.* 1995; 29:64-70.
34. Tilders, F.J.H., Schmidt, E.D. and De Goeij, D.C.E. Phenotypic plasticity of CRF neurons during stress. *Ann. N.Y. Acad. Sci.* 1993; 697:39-52.
35. Watanobe, H. and Takebe, K. Intrahypothalamic infusion with interleukin-1 beta stimulates the release of corticotropin-releasing hormone and arginine vasopressin and the plasma adrenocorticotropin in freely moving rats: a comparative perfusion of the paraventricular nucleus and the median eminence. *Neuroendocrinol.* 1993; 57:593-599.
36. Saperstein, A., Brand, H., Audhya, T., et al. Interleukin-1β mediates stress-induced immunosuppression via corticotropin-releasing factor. *Endocrinology* 1992; 130:152-159.
37. Ando, T. and Dunn, AJ. Mouse tumor necrosis factor-alpha increases brain tryptophan concentrations and norepinephrine metabolism while activating the HPA axis in mice. *Neuroimmunomod* 1999; 6:319-329.

38. Hayley, S., Brebner, K., Lacosta, S., et al. Sensitization to the effects of tumor necrosis factor-α: Neuroendocrine, central monoamine and behavioral variations. *J. Neurosci.* 1999a; 19:5654-5665.
39. Zhou, D., Shanks, N., Riechman, S.E. et al. Interleukin 6 modulates interleukin-1 and stress-induced activation of the hypothalamic-pituitary-adrenal axis in male rats. *Neuroendocrinol.* 1996; 63: 227-236.
40. Bernardini, R., Kamilaris, T.C., Calogero, A.E., et al. Interactions between tumor necrosis factor–α, hypothalamic corticotropin-releasing hormone, and adrenocorticotropin secretion in the rat. *Endocrinology* 1990; 126:2876-2881.
41. Kageyama, K., Watanobe, H., Takebe, K. In vivo evidence that arginine vasopressin is involved in the adrenocorticotropin response induced by interleukin-6 but not by tumor necrosis factor-alpha in the rat. *Neuroimmunomodul.* 1995; 2: 137-140.
42. Kobayashi, H., Fukata, J., Murakami, N., et al. Tumor necrosis factor receptors in the pituitary cells. Brain Res. 1997; 758:45-50.
43. Ritchie, P.K., Ashby, M., Knight, H.H., et al. Dopamine increases interleukin 6 release and inhibits tumor necrosis factor release from rat adrenal zona glomerulosa cells in vitro. *Eur. J. Endocrinol.* 1996; 134: 610-616.
44. Sweep, C.G., Van Der Meer, M.JM., Hermus, A.R., et al. Chronic stimulation of the pituitary-adrenal axis in rats by interleukin-1β infusion: in vivo and in vitro studies. *Endocrinology* 1992; 130:1153-1164.
45. Van Der Meer, M.J., Sweep, C.G., Pesman, G.J., et al. Chronic stimulation of the hypothalamus-pituitary-adrenal axis in rats by interleukin 1βeta: central and peripheral mechanisms. *Cytokine* 1996; 12:910-919
46. Van Der Meer, M.J.M., Sweep, C.G.J., Pesman, GJ., et al. Synergism between IL-1β and TNF–α on the activity of the pituitary-adrenal axis and on food intake in rats. *Am. J. Physiol.* 1995; 268:E551-E557.
47. Lacosta, S., Merali, Z. and Anisman, H. Influence of interleukin-1 on exploratory behaviors, plasma ACTH and cortisol, and central biogenic amines in mice. *Psychopharmacol.* 1998; 137:351-361.
48. Anisman, H., Hayley, S. and Merali, Z. Behavioral and central neurochemical consequences of cytokine challenge. In: *Neuroimmune Biology: New Foundation of Biology*, J Bienenstock, R Gorzynski, I Berczi, eds. Elsevier Science, NY, pp. 141-162, 2001.
49. Kaur, D., Cruess, D.F. and Potter, W.Z. Effect of IL-1alpha on the release of norepinephrine in rat hypothalamus. *J. Neuroimmunol.* 1998; 90:122-127.
50. Carmelia, G., Pietro, G. and De Simoni, M.G. Activation of the hypothalamic serotonergic system by central interleukin-1. *Eur. J. Pharmacol.* 1991; 209:139-140.
51. Zalcman, S., Green-Johnson, J.M., Murray, L, et al. Cytokine-specific central monoamine alterations induced by interleukin (IL)-1, IL-2 and IL-6. *Brain Res.* 1994; 643:40-49
52. Linthorst, A.C.E., Flachskamm, C., Muller-Preuss, P., et al. Effect of bacterial endotoxin and interleukin-1β on hippocampal serotonergic neurotransmission, behavioral activity, and free corticosterone levels: An in vivo microdialysis study. *J. Neurosci.* 1995; 15:2920-2934.
53. Shintani, F., Kanba, S., Nakaki, T., et al. Interleukin-1β augments release of norepinephrine, dopamine and serotonin in the rat anterior hypothalamus. *J. Neurosci.* 1993; 13:3574-3581.
54. Mohankumar, P.S. and Quadri, S.K. Systemic administration of interleukin-1 stimulates norepinephrine release in the paraventricular nucleus. *Life Sci* 1993; 52:1961-1967.

55. Mohankumar, P.S., Thyagarajan, S. and Quadri, S.K. Interleukin-1β increases 5-hydroxyindoleacetic acid release in the hypothalamus in vivo. *Brain Res. Bull.* 1993; 31:745-748
56. Kamikawa, H., Hori, T., Nakane, H., et al. IL-1beta increases norepinephrine level in rat frontal cortex: involvement of prostanoids, NO, and glutamate. *Am. J. Physiol.* 1998; 275: R803-810.
57. Merali, Z., Lacosta, S. and Anisman, H. Effects of interleukin-1β and mild stress on alterations of norepinephrine, dopamine and serotonin neurotransmission: A regional microdialysis study. *Brain Res.* 1997; 761:225-235.
58. Song, C., Merali, Z. and Anisman, H. Variations of nucleus accumbens dopamine and serotonin following systemic interleukin-1, interleukin-2 or interleukin-6 treatment. *Neuroscience* 1998; 88:823-836.
59. Kabiersch, A., Del Rey, A., Honegger, C.G., et al. Interleukin-1 induces changes in norepinephrine metabolism in the rat brain. *Brain Behav. Immun.* 1988; 2:267-274.
60. Mefford, I.N., Masters, C.F., Heyes, M.P, et al. & Eskay, R.L. Cytokine-induced activation of the neuroendocrine stress axis persists in endotoxin-tolerant mice. *Brain Res.* 1991; 557:327-330.
61. Hayley, S., Brebner, K., Staines, W.M., et al. Time-dependent sensitization effects associated with murine tumor necrosis factor-α: Behavioral and neurochemical processes. *Neuroimmunomodulation* 1999; 6:224.
62. Cho, L., Tsunoda, M. and Sharma, R.P. Effects of endotoxin and tumor necrosis factor alpha on regional brain neurotransmitters in mice. *Nat. Toxins* 1999; 7(5):187-195.
63. Pauli, S., Linthorst, A.C.E. and Reul, J.M.H.M. Tumor necrosis factor-a and interlekin-2 differentially affect hippocampal serotonergic neurotransmission, behavioral activity, body temperature and hypothalamic-pituitary-adrenocortical axis activity in the rat. *Eur. J. Neurosci.* 1998; 10:868-878.
64. Hayley, S., Ethier, K., Wall, P., et al. Immediate and protracted effects of TNF-alpha: Peripheral vs central sites of action. *Brain. Behav. Immun.* 2001; 15:155.
65. Turrin, N.P. and Plata-Salaman, C.R. Cytokine-cytokine interactions and the brain. *Brain Res. Bull.* 2000; 51:3-9.
66. Avitsur, R and, Yirmiya, R. Cytokines inhibit sexual behavior in female rats: I. Synergistic effects of tumor necrosis factor alpha and interleukin-1. *Brain Behav. Immun.* 1999; 14-32.
67. Perlstein, R.S., Whitnall, MH., Abrams, J.S., et al. Synergistic roles of interleukin-6, interleukin-1, and tumor necrosis factor in the adrenocorticotropin response to bacterial lipopolysaccharide in vivo. *Endocrinology* 1993; 132:946-952.
68. Sonti, G., Ilyin, S.E. and Plata-Salamn, C.R. Anorexia induced by cytokine interactions at pathophysiological concentrations. *Am. J. Physiol 1996;* 270, R1394-402.
69. Borowski, T., Kokkinidis, L., Merali, Z., et al. Lipopolysaccharide, central in vivo biogenic amine variations, and anhedonia. *Neuroreport* 1998; 9:3797-3802.
70. Hayley, S., Staines, W., Merali Z., et al. Changes in immunoreactivity for AVP, CRH, NOS and GFAP in the murine brain after tumor necrosis Factor-α treatment. *Soc. Neurosci.* Abstract 1999;. 25: 1121
71. Hayley, S., Lacosta, S., Merali, Z., et al. Central Monoamine and Plasma Corticosterone Changes Induced by a Bacterial Endotoxin: Sensitization and Cross-sensitization Effects. *Eur. J. Neurosci.* 2001; 13:1155-1165.
72. Rivest, S. and Rivier, C. Stress and interleukin-1 beta-induced activation of c-Fos, NGFI-B and CRF gene expression in the hypothalamic PVN - Comparison between Sprague-Dawley, Fisher-344 and Lewis rats. *J. Neuroendocrinol.* 1994; 6:101-117.

73. Rivest, S. and Rivier, C. Influence of the paraventricular nucleus of the hypothalamus in the alteration of neuroendocrine functions induced by intermittent footshock or interleukin. *Endocrinology*, 1991; 129:2049-2057.
74. Makara, G.B. and Kovacs, K.J. Lesioning of the hypothalamic paraventricular nucleus inhibits ether-induced ACTH but not prolactin release. *Neurobiology*, 1997; 5:403-411.
75. Shintani, F., Nakaki, T. and Kanba, S. Involvement of interleukin-1 in immibilization stress-induced increase in plasma adrenocorticotropic hormone and in release of hypothalamic monoamines in the rat. *J. Neurosci.* 1995; 15:1961-1970.
76. Sawchenko, P.E., Brown, E.R., Chan, R.K., et al. The paraventricular nucleus of the hypothalamus and the functional neuroanatomy of visceromotor responses to stress. *Prog. Brain. Res.* 1996; 107:201-222
77. Lee, S. and Rivier, C. Interaction between corticotropin-releasing factor and nitric oxide in mediating the response of the rat hypothalamus to immune and non-immune stimuli. *Mol. Brain Res.* 1198; 57, 54-62.
78. Imaki, T., Shibasaki, T., Hotta, M., et al. Intracerebroventricular administration of corticotropin-releasing factor induces c-fos messenger RNA expression in brain regions related to stress responses - comparison with pattern of c-fos messenger RNA induction after stress. *Brain Res.* 1993; 616:114-125.
79. Davis, M. The role of the amygdala in fear and anxiety. *Ann. Rev. Neurosci.* 1992; 15, 353-375.
80. LeDoux, J.E. Emotion circuits in the brain. *Ann. Re.v Neurosci.* 2000; 23:155-184.
81. Connor, T., Song, C., Leonard, B.E., et al. An assessment of the effects of central interleukin-1β, -2, -6 and tumor necrosis factor–α administration on some behavioural, neurochemical, endocrine and immune parameters in the rat. *Neuroscience* 1998; 84:923-933.
82. Dunn, A.J. and Swiergiel, A.H. Behavioral responses to stress are intact in CRF-deficient mice. *Brain Res.* 1999; 845:14-20.
83. Lee, H.Y., Whiteside, M.B. and Herkenham, M. Area postrema removal abolishes stimulatory effects of intravenous interleukin-1beta on hypothalamic-pituitary-adrenal axis activity and c-fos mRNA in the hypothalamic paraventricular nucleus. *Brain Res. Bull.* 1998; 46:495-503.
84. Tilders, F.J.H. and Schmidt, E.D. Interleukin-1-induced plasticity of hypothalamic CRH neurons and long-term stress hyperresponsiveness. *Ann. N.Y. Acad. Sci* 1998; 840:65-73.
85. Bartanusz, V., Jezova, D., Bertini, L.T., et al. Stress-induced increase in vasopressin and corticotropin-releasing factor expression in hypophysiotrophic paraventricular neurons. *Endocrinology* 1993; 132:895- 902.
86. de Goeij, D.C., Dijkstra, H. and Tilders, F.J.H. Chronic psychosocial stress enhances vasopressin, but not corticotropin-releasing factor, in the external zone of the median eminence of male rats: relationship to subordinate status. *Endocrinology* 1992; 131:847-53.
87. Schmidt, E.D., Janszen, A.W.JW., Wouterlood, FG., et al. Interleukin-1 induced long-lasting changes in hypothalamic corticotropin-releasing hormone (CRH) neurons and hyperresponsiveness of the hypothalamic-pituitary-adrenal axis. *J. Neurosci.* 1995; 15:7417-7426.
88. de Goeij, D.C., Jezova, D. and Tilders, F.J.H. Repeated stress enhances vasopressin synthesis in corticotropin releasing factor neurons in the paraventricular nucleus. *Brain Res.* 1992; 577:165-168.
89. de Goeij, D.C., Kvetnansky, R., Whitnall, M., et al. Repeated stress-induced activation of corticotrophin-releasing factor neurons enhances vasopressin stores and colocalization

with corticotrophin-releasing factor in the median eminence of rats. *Neuroendocrinology* 1991; 53:150-159.

90. Sawchenko, P. Evidence for differential regulation of corticotrophin-releasing factor and vasopressin immunoreactivies in parvocellular neurosecretory and autonomic-projections of the paraventricular nucleus. *Brain Res.* 1987; 437:253-263.
91. Whitnall, M.H., Key, S. and Gainer, H. Vasopressin-containing and vasopressin-deficient subpopulations of corticotropin-releasing factor axons are differentially affected by adrenalectomy. *Endocrinol.* 1987; 120:2180-2182.
92. Schmidt, E.D., Binnekade, R., Janszen, A.W., et al. Short stressor induced long-lasting increases of vasopressin stores in hypothalamic corticotropin-releasing hormone (CRH) neurons in adult rats. *J. Neuroendocrinol.* 1996; 8:703-712.
93. Finlay, J.M., Jedema, H.P., Rabinovic, AD, et al. Impact of corticotropin-releasing hormone on extracellular norepinephrine in prefrontal cortex after chronic cold stress. *J. Neurochem.* 1997; 69:144-150.
94. Gresch, P.J., Sved, A.F., Zigmond, M.J., et al. Stress-induced sensitization of dopamine and norepinephrine efflux in medial prefrontal cortex of the rat. J. Neurochem. 1994; 63:575-583.
95. Jordan, S., Kramer, G.L., Zukas, P.K., et al In vivo biogenic amine efflux in medial prefrontal cortex with imipramine, fluoxetine, and fluvoxamine. *Synapse* 1994; 18: 294-297.
96. Robinson, T.E., Angus, A.L. and Becker, J.B. Sensitization to stress: The enduring effects of prior stress on amphetamine-induced rotational behavior. *Neuropharamacol.* 1987; 26:679-691.
97. Tilders, F.J.H. and Schmidt, ED. Cross-sensitization between immune and non-immune stressors. A role in the etiology of depression? *Adv. Exp. Med. Biol.* 1999; 461:179-197.
98. Van Dijken, H.H., Van der Heyden, J.A., Mos, J, et al. Inescapable footshocks induce progressive and long-lasting behavioural changes in male rats. *Physiol. Behav.* 1993; 51:787-794.
99. Deutch, A.Y. and Roth, R.H. The determinants of stress-induced activation of the prefrontal cortical dopamine system. In *Progress in Brain Research, Vol. 85,* HBM Uylings, CG Van Eden, JPC De Bruin, MA Corner, MGP Feenstra, eds. New York, Elsevier; 367-403, 1990.
100. Jedema, H.P., Sved , A.F., Zigmond, M.J., et al. Sensitization of norepinephrine release in medial prefrontal cortex: effect of different chronic stress protocols. *Brain Res.* 1999; 830:211-217.
101. Irwin, J., Ahluwalia, P. and Anisman, H. Sensitization of norepinephrine activity following acute and chronic footshock. *Brain Res.* 1986; 379:98-103.
102. Nisenbaum, L.K., Zigmond, M.J., Sved, A.F.et al. Prior exposure to chronic stress results in enhanced synthesis and release of hippocampal norepinephrine in response to a novel stressor. *J. Neurosci.* 1991; 11, 1478-1484.
103. Bhatnagar, S., Dallman, M.F., Roderick, R.E., et al. The effects of prior chronic stress on cardiovascular responses to acute restraint and formalin injection. *Brain Res.* 1998; 797:313-320.
104. Chung, K.K., Martinez, M. and Herbert, J. c-fos expression, behavioural, endocrine and autonomic responses to acute social stress in male rats after chronic restraint: modulation by serotonin. *Neuroscience* 2000; 95:453-463.
105. Cuadra, G., Zurita, A., Lacerra, C., et al. Chronic stress sensitizes frontal cortex dopamine release in response to a subsequent novel stressor: reversal by naloxone. *Brain Res. Bull.* 1999; 48:303-308.

106. McEwen, B.S. Stress, adaptation, and disease. Allostasis and allostatic load. *Ann. N.Y. Acd. Sci.* 1998; 840: 33-44.
107. Tilders, F.J.H., DeRijk, R.H., Van Dam , A.M., et al. Activation of the hypothalamus-pituitary–adrenal axis by bacterial endotoxins: routes and intermediate signals. *Psychoneuroendocrinol.* 1994; 19, 209-232.
108. Plotsky, P.M. (1991). Pathways to the secretion of adrencorticotropin: a view from the portal. *J. Neuroendocinol.*, 3, 1-9. (Different date in text.)
109. Wang, G., Klostergaard, J., Khodadadian, M., et al. Murine cells transfected with human Hsp27 cDNA resist TNF-induced cytotoxicity. *J. Immunother. Emphasis Tumor. Immunol.* 1996; 19, 9-20.
110. Brouckaert, P., Libert, C., Everaerdt, B., et al. Selective species specificity of tumor necrosis factor for toxicity in the mouse. *Lymphokine Cytokine Res.* 1992; 11:193-196.
111. Beutler, B. and van Huffel, C. Unraveling function in the TNF ligand and receptor families. *Science* 1994; 264:667-668.
112. Brown, R.E., Stevens, D.R. and Haas, H.L. The physiology of brain histamine. *Prog. Neurobiol.* 2001; 63:637-672.
113. Brzezinska-Blaszczyk, E., Forczmanski, M., Pietrzak, A. The action of tumor necrosis factor-alpha on rat mast cells. *J. Interferon Cytokine Res.* 2000; 20: 377-382.
114. Schwartz, J.C., Arrang, J.M., Garbarg, M., et al. Histaminergic transmission in the mammalian brain. *Physiol. Rev.* 1991; 7:1-51.
115. Giovannini, M.G., Bartolini, L., Bacciottini, L., et al. Effects of histamine H3 receptor agonists and antagonists on cognitive performance and scopolamine-induced amnesia. *Behav. Brain Res.* 1999; 104, 147-155.
116. Masaki, T., Yoshimatsu, H., Chiba, S., et al. Targeted disruption of histamine H1-receptor attenuates regulatory effects of leptin on feeding, adiposity, and UCP family in mice. *Diabetes* 2001; 50, 385-391.
117. Kjaer, A., Larsen, P.J., Knigge, U., et al. Neuronal histamine and expression of corticoptropi-rleeasing hormone vasopressin and oxytocin in the hypothalamus: relative importance of H1 and H2 receptors. *Eur. J. Endocrinol.* 1998; 139: 238-243.
118. Hayley, S., Kelly, O., Anisman, H. Murine tumor necrosis factor α sensitizes plasma corticisoterone and manifestation of shock: modulation by histamine. *J. Neuroimmunol.* 2002; in press.
119. Nadeau, S. and Rivest, R. Effects of circulating tumor necrosis factor on the neuronal activity and expression of the genes encoding the tumor necrosis factor receptors (p55 and p75) in the rat brain: a view from the blood-brain barrier. *Neuroscience* 1999; 93: 1449-1464.
120. Hayley, S., Staines, W.M., Merali, Z., et al. Time- dependent sensitization of corticotropin releasing hormone, arginine vasopressin and *c-fos* immunoreactivity within the mouse brain in response to tumor necrosis factor-α. *Neuroscience* 2001; 106: 137-148.
121. Sheng, M. and Greenberg, M.E. The regulation and function of c-fos and other immediate early genes in the nervous system. *Neuron* 1990;4, 477-485.
122. Soliven, B. and Albert, J. Tumor necrosis factor modulates the inactivation of catchecholamine secretion in cultured sympathetic neurons. *J. Neurochem.* 1992; 58, 1073-1078.
123. Chang, S.L., Ren, T. and Zadina, J.E. Interleukin-1 activation of FOS proto-oncogene protein in the rat hypothalamus. *Brain Res.* 1993; 617:123-130.
124. Lee, Y. and Davis, M. Role of the hippocampus, the bed nucleus of the stria terminalis, and the amygdala in the excitatory effect of corticotropin-releasing hormone on the acoustic startle reflex. *J. Neurosci.* 1997; 17, 6434-6446.

125. Van Bockstaele, E.J., Colago, E.E. and Valentino, R.J. Amygdaloid corticotropin-releasing factor targets locus coeruleus dendrites: substrate for the co-ordination of emotional and cognitive limbs of the stress response. *J. Neuroendocrinol.* 1998; 10, 743-757.
126. Ignatowski, T.A., Noble, B.K., Wright, J.R., et al. Neuronal-associated tumor necrosis factor (TNF-α): its role in noradrenergic functioning and modification of its expression following antidepressant drug administration. *J. Neuroimmunol.* 1997; 79:84-90.
127. Mandrup-Poulsen, T., Nerup, J., Reimers, J.I., et al. Cytokines and the endocrine system. I. The immunoendocrine network. *Eur. J. Endocri.*1995; 133, 660-671.
128. Galanos, C. and Freudenberg, M.A. Mechanisms of endotoxin shock and endotoxin hypersensitivity. *Immunobiol.* 1993; 187, 346-356.
129. Berczi, I., Chow, D.A. and Sabbadini, E.R. Neuroimmunoregulation and natural immunity. *Domest. Anim. Endocrinol.* 1998; 15:273-281.
130. Breder, C.D., Hazuka, C., Ghayur, T., et al. Regional induction of tunor necrosis factor a expression in the mouse brain. *Proc. Natl. Acad. Sci.* 1994; 91:11393-11397.
131. Plata-Salaman, C.R., Oomura, Y. and Kai, Y. Tumor necrosis factor and interleukin-1 beta: suppression of food intake by direct action in the central nervous system. *Brain Res.*1998; 448, 06-114.
132. Maes, M. Major depression and activation of the inflammatory response system. *Adv. Exp. Med. Biol.*1999; 461, 25-46.
133. Berczi, I. Neuroendocrine defence in endotoxin shock (a review). *Acta. Microbiol. Hung.* 1993; 40: 265-302.
134. West, M.A., Bennet, T., Seatter, S.C., et al. LPS pretreatment reprograms macrophage LPS-stimulated TNF and IL-1 release without protein tyrosine kinase activation. *J. Leukoc. Biol.*1997; 61, 88-95.
135. Wallach, D., Holtmann, H., Engelmann, H., et al. Sensitization and desensitization to lethal effects of tumor necrosis factor and IL-1. *J. Immunol.*1988; 140, 2994-2999.
136. Pezeshki, G., Pohl, T. and Schbitz, B. Corticosterone controls interleukin-1 beta expression and sickness behavior in the rat. *J. Neuroendocrinol.* 1996; 2, 129-35
137. Hardaway, R.M. A review of septic shock. *Am. Surg.* 2000; 66:22-29.
138. Hayley, S., Merali, Z. and Anisman, H. Central monoamine activity in genetically distinct strains of mice following a psychogenic stressor: Effects of predator exposure. *Brain Res.* 2001; 892: 293-300.
139. Lacosta, S., Merali, Z. and Anisman, H. Behavioral and neurochemical consequences of lipopolysaccharide in mice: anxigenic-like effects. *Brain Res.*1999; 818, 291-303.
140. Linthorst, A.C. and Reul, J.M. Brain neurotransmission during peripheral inflammation. *Ann. N.Y. Acad. Sci.*1998; 840, 139-152
141. Nagano, I., Takao, T., Nanamiya, w., et al. Differential effects of one and repeated endotoxin treatment on pituitary- adrenocortical hormones in the mouse: role of interleukin-1 and tumor necrosis factor-alpha. *Neuroimmunomodul.* 1999; 6: 284-292.
142. Freudenberg, M.A. and Galanos, C. The metabolic fate of endotoxins. *Prog. Clin. Biol. Res.* 1988; 272:63-75.
143. Heremans, H., Van Damme, J., Dillen, C., et al. Interferon gamma, a mediator of lethal lipopolysaccharide-induced Shwartzman-like shock reactions in mice. *J. Exp. Med.* 1990; 171:1853-1869.
144. Derijk, R., Van Rooijen, N., Tilders, F.J.H., et al. Selective depletion of macrophages prevents pituitary-adrenal activation in response to subpyrogenic, but not to pyrogenic, doses of bacterial endotoxin in rats. *Endocrinology* 1991; 129:330-338.

145. Fahmi, H., Ancuta, P., Perrier , S., et al. Preexposure of mouse peritoneal macrophages to lipopolysaccharide and other stimuli enhances the nitric oxide response to secondary stimuli. Inflamm Res 1996; 45:347-353.
146. Yirmiya, R., Weidenfeld, J., Pollak, Y., et al. Cytokines, "depression due to a general medical condition," and antidepressant drugs. *Adv. Exp. Med. Biol.* 1999; 461: 283-316.
147. Musselman, D.L., Lawson, D.H., Gumnick, J.F et al. Paroxetine for the Prevention of the Depression and Neurotoxicity Induced by High Dose Interferon Alpha". *New Eng. J. Med.* 2001; 344, 961-966.
148. Heim, C., Owens, M.J., Plotsky, P.M., et al. Persistent changes in corticotrophin-releasing factor systems due to early life stress: relationship to the pathophysiology of major depression and post-traumatic stress disorder. *Psychopharmacol. Bull.* 1997; 33:185-192.
149. Meaney, M.J., Diorio, J., Francis, D., et al. Early environment regulation of forebrain glucocorticoid receptor gene expression: implications for adrenocortical responses to stress. *Dev. Neurosci.*1996; 18, 49-72.
150. Shanks, N., Windle, R.J., Perks, P.A., et al. Early-life exposure to endotoxin alters hypothalamic-pituitary-adrenal function and predisposition to inflammation. *Proc. Natl. Acad. Sci.*2000; 97, 645-650.

Chapter 12

CYTOKINE REGULATION IN MAJOR DEPRESSION

Ziad Kronfol
Department of Psychiatry, University of Michigan Health System, Ann Arbor, Michigan

INTRODUCTION

Two decades have now elapsed since the first reports of immune abnormalities in patients with major depression have been published (1-4). While these early reports were generally well received, there was a good deal of skepticism regarding their clinical significance (5). With the very rapid growth of research in the field of psychoneuroimmunology that followed (6) new meaning and new insights have been added to the neuroimmunology of affective disorders. More specifically, the explosion in cytokine research over the last decade or so and the role these cytokines may play in various neurochemical, neuroendocrine and behavioral phenomena have offered new possibilities for a role for cytokines in major depression. This review will first examine the evidence for immune dysregulation in major depression in general, and will then focus on cytokine dysregulation in particular. A critical review of the "cytokine hypothesis" of major depression will then follow.

1. DEPRESSION AND THE IMMUNE SYSTEM

1.1 Summary of existing findings

The rationale for the early studies of immune regulation in major depression can be summarized as follows: 1) depression is a form of stress and stress has been associated with immune dysregulation, particularly immune suppression (7); 2) depression is associated with hyperactivity of the hypothalamic-pituitary adrenal (HPA) axis, resulting in hypersecretion of cortisol (8), an immunosuppressive hormone and 3) depression is associated with increased frequency of immune-related disorders such as infection (9), allergy (10), autoimmune diseases (11) and cancer (12). The hope was that a specific immune abnormality would be found that could tie these findings together and explain the relation between depression and at least one or two medical conditions. The possibility that the immune disturbance would be related to the pathophysiology of depression was not directly entertained at that early stage. Several investigators have therefore explored different facets of the immune system in patients with major depression. The results revealed a wide array of immunological abnormalities in patients with major depression. These can be grouped into three major categories: 1) abnormalities of leukocytes trafficking and lymphocyte subsets; 2) abnormalities of leukocyte functions and 3) abnormal humoral factors. A meta-analytic review of the literature, using both a fixed effects analysis that allows generalization to other subjects like those in the reviewed studies and a more conservative random effects analysis that allows generalization to future studies has recently been conducted (13). The results are summarized in Table 1.

Table 1. Summary of Immune Abnormalities in Major Depression

Abnormalities of Leukocyte Trafficking and Lymphocyte Subsets	Leukocytosis*, neutrophilia*, lymphocytopenia*, decreased NK cell counts and T-cell proportions, increased CD_4/CD_8 ratios.*
Abnormal Leukocyte Functions	Decreased neutrophil phagocytosis, decreased lymphocytic responses to the mitogens PHA*, Con A* and PWM*, reduced NK cytotoxicity.*
Abnormal Humoral Factors	Increased IgM, increased PGE_2*, increased IL-6 levels*, increased sIL-2R, increased acute phase proteins (eg. haptoglobin*, α_1 acid glycoprotein, α_1 antitrypsin,)

*confirmed by random effect meta-analysis.

Results of leukocyte trafficking in patients with major depression reveal an increase in the total number of circulating leukocytes accompanied by an increase in the absolute and relative (percentage) number of neutrophils and a decrease in the absolute and relative (percentage) number of lymphocytes (3,14,15). For monocytes, the absolute numbers increased, while the relative (percentage) numbers decreased (3,15). Results of lymphocyte subset analysis indicates no changes in the absolute number of B cells, CD_3+ cells, CD_4+ cells and CD_8+ cells (14). The absolute numbers of NK cells is slightly decreased (16). As for the relative (percentage) numbers, B cells increased while CD_3 cells decreased slightly (17). The percentage of CD_4+

cells increased while the percentage of CD_8 cells did not change. The ratio of CD_4/CD_8 cells increased (18). The percentage of NK cells did not change. Both the absolute and relative number of activated lymphocytes $CD_{25}+$ and HLA-DR increased (17).

Results of leukocyte function indicate a mild decrease in neutrophil phagocytosis (19) and moderate decreases in the lymphocyte responses to stimulation with the mitogens Phytoheonagglutinin (PHA), Concanavalin A (ConA) and Pokeweed Mitogen (PWM) (1,2,14,20). NK cytoxicity is also moderately decreased in the depressed group (21,22). IL-2 production is not changed (23).

Assessment of humoral factors reveals a decrease in total serum protein, a decrease in serum albumin and an increase in acute phase proteins, including haptoglobin, α_1-acid glycoprotein and α_1-antitripsin (24). IgM concentrations are increased, while IgG and IgA are unchanged (14). Prostaglandin PGE_2 concentration is increased (25). Plasma levels of IL-6 and soluble IL-2 receptors (sIL-2R) are also increased (26).

In summary, there are numerous reports of immunological abnormalities in patients with major depression including abnormalities in leukocyte trafficking, lymphocyte subsets, leukocyte function and humoral factors. The clinical significance of these findings however remains unclear. While some of these results have at times been difficult to interpret (e.g., evidence of both immune activation as evidenced by increased levels of IL-6 and haptoglobin and immune suppression as evidenced by decreased lymphocyte NK cytotoxicity), it is now well known that immune regulation is a complex phenomenon and that both immune activation and immune suppression can occur simultaneously and can be associated with stress (27).

1.2 Proposed Mechanisms

Because these immunological findings were at first believed to be strictly related to the well-established neuroendocrine abnormalities seen in depressive illness (28,29), many of the early studies addressing mechanisms of immunosuppression in major depression focused on dysregulation of the HPA axis and its relation to immune function. In a study of 40 patients with major depression and 37 matched controls, we found a clear association between cortisol values and the traffic of leukocytes: higher cortisol values were associated with higher percentages of circulating neutrophils and lower percentages of circulating lymphocytes (14). The association between measures of lymphocyte function and measures of HPA activity however has been more difficult to characterize. In separate studies of patients with major depression, we found no significant associations between lymphocyte responses to mitogenic stimulation on the one hand and measures of 24-hour urinary-free cortisol (30) or results of the dexamethasone suppression test on the other (31). Darko et al. found a significant positive correlation between NK cytotoxicity and plasma β-endorphin levels in depressed patients but not in healthy controls (32). Irwin and colleagues found a significant negative correlation between neuropeptide Y concentrations and NK activity in a group of depressed patients (33). We reported a negative association between levels of the complement component C_4 and lymphocytic responses to the mitogens PHA, Con A and PWM (14). There are also reports of an association between measures of HPA activity and cytokine secretion. In a study of 28 inpatients with major or minor depression Maes and colleagues found a statistically significant positive correlation between mitogen-induced IL-1 β production and post DST cortisol values. They concluded that IL-1β oversecretion may contribute to HPA axis hyperactivity in major depression (34). More recent research indicates that the relation between neuroendocrine and immune regulation in general and in depressed patients in particular is quite complex and involves several interrelated systems that include cytokines, hormones and neurotransmitters (35)

1.3 Clinical Applications

One of the most pressing and yet unresolved issue in clinical psychoneuroimmunology is the relation between stress-induced immuno-suppression and health status, particularly in disorders that are thought to be of immune and/or inflammatory origin (36). While epidemiological studies suggest a possible association between stress (or depression) and disease onset or disease progression, the role of stress-induced immune abnormalities in explaining these changes in health status have not been easy to document. As an illustration, we will briefly review pertinent literature summarising the relation between depression, immune dysfunction and the following disorders: cancer, infection, AIDS and coronary artery disease.

1.3.1 Cancer

Several epidemiological studies have noted an association between depressed mood and an increased risk of developing cancer (37,38). These findings however have not always been confirmed (39). A meta-analytic review concluded that depression was associated with a small though statistically significant increase in the risk of cancer (12). Another controversy in that domain pertains to the effects of psychosocial intervention in the progression of cancer. Early reports have noted an increase in survival rate among women with breast cancer who participated in semi-structured group psychotherapy (40). More recent studies however have not been able to confirm these findings (41). Furthermore, the relation between depression, immune dysregulation and cancer progression has rarely been addressed. Levy and colleagues (1987) reported an association between stress factors, a decline in natural killer cell activity and poor prognosis in patients with breast cancer (42). Unfortunately, these findings have not been replicated nor has there been any serious effort to explore the complex relationship between depression, immunity and cancer progression.

1.3.2 Infection

Infection occurs as a result of complex interactions between an infectious agent and the host. Animal studies show that inescapable shock, an animal model of depression, increases susceptibility to viral diseases such as herpes simplex and influenza viruses by altering immune function (43). Human studies however have been more difficult to conduct. In an elegant study to address these issues, Cohen and colleagues experimentally inoculated healthy volunteers with different strains of the common cold viruses following an extensive assessment of their psychological profiles (9). They found significant associations between the frequency and severity of infections and the levels of depression and stress. In another study, the same group of researchers experimentally inoculated healthy volunteers with the influenza A virus and monitored upper respiratory symptoms along with nasal levels of IL-6. As predicted, the authors found significant associations between psychological stressors, severity of upper respiratory symptoms and nasal levels of IL-6. They concluded that IL-6 may play a role as mediator between stress and illness (44). In a different line of research, Kiecolt-Glaser and colleagues reported that spousal caregivers of dementia patients (a model of chronically stressed individuals) had a poorer antibody response to influenza vaccination than non-stressed age-and-sex-matched controls (45). Irwin and colleagues found that the frequency of cells in the peripheral blood capable of proliferating to the Varicella Zoster Virus were significantly decreased in patients with major depression compared to matched healthy controls (46). The relation to clinical Herpes Zoster or shingles however remains to be determined.

1.3.3 AIDS

A good deal of research has recently been conducted to address the relation between psychosocial factors and HIV progression. The general hypothesis here has been that levels of stress and/or depression, which are usually elevated in patients with HIV infection, will hasten the progression to AIDS and adversely impact the course of this illness. The results however have been inconsistent. Lesserman and colleagues found a significant association between severe stress, depressive symptoms and declines in several lymphocyte subsets in men infected with HIV over a two-year period (47). The same authors have also reported that more stress, more depression and less social support may accelerate the rate of HIV progression to AIDS. (48). As to possible associations between depressive symptoms and mortality among AIDS patients, Mayne and colleagues reported a small but statistically significant increase in mortality among AIDS patients with chronic depressive symptoms (49). Burrack and colleagues found a significant negative association between depressive symptoms and CD_4+ cells but not with morbidity or mortality (50). In contrast to these reports, Lykestos et al (1993) found no relation between depressive symptoms, CD_4+ cells, morbidity or mortality in HIV infected patients (51).

1.3.4 Coronary Artery Disease

Although coronary artery disease has not traditionally been considered an immunological disorder, there is strong recent evidence that inflammatory processes play an important role in the pathophysiology of artherosclerosis (52). Increased levels of systemic markers of inflammation such as C-reactive proteins and fibrinogen have been associated with increased cardiac morbidity and mortality both in patients with coronary artery disease and apparently healthy individuals (53,54). Increased levels of the pro-inflammatory cytokines IL1β, IL6 and IL-8 have also been reported in unstable angina (58), with reports of an association between IL-6 and mortality in this disease (55-57). Epidemiological studies have shown that depression is common among patients recovering from a myocardial infarction, and that depressed patients have a higher risk of cardiac mortality in follow-up studies (59). Furthermore, several epidemiological studies seem to support the notion that depression increases the risk of developing coronary artery disease both in men and women (60). In spite of these findings, a clear link between depression, immune abnormalities, and the onset and course of coronary artery disease has not been firmly established.

In conclusion, various immunological abnormalities have been reported over the last two decades in patients with major depression. These include abnormalities in leukocyte trafficking, leukocyte function, as well as evidence of immune activation. Attempts have been made to explain these findings in relation to neuroendocrine dysregulation, particularly HPA dysfunction. The clinical implications have also been addressed and the relation of depression to immune-related disorders such as infection, cancer and AIDS has been explored. While several of the early research questions have been addressed, the answers provided have raised many more questions. The need to press on with fresh new and testable hypotheses continues.

2. THE CYTOKINE HYPOTHESIS OF MAJOR DEPRESSION

2.1 Background and Hypothesis

Early studies of immune regulation in patients with major depression, relying mostly on assays of lymphocyte functions such as lymphocyte responses to mitogen stimulation and lymphocyte NK cell activity, pointed in general in the direction of immune suppression. These results were confirmed by several laboratories as indicated by the meta-analytic reviews. The magnitude of the change however was relatively small and the clinical significance of these findings was difficult to determine. Furthermore, a general hypothesis regarding the pathophysiology of depression that could incorporate these and other findings was essentially lacking (5). The decade of the 90's however has witnessed a more steady stream of articles suggesting immune activation in patients with major depression, particularly an increase in the secretion of specific cytokines and acute phase proteins (24,61,62). This was coupled with an explosion of basic research in the field of psychoneuroimmunology, stressing the important role of cytokines as neuromodulators within the brain, and emphasising their respective effects on neurotransmitter and neurohormone regulation (6,63). The behavioral effects of cytokines, including effects on sleep, appetite, sexual behavior, memory, and motor activity were also being recognized (64). All of this has helped in the development of a "cytokine hypothesis" of major depression. The hypothesis was first formulated by Smith in 1991 as the "macrophage hypothesis" of major depression (65). It was then developed further by Maes and colleagues (66). More recently, Dantzer and other researchers have organized two international conferences devoted to the discussion of the role of cytokines in major depression. The hypothesis has been gaining momentum ever since.

But what is the hypothesis about? In its simplest form (figure 1), it suggests that cytokines play a key role in the pathophysiology of major depression. Pro-inflammatory cytokines such as IL-1β, IL-6 and/or TNF are increased as a result of stress, either physical (e.g., virus) or psychological

Important Biological Features of Depression	BRAIN Effects of Cytokines	Clinical Symptoms of Depression (DSM-IV)
▪ Changes in NE, 5HT, DA metabolism ▪ HPA-axis activation: Hypercortisolemia/DST non-suppression ▪ Changes in sleep EEG architecture ▪ Leukocytosis, neutrophilia ▪ ↓ NK cell activity ▪ ↑ haptoglobin ▪ ↑ IL-6 ▪ hypofrontal metabolism on PET brain imaging	▪ activation of cytokine pathways ▪ changes in neurotransmiter turnover (e.g., NE, 5HT, DA) ▪ increased CRH secretion/HPA axis activation ▪ changes in sleep EEG architecture ▪ behavioral changes: sadness, irritability, anxiety, fatigue, decreased social and sexual activity, increased sleep, decreased food intake, cognitive impairment. ▪ changes in cerebral blood flow distribution?	▪ sadness ▪ anhedonia ▪ sleep disturbance ▪ appetite disturbance/ weight loss ▪ psychomotor retardation ▪ fatigue, loss of energy ▪ cognitive impairment ▪ feelings of worthlessness/guilt ▪ suicidal ideation

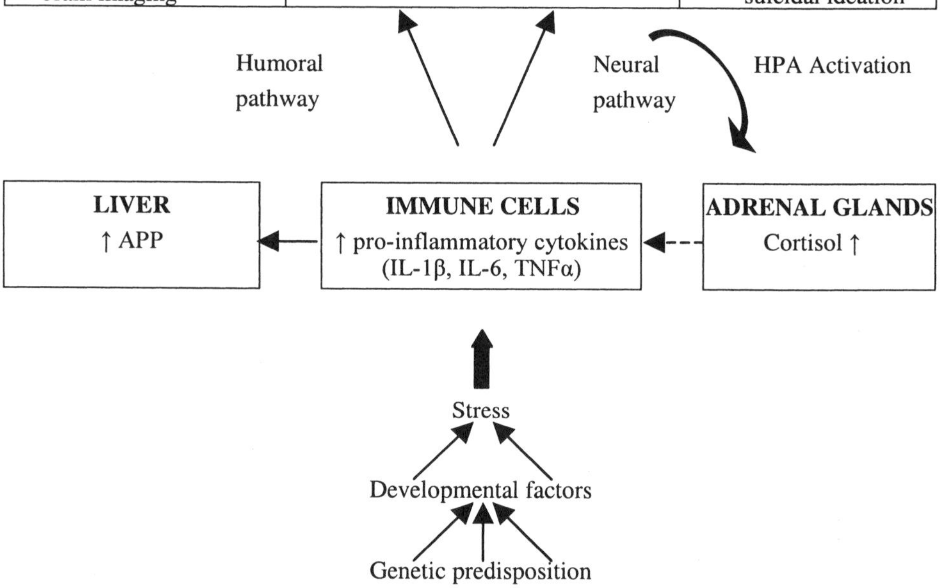

Figure 1. Cytokine hypothesis of major depresson. Stress (physical or psychological) increases the secretion of pro-inflammatory cytokines in genetically and/or developmentally predisposed individuals. Excessive cytokine secretion will directly or indirectly affect the brain resulting in biological and behavioral changes (middle column) that mimic the biological (first column) and clinical (last column) features of depression. Other organs are also involved (e.g., liver, adrenals). The adrenals in particular play an inhibitory role through negative feedback mechanisms (broken arrow). APP: acute phase proteins; DST: dexamethasone suppression test; NE: norepinephrine; 5HT: serotonin; DA: dopamine; NK: natural killer; CRH: corticotropin-releasing hormone; HPA: hypothalamic-pituitary-adrenal; EEG: electroencephalogram.

(e.g., divorce or bereavement), This increase in cytokines is conveyed to the brain directly or indirectly through humoral or neural pathways. Increased cytokine signaling in the brain will result in neurochemical changes (such as changes in serotonin and/or norepinephrine transmission), neuroendocrine changes (such as increased HPA activity) and behavioral changes such as sleep and appetite disturbance, social withdrawal and decreased motor activity. In subsequent, more refined versions of the hypothesis, the changes in cytokine levels were described as state-dependent, not trait-dependent. Furthermore, antidepressant medications were postulated to work, at least in part, by down-regulating excessive cytokine secretion. In the following paragraphs, we will examine the evidence for these assertions and present a critical review of the hypothesis as it stands today.

2.2 Neurochemical/Neuroendocrine Factors

As indicated in Chapter 4 of this volume, cytokines in the brain have been associated with a variety of neurochemical changes including changes in the metabolism of norepinephrine, serotonin and dopamine, neurotransmitters that have been implicated in the pathophysiology of major depression. IL-1 for instance has been shown to increase the brain concentrations of norepinephrine and serotonin but not dopamine. IL-2 increases the levels of norepinephrine and dopamine but not serotonin. IL-6 increases the levels of serotonin only. TNFα increases the levels of norepinephrine but only when used in high doses. IFNγ had no measurable effect on the turnover of norepinephrine, dopamine or serotonin (67). Thus, increases in cytokines brought about by psychosocial stress or virus can lead to neurochemical changes in key portions of the brain that affect mood and vegetative functions, and possibly leading to major depression.

Cytokines have also been shown to affect neuroendocrine regulation (see chapter 5 of this volume). Specifically, intraperitoneal or intracerebroventricular injections of IL-1β has been associated with activation of the HPA axis and increases in the secretion of CRH, ACTH and cortisol (68). HPA activation has been one of the most replicated findings in the psychobiology of major depression, and a defect in feedback regulation, perhaps at the glucocorticoid receptor level, has been postulated (69). It is therefore possible that excessive secretion of cytokines, possibly in relation to stress, might activate the HPA axis, probably through an increase in the secretion of CRH (68). Both an increase in cytokines (26,61) and an increase in HPA activity (28,29) are common in major depression. Furthermore, the increases in at least one cytokine (IL-Iβ) and cortisol are significantly and positively correlated (34). Similar findings have also been reported for IL-6 (70).

2.3 Behavioral Factors

One of the most remarkable features of brain cytokines, it has recently been demonstrated, is their ability to affect the behavior of the organisms at large, particularly during the course of an infection, resulting in what has been called "sickness behavior" (71). Animal studies show that rodents injected systemically or centrally with pro-inflammatory cytokines or cytokine-inducing agents demonstrate a series of behavioral effects such as increased sleep, decreased food and water intake, decreased exploratory behavior, decreased social and sexual behavior, decreased preference for saccharin, decreased brain self-stimulation and impaired learning and memory (72-76). These activities, it has been argued, allow the animal to "conserve" energy and devote all available resources to combating the infection. Some researchers have noted the resemblance between "sickness behavior" in animals and the core symptoms of major depression, such as anhedonia, sleep disturbance, appetite disturbance, psychomotor retardation, social withdrawal, decreased libido and impaired concentration and memory (76). Sleep disturbance in depression can take the form of insomnia or hypersomnia. Insomnia has been described as early, middle or late depending on whether it tends to occur during the early, middle or late part of the night. The most common form of insomnia in major depression is late insomnia with early morning awakening. Hypersomnia has been reported mostly among bipolar patients and patients with atypical depression (77). Sleep EEG studies of patients with major depression have confirmed the sleep disturbance in these patients and revealed abnormalities in sleep architecture such as shortened Rapid-Eye-Movement (REM) latency and a decrease in sleep stages 3 and 4 (delta sleep) (78). Appetite disturbance in patients with major depression has usually been in the form of decreased caloric intake, often resulting in weight loss although in some forms of depression (bipolar or atypical) patients tend to eat more, particularly high starch/high carbohydrate food items, with resulting weight gain (79). Investigators have also noted the similarities between anhedonia, the hallmark of clinical depression, on the one hand and animal behaviors such as decreased preference to saccharine and decreased brain self-stimulation on the other (74). Social isolation, decreased sexual behavior and decreased learning and memory have also been observed in patients with major depression as well as in rodents injected with cytokines (76).

Another line of evidence in support of the "cytokine hypothesis" of major depression has been coming from the reported adverse effects associated with the use of cytokines in the treatment of several medical conditions, notably infection or cancer. One such treatment is IFN-α which is now often used in treating patients with malignant melanoma or patients with Hepatitis

C. Treatment of these patients with IFN-α has been associated with symptoms of clinical depression in 30-40% of patients. These symptoms include sadness, irritability, fatigue, lack of energy, lack of motivation, feelings of hopelessness, guilt and occasional suicidal ideation (80). Rare incidents of completed suicide have also been reported (81). IFN-α treatment has also been associated with cognitive impairment resulting in decreased attention span, spatial orientation and short-term memory (82). These effects appear within 2-3 weeks following the initiation of treatment. They seem to resolve with the discontinuation or the termination of treatment. It is unclear which patients are at risk for the cytokine-induced depressive syndrome. A recent report suggests that the presence of depressive symptoms at baseline may be a predictor of cytokine-induced clinical depression (83). There is also some preliminary evidence that pre-treatment with an antidepressant such as paroxetine in high-risk patients can prevent or abort the development of depressive symptoms (84). Other cytokines have also been used in the treatment of multiple sclerosis and advanced malignancies. Results with IFN-β and IL-2 suggest that these cytokines have significant psychiatric adverse effects, resembling or possibly exceeding those associated with IFN-α (85).

2.4 Effects of Antidepressant Medications

Antidepressant medications such as tricyclic antidepressants (TCA) and selective serotonin reuptake inhibitors (SSRI) have been used with great success to treat episodes of major depression. It is estimated that roughly 70-80% of patients on these medications do show a clinically significant therapeutic response. While the mechanism(s) of action of these medications have not been completely elucidated, the bulk of the evidence favors enhanced neurotransmission and/or receptor stimulation directly affecting specific neurotransmitters such as norepinephrine, serotonin or dopamine(86). It is largely because of such mechanisms of action for the most common antidepressant medications that depression hypotheses have usually implicated a dysregulation of one or more of these neurotransmitters as a key factor in the etiology and/or pathophysiology of major depression.

How does the effect of antidepressant medications influence the "cytokine hypothesis" of major depression? From a theoretical standpoint, if depression is associated with an increase in the secretion of specific cytokines, and if this increase is a state marker of depression, then treatment with antidepressant medications would be expected to decrease the levels of the specific cytokine(s). Furthermore, the decrease in the levels of cytokine would be expected to correlate with the therapeutic response. Unfortunately, the number of studies addressing these issues has been very limited. Among the studies that have been reported, the results have been mixed. In 1994, Weizman and colleagues reported a reduction in the production of the cytokines IL1β, IL-2 and IL-3 in patients with major depression. Treatment with chlomipramine, a tricyclic antidepressant, increased the levels of cytokines (87). In 1995, Sluzewska and colleagues reported an increase in plasma levels of IL-6 in depressed patients. Treatment with fluoxetine, an SSRI, decreased the levels of IL-6 in 6 out of 22 patients (88). Landmann et al, on the other hand, failed to demonstrate any change in the production of TNF in their depressed patients or any effect of treatment with meclobemide, a monoamine oxidase inhibitor MAOI (89). There are also occasional reports of the effects of antidepressant medications on cytokine production in experimental animals. Kubera and colleagues reported an increase in the secretion of IL-1 and IL-2 by splenocytes in rats subjected to a chronic stress model of depression. Treatment with Imipramine, a TCA, reversed these effects (90). Suzuki reported an increase in IL-1β and IL-1 receptor antagonist mRNA by chronic treatment with sertraline (SSRI), imipramine (TCA) and maprotiline (MAOI). The areas affected included the hypothalamus, hippocampus, frontal cortex and brain stem (91).

2.5 Immunological Evidence

No matter how powerful and attractive the arguments in favor of a "cytokine hypothesis" of major depression, crucial support for or against the hypothesis has to come from the immunological evidence. Table 2 summarizes the results of serum and/or CSF concentrations of cytokines and/or cytokine receptors in patients with major depression.

As indicated in this summary, serum IL1β levels have been reported to be increased by some investigators (92), but not by others (93,94). Similarly, IL-1Ra levels have also been shown to be increased by Maes et al (92), but not by Kubera and colleagues (96) or Kronfol and colleagues (94). Cerebrospinal fluid levels of IL-1β have been reported to be increased (95). Serum Interleukin-2 levels have rarely been assessed in depressed patients, but soluble IL-2 receptors (sIL-2R) measures have been found to be increased (26,97). Serum IL-6 have been shown to be increased by several groups (26,97-99), but not by others (93,94,96,100). Cerebrospinal fluid levels of IL-6 have been reported to be decreased (95). As for concentrations of soluble IL-6 receptors (sIL-6R), they have been reported to be increased again by the same group (26,97) who also reported higher levels of IL-1β, IL-1Ra, IL-6 and sIL-2R. Serum TNFα concentrations have been reported to be increased (101,102), or decreased (100) in depressed patients compared to controls. TNF soluble receptors ($TNFsr_1$ and $TNFsr_2$) have been found to be similar in depressed patients and normal controls

(94,100). Other cytokine concentrations that have been measured in depressed patients and found to be similar to controls include IL-10 (96) and IFNγ (89).

Table 2. Serum and CSF concentrations of cytokines and/or soluble cytokine receptors in patients with depressive illness compared to healthy controls.

IL-1:	IL-1β ↑ (92), IL-1β ns (93), IL-1β ns (94); IL-1β ↑ in CSF (95) IL-1Ra ↑ (92), IL-1Ra ns (96) IL-1Ra ns (94).
IL-2:	sIL-2R ↑ (26), sIL-2R ↑ (97).
IL-6:	IL-6 ↑ (26), IL-6 ↑ (97), IL-6 ↑ (98), IL-6 ↑ (99), IL-6 ns (93), IL-6 ns (94); IL-6 ns (96), IL-6 ns (100), sIL-6R ↑(97), sIL-6R ↑ (26), IL-6 ↓ in CSF (95).
IL-10:	IL-10 ns (96).
IFN:	IFNγ ns (89).
TNF:	TNFα ↑ (101), TNFα ↑ (102), TNFα ↓ (100), TNFsr ns (100), TNFsr ns (94).

↑ concentrations significantly higher than controls.
↓ concentrations significantly lower than controls.
ns concentrations not significantly different from controls.
Numbers in parentheses refer to bibliography at the end of the chapter

Perhaps more important than the studies showing serum levels of cytokines in depressed patients and normal controls are the studies comparing ex-vivo stimulated cytokine production in these two groups. Stimulated cytokine production by cultured leukocytes is usually a better indicator of cytokine secretion and therefore more likely to provide useful information. As shown in Table 3, most cytokine production assays show that the levels of secreted cytokines are either increased or unchanged in the depressed group. More specifically, studies conducted by Maes and colleagues and using peripheral blood mononuclear cells in culture with PHA as the stimulus show an increase in the production of the pro-inflammatory cytokines IL-1β and IFNγ (34,61). Seidel and colleagues, using whole blood assays and PHA as the stimulus were only able to confirm the increase in IFNγ but not the IL-1β (23). Kronfol and colleagues, using whole blood assays and both PHA and LPS as stimuli, found a marginal increase in the secretion of IL-1β (with LPS only), a significant increase in the secretion of IL-6 (with LPS only), and no changes in the secretion of TNFα (94). Only one study reported a decrease in the production of IL-1β, IL-2 and IL-3 (87). The reasons for the discrepancies in the results are unclear. They most likely include clinical factors such as age, gender, ethnicity, medication status, nature and severity of depressive symptoms, chronicity of illness and different depression subtypes, as well as methodological factors such as sample size, statistical analysis, blood processing protocols, culture conditions and assay techniques.

Table 3. Cytokine production in stimulated leukocyte cultures of blood samples obtained from depressed patients and healthy controls.

Cytokine	Tissue Source	Stimulus	Results	References
IL1β	PBMC	PHA	↑	Maes et al (34),
IL1β	PBMC	LPS	↓	Weizman (87)
IL1β	WBA	PHA	-	Seidel (23)
IL1-β	WBA	LPS	-	Kronfol (94)
IL1β	WBA	PHA	↑*	Kronfol (94)
IL-2	PBMC	LPS	↓	Weizman (87)
IL-2	WBA	PHA	-	Seidel (23)
IL-3	PBMC	LPS	↓	Weizman (87)
IL-6	WBA	PHA	-	Seidel (23)
IL-6	WBA	PHA	-	Kronfol (94)
IL-6	WBA	LPS	↑	Kronfol (94)
IL-10	WBA	PHA	-	Seidel (23)
TNFα	PBMC	LPS	-	Landmann (89)
TNFα	WBA	LPS, PHA	-	Kronfol (94)
IFNγ	PBMC	PHA	↑	Maes (61)
IFNγ	WBA	PHA	↑	Seidel (23)

PBMC peripheral blood mononuclear cells, WBA whole blood assay; PHA-phytohemagglutinin; LPS: lipopolysaccharide; ↑ increased compared to controls, ↓ decreased compared to controls, - no significantly different results; * non-significant trend.

2.6 Critique

The "cytokine hypothesis" as described above could be, at first glance, real attractive. Many behavioral and vegetative symptoms of depression such as a decline in pleasurable activities, social isolation, sleep and appetite disturbances and decreased motor activity have been seen in animals injected with LPS or cytokines. Depressive mood and cognitive impairment have been observed in human volunteers experimentally injected with LPS as well as patients who receive treatment with cytokines such as IFNα for conditions such as malignant melanoma or hepatitis C. The neuroendocrine changes in depression, characterised by HPA axis activation, and the neurochemical changes, characterised by altered metabolism of NE, 5HT or DA in key areas of the brain, have also been associated with the brain actions of cytokines. Stress, a frequent trigger for episodes of depression, has been shown to increase cytokine secretion under certain conditions. Lastly, there are occasional reports of a decrease in cytokine production in conjunction with the use of specific antidepressant medications. Most of these factors, however, represent circumstantial evidence for a role for cytokines in major depression. Direct evidence would ideally be provided by actual measurement of cytokines and/or soluble cytokine receptors in specific areas of the brain. However, since the technology is not yet available, confirmation of the hypothesis would have to rely on measurements of cytokines and/or soluble cytokine receptors in body fluids such as plasma or CSF which may not necessarily reflect cytokine activities in key areas of the brain. Ex-vivo cytokine production assays would also be particularly important and useful since the concentrations of circulating cytokines at basal conditions may often be difficult to detect. Unfortunately, as shown in the previous section on immunological evidence, the immunological data are inconsistent and as such cast some doubt on the validity of the cytokine hypothesis. Most of the results showing support for the cytokine hypothesis come from a very small number of laboratories. A number of studies have not been able to confirm these results or produced different findings. Thus, further confirmation by independent investigators controlling as much as possible for clinical and methodological factors is critical for the cytokine hypothesis to be universally accepted as a viable and serious hypothesis. But even with more data in support of an activation of monocytes and lymphocytes in patients with major depression, there are several challenges to this hypothesis still lying ahead. On the clinical level, depression is most commonly associated with insomnia, not hypersomnolence, and sleep EEG records on these patients reveal a decrease in delta sleep (stage 3 and 4) unlike the findings usually associated with cytokines whereby delta sleep is increased. Furthermore, according to recent reports, cytokine abnormalities are most often reported in the milder forms of depression, i,.e., dysthymia, where vegetative signs of depression such as sleep and appetite disturbances are not very prominent and signs of HPA activation are usually lacking (103).

On the scientific level, while some studies have shown that treatment of animals with antidepressant drugs could prevent the reduction in pleasurable activities induced by LPS (104), not all studies agree (105). Furthermore, because of the many similarities in the endocrine and behavioral effects of

IL-1 and CRH, and because IL-1 stimulates the release of CRH, researchers need to clearly demonstrate that the effects of cytokines in the brain are not mostly due to CRH release, and that the "cytokine hypothesis" of major depression is not in fact another version of the "CRH hypothesis" in disguise (105).

3. CONCLUSION

It is now quite clear that interactions between the brain and the immune system occur on a regular basis and are part of a homeostatic mechanism to keep the organism healthy. Whether these neural-immune interactions play a role in the etiology/pathogenesis of psychiatric disorders in general, and affective disorders in particular, remains somewhat controversial. In the present chapter, we briefly reviewed existing evidence for immune abnormalities in patients with major depression along with postulated mechanisms and possible clinical implications. We then discussed the central role that cytokines may play in the etiology/pathogenesis of some forms of depressive illness, and presented a critical review of the "cytokine hypothesis" of major depression.

In conclusion, there is strong evidence, both direct and indirect, for a role for cytokines in the etiology of major depression. The evidence, though quite compelling, comes mostly from a very small number of laboratories and has not been universally replicated. Future studies should examine a broader set of cytokines, their soluble receptors, and possible natural antagonists. These cytokines should be assessed in relation to clinical variables such as sleep and appetite disturbance, anhedonia, decreased motivation and impaired cognition. Cytokines should also be assessed in relation to laboratory markers of depression such as sleep EEG and HPA challenge tests. Longitudinal studies should address the important question of whether cytokine abnormalities in major depression are state or trait markers of depression. Lastly, the issue of cytokine regulation in relation to treatment response should be examined. If cytokines are indeed found to play a major role in many of these factors, then the development of new pharmaceutical agents directed towards cytokine activity should be introduced and clinical trials should be started. This would certainly represent a major departure from the way we presently understand and treat clinical depression.

REFERENCES

1. Kronfol, Z., Silva, J., Greden, J., et al. Impaired lymphocyte function in depressive illness. *Life Sci.* 1983; 33: 241-147.
2. Schleifer, S.J., Keller, S.E., Meyerson, A.T., et al. Lymphocyte function in major depressive disorder. *Arch. Gen. Psychiat.* 1984; 41: 484-486.
3. Kronfol, Z., Turner, R., Nasrallah, H., et al. Leukocyte regulation in depression and schizophrenia. *Psychiat. Res.* 1984; 13: 13-18.
4. Schleifer, S.J., Keller, S.E., Siris, S.G., et al. Depression and immunity: lymphocyte function in ambulatory depressed patients, hospitalized schizophrenic patients and patients hospitalized for herniorrhaphy. *Arch. Gen. Psychiat. 1985; 42: 129-133.*
5. Stein, M., Miller, A.H., Trestman, R.L. Depression, the immune system, and health and illness. *Arch. Gen. Psychiat.* 1991; 48: 171-177.
6. Ader, R., Felten, D., Cohen, N. (Eds.) *Psychoneuroimmunology, 3rd edition.* New York, 2001.
7. Schleifer, S., Keller, S., Camerino, M., et al. Suppression of lymphocyte stimulation following bereavement. *JAMA* 1983; 250: 374-377.
8. Carroll, B., Curtis, G., Mendels, J., Cerobropspinal fluid and plasma free cortisol concentrations in depression. *Psychol. Med.* 1976; 6: 235-244.
9. Cohen, S., Tyrell, D.A.J., Smith, A.P. Psychological stress and susceptibility to the common cold. *New Engl. J. Med.* 1991; 325: 606-612.
10. Nasr, S., Altman, E., Meltzer, H. Concordance of atopic and affective disorders. *J. Affect. Dis.* 1981; 3: 291-296.
11. Schiffer, R., Babigian, H. Behavioral disorders in multiple sclerosis, temporal lobe epilepsy and amyotrophic lateral sclerosis. *Arch. Neurol.* 1984; 41: 1067-1069.
12. McGee, R., Williams, S., Elwood, M. Depression and the development of cancer: a meta-analysis. *Soc. Sci. Med.* 1994; 38: 187-192.
13. Zorrilla, E.P., Luborsky, L., McKay, J.R., et al. The relationship of depression and stressors to immunologic assays: a meta-analytic review. *Brain, Behav. Immun.* 2001; 15: 199-226.
14. Kronfol, Z., House, J.D. Lymphocyte mitogenesis, immunoglobulin and complement levels in depressed patients and normal controls. *Acta Psychiatrica Scand.* 1989; 80: 142-147.
15. Maes, M., Van der Planken, M., Stevens, W.J., et al. Leukocytosis, monocytosis and neutrophilia: hallmarks of severe depression. *J. Psychiat. Res.*, 1992; 26: 125-134.
16. Evans, D.L., Folds, J.D., Petitto, J.M., et al. Circulating natural killer cell phenotypes in men and women with major depression. *Arch. Gen. Psychiat.* 1992; 49: 388-395.
17. Maes, M., Lambrechts, J., Bossman, E., et al. Evidence for a system immune activation during depresion: results of leukocyte enumeration by flow cytometry in conjunction with monoclonal antibody staining. *Psychol. Med.* 1992; 22: 45-53.
18. Muller, N., Ackenheil, M., Hofschuster, E. Altered T-cell number and reduced suppressor cell activity in patients with affective psychosis. In: *Interactions Among Central Nervous System, Neuroendocrine and Immune Systems*, Hadden, W., Masek, K., Nistizo, G. (Eds.). Rowe: Pythegora Press, 1989; pp. 385-394.
19. McAdams, C., Leonard, B.E. Neutrophil and monocyte phagocytosis in depressed patients. *Progr. Neuropsycopharmacol.* 1993; 17: 971-984.
20. Schleifer, S.J., Keller, S.E., Bond, R.N., et al. Major depressive disorder and immunity: role of age, sex, severity, and hospitalization *Arch. Gen. Psychiat.* 1989; 46: 81-87.
21. Kronfol, A., Nair, M., Goodson, J., et al. Natural killer cell activity in depressive illness: preliminary report. *Biol. Psychiat.* 1989; 26: 753-756.

22. Irwin, M., Lacher, U., Caldwell, C. Deprssion and reduced natural killer cytotoxicity: a longitudinal study of depresed patients and control subjects. *Psychol. Med.* 1992; 22: 1045-1050.
23. Seidel, A., Arolt, V., Hunstiger, M., et al. Cytokine production and serum proteins in depression. *Scand. J Immunol.* 1995; 41: 434-538.
24. Maes, M. Acute phase protein alterations in major depression: a review. *Rev. Neurosci.* 1993; 4: 407-416.
25. Calabrese, J.R., Skwerer, R.G., Barna, B., et al. Depression, immunocompetence, and prostaglandins of the E series. *Psychiat. Res.* 1986; 17: 41-47.
26. Maes, M., Meltzer, H., Bosmans, E. Increased plasma levels of interleukin-6, soluble interleukin-6, soluble interleukin-2 and transferin receptors in major depression. *J. Affect. Dis.* 1995; 34: 301-309.
27. Biondi, M. Effects of stress on immune functions: an overview. In: *Psychoneuroimmunology, 3rd Edition*, Ader, R., Felten, D., Cohen, N. (Eds.). New York: Academic Press, 2001; pp. 189-226.
28. Carroll, B.J., Curtis, G., Mendels, J. Neuroendocrine regulation in depression I. Limbic system adrenocortical dysfunction. *Arch. Gen. Psychiat.*1976; 33: 1039-1044.
29. Carroll, B., Curtis, G., Davies, B., et al. Urinary free cortisol excretion in depression. *Psychol. Med.* 1976; 6: 43-50
30. Kronfol, Z., House, J.D., Silva, J., et al. Depression, urinary free cortisol excretion and lymphocyte function. *Brit. J. Psychiat.* 1986; 148: 70-73.
31. Kronfol, Z., House, J.D. Depression, hypothalamic-pituitary-adrenocortical activity and lymphocyte function. *Psychopharmacoll. Bull. 1985; 21: 476-478.*
32. Darko, D.F., Irwin, M.R., Risch, S.C., et al. Plasma beta-endorphin and natural killer cell activity in major depression: a preliminary study. *Psychiat. Res.* 1992; 43: 111-119.
33. Irwin, M., Brown, M., Patterson, T., et al. Neuropeptide Y and natural killer cell activity: findings in depression and Alzheimer's caregiver stress. *FASEB J.* 1991; 5: 3100-3107.
34. Maes, M., Bosmans, E., Meltzer, H.Y., et al. Interleukin-1β: a putative mediator of HPA-axis hyperactivity in major depression? *Amer. J Psychiat.* 1993; 150: 1189-1193.
35. Kronfol, Z., Nair, M., Zhang, Q., et al. Circadian immune measures in healthy volunteers: relationship to hypothalamic-pituitary-adrenal axis hormones and sympathetic neurotransmitters. *Psychosom Med.* 1997; 59: 42-50.
36. Cohen, S. and Miller, G. Stress, immunity and susceptibility to upper respiratory infection. In: *Psychoneuroimmunology (3rd edition)* 2001, Ader, R., Felten, D., Cohen, N. (eds.), Academic Press, New York, pp. 499-509.
37. Friedman, G.A. Psychiatrically-diagnosed depression and subsequent cancer. *Cancer Epidemiol. Biomarkers Prev.* 1994; 3: 11-13.
38. Penninx, B., Guraluik, M., Pahor, M., et al. Chronically depressed mood and cancer risk in older persons. *J. Natl. Cancer Inst.* 1998; 90: 1888-1893.
39. Zonderman, A., Costa, P., McCrae, R. Depression as a risk for cancer morbidity and mortality in a nationaly representative sample. *JAMA* 1989; 262 1191-1195.
40. Spiegel, D., Bloom, J.R., Kraemer, H.C., et al. Effect of psychosocial treatment on survival of patients with metastatic breast cancer. *Lancet* 1989; 2: 888-891.
41. Goodwin, P., Leszcz, M. Ennis, M., et al. The effect of group psychological support on survival in metastatic brain cancer. *New Engl. J. Med.* 2001; 345: 1719-1726.
42. Levy, S.M., Harberman, R.B., Lippman, M., et al. Correlation of stress factor with sustained depression of natural killer cell activity and predicted prognosis in patients with breast cancer. *J. Clin. Oncol.* 1987; 5: 348-353.

43. Sheridan, J., Dobbs, C., Brown, D., et al. Psychoneuroimmunology: stress effects on pathogenesis and immunity during infection. *Clin. Micros. Rev.* 1994: 7: 200-212.
44. Cohen, S., Doyle, W.J., Skoner, D.P. Psychological stress, cytokine production, and severity of upper respiratory illness. *Psychosom. Med.* 1999; 61: 175-180.
45. Kiecolt-Glaser, J., Glaser, R., Gravenstein, S., et al. Chronic stress alters the immune response to influenza virus vaccine in older adults. *Proc. Natl. Acad. Sci. USA* 1996; 93: 3043-3047.
46. Irwin, M., Costlow, C., Williams, H., et al. Cellular immunity to varicella-zoster virus in depression. *J. Infect. Dis.* 1998; 178: 104-108.
47. Leserman, J., Petitto, J.M., Perkins, D.O., et al. Severe stress, depressive symptoms, and changes in lymphocyte subsets in human immunodeficiency virus-infected men. *Arch. Gen. Psychiat.* 1997; 54: 279-285.
48. Lesserman, J., Jackson, E., Petitto, J., et al. Progression to AIDS: the effects of stress, depressive symptoms and social support. *Psychosom. Med., 1999, 61: 397-406*
49. Mayne, T.J., Vittinghoff, E., Chesney, M.A., et al. Depressive affect and survival among gay and bisexual men infected with HIV. *Arch. Intern. Med.* 1996; 156: 2233-2238.
50. Burack, J.H., Barrett, D.C., Stall, R.D., et al. Depressive symptoms and CD4 lymphocyte decline among HIV-infected men. *JAMA* 1993; 270: 2568-2573.
51. Lyketsos, C.G., Hoover, D.R., Guccione, M., et al. Depressive symptoms as predictors of medical outcomes in HIV infection. *JAMA* 1993; 270: 2563-2567.
52. Ross., R. Atherosclerosis: an inflammatory disease. *N. Engl. J. Med.* 1999; 340: 115-126.
53. Mendall, M.A., Strahan, D.P., Butland, B.K., et al. C-reactive protein: relation to total mortality, cardiovascular mortality and cardiovascular risk factors in men. *Eur. Heart J.* 2000; 21: 1584-1590.
54. Ridker, P.M., Kennekens, C.H. Buring, J.E., et al. C-reactive protein and other markers of inflammation in the prediction of cardiovascular disease in women. *N. Engl. J. Med.* 2000; 342: 836-843.
55. Biasucci, L.M., Vitelli, A., Liuzzo, G., et al. Elevated levels of interleukin-6 in unstable angina. *Circulation.* 1996; 94: 874-877.
56. Biasucci, L.M., Liuzzo, G., Fantuzzi, G., et al. Increasing levels of interleukin (IL)-1Ra and IL-6 during the first 2 days of hospitalization in unstable angina are associated with increased risk of in-hospital coronary events. *Circulation. 1999;* 99: 2079-2084.
57. Simon, A.D., Yazdani, S., Wang, W., et al. Circulating levels of IL-1β, a prothrombotic cytokine, are elevated in unstable angina vs. stable angina. *J. Thromb. Thrombolysis* 2000; 9: 217-222.
58. Lindmark, E., Diderholm, E., Wallentin, L., et al. Relationship between interleukin-6 and mortality in patients with unstable coronary artery disease. *JAMA* 2001; 286: 2107-2113.
59. Frasure-Smith, N., Lesperance, F., Talajic, M. Depression following myocardial infarction: impact on 6-month survival. *JAMA* 1993; 270: 1819-1825.
60. Glassman, A.H., Shapiro, P.A. Depression and the course of coronary artery disease. *Am. J. Psychiat.* 1998; 155: 4-11.
61. Maes, M., Scharpe, S., Meltzer, H.Y., et al. Increased neopterin and interferon gamma secretion and lower L-troptophan levels in major depression: Further evidence for immune activation in severe depression. *Psychiat. Res.* 1994: 54: 143-160.
62. Dantzer, R., Wollman, E., Yirmiya, R., (Eds.) *Cytokines, Stress and Depression.* 1999, New York: Kluwer Academic Press.

63. Dantzer, Bluthe, R.M., Castanon, N., et al. Cytokine effects on behavior. In: *Psychoneuroimmunology, 3rd Edition,* Ader, R., Felten, D., Cohen, N. (Eds.). San Diego: Academic Press, 2001; pp. 703-727.
64. Smith. R.S. The macrophage theory of depression. *Med. Hypoth.* 1991; 35: 298-306.
65. Maes, M., Smith, R., Scharpe, S. The monocyte-T lymphocyte hypothesis of major depression. *Psychoneuroendocrinol.* 1995; 20: 111-116.
66. Dunn, A.J., Vickers, S.L. Neurochemical and neuroendocrine responses to Newcastle disease virus administration in mice. *Brain Res.* 1994; 645: 103-112.
67. Dunn, A.J., Wang, J.P., Ando, T. Effects of cytokines on central neurotransmission: comparison with the effects of stress. In: *Cytokines, stress and depression,* Dantzer, R., Wollman, E.E., Yirmiya, R., (Eds.) New York: KluwerAcademic/Plenum Publishers, 1999; pp. 117-127.
68. Sapolsky, R., Rivier, C., Yamamoto, G., et al. Interleukin-1 stimulates the secretion of hypothalamic corticotropin-releasing factor. *Sci.* 1987; 238: 522-524.
69. Young, E.A., Haskett, R.F., Murphy-Weiberg,V., et al. Loss of glucocorticoid fast feedback in depression. *Arch. Gen. Psychiat.* 1991; 68: 693-699.
70. Maes, M., Scharpe, S., Meltzer, H. Relationship between interleukin-6 activity, acute phase proteins, and HPA-axis function in severe depression. *Psychiat. Res.* 1993; 49: 11-27.
71. Dantzer, R. Cytokine-induced sickness behavior: where do we stand? *Brain, Behav. Immunit* 2001; 15: 7-24.
72. Krueger, J., Majde, J. Microbiol products and cytokines in sleep and fever regulation. *Crit. Rev. Immunol.* 1994; 14: 355-379.
73. Plata-Salaman, C.R. Brain mechanisms in cytokine-induced anorexia. *Psychoneuroendocrinol.* 1999; 24: 25-41.
74. Kent., S., Bluthe, R.M., Kelley, K.W., et al. Sickness behavior as a new target for drug development. *Trends Pharmacol. Sci.* 1992; 13: 24-28.
75. Gibertini, M., Newton, C., Friedman, H., et al. Spatial learning impairment in mice infected with Legionella pneumophila or administered exogenous interleukin-1-beta. *Brain. Behav. Immun.* 1995; 9: 113-128.
76. Yirmiya, R. Endotoxin produces a depressive-like episode in rats. *Brain. Res.* 1996; 711: 163-174.
77. Reynold, C., Collin, J., Kupfer, D. Sleep and affective states. In: *Psychopharmacology, the Third Generation of Progress.* Meltzer, H. (ed.) 1987, Raven Press, New York, pp. 647-654.
78. Kupfer, D., Reynold, C., Grochocinski, V., et al. Aspects of short REM latency in affective states: A revisit. *Psychiat. Res.* 1986; 17: 49-59.
79. Casper, R., Redmond, E., Katz, M. Somatic symptoms in primary affective disorder: presence and relationship to the classification of depression. *Arch. Gen. Psychiat.* 1985; 42: 1098-1104.
80. Renault, P.F., Hoofnagle, J.H., Park, H. et al. Psychiatric complications of long term interferon alpha therapy. *Arch. Intern. Med.* 1987; 147: 1577-1580.
81. Janssen, H.L., Brouwer, J.T., van der Mast, R.C., et al. Suicide associated with alfa-interferon therapy for chronic viral hepatitis. *J. Hepatol.* 1994; 21: 241-243.
82. Meyers, C., Scheibel, R., Forman, A. Persistent neurotoxicity of systemically-administered interferon-alpha. *Neurology* 1991; 41: 672-676.
83. Capuron, L., Ravaud, A. Prediction of the depressive effects of interferon alfa therapy by the patient's initial affective state. *N. Engl. J. Med.* 1999; 340: 1370.
84. Musselman, D.L., Lawson, D.H., Gumnick, J.F., et al. Paroxetine for the prevention of depression induced by high dose interferon-alfa. *N. Engl. J. Med.* 2001; 344: 961-966.

85. Capuron, L., Ravaud, A., Dantzer, R. Timing and specificity of the cognitive changes induced by interleukin-2 and interferon-α treatments in cancer patients. *Psychosom. Med.* 2001; 63: 376-386.
86. Blier, P. Crosstalk between the norepinephrine and serotonin systems and its role in the antidepressant response. *J. Psychiat. Neurosci.* 2001; 26: 3-10.
87. Weizman, R., Laor, N., Podliszewski, E., et al. Cytokine production in major depressed patients before and after cholmipramine treatment. *Biol. Psychiat.* 1994; 25: 42-47.
88. Sluzewska, A., Rybakowski, J.K., Laciak, M., et al. Interleukin-6 serum levels in depressed patients before and after treatment with fluoxetine. *Annals NY Acad. Sci. 1995; 762: 474-477.*
89. Landmann, R., Schaub, B., Link, S., et al. Unaltered monocyte function in patients with major depression before and afer three months of antidepressive therapy. *Biol. Psychiat.* 1997; 41: 675-681.
90. Kubera, M., Symbirse, A., Basta-Kaim, A., et al. Effect of chronic treatment with imipramine on interleukin 1 and interleukin 2 production by splenocytes obtained from rats subjected to a chronic mild stress model of depression. *Pol. J Pharmacol. 1998;* 48: 503-506.
91. Suzuki, E., Shintani, F., Kanba, S., et al. Induction of interleukin-1β and interleukin-1 receptor antagonist mRNA by chronic treatment with various psychotropics in widespread area of rat brain. *Neurosci. Lett.* 1996; 215: 201-204.
92. Maes, M., Bosmans, E., DeJongh, R., et al. Increased serum IL-6 and IL-1 receptor antagonist concentrations in major depression and treatment resistent depression. Cyto*kine* 1997; 9: 853-858.
93. Brambilla, F., Maggione, M. Blood levels of cytokines in elderly patients with major depressive disorder. *Acta. Psychiatrica Scand.* 1998; 97: 309-331.
94. Kronfol, Z., Aziz, M., Remick, D. Increased LPS-stimulated levels of IL-6 in patients with major depression. *Am. Psychiat. Assoc. New Res. Abs. 273*, Philadelphia, 2002.
95. Levine, J., Barak, Y., Chengappa, K.N. et al. Cerebrospinal cytokine levels in patients with acute depression. *Neuropsychobiol.* 1999; 40: 171-176.
96. Kubera, M., Keniz, G., Bosmans, Et., et al. Plasma levels of interleukin-6, interleukin-10 and interleukin-1 receptor antagonist in depression: comparison between the acute state and after remission. *Po. J. Pharmacol.* 2000; 52: 237-241.
97. Sluzewska, A., Rybakowski, J., Bosmans, E., et al. Indicators of immune activation in major depression. *Psychiat. Res.* 1996; 64: 161-167.
98. Frommberger, U.H., Bauer, J., Haselbauer, P. et. al. Interleukin-6 (IL-6) plasma levels in depression and schizophrenia: comparison between the acute state and after remission. *Eur. Arch Psychiat.*1997; 247-233.
99. Lu, S.J., Shiah, I.S., Yatham, L.N., et al. Immune-inflammatory markers in patients with seasonal affective disorder: effects of light therapy. *J. Affect. Dis.* 2001; 63: 27-34.
100. Haack, M., Hinze-Selch, D., Fenzel, T., et al. Plasma levels of cytokines and soluble cytokine receptors in psychiatric patients upon hospital admission: effects of confounding factors and diagnosis. *J. Psychiat. Res.* 1999; 33: 407-418.
101. Mikova, O., Yakimova, R., Bosmans, E., et al. Increased serum tumor necrosis fator alpha concentratios in major depression and multiple sclerosis. *Eur. Neuropsychopharmacol.* 2001; 11: 203-208.
102. Lanquillon, S., Krieg, J.C., Bening-Abu-Shach, U., et al. Cytokine production and treatment response in major depressive disorder. *Neuropsychopharmacol.* 2000; 22: 370-379.

103. Anişman, H., Ravindran, A., Griffith, J., et al. Endocrine and cytokine correlates of major depression and dysthymia with typical or atypical features. *Mol. Psychiat.* 1999; 4: 182-188.

104. Yirmiya, R., Weidenfeld, J., Polleck, Y., et al. Cytokines, "depression due to general medical condition" and antidepressant drugs. In: *Cytokines, Stress and Depression*, Dantzer, R., Wollman, E., Yirmiya, R. (eds.) 1999, Kluwer Academic Press, New York, pp. 283-316.

105. Connor, T.J., Leonard, B.E., Depression, stress and immunological activation: the role of cytokines in depressive disorders. *Life Sci.* 1998; 62: 583-606.

Chapter 13

CYTOKINE REGULATION IN ANXIETY DISORDERS

Francesca Brambilla
Dipartimento Di Salute Mentale-Ospedale Sacco-Milano-Italy

1. INTRODUCTION

The functional connections between the Central Nervous System (CNS), the Immune System (IS) and the Neuroendocrine System (NES), with their reciprocal stimulatory-inhibitory links, are one of the bases for the capacity of animals and humans to adapt correctly and to survive to internal and external stimuli, to changes in the environment or in the internal milieu, and to physical and psychological insults. Impaired functions and interrelationships between the CNS, IS and NES may be one of the bases for the development of psychoneurological, neuroendocrinological and immune pathologies. Whether or not CNS, IS and NES impairments are, respectively, primary or secondary events in the pathogenesis of immunopathies, endocrinopathies or psychopathies is still a question of debate.

There are multiple IS and NES alterations in humans in the course of mental disorders, but their etiopathogenetic significance, however, is still questionable. Data in anxiety disorders (AD) are especially relatively scanty and rather contradictory.

To define and discuss the aspects of immune function in AD it is necessary to briefly mention some data about CNS, NES and IS activities in the course of or as a consequence of acute and chronic psychologically stressful life events. In fact, psychological stresses precede AD, possibly being one of their causes, and are often followed by anxiety reactions of variable severity. In addition, the stress-induced biological impairments mimic some of the organic abnormalities seen during AD, thus making the interconnection between stress-related reactions and AD even more strong. Acute or chronic psychological stresses are reported to be followed by impairments of CNS, IS and NES, the latter two possibly being the consequences of the former, in the frame of a continuum which ends in an incapacity to adapt rapidly , physically or psychologically, to subsequent stresses that are not even severe but pertain to relatively normal every day life events.

Even though the aim of this chapter is to present and discuss data about the secretion and function of cytokines in AD, some general data about cell-mediated and humoral immunological aspects related to stress-linked events must be briefly reported. Cytokines are known to modulate the complex systems of interconnection between cell-mediated immunity, humoral immunity and acute-phase protein secretion, regulating the different phases of the immune response. Thus, it is impossible to study cytokine secretion and function without knowing the situation of the other immune processes.

Before proceeding to the presentation of aspects of cytokine secretion in AD, it is important to recall that cytokines are divided into pro-inflammatory (Interleukin-1β, Interleukin-2, Interleukin-6, Interleukin-8, Tumor Necrosis Factor-α, Interferon-γ), anti-inflammatory (Interleukin-4, Interleukin-8, Interleukin-10, Interleukin-13) and hematopoietic (Interleukin-3, Interleukin-5), according to their primary effects. Thus, alterations of one or another cytokine group may have markedly different effects on brain function and therefore be of different importance for modulation of mental disorders.

2. EFFECTS OF PSYCHOLOGICAL STRESS ON IMMUNE FUNCTION

It has been anecdotally reported, and then demonstrated scientifically, that acute and chronic psychologically stressful events often precede and modulate the development of infectious diseases, of allergies and of cancer, by inducing alterations in the immune function (1). These impairments involve all the activities of the IS, which are first acutely and briefly stimulated and then mostly depressed, even though the *consecutio temporum* and the consistency of these processes have not always been confirmed (2).

Stressors are not always very severe and easily recognizable physical and psychological insults, but include persistent daily harassment, life changes and situations which may not be seen as harmful as themselves but are perceived as such by specific individuals. Subjective capacities for confronting and coping with new or persistent life aspects and direct cognitive appraisal of stimuli may turn normal every-day life situations into events which are lived positively or negatively by specific individuals only, with obviously different effects on the CNS, NES and IS functions. This aspect is particularly important in relation to the pathogenesis of AD since the process of cognitive appraisal of a life situation determines the direction and intensity of the stress-related emotions (3). Psychological factors, such as temperament traits, personality characters, mood states, defense mechanisms, availability of social support, perceived controllability and capacity of coping with the situations are mandatory modulators of the type of CNS, NES and IS responses observed in individuals or in groups of individuals, influencing the development of specific psychoimmunoendocrine pathologies.

Preliminary studies in humans suggested that psychological stresses (including bereavement in widows or widowers, caregiving mothers of low birthweight infants, divorce, separation, unemployment, war attacks, war imprisonment, earthquakes) are associated with depressed immune functions (4-6). More recently, it has been observed that acute psychological stresses are followed in normal humans by transient increases in leukocyte and specifically in monocyte counts, in Natural Killer (NK) cell numbers and activity, in T suppressor/cytotoxic cell numers. Contemporaneously, there is a moderate lymphopenia, reduced CD4/CD8 ratio, decreased numbers of T helper cells, blunted lymphocytic responses to *in vitro* mitogen stimulation, and decreased skin test reactivity (2,7-16). As for humoral immune parameters, increased salivary levels of immunoglobulin A (IgA), IgM, IgG, α2-M, C3 and C4 complement factors, acute phase (AP) protein-α, macroglobulin, haptoglobin and α1 acid glycoprotein are increased (13,17-18). The immune alterations seemed to correlate with the stress-perception expressed by the probands (13), with the mental efforts of the subjects to confront the stress and with its controllability (16).

Chronic psychological stresses in humans induce the same phenomena, but occasionally increased numbers and activity of NK cells and lower than normal production of soluble IgA have been reported. (2,19-20).

All these phenomena might be primary direct responses to the real or lived-as-real stressful situations; but may also be secondary to changes in CNS and/or NES functions, since it is well known that stress modifies the secretion of neurotransmitters, including serotonin (5-HT), noradrenaline (NE), dopamine (DA), epinephrine (EPI), gammaaminobutyric acid (GABA), of neuropeptides such as corticotropin-releasing hormone (CRH), thyrotropin-releasing hormone (TRH), growth hormone-releasing hormone (GHRH), somatostatin (SRIF), endogenous opioids and especially of β-endorfin (β-EP), neuropeptide Y (NPY), cholecystokinin (CCK), of neurosteroids, and of peripheral hormones such as ACTH, cortisol (CORT), growth hormone (GH) and prolactin (PRL), which modulate the immune function (21).

Coming to studies of cytokine production and concentrations in body fluids, there are consistent data in the literature showing that in experimental animals psychological stressors increase the central and peripheral secretion of pro-inflamatory cytokines and their receptors, including Interleukin-6 (IL-

6), IL-6 messenger RNA (in the midbrain), Interleukin-1 (IL-1) and Tumor Necrosis Factor-α (TNF-α). This increase may be subjected to learning processes, such as classical conditioning and sensitization (22), since early life stresses may sensitize the organism to respond with increased inflammatory-like reactions to subsequent restressors. In other words, the immune system is dependent on "sensitization", meaning that psychological and physical stresses may result in long-lasting alterations of immune responses, which will worsen with the passage of time.

Data for humans are relatively few and rather contradictory. Increased production of the pro-inflamatory IL-1β, IL-1β messenger RNA (in the hypothalamus), IL-1β receptor antagonist (IL-1βRa), IL-6, TNF-α, Interferon-γ (IFN-γ), and of theanti-inflammatory Interleukin-10 (IL-10) have been reported as a consequence of psychological stresses, and are positively correlated with the perception of the stress and of the stress-induced changes (1, 14, 23-30). Maes and coworkers (26) observed that stress-linked increases in leucocytes and neutrophils and activation of T cells (CD2 +, CD26 +, CD2+, HLA-DR +) are correlated with the hypersecretion of pro-inflamatory cytokines (IL-1β, IL-6, TNF-α, INF-γ). It has also been reported, however, that psychological stresses do not change plasma concentrations of IL-1β and IL-6 or reduce INF-γ production (23,24,29,31).

The reason for the discrepancy between these data is not known, possibly resulting from the selection of the type of stressors used, the protocols of examination and the type of subjects included in the studies. In this regard, it is worth mentioning that subjects with high stress-induced anxiety had high INF-γ and low IL-10 and Clara Cell (CC-16) concentrations, while subjects with low stress-induced anxiety had the opposite (26,32). In addition, cytokine secretion in response to stressful events seemed to correlate with loneliness (29). This shows that the stressor itself is not as important as the type of subjective adaptation to it, possibly in relation to individual temperament traits. It is also worth to mention that in some cases anticipation of a stressful event (surgery, school examination) induces increased IL-1β and IL-1β-Ra production (25,30), even though the phenomenon has not always been confirmed (31).

All together, these findings suggest that events which are stressful or are perceived as stressful are mostly accompanied by activation of the pro-inflamatory response system (23,24,26-28,32).

Pro-inflamatory cytokines, apart from their efficacy in making the animal and human body to adapt physically to positive or negatve stimuli, induce changes in mental functions, again possibly as expressions of adaptation to external and internal events. In experimental animals these changes include anedonia, suppression of social behavior and social withdrawal, anxiety, irritability and depression, lethargy, anorexia, suppression of drinking and of sexual behavior, impaired learning and memory, decreased brain self-stimulation, decreased body care activity (33,34). In humans, administration of interferons for therapeutic use and infection-related hypersecretion of IL-1β, IL-6 and TNF-α induces appearance of the same symptomatology, which is described as "sickness behavior" (34).

In parallel to induction of the immune responses, acute and chronic stressors stimulate the secretion of hypothalamo-pituitary-adrenal hormones (CRH, ACTH, CORT, endogenous opioids, dehydroepyandrosterone = DHEA), of PRL, GH and of multiple neurosteroids and neuropeptides. Simultaneously, the CNS and the peripheral sympathetic nervous system react to stressors with an increased secretion of NE, EPI, 5-HT, and DA. The NES changes have complex modulatory effects on IS function, in part inhibiting or modifying its responses to stimuli (CRH, ACTH, CORT), in part stimulating or just protecting them (GH, PRL, opioids, DHEA). Peripheral hypersecretion of catecholamines, in particular NE, has been reported as possibly being responsible for the blunted NK cell activity and mitogen-induced lymphocyte proliferation observed during these events (35-37). Key mediators of the inhibitory effects of NE on cell-mediated immunity are IFN-γ, IL-10, IL-6 and TNF-α (38-40). NE and NE-agonists may inhibit or stimulate *in vivo* and *in vitro* the production of IL-6, IL-6 mRNA, TNF-α, IL-1, and IL-2 (40-42). Serotonin and dopamine have been reported to stimulate the proliferative activity of splenocytes and the secretion of IL-6, IL-1, IL-2 and TNF-α (43). In turn, cytokines administered centrally or peripherally to experimental animals increase NE, DA and 5-HTconcentrations in the hypothalamus and in different brain areas (44). On the other hand, it is well known that cytokines stimulamulate HPA secretion (45) which, in turn, inhibit 5-HT and stimulate NE and DA functions (46).

The CNS, NES and IS responses to acute and chronic stressors are physiological events as long as they are temporally concluded and vanish with the end of the stress itself. But when the capacity of one of the three systems, or of all of them, to adapt to stressful stimuli is lost because of genetic-linked impairments or of previously prolonged or repeated stressful events, changes in immune functions become pathologic and can interfere with the appearance of psychiatric disorders, through the simultaneous effects they exert on the NES and on the CNS.

The data reported above are rather contradictory, pointing out that stress modulates the immune function, and especially cytokine secretion, both by inhibiting and stimulating its activities and production. This apparent contradiction clearly results from many factors, some technical (acute or chronic nature of the stressors and their severity, type of immune variable taken into consideration, protocols of investigations, methods of assay, *in vitro or in vivo* examinations, human or animal experiments, influence of age and gender) and some linked to personality characters of the probands and to their capacity for coping with and adapting to the stressful situations. Even taking these biases into consideration, it is possible to suggest that both pro-inflamatory and anti-inflammatory cytokine concentrations increase under acute or prolonged psychological stresses, the prevalence of the pro- or anti-inflammatory cytokine secretion depending mostly from the stress perception, stress-induced anxiety and lack of external support. In these cases, it is possible that an unbalance between pro- and antinflammatory cytokine secretion would be responsible for the occurrance of physical and psychological damages.

The fact that stressful situations which stimulate the hypothalamo-pituitary-adrenal (HPA) axis and central and peripheral NE and DA secretions, which are known to inhibit immune function in general and cytokine production in particular, do not seem to block immune activity is not inexplicable. As mentioned above, multiple factors intervene in immune modulation, some stimulating and some inhibiting it, most of the former and of the latter being hypersecreted under stress. It is only logical, therefore, that the combination of stimulatory and inhibitory influences may obscure the effects of each one of them, ending in results complex and difficult to explain.

3. IMMUNE FUNCTION IN ANXIETY DISORDERS WITH SPECIFIC REFERENCE TO CYTOKINE SECRETION

3.1 Subclinical Anxiety

Patients with subclinical anxiety show decreased lymphocyte responses to mitogen stimulation (47-49), decreased salivary IgA concentrations (50), decreased monocyte numbers and CD4/ CD8 ratio, with reduced NK cell activity (51).

Data on cytokine secretion and activity in subclinially anxious subjects are very scanty. Examination of 497 18-19-year-old university students revealed that the concentrations of the pro-inflamatory IL-1β and IL-2 are lower than normal in subjects with high subclinical anxiety scores, who tend to give more negative interpretations of stressful events. This occurred in parallel with subclinical activation of the hypothalamo-pituitary-adrenal (HPA) function, even though there was no significant correlation between NES and IS impairments (48). The biological mechanism underlying the blunted ,cytokine production in these patients is not clear. Increased secretion of NE and DA, which has been seen to occur in stressful situations, might be responsible for both the inhibition of cytokine production and the activation of the HPA function. It is worth mentioning, however, that Koh et al. (52-54) observed increased IL-2 production, positively correlating with the anxious symptomatology.

These data suggest that in subclinical anxiety there are alterations of immune function, and in particular an impaired cytokine secretion, without the possibility, at the moment to differentiate between activation and inhibition of the IS.

3.2 Generalized Anxiety

It has been reported that patients with Generalized Anxiety Disorder (GAD) develop infections more frequently than normal non-anxious subjects, which suggests that immune defenses are impaired in the course of the disease (55).

GAD patients have increased numbers of monocytes, CD13 + and CD8 +, decreased numbers and activity of NK cells and blunted proliferative responses to mitogens (56).

In 16 GAD patients, phytohemoagglutinin (PHA)-induced production of IL-2 is blunted (54), that of INF-γ is normal (56) and concentrations of IL-2-receptors (IL-2-R) on T-lymphocytes stimulated *in vitro* with anti-CD3 are reduced, correlating negatively with the degree of anxiety (55). These data seem to overlap with those obtained in subjects with subclinical anxiety.

In day 1-3 postpartum women with " the blues" and pathological anxiety, blood concentrations of IL-6, IL-6 receptor (IL-6-R), Leukemia Inhibitory Factor receptor (LIF-R), IgP 130 (the IL-6 signalling protein) and IL-1 receptor antagonist (IL-1-Ra) are significantly higher than in controls, and are correlated with both anxious and depressive symptomatology, these impairments being indicative of an increased inflammatory capacity (44). The difference in cytokine concentrations between pure GAD patients and postpartum GAD subjects with a comorbidity for depression may be linked to the presence in the latter of depression, since for Major Depressive Disorders (MDD) pro-inflamatory cytokine concentrations higher than normal have been repeatedly reported (22).

All together, the state of secretion of pro-inflamatory cytokines and of their antagonists and the activity of their receptors is contradictory, since they have been simultaneously reported to be normal, decreased or increased in GAD.

3.3 Phobias

Social phobias, agoraphobia and other phobias have been reported to be associated with normal cell counts of CD3 +, CD4+, CD8+, CD19+, CD57+, normal CD4+/CD8+ ratio, and increased values of monocytes and NK cells (57-61). However, NK cell numbers seem to be increased only in patients with normal levels of trait worry, while those with high levels of trait worries showed blunted peripheral NK cell response to the exposure to the phobic stimulus.

IL-2 and IL-2R serum levels were reported to be normal in 15 patients with generalized subtype of social phobia, suggesting that these subjects do not have a T-lymphocyte activation (62).

Obviously, these data are too few to enable us to make any suggestion.

3.4 Panic Disorder

Data about immune parameters in patients with Panic Disorder (PD) are rather contradictory. Total numbers of lymphocytes, leukocytes, T and B cells, CD4+, CD8+ , CD4 +/CD8+ ratio, CD56 +, CD57 +, NK cell, and lymphocyte responses to mitogen stimulation have been reported to be mostly normal in PD patients, but there are also data showing alterations of the above-mentioned immune parameters, showing both an activation and an inhibition of the system. Serum levels of immunoglobulins generally show increased values of IgA, normal of IgM, IgG and IgE (41, 54, 63-70).

Studies of cytokine concentrations and production have also shown contradictory results. Plasma levels of the pro-inflamatory cytokine IL-1β have been reported by Brambilla et al. (71) to be increased both before and after 1 month of successful alprazolam therapy in 10 patients with PD and agoraphobia. IL-2 plasma concentrations were also reported to be higher than normal in 36 patients with PD and agoraphobia examined by Rapaport et al. (59). On the contrary, Rapaport et al. (62) found normal plasma concentrations of IL-1β, IL-1α, IL-2R and Brambilla et al. (72) normal values of TNF-α both before and after one month of alprazolam therapy. The contradictions between data are difficult to explain, since patients selected from Brambilla et al. (71,73) and Rapaport et al. (62) had super-imposable demographic and clinical characteristics. The only difference was in the type of assay, Brambilla et al. using a radioimmmuno-assay and Rapaport et al. a competitive enzyme immunoassay.

In vitro production of the pro-inflamatory IL-2 and of the hematopoietic IL-3 from peripheral blood mononuclear cells (PBMC) of PD patients is normal in unstimulated conditions and blunted after PHA stimulation. A negative correlation was found between IL-3 production and severity of anxiety (54,74).

3.5 Obsessive-Compulsive Disorders

The data on immune function of patients with Obsessive-Compulsive Disorder (OCD) are relatively scanty and contradictory.

That OCD has an autoimmunoetiology has been suggested on the basis of a connection of the disease with Sydenham's chorea, rheumatic fever, or herpes virus infection (75-79). However, no changes in T-lymphocyte subsets or in concentrations of transferrin, an acute phase protein, have been reported (80). On the contrary, Marazziti et al. (81) observed increased CD4+ and CD8+ lymphocyte numbers in OC patients.

Concentrations of the pro-inflamatory cytokine IL-1β were been reported to be normal in 42 OC patients of Maes et al. (80) and Monteleone et al. (82), while were lower than normal in a group of 27 patients investigated by Brambilla et al. (83). Serum IL-6, soluble IL-6 receptor (sIL-6R), and soluble IL-2 receptor (sIL-2R) concentrations were found normal in the patients of Maes et al. (80) and Monteleone et al. (82), while TNF-α values were lower in 41 OC subjects investigated by Brambilla et al. (83) and Monteleone et al. (82). Correlation between cytokine and cortisol values was expected but not not observed by Monteleone et al. (82). What is intriguing is the fact that the normal cytokine concentrations of Maes patients correlated with the severity of compulsive, but not obsessive, symptomatology, which did not occur for Monteleone and for Brambilla patients. Moreover, even though being normal, sIL-2R concentrations correlated negatively with higher than normal cortisol values in Maes patients. This suggests that the regulation of the NES and IS relationship is altered in OCD.

In vitro unstimulated production of the pro-inflamatory cytokines IL-1β, IL-2 and the hematopoietic cytokine IL-3 by peripheral blood mononuclear cells (PBMC) were normal (84), while values of the pro-inflamatory INF-γ were higher than normal, correlating positively with high scores of somatization (85).

3.6 Post-Traumatic Stress Disorder

The immune aspects of Post-Traumatic Stress Disorder (PTSD) are presented and discussed in chapter 17. However, this presentation to be complete we must briefly report here some data.

Skin responsiveness to various antigens is increased in PTSD patients (86-87). NK cell cytotoxicity and CD2+, CD4+, CD8+, CD16+, CD20+ and

CD56+ counts are decreased in correlation with the severity of the PTSD symptomatology (88), or are normal (87, 89). Increased NK cell activity was detected by Laundelager et al. (89). Finally, increased counts of total leukocytes and lymphocytes, T-cells, CD4 and CD8 have been reported (87) The lymphoproliferative response to mitogen stimulation has been reported as normal or as reduced (90).

In 19 Israeli male army veterans with PTSD, IL-1β values were increased, while those of sIL-2R were not. IL-1β values correlated positively with the duration of the illness, but not with its severity, with cortisol secretion, with anxiety or depressive symptomatology or with alexithymia scores (91).

Soluble IL-1-receptor antagonist (sIL-1Ra), IL-6, sIL-6R and IL-2 receptor (IL-2R) concentrations are normal in PTSD [13 cases of Maes et al., (92), 25 cases of Muller et al., (93)]. Concentrations of Clara cells (CC16) and glycoprotein 130, which are IL-6 dependent and transduce IL-6 signals, are also normal (92). The data of Maes et al. and Muller et. al. (92-93) seem to exclude the existence of cytokine alterations in PTSD, which is certainly intriguing since there have been reports of impairments of neurotransmitters and neurohormones which modulate immune function.

4. PHARMACOLOGICAL TREATMENT OF ANXIETY DISORDERS: EFFECTS ON IMMUNE FUNCTIONS

It has been suggested that psychopharmacological therapies influence immune functions, partly through their capacity to modulate brain neurotransmitter secretion and receptor activities, which in turn influence multiple immune parameters, and partly by acting directly on receptors located in central and peripheral immune tissues. These phenomena have been demonstrated in experimental animals and in humans for two classes of psychotropic drugs that have been widely and effectively used in AD, that is anxioyitic benzodiazepines (BDZs) and antidepressants (tricyclic antidepressants, heterocyclic antidepressants and specific serotonin reuptake inhibitors SSRI), acting as anxyolitic possibly because of their 5-HT modulatory capacity .

4.1 Benzodiazepines

Benzodiazepines (BDZs) influence the function of the immune system by acting on the benzodiazepine receptors located on the membranes of immune cells (monocytes, neutrophils, B cells, NK cells, CD4 +,CD8+), in both experimental animals and humans (94,95). BDZs stimulate chemotaxis and phagocyte defenses, decrease the response to mitogen stimulation, decrease CD3+, CD4+ , CD16+, CD57+, NK cell counts and enhance NK cell activity (96,97). In experimental animals BDZs reverse or attenuates the stress-induced immunosuppression (98-100).

Production of the pro-inflamatory cytokines IL-1β, IL-6, TNF-α and of granulocyte/macrophage colony stimulating factor have been observed to be modulated by BZDs in both the peripheral immune cells and in macrophages in the brain of rats (101,102).

In nonpsychotic and nondepressed humans chronically using BDZs, CD3+, CD4+, CD16+, NK cell counts are lower than normal and NK cell activity is increased (100).

Data about cytokine regulation by BDZs in human anxiety disorders are missing, apart from those of Brambilla et al. (71,72) obtained for patients with PD, in whom the higher than normal values of IL-1β and the normal values of TNF-α observed before therapy were not modified by a clinically succesful one month administration of full therapeutic doses of alprazolam.

4.2 Antidepressants

Tricyclic antidepressants, heterocyclic antidepressants and SSRI, which have all been widely and often successfully administered in the treatment of AD, have been shown to inhibit multiple immune fuctions. Administration of SSRI antidepressants, in particular fluoxetine and citalopram, to experimental animals inhibits the stress-induced peripheral IL-1 and IL-2 production, suppress IL-4 production, and increases IL-6 and IL-10 secretion. In the brain, they stimulate IL-1 and IL-1Ra secretion (103,104) .

It has been reported anecdotally that patients on fluoxetine therapy have reactivation of herpes simplex infection and development of cutaneous pseudolymphoma lesions (103).

In vitro studies with blood obtained from normal humans have shown that the tricyclic antidepressants chlomipramine and imipramine inhibit the mitogen-stimulated T-lymphocyte or monocyte secretion of IL-2, IL-1β, TNF-α, INF-γ, and stimulate IL-10 secretion (105). The heterocyclic antidepressant trazodone reduces the release of INF-γ and increases the secretion of IL-10 (106). The SSRI sertraline and citalopram suppress the IL-2 secretion by stimulated T-lymphocytes (105).

Studies *in vivo* in humans regard only the effects exerted on depressed patients, in whom the pathologically elevated cytokine secretion is inhibited by SSRI treatment (107), and in psychosomatic disorders, but no data have been reported for anxious patients.

5. CONCLUSIONS

The data in the literature on secretions and functions of cytokines, cytokine receptors and cytokine antagonists in the course of Anxiety Disorders are scanty and too controversial to allow us to draw definite conclusions. Normal, increased or decreased basal and stimulated concentrations in body fluids and *in vitro* production of IL-1β, IL-1β-Ra, IL-1α, IL-6, IL-6R, IL-6-dependent IGP 130, TNF-α, INF-γ, IL-2, IL-2R, IL-3, IL-10, LIF R have been found contemporaneously in various studies. Sometimes the alterations of pro-inflammatory cytokines, their receptors and signal transducers were specular to those of anti-inflammatory cytokines or of pro-inflammatory receptor antagonists, as expected, suggesting the existence of an inflammatory-like or an anti-inflammatory like situation in specific AD. But mostly they were not, the alterations going in the same direction for both groups of the immune parameters, suggesting the contemporary presence of pro-inflammatory and anti-inflammatory situations, being possibly compensatory phenomena. Therefore, the significance of alterations when present is difficult to assess and its etiopathogenetic meaning even more difficult to understand.

During and after stressful situations pro-inflamatory and anti-inflammatory cytokine secretion has been generally found to be normal, activation of pro-inflammatory cytokine occurring only when the subjects are not able to cope with the stressful events or have specifically negative approaches to them or do not have social support or there is development of marked, even though still not pathological, anxiety. This observation suggests that IS responses to stressful events are not linked to the magnitude of severity of the stressors but rather to the presence in the probands of specific temperament traits which make them less able to cope with external and internal stimuli. The fact that in ADs, which have often been considered to be stress-like psychological disorders occurring in individuals with predisposing temperaments, this activation does not always occur, whatever the severity of the anxiety disorders, suggests either that ADs are not simple stress disorders and the consequence of stressful life events, or that their psychobiological background is much more complex than that pertaining to usual responses to stress. Only GAD data partly overlap with those of stress, in that the IL-6 system (including IL-6R and IGP-130) is activated in GAD patients as in normal subjects under stress. However, the data for the other cytokines and their receptors do not confirm this aspect, which seems to be confined only to the IL-6 system, suggesting that there is a disconnection between the secretions and functions of the cytokines. In fact, it has been amply demonstrated in experimental animals and normal humans that the secretion of pro-inflammatory and of anti-inflammatory cytokines, their receptors and receptor antagonists is regulated as a cascade of steps, the secretion of one element inducing the stimulation or inhibition of the others. In the case of ADs this does not seem to occur, and it can be tentatively proposed that this might be the pathology specific for these disorders, and not just an excess or a defect of IS function. Like the disconnection between cytokine parameters, in AD there is also a disconnection between CNS, NES and IS functions. In fact, the normal agonistic-antagonist relationship between neurotransmitters, neurohormones and cytokines, which is specific for stress reactions in experimental animals and normal humans, is mostly lost in AD. Typical of this is the lack of correlation between cytokine secretion and HPA function repeatedly reported for AD patients.

As has been observed for cytokine secretion, the other immune parameters, reported very briefly here, tend also to show different results in the different AD. Inhibition of IS is mostly present in subclinical anxiety and in PTSD, activation in Phobias, normality or an activation in PD and normal functions in OCD.

At this point, however, it must be mentioned that the contradictions between results of different studies, which have shown both normal and pathological cytokine secretions in the same type of AD pathologies, could be due to the heterogeneity of the populations investigated, including differences in age and gender of the subjects, duration and severity of the diseases, type and duration of previous pharmacological treatments, differences in personality characters, presence of comorbidities, all factors which can modify the secretion of cytokines. Taking the age factor into consideration, for instance, most of the time the populations studied were a mixture of subjects of different ages and sometimes the controls used were not matched strictly for age with the patients, when it is well known that immune function, and specifically cytokine secretion, is strongly influenced by age (108). Similarly, the gender of the patients has not generally been taken into consideration, and in most of the studies it was not even mentioned whether or not the patients were men or women, or how many were men or women. Gender has been reported to influence immune function (109), even though no specific data on cytokine secretion have been reported. Moreover, in women the fluctuations of the hormonal milieu related to the ovarian cycle and/or menopause may also influence cytokine secretion (110), but no data were included in the studies mentioned above on the phases of the cycle and/or menopausal state during which the investigations were conducted.

Most of the time the duration of anxiety disorders and the severity of anxiety have not been reported in the various studies, even though it has been shown that during acute, long-lasting or chronic stressful situations the resulting immune function correlates with the severity of stress-induced anxiety.

It has been demonstrated that pharmacological manipulations (including not only anti-anxiety and antidepressant drugs but also anti-psychotics, antibiotics and hormones) can modify IS functions. Sometimes, but not always, it was stated that the AD patients under investigation were drug free, but no history of previous pharmacological treatments was included.

Finally, differences in personality characters may be responsible for the development of specific biochemical and immunological derangements which in turn might underlie one and not another AD.

Future researches should take into investigation all these parameters, to make the population investigated more compact, with the hope of producing more consistent results.

Even after taking into consideration all these biases, it seems possible to state that the data of the literature reveal the presence in AD of immune alterations whose definition of type and etiopathogenetic significance still need further clarification. Since we investigated AD patients during the course of and not, obviously, before or just at the beginning of the diseases, it is possible that we missed etiopathogenetically significant alterations of immune parameters, and just measured the results of the evolution of the pathological processes. Moreover, most of the data were obtained just once during the course of the diseases and not repeatedly during and after their end, when it is well known that IS function is not static but highly dynamic. At the moment it is clear that, even though all these diseases are classified as AD, their IS backgrounds are different in each disorder. What the significance of these differences is, whether or not they are primary or secondary for the development of the disease, and whether or not they are specific for each disease or linked to temperament traits is, for the moment, impossible to say. It is evident that the data obtained up to now must be considered only preliminary, and need further investigations to clarify their significance.

REFERENCES

1. Cohen, S., Doyle, W.J. and Skoner, D.P. Psychological stress, cytokine production and severity of upper respiratory illness. *Psychosom. Med.* 1999; 61: 175-180
2. Biondi, M. "Effects of stress on immune function: an overview". In *Psychoneuroimmunology, 3rd Edition*, Adler, R., Felten, D., Cohen, N. (eds.), Academic Press, 2001, pp 189-226
3. Lazarus, R. Coping theory and research, past, present and future. *Psychosom. Med.* 1993; 55: 234-247 -

4. Dantzer, R. and Mormede, P. "Psychoneuroimmunology of stress". In *Stress, the immune system and psychiatry.* B.F. Leonard, K. Miller, eds, Chichester, Wiley, 1995, pp 48-83.
5. Gennaro, S., Fehder, W., Nuamah, I.F., et al. Caregiving to very low birthweight infants: a model of stress and immune response. *Brain Behav. Immun.* 1997; 11: 201-215
6. Solomon, G.F., Segerstrom, S.C., Grohr, P., et al. Shaking up immunity: psychological and immunological changes after a natural disaster. *Psychosom. Med.* 1997; 59: 114-127
7. Arnetz, B.B., Wasserman, J., Petrini, B., et al. Immune function in unemployed woman. *Psychosom. Med.* 1987; 49: 3-11
8. Kiecolt Glaser, J.K., Fisher, L.D., Ogrocki, P., et al. Marital quality, marital disruption, and immune function. *Psychosom. Med.* 1987; 49: 13-34
9. Fittshen, B., Schulz, H., Raedler, A., et al. Changes of immunological parameters in healthy subjects under examination stress. *Int. J. Neurosci.* 1990; 51: 241-242
10. Brosschot, J.F., Benshop, R.J., Godaert, G.L.R., et al. Effects of experimental psychological stress on distribution and function of peripheral blood cells. *Psychosom. Med.* 1992; 54: 394-406
11. Dekaris, D., Sabioncello, A., Mazuran, R., et al. Multiple changes of immunological parameters in prisoners of war. *J. Am. Ment. Ass.* 1993; 270: 595-599.
12. Weiss, D.W., Hirt, R., Tarcic, N., et al. Studies in psychoimmunology: psychological, immunological, and neuroendocrinological parameters in Israeli civilians during and after a period of Scud missile attack. *Behav. Med.* 1996; 22: 5-14
13 Maes, M.., Hendriks, D.., Van Gastel, A., et al. Effects of psychological stress on serum immunoglobulin, complement and acute phase protein concentrations in normal volunteers. *Psychoneuroendocrinol.* 1997; 22: 397-409
14. Maes, M., Von Bockstaele, D., Van Gastel, A., et al. Influence of psychological stress on leukocyte subset distribution in normal humans: evidence for interrelated immunosuppression and T cell activation. *Neuro psychobiol.* 1999; 39: 1-9
15. Breznitz, S., Ben Hur, H., Berzon, Y., et al. Experimental induction and termination of acute psychological stress in human volunteers, effects of immunological, neuroendocrine, cardiovascular, and psycological parameters. *Brain Behav. Immun.* 1998; 12: 34-52
16. Peters, M.L., Godaert, G.L.R., Ballieux, R.E., et al. Immune response to experimental stress: effects of mental effort and uncontrollability. *Psychosom. Med.* 1999; 61: 513-524
17. Jemmott, J., Borysenko, M. and Chapman, R. Academic stress, power motivation and decrease in secretion rate of salivary secretory immunoglobulin A. *Lancet* 1988; i: 1400-1402
18. Bosch, J.A., Brand, H.S., Ligtenberg, T.J.M., et al. Psychological stress as a determinant of protein levels and salivary-induced aggregation of streptococcus gordonii in human whole saliva. *Psychol. Med.* 1996; 58: 374- 382
19. Glaser, R., Kiecolt-Glaser, J.K., Stout, J.C., et al. Stress-induced impairments in cellular immunity. *Psychiat. Res.* 1985; 16: 233-239
20. Backen, E.A., Manuck, S.B., Marsland, A.L., et al. Lymphocyte subset and cellular immune responses to a brief experimental stressor. *Psychosom.Med.* 1992; 54: 673-679
21. Dantzer, R. and Kelley, K.W. Stress and immunity: an integrated view of relationships between the brain and the immune system. *Life Sci.* 1989; 44: 1995-2008

22. Mąes, M. Psychological stress, cytokines and the inflammatory response system. *Current Opinion in Psychiatry* 1999; 12: 695-700
23. Dobbin, J.P., Harth, M., Mc Cain, G.A., et al. Cytokine production and lymphocyte transformation during stress. *Brain, Behav., Immun.* 1991; 5: 339-348
24. Duguè, B., Lappanen, E.A., Teppo, A.M., et al. Effects of psychological stress on plasma interleukins-1 beta and 6, C-reactive protein, tumor necrosis factor alpha, antidiuretic hormone and serum cortisol. *Scand. J. Clin. Lab. Invest.* 1993; 56: 555-561.
25. Schulte, H.M., Bamberger, C.M., Elsen, H., et al. Systemic interleukin-1α and interleukin-2 secretion in response to acute stress and to corticotropin-releasing hormone in humans. *Eur. J. Clin. Invest.* 1994; 24: 773-777
26. Maes, M., Song, C., Lin, A., et al. The effects of psychological stress on humans: increased production of pro-inflammatory cytokines and a Th-1-like response in stress-induced anxiety. *Cytokine* 1998; 10: 313-318
27. Connor, T.J. and Leonard, B.E. Depression, stress and immune activation: the role of cytokines in depressive disorders. *Life Sci.* 1998; 62: 583-606
28. Mayer, S.F. and Watkins, L.R. Cytokines for psychologists: implications of bidirectional immune-to-brain communication for understanding behavior, mood and cognition. *Psych logical Rev.* 1998; 105: 83-107
29. Marshal, G.D., Agarwal, S.K., Loyd, C., et al. Cytokine dysregulation associated with exam stress in healthy medical students. *Brain, Behav. Immun.* 1998; 12: 297-307
30. Song, C., Kenis, G., Van Gastel, A., et al. Influence of psychological stress on immune-inflammatory vartiables in normal humans. Part II. Altered serum concentrations of natural anti-inflammatory agents and soluble membrane antigens of monocytes and T lymphocytes. *Psychiat. Res.* 1999; 85: 293-303
31. Lacey, K., Zaharia, M.D., Griffiths, J., et al. A prospective study of neuroendocrine and immune alterations associated with the stress of an oral academic examination among graduate students. *Psychoneuroendocrinol.* 2000; 25: 339-356
32. Maes, M., Song, C., Lin, A., et al. Immune and clinical correlates of psychological stress-induced production of interferon-γ and IL-10 in humans. In *Cytokine, stress and immunity.,*Plotnikoff N.P., Faith R.E., Murgo A.J., Good R.A., eds. Boca Raton, Raven Press, 1998, pp 39-50.
33. Dobkin, P.L., Fortin, P.R., Joseph, L., et al. Psychosocial contributors to mental and physical health in patients with systemic lupus erythemathosous. *Arthritis Care and Research* 1998; 11: 23-31.
34. Dantzer, R. Cytokine-induced sickness behavior: where do we stand? *Brain, Behav. Immun.* 2001; 15: 7-24
35. Croiset, G., Heijnen, C.J., van der Wal, W.E., et al. A role for the autonomic nervous system in modulating the immune response during mild emotional stimuli. *Life Sci.* 1990; 46: 419-425
36. Sgoutas-Emch, S., Cacioppo, J.T., Uchino, B.N., et al. The effects of an acute psychological stressor on cardiovascular, endocrine and cellular immune response: a prospective study of individuals high and low in heart rate reactivity. *Psychoneuro-endocrinol.* 1994; 31: 264-271
37. Shimizu, N., Kaizuka, Y., Hori, T., et al. Immobilization increases norepinephrine releases and reduces NK cytotoxicity in spleen of conscious rat. *Am. J. Physiol.* 1996; 271: 537-544
38. Cavaillon, J.M. *Les Cytokines*, Cavaillon J.M., ed., Paris, Masson, 1996

39. Cole, S.W., Korin, Y.D., Fahey, J.L., et al. Norepinephrine accelerates HIV replication via protein kinase A-dependent effects on cytokine production. *Immunol.* 1998; 161: 610-616
40. Maes, M., Lin, A., Kenis, G., et al. The effects of noradrenaline and alpha-2 adrenoceptor agents on the production of monocytic products. *Psychiat. Res.* 2000; 96: 245-254
41. Andreoli, A., Keller, S.E., Rabaeus, M., et al. Immunity, major depression and panic disorder comorbidity. *Biol.Psychiat.* 1992; 31: 896-908
42. Kronfol, Z and Remick, D.G. Cytokines and the brain: implications for clinical psychiatry. *Am J. Psychiat.* 2000; 157: 683-694
43. Solzynski, D., Kozak, W., Conn, C.A., et al. Beta-adrenoceptor antagonists suppress elevation in body temperature and increase in plasma IL-6 in rats exposed to open field. *Neuroendocrinol.* 1996; 63: 459-467
44. Dunn, A.J. "Interactions between the nervous system and the immune system: implications for psychopharmacology". In *Psychopharmacology: the fourth generation of progress.* F.E. Bloom, D.J. Kupfer, eds, New York, Raven Press, 1995.
45. Besedowsky, H., Del Rey, A., Sorkin, E., et al. Cytokines as modulators of the hypothalamus-pituitary-adrenal axis. *J. Steroid Biochem. Mol. Biol.* 1991; 40: 613-618
46. Heim, C. and Nemeroff, C.B. The role of childhood trauma in the neurobiology of mood and anxiety disorders: preclinical and clinical studies. *Biol. Psychiat.* 2001; 49: 1023-1039
47. Linn, B.S., Linn, M.W. and Jensen, J. Anxiety and immune responsiveness. *Psychol. Reports* 1981; 49: 969-970
48. Zorrilla, E.P., Redei, E. and De Rubeis, R.J. Reduced cytokine levels and T-cell function in healthy males: relation to individual differences in subclinical anxiety. *Brain, Behav. Immun.* 1994; 8: 293-312
49. Gonzales Quijano, M.I., Martin, M., Millan, S., et al. Lymphocyte response to mitogens: influence of life events and personality. *Neuropsychobiol.* 1998; 38: 90-96
50. Graham, N., Bartohlomeuse, R., Taboonpong, N., et al. Does anxiety reduce the secretion rate of secretory IgA in saliva? *Med. J. Australia* 1988; 148: 131-133
51. Fredrikson, M., Furst, C.J., Lekander, M., et al. Trait anxiety and anticipatory immune reactions in women receiving adjuvant chemotherapy for breast cancer. *Brain Behav. Immun.* 1993; 7: 79-90
52 Koh, K.B. Exam stress enhances lymphocyte proliferation. *14th World Congress of Psychosom. Med.* 1997 (abstract).
53 Koh, K.B. Emotion and immunity. *J. Psychosom. Res.* 1998; 45: 107-115
54 Koh, K.B. and Lee, B.K. Reduced lymphocyte proliferation and interleukin-2 production in anxiety disorders. *Psychosom. Med.* 1998; 60: 479-483
55. La Via, M.F., Workman, E.W., Lydiard, R.B. Subtype response to stress-induced immunodepression. *Funct. Neurol.* 1992; 7: 19-22
56. Castilla Cortazar, I., Castilla, A. and Gurpegui, M. Opioid peptides and immunodysfunction in patients with major depression and anxiety disorders. *J. Physiol. Biochem.* 1998; 54: 203-216
57. Covelli, V., Munno, I., Altamura, M., et al. Administration of thymopentin to patients with phobic disorders improves depressed phagocytic functions. *Immunopharmacol. Immunotoxicol.* 1990; 12: 619-631
58. Tancer, M.E., Folds, J. and Evans, D.L. Natural Killer cell levels in social phobia: a pilot study. *Biol. Psychiat.* 1992; 31: 236A (abstract)
59. Rapaport, M.H. and Stein, M.B. Serum cytokine and soluble interleukin-2 receptors in patients with panic disorder. *Anxiety* 1994; 1: 22-25

60. Gerritsen, W., Heijnen, C.J., Wiegant, V.M., et al. Experimental social fear: immunological, hormonal and autonomic concomitants. *Psychosom. Med.* 1996; 58: 273-286
61. Segerstromm S.C., Gloverm D.A., Craskem M.G., et al. Worry affects the immune response to phobic fear. *Brain, Behav. Immun.* 1999; 13: 80-92
62. Rapaport, M.H. and Stein, M.B. Serum interleukin-2 and soluble interleukin-2 receptor levels in generalized social phobia. *Anxiety* 1994; 1: 50-53
63. Schleifer, S.J., Keller, S.E., Scott, B.J., et al. Lymphocyte function in panic disorder. *Biol. Psychiat.* 1990; 27, 66A (abstract)
64. Tondo, L., Pani, P.P., Manconi, P.E., et al. Lymphocyte subset in Panic Disorder: a controlled study. *Biol. Psychiat.* 1991; 29(11S): 588 (abstract)
65. McDaniel, J.S., Risby, E.D., Stipetic, M., et al. Natural killer cell activity in patients with panic disorder. *Anxiety* 1991; 1: 192-195
66. Marazziti, D., Ambrogi, F., Vanacore, R., et al. Immune cell imbalance in major depressive and panic disorder. *Neuropsychobiol.* 1992; 26: 23-26
67. Brambilla, F., Bellodi, L., Perna, G., et al. Psychoimmunoendocrine aspects of panic disorder. *Neuropsychobiol.* 1992; 26: 12-22
68. Perini, G.I., Zara, M., Carraro, C., et al. Psychoimmunoendocrine aspects of panic disorders . *Human Psychopharmacol.* 1995; 10: 461-465
69. Surman, O.S., Williams, J., Sheean, D.V., et al. Immunological response to stress in agoraphobia and panic attaks. *Biol. Psychiat.* 1996; 21: 768-774
70. Rapaport, M.H. Circulating lymphocyte phenotypic surface markers in anxiety disorder patients and normal volunteers. *Biol.Psychiat.* 1998; 43: 458-463
71. Brambilla, F., Bellodi, L., Perna, G., et al. Plasma interleukin-1 beta concentrations in panic disorder. *Psychiat. Res.* 1994; 54: 135-142
72. Brambilla, F., Bellodi, L. and Perna, G. Plasma levels of Tumor Necrosis Factor-α in patients with panic disorder: effects of alprazolam therapy. *Psychiat. Res.* 1999; 89: 21-27
73. Brambilla, F., Picotti, G.B., Bondiolotti, G.P., et al. Neurochemical-hormonal investigation of panic disorder. Effects of alprazolam therapy. In *Stress and related disorders.* A.R. Genazzani, G. Nappi, F. Petraglia, A. Martignoni., eds., Cansterton, Parthenon Press, 1991, pp 175-182.
74. Weizman, R., Laor, N., Wiener, Z., et al. Cytokine production in panic disorder patients. *Clin. Neuropharmacol.* 1999; 22: 107-109
75. Murphy, T.K., Goodman, W.K., Fudge, M.W., et al. Lymphocyte antigen D8/17: a peripheral marker for childhood-onset obsessive-compulsive disorder and Tourette's Syndrome? *Am. J. Psychiat.* 1997; 154: 402-407
76. Khanna, S., Gokul, B.N., Reddy, P.L., et al. Humoral immune function in obsessive-compulsive disorder. *Indian J. Psychol. Med.* 1990; 13: 31-37
77. Khanna, S., Ravi, V., Shenoy, P.K., et al. Cerebrospinal fluid viral antibodies in obsessive-compulsive disorder in an Indian population. *Biol. Psychiat.* 1997; 41: 883-890
78. Swedo, S.E., Rapoport, J.L., Cheslow, D.L., et al. High prevalence of obsessive-compulsive symptoms in patients with Sydenham's Chorea. *Am. J.Psychiat* 1989; 146: 246-249
79. Swedo, S.E., Leonard, H.L., Garvey, M., et al. Pedriatic autoimmune neuropsychiatric disorders associated with streptococcal infections: clinical description of the first 50 cases. *Am. J. Psychiat.* 1998; 155: 264-271

80. Mąes, M., Meltzer, H. Y. and Bosmans, E. Psychoimmune investigation in obsessive-compulsive disorder: assays of plasma transferrin, IL-2 and IL-6 receptor, and IL-1beta and IL-6 concentrations. *Neuropsychobiol.* 1994; 30: 57-60
81. Marazziti, D., Presta, S., Pfanner, C., et al. Immunological alterations in adult obsessive-compulsive disorder. *Biol. Psychiat.* 1999; 46: 810-814
82. Monteleone, P., Catapano, F., Fabrazzo, M.,. Decreased blood levels of tumor necrosis factor-alpha in patients with obsessive-compulsive disorder. *Neuropsychobiol.* 1998; 37: 182-185
83. Brambilla, F., Perna, G., Bellodi, L., et al. Plasma interleukin-1beta in obsessive-compulsive disorders. *Biol. Psychiat.* 1997; 42: 976-981
84. Weizman, R., Laor, N., Barber, Y., et al. Cytokine production in obsessive-compulsive disorder. *Biol. Psychiat.* 1996; 40: 908-912
85. Surkina, I., Koptelov, O., Moukhine, A., et al. Immune function in patients with stress related mental disorder. *10th Congress Psychiatry*, Madrid, Spain 1996 (abstr.), p. 243.
86. Watson, I.P.B., Muller, H.K., Jones, I.H., et al. Cell-mediated immunity in combat veterans with post-traumatic stress disorder. *Med.J. Australia* 1993; 159: 513-516
87. Boscarino, J. and Chang, J. Higher abnormal leukocyte and lymphocyte counts 20 years after exposure to severe stress: research and clinical implications. *Psychosom. Med.* 1999; 61: 378-386
88. Ironson, G., Wynings, C., Schneiderman, N., et al. Posttraumatic stress symptoms, intrusive thoughts, loss, and immune function after hurricane Andrew. *Psychosom. Med.* 1997; 59: 128-141
89. Laundeslager, M.L., Aasal, R., Adler, L., et al. Elevated cytotoxicity in combat veterans with long-term pos-traumatic stress disorder: preliminary observations. *Brain, Be hav. Immun.* 1998; 12: 74-79
90. Everson, M.P., Kotler, S. and Blackburn, W.D. "Stress and immune dysfunction in Gulf war veterans". In *Neuroendocrine Immune Basis of the Rheumatic Diseases.* M. Cutolo, A.T. Masi, J.W. Bijlsma, et al., eds. New York: The New York Academy of Sciences, 1999.
91. Spivak, B., Shohat, B., Mester, R., et al. Elevated levels of serum interleukin-1beta in combat-related postraumatic stress disorder. *Biol.Psychiat.* 1997; 42: 345-348
92. Maes, M., Lin, A., Delmeire, L., et al. Elevated serum interleukin-6 (IL-6) and IL-6 receptor concentrations in posttraumatic stress disorder following accidental man-made traumatic events. *Biol.Psychiat.* 1999; 45: 833-839
93. Muller, R.J., Sutherland, A.G., Hutchison, J.D., et al. C-reactive protein and interleukin 6 receptor in post-traumatic stress disorder: a pilot study. *Cytokine* 2001; 13: 235-255
94. Zavala, F., Haumont, J. and Lenfant, M. Interaction of benzodiazepines with mouse macrophages. *Europ. J. Pharmacol.* 1984; 106: 561-566
95. Canat, X., Carayon, P., Bouaboula, M., et al. Distribution profile and properties of peripheral-type benzodiazepine receptors on human hemopoietic cells. *Life Sciences* 1992; 52: 107-118
96. Ruff, M.R., Pert, C.B., Weber, R.J., et al. Benzodiazepine receptor-mediated chemotaxis of human monocytes. *Science* 1985; 229: 1281-1283
97. Zavala, F. and Lenfant, M. Peripheral benzodiazepines enhance the respiratory burst of macrophage-like P388D1 cells stimulated by arachidonic acid. *Int. J. Immunopharmacol.* 1988; 9: 269-274

98. Fręire Garabal, M., Belmonte, A., Orallo, F., et al. Effects of alprazolam on T-cell immunosuppressive response to surgical stress in mice. *Cancer Letters* 1991; 58: 183-187

99. Freire Garabal, M., Couceiro, J., Balboa, J.L., et al. Effects of diazepam on the resistance, on the development of immunity and on the passive transfer of immunity to Listeria monocytogenes in mice submitted to surgical stress. *Acta Ther.* 1991; 17: 355-362

100. Lechin, F., van der Dijs, B., Vitelli-Flores, G., et al. Peripheral blood immunological parameters in long-term benzodiazepine users. *Clin. Neuropharmacol.* 1994; 17: 63-72

101. Bender, A.S. and Hertz, L. Pharmacological evidence that the non-neutral dizepam bindings in primary cultures of glial cells is associated with a calcium channel. *Europ. J. Pharmacol.* 1985; 110: 287-288

102. Taupin, V., Toulmond, S., Serrano, A., et al. Increase in IL-6, IL-1 and TNF levels in rat brain following traumatic lesion: influence of pre- and posttraumatic treatment with Ro54864, a peripheral type (p site) benzodiazepine ligand. *J. Neuroimmunol.* 1993; 42: 177-185

103. Kubera, M., Simbirtsev, A., Matheson, R., et al. Effects of repeated fluoxetine and citalopram administration on cytokine release in C57BL/6 mice. *Psychiat. Res.* 2000; 95: 255-266

104. Maes, M. The immunoregulatory effects of antidepressants. *Human Psychopharmacol.* 2001; 16: 95-103

105 Xia, Z., De Pierre, J.W. and Nassberger, L. Tricyclic antidepressants inhibit IL-6, IL-1β and TNF-α release in human blood monocytes and IL-2 and interferon-γ in T cells. *Immuno pharmacol.* 1996; 34: 27-37

106. Song, C. The interaction between cytokines and neurotransmitters in depression and stress: possible mechanism of antidepressant treatments. *Human Psychopharmacol.* 2000; 5: 199-211

107. Sluzewska, A., Rybakowski, J.K., Laciak, M., et al. Interleukin-6 serum levels in depressed patients before and after treatment with fluoxetine. *Ann. NY Acad. Sci.* 1995; 762: 474-476

108. Weksler, M.E. Senescence of the immune system. *Med. Clin. North Am.* 1983; 67: 263- 272

109. Oyeynka, G.O. Age and sex difference in immunecompetence. *Gerontology* 1984; 30: 188-195

110. Komorowski, J. and Stepien, H. FSH and LH induce Interleukin-6 (IL-6) release from human peripheral blood monocytes cultures *in vitro.* A dose response study. *Horm. Metab.* 1994; 26: 438-439

Chapter 14

CYTOKINES AND POST TRAUMATIC STRESS DISORDERS

Dewleen G. Baker[1], Thomas D. Geracioti, Jr.[1], John W. Kasckow[1], Emmanouil Zoumakis[2], George P. Chrousos[2]
[1]Cincinnati Veterans Affairs Medical Center and Department of Psychiatry, University of Cincinnati College of Medicine, Cincinnati, Ohio; [2]Pediatric and Reproductive Endocrinology Branch, NICHD, NIH, Bethesda, Maryland

1. INTRODUCTION

Posttraumatic stress disorder (PTSD) is unique among psychiatric syndromes in that a state of threatened homeostasis, or stress, is a known direct precipitant for the disorder. Thus, an essential diagnostic criterion for PTSD is exposure to an environmental traumatic event or stressor, during which the individual subjectively experiences extreme fear for self or someone else, feels helplessness and often a sense of horror, emotions that initiate a strong and prolonged psychophysiologic response. Although predisposing vulnerabilities are likely (1-3) the inability to fully re-establish pre-stress homeostasis and the subsequent post-stress maladaptive accommodation result in PTSD. The symptoms of PTSD, once established, often persist for years. The post-trauma emotional changes include persistent re-experiencing of the adverse event through intrusive thoughts and nightmares, emotional numbing, avoidance of stimuli that might be a reminder of the event, memory and cognitive difficulties, as well as a state of increased arousal experienced as difficulty sleeping, irritability, increased startle and hypervigilence. The disorder is associated with well-documented physiologic abnormalities in the major stress response systems, the hypothalamic-pituitary-adrenal (HPA) axis and sympathetic nervous system (SNS) (4-6) as well as other brain systems (7) for review.

Evidence for specific abnormalities in the major stress response systems in PTSD introduces the possibility that components of the immune system participate in the maladaptive response, since the brain and immune systems are functionally linked (8). The HPA axis through neuroendocrine humoral outflow by way of the pituitary and adrenal glands, the autonomic nervous system through direct efferent SNS innervation of lymphoid tissues, and the cytokines, pleotropic humoral immune messengers with endocrine, paracrine and autocrine effects, are active participants in this functional link. The wealth of evidence for bi-directional communication between the major neuroendocrine stress-response systems, i.e. the hypothalamic-pituitary-axis and sympathetic nervous system, and the cytokines has been amply reviewed by Andrew Miller in this publication, as well as by others (9,10).

Although stress is the direct precipitant of PTSD, and thought to be involved in anxiety, depression, and the chronic pain and fatigue syndromes as well, the relations between stress, anxiety, depression, the fibromyalgia and chronic fatigue syndromes and PTSD have not yet been fully elucidated (11). Traumatic stress, itself, can be followed by depression, PTSD, or a co-morbid state in which symptoms of both are expressed (12,13). Nearly 50% of patients with chronic PTSD experience intermittent co-morbid major depressive episodes, and 21 % of veterans in a recent study met diagnostic criteria for fibromyalgia, adding to the overall morbidity and diminished health-care quality of life associated with PTSD (14-17).

There is increasing evidence for abnormal central nervous system (CNS) regulation of both the HPA axis and the SNS in all of these disorders. In depression and PTSD, CSF studies provide direct evidence of CNS CRH activation and catechoaminergic hyperactivity (5,6,18). In fibromyalgia, CSF studies show noradrenergic hypofunction and based on indirect evidence the CRH neuron has been hypothesized to be either hyper- and hypoactive (19-21).

Disorder-specific differences in peripheral stress hormone secretion are also evident. Adrenocorticotropin (ACTH) and cortisol are secreted in excess in melancholic depression, but are normal or low in PTSD, and chronic pain and fatigue syndromes (4,5,11,13,18,22). Measurement of both basal and stress-related urinary and plasma catecholamine levels indicate excess secretion in depression and PTSD, but normal or low secretion in the chronic pain and fatigue syndromes (20,23-27).

These differences in HPA axis and SNS regulation could result in, or to be the result of, differences in neuroimmunomodulation. The complexity of the cytokine-nervous system interaction offers ample opportunity for neuroimmunomodulation at multiple levels and at many sites. These occur at the level of the cell through modulation of the signaling systems, including the levels of gene transcription or translation, and at the brain, pituitary, adrenal or immune tissues (28-31). Consequently, altered cytokine physiology has been hypothesized in PTSD, as well as in depression and fibromyalgia (32-35). The cytokine alterations are conceived as primary and/or secondary and thought to be potential mediators of long-term health consequences (34,36).

There are a number of mechanisms via which cytokines could contribute to the pathophysiology of PTSD and other post-stress syndromes. The contribution, if primary, could be thought of as acting on a genetic background of an increased vulnerability to these disorders, possibly a result of genetic polymorphisms in genes of the various cytokine cascades that participate in neural growth, development and/or function (37) or by participation in tissue specific sensitization prior to, and at the time of the index trauma event (38). For more information about sensitization, please refer to Chapter 11 of this book. Post-trauma, cytokines could participate in symptom expression either directly by effects on neurons or indirectly through effects on neurotransmitters, neuropeptides or other cytokines (39,40). Alternatively, as secondary participants, cytokines in the CNS might serve as neural protectors or may promote neural inflammation. In the periphery, they might mediate secondary health vulnerabilities, such as heart disease through inflammatory mechanisms (36).

Although cytokine participation in the pathophysiology of PTSD is an attractive hypothesis, there are no conclusive data to support it. There is preliminary evidence for cytokine dysregulation in humans but, to date, most published reports have been limited to the measurements of pro-inflamatory cytokines, TNFα, IL-1 and IL-6 and/or their receptors. Consequently, we will begin this chapter with a brief overview of cytokine physiology and their actions in the CNS as well as a discussion of possible models for involvement of cytokines in the pathophysiology of stress and PTSD. Subsequently, we will present an overview of immune studies in PTSD, followed by a synthesis of these results within the context of current pre-clinical research.

2. PRE-CLINICAL STUDIES: MODEL FOR CYTOKINES IN PTSD

2.1 Cytokines in the Central Nervous System

It is now recognized that that many, if not most cytokines and their receptors are constitutively expressed by and act upon neurons in central nervous system under normal conditions as well as in pathologic states (41). In the CNS, they are synthesized by glial, vascular, immune as well as by nerve cells. Additionally, there is evidence for peripheral to brain cytokine transport by a number of possible mechanisms: through active transport across the blood-brain barrier, through the circumventricular organs that lack a blood-brain barrier, and perhaps through afferent signaling via the autonomic afferent nerves. (42-47). Changes in cytokine levels in the CNS can have distal effects in the periphery, and *vice versa*, the peripheral cytokine compartment can influence CNS cytokine secretion. Yet, an understanding of cytokine production and regulation in each compartment will be critical to fully understanding their role in stress-related disorders. And, neither plasma nor CNS data alone are likely to be sufficient for fully understanding their effects in PTSD.

Under physiologic conditions, cytokines in the CNS are produced by and act directly upon neuronal cells to increase or decrease firing, causing changes in physiology and behavior critical to survival of the organism, such as sleep, motivation, appetite, and cognition (41). Alternatively, as products of nonneuronal brain cells, such as astrocytes, microglia, macrophages, mast and endothelial cells, cytokines play a role in the development or progression of pathologic states by provoking and maintaining an inflammatory response (41).

Cytokines are released in a cascade-like fashion, and often have multiple overlapping or antagonistic functions resulting in cytokine-cytokine interactions that can ultimately have an output that can be additive, synergistic or antagonistic. Turrin and Plata-Salaman point out that a holistic approach is warranted in the study of brain-cytokine interactions because the end result of cytokine signaling in the CNS is the product of highly complex modulatory interactions between neurotransmitters, neuropeptides, hormones and other cytokines as well as negative feedback regulation through autocrine, paracrine or hormonal mechanisms (41). Therefore, the sum product, rather than individual levels may best represent the physiologic and behavioral effects. Consequently, data obtained on cytokine levels of individuals with PTSD will need to be interpreted with caution and within this context.

2.2 Cytokines in Mental Health

Research on the relations between the immune system and psychiatric disorders began over a century ago. In 1927, the Nobel Prize was granted to psychiatrist Julius Wagner-Jauregg, for his success in altering the course of a psychiatric illness through use of an infectious agent. (34,48). The opportunity to productively pursue this line of inquiry has advanced considerably over the last century with the discovery and isolation of two key mediators: the stress neurohormones CRH (49) and norepinephrine (50) along with proof of the bi-directional communication between the brain and cytokine messengers (8,51). Over the past two decades, the detection of an ever-increasing number of cytokines and the explosively expanding research into their functions provides a framework for a more sophisticated understanding of their effects on behavior and their interactions with hormones of the stress axis, as likely occurs in PTSD. A core assumption of this work, predicated on evidence of bi-directional communication between the immune and nervous systems, is that physical and psychological stress activate the same immune-brain circuits, despite the relative independence of the central and peripheral compartments (52).

Even though over one hundred cytokines have been described, recent research has focused on the CNS effects of tumor necrosis factor-alpha (TNF-α), interleukin-1, and interleukin-6, three pro-inflammatory cytokines that are released in a cascade-like fashion in response to infection or tissue injury. Pre-clinical studies have demonstrated that in the CNS these cytokines influence core and critical behaviors, such as cognition, motivation, social exploration, sleep, and appetite in addition to producing changes in physiologic parameters such as body temperature, modulation of immune response and metabolism. These behavioral and physiologic effects can be attenuated or reversed by blocking the cytokine activity by various means (39,41,53).

The group of physiologic and behavioral changes generated by increased levels of CNS IL-1, termed "sickness behaviors" share similarities to those seen in major depression, thereby providing one of the lines of evidence leading to the hypothesis that they are CNS mediators of some symptoms of this syndrome (34,53). The "sickness behavior" hypothesis is a plausible explanation for major depression observed in individuals after chronic stress, or perhaps even after an exposure to major traumatic stress (12). However, it does not offer an accurate model for cytokine involvement in PTSD, which has a different cluster of presenting symptoms and different SNS and HPA axis dynamics than depression. Thus, a different model for PTSD will need to be developed.

2.3 Biologic model for cytokines in PTSD

2.3.1 General

Like the cytokine model of depression, a cytokine model of PTSD is predicated on a number of not fully proven assumptions: First, that stress, or in the case of PTSD, traumatic stress, up-regulates CNS cytokine activity in the brain structures that are directly or indirectly involved in emotional processing. Second, that at the CNS level of up-regulation attained, these messengers are integrated into the relevant stress circuitry, and participate with neurotransmitters and stress hormones such as CRH and NE via standard receptor and intra-cellular mechanisms to modulate the stress response. And third, that the cytokine participation can result in long-term changes in stress-responsivity and behavior. There are some pre-clinical data to support these assumptions and to support at least a preliminary model for cytokine involvement in PTSD.

2.3.2 Pre-clinical studies: Stress-related cytokine induction in the CNS

Since traumatic stress evokes intense emotion, the physiologic processing of the event is likely to involve limbic structures (54,55) through which cortical influences upon the hypothalamus and brain stem are carried by connections between the neocortex, hippocampus, and the amygdala (55). A key regulator of the behavioral and physical adaptation to the stress event is CRH, which, participating with NE in a positive feedback circuit coordinates this autonomic and behavioral response (9,56). The interconnections and various parts of this neurocircuitry, and the participating neurotransmitters, neuropeptides, and hormones have been described (9,55,57). Cytokines and cytokine receptor expression has been demonstrated at multiple points along this pathway, providing a potential substrate for interaction (58,59).

Recent studies of pathways, from the cerebral cortex to limbic and deep temporal lobe structures, including the hippocampus, further clarify the anatomic framework within which CNS NE with CRH and cytokines may interact to integrate the cognitive and behavioral aspects of stress (60). HPA axis excitation is propelled by stress-specific circuits connecting directly (brainstem) or indirectly to the parvocellular paraventricular nucleus (PVN) of the hypothalamus, although details of cytokine modulation of CRH activity in the PVN continue to clarified (57). It is known that IL-1 can increase NE metabolism in the hypothalamus, hippocampus, brain stem and spinal cord as demonstrated by an increase in the ratio of 3-methoxy-4-hydroxyphenyl-ethylene glycol (MHPG)/NE ratio (61,62). Evidence that these catecholaminergic neurons are involved in the HPA response to IL-1 through interconnective pathways with CRH in the PVN is provided by the ability to block the HPA axis response by surgical interruption of the noradrenergic innervation of the CRH-producing neurons in the PVN or by administration of NE antagonists (63-65).

A body of evidence, thus, supports a crucial role of catecholaminergic neurons in stressor- and cytokine (IL-1)- induced HPA axis activation, with stressor-specific modulatory input from the amygdala and other brain structures (57,66,67). Therefore, a credible pathway exists for cytokine modulation of the conditioned immune responses that have been demonstrated in laboratory animals and for hypothesizing cytokine involvement in the long-term mental health changes that are observed after traumatic stress.

More of an issue of debate is whether an acute or repeated psychogenic stress, such as that produced by a traumatic event of the type described by PTSD patients, can induce cytokine expression at critical locations in the neurocircuitry subserving emotional stress. Confirmatory data are emerging, but are still incomplete, and not wholly consistent. In 1991, immobilization, or physical restraint of rats was shown to increase IL-1β mRNA in the hypothalamus, but not in other important brain regions of the stress neurocircuitry (68). Later, inescapable shock (of the tail) was revealed to increase IL-1β protein in various brain regions, including the hypothalamus and hippocampus, but only in adrenalectomized, not intact, rats (69). Consistent with the idea that cytokine changes contribute to other stressor-provoked neurochemical changes, the central administration of IL-1 receptor antagonist (IL-1Ra) prevented the stressor elicited HPA axis changes (70). It also prevented the release of hypothalamic NE, dopamine and serotonin (71). Similarly, IL1-Ra prevented the development of behavioral disturbances ordinarily provoked by an uncontrollable stressor (72).

Recently, cortisol-modulated IL-1β protein induction was demonstrated in the brains and the pituitaries of intact (non-adrenalectomized) animals subjected to immobilization stress (69). The stress-induced increases in IL-1β protein were present in all brain regions assayed; the hypothalamus, nucleus tractus solitarius, cerebellum, hippocampus and posterior cortex. Moreover, there was site-specific glucocorticoid modulation; the stress-induced IL-1β protein production was suppressed by cortisol in the hypothalamus and pituitary, but not in the nucleus tractus solitarius. Stress-induced enhancement of brain IL1-β has also been observed in rats subjected to a less intense and more naturalistic form of stress, social isolation, in which animals accustomed to group housing were placed in an individual cage for 4 hours (73). The increases in IL-1β protein were found in the hippocampus and posterior cortex, but not in the pituitary gland or hypothalamus of the isolated rats. These animals had impairment in contextual fear conditioning that could be blocked by administration of IL-1Ra.

However, neither acute nor chronic exposure of rats to a feared predator, the ferret, a stress that seems intuitively similar to stresses that precipitate PTSD, induced measurable interleukin-1β mRNA. Nor did it induce any other of a number of cytokines, cytokine receptors, or neuropeptide mRNAs in brain regions thought to be involved in emotional processing, the pre-frontal cortex, amygdala, hippocampus, or hypothalamus (74). The authors suggested that the naturalistic stressor that they used, as opposed to psychogenic stressors such as immobilization stress or tail shock, activated homeostatic mechanisms such as glucocorticoid release that acted to preclude the brain cytokine alterations (74).

2.3.3 Pre-clinical trauma models: relations to clinical presentation

In view of the reported differences in CNS cytokine induction by different pre-clinical stress models, a full discussion of their potential analogy to the trauma event, an essential diagnostic criterion for PTSD, is warranted. The pre-clinical models of stress while critical to an understanding of the biology of the stress response, are limited in that none provides insight into the cognitive and emotional processing that result in the stress-induced physiologic and behavioral effects. At first glance, the paradigm of exposing a rat to its natural predator, the ferret, seems like an ideal model for the study of cytokine induction and sensitization in PTSD. However, the model may not provide a sufficiently intense or unanticipated event. The stressor was intense enough to elicit a startle response, a reduction in exploratory behaviors, and an HPA axis response equivalent to that seen in immobilization stress. However, there was other evidence, such as startle habituation, HPA axis adaptation, and partial behavioral habituation that suggests the rat did not experience an overwhelming and unmanageable level of stress. This time-related attenuation of stress-induced physiological and behavioral effects is evidence that the rat maintained a stress-response plasticity typically lacking after trauma exposure in humans with PTSD (75-77).

By analogy, men in combat, or women in domestic violence situations, often describe enduring high levels of stress for long periods of time prior to the seminal stress event without experiencing noticeable psychological or behavioral change. The change only occurs following the emotionally overwhelming seminal event. A dramatic example was given by a combat veteran who described months of intense combat, reported as very stressful, but throughout which he endured without perceived emotional or behavioral change. During the difficult time, his sense of integrity was maintained, in part by close association with a comrade, with whom he shared the stressful and dangerous situation. His psychological integrity ended, and his symptoms began after the seminal event, which occurred during a battle in which his friend's head was shot off, flew back and hit him in the chest. It was during this event that he first had the experience of intense physiologic arousal and dissociation typically described by people who recall a traumatic event. The dissociative phenomenon, such as time distortion, de-realization, and depersonalization perhaps herald recruitment of rarely manifested sensory pathways and intracellular processes. A naturalistic stress, such as a ferret walking near a cage may be insufficient to invoke these atypical CNS events.

The stress models that involve some degree of atypical handling, stress or injury, that do induce a CNS cytokine response, may not be inappropriate as models for cytokine involvement in for PTSD. As frequently as not, a trauma event severe enough to precipitate the symptoms of PTSD involves either intense physical exertion in efforts to maintain personal safety, or physical injury when those efforts fail. Burns and physical injuries, both of which would induce cytokines independently, are frequent components of traumatic stress involving domestic violence, rape, accidents and torture. And, high levels of physical exertion have been associated with cytokine induction (78). In these cases, cytokine messengers, via cross-sensitization, may enhance sensitization induced by the emotional reaction to the event itself. This cross-sensitization might explain the high incidence of PTSD in Vietnam veterans who were physically injured and the relatively high rates of PTSD in patients on burn units (79,80).

2.3.4 Cytokine sensitization and cross-sensitization

Despite some inadequacies and discrepancies, pre-clinical studies provide evidence for CNS cytokine induction by psychogenic stressors analogous to those described by patients with PTSD, in brain regions where they could participate in behavioral sensitization. Another series of pre-clinical studies provides evidence that sensitization to single stressor (footshock) does occur, that a cytokine (IL-1) can enhance that sensitization, and that cross-sensitization between a cytokine (IL-1) and a non-immune stressor (amphetamine) can also occur.

Changes in physiologic (long-lasting changes in HPA axis secretagogues) and behavioral (enhanced fear) parameters are commonly used paradigms to demonstrate sensitization. In separate experiments, van Dijken *et al* were able to show subsequent enhanced physiologic and behavioral responses after exposure to a single psychogenic stressor (foot shock) (81). Subsequently, Schmidt *et al* demonstrated cytokine-induced sensitization to a psychogenic stressor. Administration of IL-1 one to eleven days prior to exposure to either the same (cytokine) or a different (footshock) stressor was accompanied by an increased HPA axis response to the second stressor, and an enhancement of vasopressin immunoreactivity in the external median eminence (82).

Since amphetamine administration was later shown to induce behavioral sensitization, the research group hypothesized that similar mechanisms may underlie HPA axis sensitization by stress, IL-1 or amphetamine, and are likely to involve functional alterations in the CRH neurons in the PVN of the hypothalamus, as well as the brain stem catecholaminergic neurons (38). A crucial role for the brain stem catecholaminergic neurons had already been demonstrated in stressor and IL-1-induced HPA axis activation (83). This research group showed that single exposure to amphetamine or IL-1 induced cross-sensitization of HPA axis response after 11 and 22 days, but not after 42 days, along with a long-lasting increase in stimulus-secretion coupling in brainstem noradrenergic neurons that innervate the PVN. However, some functional sensitization-related differences between amphetamine and IL-1 at the level of the PVN CRH neuron were revealed. Whereas increased AVP signaling appeared to play a role in IL-1 induced sensitization, it did not do so in amphetamine-induced, or apparently in social-defeat induced sensitization of the HPA axis(84).

Thus, not only has IL-1 been shown to induce long-term sensitization; similar sensitization can be induced by other cytokines (TNF-a), the sensitization resulting in subsequent increased reactivity to an emotional stressor (a novel environment). Moreover, this reactivity is associated with increased intrahypothalamic NE release. (85)

2.3.5 Glucocorticoid feedback at the PVN: A model for PTSD

Glucocorticoids, produced in the periphery by the adrenal gland, either directly, or indirectly via input from other brain areas, impact upon the same CRH/AVP-producing cells in the PVN of the hypothalamus that are the putative agents of sensitization described above. Optimally, glucocorticoids serve to appropriately limit the stress sensitization, returning the organism to its post-stress homeostatic state, as occurs in normal healthy individuals. An area of intense investigation has been clarifying the abnormalities in the HPA feedback circuit in disorders like PTSD, the fibromyalgia/chronic fatigue syndromes, and melancholic depression, since each has HPA axis feedback dynamics that differ from the other conditions and from those of normal healthy individuals (4,9,11,20). In fact, the relatively low cortisol output associated with PTSD, perhaps due to inherited genetic polymorphisms, appears to confer a vulnerability to the disorder if a trauma occurs (1,2).

Two recently published pre-clinical studies may shed light on the interplay between the characteristics of the stressor, glucocorticoid production, and physiologic sensitization at the level of the CRH neuron in the PVN (86,87). First, Marti *et al* showed that the aspects of the stressor, such as intensity, alter dynamics of glucocorticoid output to a later stressor of the same type. Based on the serendipitous discovery that a single acute (immobilization) stress in rats did not alter the initial intensity of their corticosterone response, but that it hastened their post-stress corticosterone return to baseline, Marti *et al* set up a series of experiments to further define the characteristics of this alteration in HPA axis dynamics. The investigators determined that a faster post-stress glucocorticoid recovery to baseline was dependent upon the intensity, not the duration of the stress event, and that this rapid post-stress glucocorticoid return to baseline did not appear immediately, but consolidated over time (86). The fast HPA axis recovery is in many ways analogous to the post-challenge cortisol dynamics in PTSD (4).

The other pre-clinical study suggests how the glucocorticoid dynamics, such as early post-stress return of cortisol to baseline might alter CRH and AVP expression in the PVN so as to enhance sensitization and diminish habituation to the same stress. Pinnock and Herbert explored the effects of regulating corticosterone levels on PVN neuron adaptation while exposing male rats to repeated restraint stress. (87). In intact rats, acute restraint stress increased the expression of CRH mRNA while AVP mRNA expression was no different from that of control rats. Repeated exposure resulted in habituation of CRH expression, whereas AVP mRNA increased above that seen in either non-stressed or acutely stressed animals. In adrenalectomized rats with replacement corticosterone tablets that replicated the daily trough, basal CRH levels were raised, but a response to acute stress was still observed. However, the habituation that was seen in normal animals subjected to repeat stress was prevented, so that CRH mRNA levels continued to be high after repeat restraint stress. By contrast, the AVP response to both acute and repeated stress was unaltered in the corticosterone rats compared to controls. Replacement corticosterone pellets, approximating blood levels equivalent to the daily maximum had the opposite effect of the low replacement dose pellets. There was no change in CRH mRNA expression following either acute or repeated restraint compared to control rats, but the expected increase in AVP following repeated restraint was prevented.

Thus, the investigators showed the importance of glucocorticoids in modulating the adaptive pattern of CRH and AVP expression in the PVN. For each (CRH and AVP), the "set point" differed; for CRH, experiencing higher levels was necessary for subsequent adaptation to repeated restraint to occur, whereas for AVP a return to lower levels was necessary to allow this peptide to respond to repeated stress. Therefore, the diminished glucocorticoid response and rapid post-stress recovery to baseline observed in PTSD would serve to optimize and maintain the stress response to the same stressor, thereby maximizing and maintaining sensitization (88). This impediment of habituation would serve to promote the persistently high central SNS activity that has been observed in the disorder (6).

2.3.6 Cytokine effects on behavior

The above pre-clinical studies support the notion that cytokines, such as IL-1, can be induced by psychogenic stress, and that cytokines participate in sensitization and cross-sensitization at the level of the hypothalamic PVN, producing changes within the CRH-producing PVN neurons. Glucocorticoid feedback dynamics similar to those observed in PTSD, may be permissive of such sensitization. And under concurrent modulation from descending pathways from limbic structures, such as the amygdala, the increase in CRH expression in the PVN results in long-term activation of sympathetic brain stem nuclei. (66,67). These long-term changes in the SNS are hypothesized to cause physiologic changes such as increases in temperature and alterations in peripheral immune cell activities, since an increase or decrease in NE transmission can modulate these effects (89,90).

And, whereas the hypothalamic CRH system is important in modulating the neuroendocrine and metabolic responses to stress, the behavioral responses are likely to be modulated by interactions between cytokines and stress hormones (CRH and NE) in the amygdala and other extra-hypothalamic structures. Cytokines may modulate these changes by effects on CRH transmission, since there is evidence that IL-1, IL-2, and IL-6 can cause changes in CRH secretion, and that IL-6 can alter levels of CRH binding protein (91). Both CNS CRH and NE are increased in PTSD and CNS levels of NE in particular correlate highly with PTSD symptoms (5,6). Therefore, cytokine modulation of CRH and CRH binding protein or NE, directly, may have significant behavioral consequences in the disorder.

IL-1 might also directly, or indirectly though interactions with NE, NMDA and other transmitters effect long term potentiation (LTP) in the hippocampus, thereby playing a role in memory storage during or after the traumatic event in PTSD (60,92-94). Recently, anxiety and cognitive/memory disturbances, symptoms similar to those seen in PTSD, have been demonstrated in humans, correlated with an increase in plasma cytokine levels induced by a very low dose of LPS, and presumed to ultimately be the result of CNS cytokines induced by peripheral IL-1 and TNF-α. (39).

In this cytokine model of PTSD, genetic polymorphisms that impact on the induction of cytokine receptors and their antagonists might also have effects on the putative cytokine-modulated behavioral manifestations. IL-1 elevates blood and CSF levels of IL-6 and IL-1Ra, in addition to inducing its own receptor (95). Relative ratios of IL-1/IL-Ra may impact on sensitization, and on post-stress behavioral manifestations, such as "sickness behavior", perhaps providing an explanation of differences in post-trauma clinical presentation (12,58). By analogy, imbalances in production may occur in the CNS in the same way that an IL-1/IL-Ra imbalance is hypothesized in the periphery to be a factor in the initiation or perpetuation of rheumatoid synovitis (96).

Thus, in summary, we conclude that pre-clinical investigations provide data to support the hypothesis that stress induces CNS cytokines and that these cytokines participate, via sensitization and cross-sensitization, in the behavioral and physiological changes in PTSD. Furthermore, it is possible that the dynamics of the HPA axis feedback in some individuals genetically prone to getting PTSD may be permissive to this process.

3. OVERVIEW: CLINICAL STUDIES

3.1 Summary of results

The first studies of immune function in PTSD were published barely a decade ago and focused on peripheral measures of cellular immune reactivity (See table 1) (97-102). Within the past three years a number of publications and presentations have provided basal measurements of cytokines and cytokine receptor concentrations in patients with PTSD (See table 2) (103-108). Most of these studies have focused on pro-inflammatory cytokines, but a few have measured other humoral messengers, as well, including the anti-inflammatory cytokine, IL-10 and the cytokine-like protein, leptin.

These studies have used a number of different research paradigms. In the majority, single time point, basal plasma cytokine or soluble receptor measurements were obtained. Both plasma and cerebrospinal fluid (CSF) data are available from two of the studies, thereby allowing a comparison of central and peripheral cytokine activity (104,106).

In addition to data from the basal studies, results of in vivo and vitro challenge studies have been reported (107,108). In these studies, patients were challenged with dexamethasone, and plasma cytokine levels were measured, or lymphocytes taken from PTSD patients and normal controls were exposed, in vitro, to various concentrations of lipopolysaccaride or dexamethasone.

3.2 Methodologic issues

Characteristics of cytokines that are of interest in mental health, i.e. their responsiveness to the internal environment and their exquisitely tuned cross-

Table 1: Results of studies of Immune Cells in PTSD

Author	Year	Subjects	Number	Cells Studied	Results	Author's Conclusions
Burgess-Watson[97]	1993 & 1995	Combat Veterans Veteran and Civilian Controls	25	Response to CMI multi-test (uses standardized doses of 7 different antigens)	Higher CMI Sum Test Scores	Combat Veterans have enhanced cell-mediated (delayed-type immunity)
Laudenslager[98]	1997	Combat Veterans	10 PTSD 9 NC, combat vets (9/10 comorbid MDD)	WBC Counts NK Cell Activity Lymphycyte Subtyping	Increase NK cell cytoxicity in PTSD	PTSD, even with comorbid depres-sion associated with an increase in natural cytotoxicity
Ironson[99]	1977	Civilians 1-4 mnths after a natural disaster	180 33% with PTSD Symptoms	WBC Counts NK Cell Activity CD4, CD8, CD56	Decrease in NK Cell cytoxicity in PTSD	Intrusive thoughts and PTSD related to lower NKCC
Delahanty[200]	1997	Civilians 2-6 months after clean up at crash site	159 workers 41 controls	NK Cell Activity and Number	NK Cell activity positively correlated with IES intrusion scores	Intrusive thoughts related to higher NKCC
Boscarino[100]	1999	Vietnam Combat Veterans	286 PTSD 274 Anxiety 192 DD 2190 Total	WBC Count Total lymphocyte T-cell, CD4, CD8, Reactive CMI	Increased WBC lymphocy tes T-cell CD4	Immune abnormalities
Wilson[101]	1999	Females with child-hood sexual abuse	10 PTSD 10 NC	Leukocyte Subtypes	Ratio of CD45RO+ to CD45RA+ lymphocyte s increased	These findings suggest presence of increased lymphocyte activation in PTSD
Kawamura[102]	2001	Japanese male civilians	12 PTSD* 48 NC	NK cell activity # lymphocytes # CD_3, CD_4 and CD_8 cells IL-4	See text	Immuno-suppression in men with history of PTSD; Immunoactivation in current PTSD

*12 with history of PTSD; 3 with current diagnosis of PTSD

Table 2: Results of studies of Cytokine Production in PTSD

Author	Year	Subjects	Number	Sample	Cytokines	Results	Author's Conclusions
Spivak[103]	1997	Combat Veterans	19 PTSD 19NC	Basal Plasma	IL-1β SIL-2R	IL-1β increase SIL-2R no change	IL-1β plasma levels are high in PTSD and positively correlated with length of illness
Hagan[104]	1999	Combat Veterans	10 PTSD 10 NC	Basal Serial * Plasma Basal Serial * CSF	Leptin Leptin	No group difference in either plasma or CSF levels CSF leptin positively correlated with CSF dopamine levels and UFC.	No effect on leptin levels of PTSD Association between CSF leptin, dopamine and UFC suggests additional roles of leptin in neuroendocrine regulation
Maes[105]	1999	Civilians Trauma events Severe accidents	13 PTSD 32 NC	Basal Plasma	IL-6 sIL-6R sgp 130	Increased IL-6 Increased sIL-6R in subjects with comorbid MDD	PTSD is associated with increased IL-6 signaling ♦
Baker[106]	2001 (in press)	Combat Veterans Combat Veterans	11 PTSD 8 NC 5 PTSD 5 NC	Basal Serial* Plasma Basal CSF	IL-6 IL-6	No group difference Increased CSF IL-6	No increase in plasma IL-6 ♦ CSF levels of IL-6 increased in PTSD♦
Wong[107]	2000[(a)] (abstract)	Combat Veterans And Civilians	14 PTSD 8 NC	Basal Plasma DST Challenge Plasma Increase In Vitro LPS stimulated lymphocytes	IL-6 IL-2 SIL-2R IL-6	No Group difference No change in IL-6 levels post DEX Blunted IL-6 response to LPS	No increase in plasma IL-6 ♦ PTSD is associated with a diminished ability to respond to endotoxin.
Wong[108]	2000 [(b)] (abstract)	Combat Veterans and Civilian Controls	14 PTSD 8 Combat Vets, 6 Civilians, 8 NC	Basal Plasma DST Challenge	IL-10 IL-10	IL-10 increase in civilians. No change in IL-10 post- DEX increase	Increased IL-10 in civilians with PTSD

*6 time points per subject hourly between (11AM-5PM)

♦ See text

modulation with the nervous system complicate their accurate measurement in clinical studies. Differences in cytokine concentrations over time, such as might occur through circadian variation, and differences in cytokine production resulting from physiologic variables of such as age, gender, body mass index or use of medications introduce sample variability. Given the inconsistencies in data gathered in mental health studies to date, a brief review of physiologic factors that alter their production is warranted.

Depending on the cytokine involved, intra-individual plasma cytokine levels have been shown to vary substantially over time. Cava *et al* measured plasma cytokine levels of IL-2R and IL-6 monthly over a six-month interval. Whereas the intra-individual correlations of IL-2R were relatively stable, the within-subject variability of IL-6 was high. Consequently, the authors recommended that more than one measurement of this cytokine would improve accuracy(109). Additionally, levels of IL-1 and IL-6 may vary during the day, since circadian variation in plasma levels has been reported for both cytokines (110-114).

Of cytokines presented in this chapter, plasma TNFα, leptin, IL-6 and perhaps IL-1Ra concentrations positively co-vary with body mass index (BMI) (115-117). Additionally, plasma IL-6 increases with age (106,118-123) . Smoking status has also been reported to alter plasma levels of a number of the cytokines studied in PTSD, including leptin, IL-6 and IL-2R but we failed to find such an association for leptin or for IL-6 in our cohort (104,106,122,123).

Additionally gender and menstrual status can influence cytokine production since sex steroids and cytokines directly and indirectly modulate each other (123-127).

Therefore, close attention to controlling for these non-study related sources of variability is indicated. Inconsistencies between studies may be a result, at least in part, of a failure to do so (123).

3.3 Basal studies: Central Nervous System

3.3.1 Interleukin 6: Overview

When fully processed, IL-6 is a 184-amino-acid peptide that achieves its effects through a ligand-specific receptor that is active in both membrane bound (IL-6R) and soluble (sIL-6R) forms. Membrane-bound IL-6R in conjunction with glycoprotein 130 (gp-130) forms a high affinity bond with IL-6 to induce signal transduction. However, when gp-130 is produced within a cell, cells without membrane bound IL-6 receptors can also respond to IL-6 presenting as part of an IL-6 soluble receptor complex (128). Alternatively, circulating sIL-6R and gp-130 units serve as a highly efficient buffering system for IL-6 activity (129).

Many peripheral cell types produce IL-6, which it is involved in plethora of activities, including regulation of inflammation, immunity, hematopoesis, bone metabolism and reproduction (122,130). In the CNS IL-6 is produced by glial cells and astrocytes and perhaps by neuronal cells (131-133). Among its many CNS functions is neuronal cell proliferation and differentiation (122,130,131,134).

While peripheral-CNS communication occurs, there is experimental and clinical evidence for differential IL-6 regulation between the two compartments (95,135,136). IL-6 functions effectively as a messenger between the brain and the rest of the body during stress and immune challenge, but many details of the specific method(s) and regulation of the peripheral-CNS cytokine signaling remain to be determined.

3.3.2 IL-6: CSF Concentrations in PTSD

Data from our recent study of IL-6 in PTSD, although preliminary, suggests a significant increase in CSF IL-6 concentration in the patients with PTSD, as compared to the normal control subjects (106). The correlation between CSF and plasma concentrations was significant and strongly positive in the normal subjects, but was negative and non-significant in the patients with PTSD, perhaps reflecting an increase in production of IL-6 in the brain compartment in the latter group (Table 3). These and other data provide further evidence that CNS and peripheral IL-6 are differentially regulated in humans (135,137,138).

Table 3. Correlation between CSF and plasma levels.

Variable		Plasma	N	CSF	N	Correlation	Significance
IL-6 pg/ml	Group	7.23 (0.86)	19	19.28 (2.94)	10	0.33	n.s.
	PTSD	8.04 (1.23)	11	24.00 (3.04)	5	-0.67	n.s.
	Control	6.11 (1.12)	8	14.56 (2.83)	5	0.92	0.08
AM Leptin ng/ml	Group	4.53 (0.70)	19	0.071 (0.023)	19	0.63	< .002*
	PTSD	5.55 (0.78)	11	0.074 (0.035)	11	0.86	<.001
	Control	3.14 (1.13)	8	0.067 (0.028/)	8	0.41	n.s.
PM Leptin ng/ml	Group	3.56 (0.48)	19	0.47 (0.017)	19	0.88	<.001*
	PTSD	4.19 (0.57)	11	0.064 (0.027)	11	0.88	< .001
	Control	2.70 (0.78)	8	0.024 (0.014	8	0.83	< .05

*Hagan[104]

Interestingly, the high CSF IL-6 concentrations in PTSD contrast with low CSF IL-6 levels observed in subjects with major depression (139,140). Ultimately, these differences in CSF IL-6 concentration may be a factor in differential symptom expression. As noted previously, anxiety and cognitive disturbances correlate with increased plasma cytokine levels induced by very low dose LPS. Such an increase in plasma cytokines would likely result in an increase in CNS cytokines, which could directly or indirectly modulate such symptoms (39). Increased levels of central IL-6 in PTSD may protect against anxiety and cognitive/memory symptoms, perhaps through an increase in CRH binding protein, since our data show a negative correlation between CSF IL-6 concentration and PTSD symptoms as measured by the Clinician Administered PTSD Scale (CAPS) (91).

It is also possible that IL-6 serves a neuro-protective role for CNS noradrenergic neurons, which overproduce NE in PTSD (6). IL-6 has been reported to enhance the survival of sympathetic neurons by diminishing neuronal excitability or by induction of nerve growth factors (132,141). These neuroprotective actions might be via induction of adenosine, an inhibitory neuromodulator that inhibits transmitter release and induces hyperpolarization. In vivo, IL-6 has been shown to induce a concentration and time-dependent up-regulation of adenosine A_1 receptor mRNA (142).

3.3.3 Leptin: Overview

Leptin is a phylogentically conserved, 16kD cytokine-like protein that exerts pleiotropic effects on many different cell types. It has wide ranging roles directed toward the overall integration of metabolism and regulation of related physiological functions under basal conditions and during immune challenge (143,144). Like, IL-6, leptin utilizes gp130, and the Jak-stat signal transducer and activator of transcription (STAT) and MAP kinase pathways (143,145).

Leptin receptors, of which there are two major isoforms (ObRa and ObRb) were originally thought to be concentrated primarily in the CNS, but are now realized to have a widespread tissue distribution throughout the body (146,147). Both leptin and its receptor share structural and functional similarities with the IL-6 family of cytokines (148).

In humans, leptin is a participant in the immune response, and plays an important role in the regulation of metabolism and energy homeostasis. Additionally, it may serve as a regulator of the hypothalamic-pituitary-gonadal axis (149).

There is cross-modulation between the SNS and leptin. Central infusion of leptin has been shown to decrease food intake while activating the sympathetic nervous system in nonhuman primates, and reciprocally in the periphery the SNS participates in regulation of leptin secretion (150,151). Furthermore, catecholamines modulate the entry of leptin into the brain via an alpha-1 adrenergic mechanism (46).

The CSF and circulating peripheral levels of leptin are positively correlated, with the constant ratio of CSF to plasma leptin, although the set-point of this ratio varies, dependent upon the body mass index (152).

There is also cross-regulation between leptin and the IL-1 family of cytokines (153). In fact, IL-1 may mediate the actions of leptin on food intake and body temperature in the CNS (154).

3.3.4 Leptin: CSF (and plasma) Concentrations in PTSD

No significant differences were found in either CSF or plasma leptin concentrations between the PTSD and normal volunteer groups in our serial CSF study (104). CSF and plasma leptin correlations were high and similar in the both PTSD and normal volunteer study groups (see table 3). Given the correlation coefficients (above 60% in the AM, and above 80% in the PM), plasma levels could serve as a surrogate for CSF levels when assaying for leptin.

Not surprisingly, leptin significantly correlated with BMI, as did a number of other neuroendocrine variables measured. However, even after controlling for BMI, a strong correlation was found between CSF leptin and urinary free cortisol (UFC) concentrations, implying a possible CNS role for leptin in HPA axis regulation. Some prior data support a role of leptin in HPA axis stimulation, but other evidence does not (149,155). However, we found no relation between leptin and CRH, nor between CRH and UFC in this study (104). Possible explanations for the failure to find either relation may be that the association between leptin and UFC may not be mediated by CRH, or alternatively, that the CRH measurements in the CSF were a product of predominately extra-hypothalamic sources, therefore obscuring the relations between leptin and CRH and CRH and UFC.

Finally, we noted a strong positive correlation between plasma leptin and CSF dopamine levels. This relationship may imply additional roles for leptin in CNS neuroendocrine regulation that may contribute to central modulation of food intake and energy balance. Further support for a leptin-dopamine interaction, and for a role of leptin in food intake was provided by a pre-clinical study published in the past year. Investigators in this study observed that food restriction enhanced conditioned-place-preference in rats can be reversed by infusion of either leptin or a dopamine receptor agonist (156).

3.4 Basal Studies: Periphery

3.4.1 Interleukin 1: Overview

In the periphery interleukin-1, like IL-6, is synthesized by a number of cell types, including macrophages (157). CNS IL-1 producing cells include cerebrovascular cells, microglia, astrocytes and neurons (41). Peripheral and central actions of IL-1 are numerous, and include behavioral and neuromodulatory effects. IL-1 is a key a mediator of the response to inflammation and stress, as discussed earlier in this chapter.

Interleukin-1 is a family of three closely related proteins that are products of separate genes. The agonists, IL-1α and IL-1β are believed to exert identical actions through binding to a the cell surface receptor, IL-1RI, which for signal transduction requires association with an accessory protein, AcP (158). A second receptor, IL-1RII also binds IL-1, but lacks an intracellular domain, so does not initiate signal transduction. IL-1 receptor antagonist (IL-1Ra) is a highly selective competitive receptor antagonist that binds to IL-1-RI, but fails to trigger the association between IL1RI and AcP, necessary for subsequent signaling, thus blocks all actions of IL-1α and IL-1β. All IL-1 ligands are formed as precursors that are biologically active, and must be cleaved by an enzyme ICE, or caspase 1, which with the other members of the IL-1 family participates in autocrine and paracrine regulation of IL-1 signaling (159).

CNS levels of IL-1 in the unstimulated state are low, leading to uncertainty as to whether IL-1 plays a physiologic role in the normal healthy brain. Knock-out mice lacking various components of the IL-1 family of cytokines, develop normally if not subjected to pathological challenge (160). However, there is evidence for IL-1 physiologic regulation in sleep (161). In addition, IL-1 may modulate synaptic plasticity, specifically inhibiting the development of long term potentiation (LTP) in the hippocampus (94,122).

3.4.2 IL-1: Plasma Concentrations in PTSD

In the only published study of IL-1 in PTSD patients to date, Spivak *et al* reported increased plasma IL-1β concentrations in PTSD (103). These results, while unreplicated, are consistent with a hypothesis of immune activation, probably by sympathetic stimulation of peripheral immune tissues or by the decreased cortisol production by the adrenal gland (89,90,162).

These data showing increased plasma levels of IL-1 in PTSD would be consistent with human studies demonstrating CNS hyperactivity, and with data from a series of pre-clinical studies. In these studies, rats receiving a prolonged intracerebroventrically (ICV) infusion of CRH, were found to have an increase in immune responsivity, evidenced by splenocyte activation and abnormal cytokine expression (163). The rats not only showed increased IL-1β and interleukin-2 receptor antagonist (IL-2Ra) mRNA expression in splenic lympocytes; they also showed abnormal responses to immune challenge (intraperitoneal infusion of LPS), including blunting of the fever response and higher plasma levels of IL-1β and IL-6, as well as behavioral abnormalities (163). The thermogenic and immune changes seen in these animals was assumed to be mediated by the SNS, since there was prior evidence of their modulation by an increase or attenuation of NE transmission (89,90,162).

IL-1Ra has also been measured in subjects with PTSD (164). Unfortunately a statistical comparison of IL-1Ra concentrations, between study groups (PTSD vs normal control) was not explicitly presented in the text of the article published (164). However, the author did divide PTSD subjects into those without and those with major depression, presenting the data in table form. In comparison to the study volunteers with PTSD alone, study volunteers with co-morbid PTSD and major depression had lower IL-1Ra concentrations; levels were .57 (.096) versus .34 (0.33) ng/ml, respectively. The mean level for comparison control subjects was .26 (0.26ng/ml) (164).

Since IL-1 levels were not assessed in this study, the ratio of IL-1Ra to IL-1 could not be determined. However, it is possible that IL-1Ra, by blocking the activity of IL-1, might block its behavioral effects, such as depressive-like symptoms. Plasma levels of IL-1Ra may be higher in the study subjects with PTSD alone because of higher ambient levels of NE in PTSD patients without depression. Adrenaline infusion has been shown to increase both IL-1Ra and IL-6 plasma concentration (165). And, Yehuda *et al.* found higher levels of NE in PTSD subjects without depression than with co-morbid depressive illness and a negative correlation between plasma NE and Hamilton-D (depression) scores (166). It is possible that higher levels of IL-1Ra are protective against the depressive symptoms, and that the co-morbid patients have a relatively lower IL1-Ra/IL-1 ratio due either to diminished secretion because of slightly lower plasma NE, or because of other factors such as genetic variation or peripheral hormone balance.

The relationship between relative levels of plasma IL-1Ra and catecholamines and the presence or absence of co-morbid depression in PTSD awaits future study. However, there is evidence that CNS up-regulation of IL-1Ra activity results in an antidepressant effect. Antidepressants are known to attenuate LPS-induced sickness behavior, probably by induction of anti-inflammatory cytokines (52,167,168).

Furthermore, in a single study of the effects of antidepressants on pro-inflammatory cytokine production in the brain, Suzuki *et al* gave various antidepressants and found that they induced IL-1 and IL-1Ra production, with relatively larger increases in IL-1Ra. The time profile of IL-1Ra induction coincided with the therapeutic efficacy (169).

3.4.3 IL-6: Plasma Concentrations in PTSD

Results of plasma IL-6 concentrations in PTSD have been inconsistent. Maes, et al measured IL-6 in 45 civilian subjects, 13 with PTSD and 32 normal volunteers, and found increased levels of both IL-6 and the soluble IL-6 receptor (sIL-6R). In contrast Wong et al measured IL-6 at the same time of day in 30 subjects, 22 with PTSD and 8 normal control subjects, but found no group difference (107). Our research group measured plasma IL-6 across six time points, from 11:00 a.m. until 5:00 p.m., in conjunction with a serial CSF study of IL-6, and we found no significant difference between groups over this time period (106).

However, results of our analysis did show a group difference in IL-6 concentration over time; the control group plasma IL-6 levels were lower at the beginning of the collection period, increasing over time, whereas the IL-6 concentrations of the veterans with PTSD were consistently higher, especially in the early hours of the collection period. An analysis of a single time point, at the 11:00 a.m. would have shown a significant group difference, with PTSD subjects having higher plasma IL-6 levels. Thus, while we failed to find a statistically significant increase in plasma IL-6 in PTSD patients, further study may be warranted to determine if plasma IL-6 levels are normal in this disorder throughout the 24-hour circadian period. It is plausible that the Maes *et al* single time point IL-6 determination, nearly 4 hours earlier (at 7:45 a.m.) than our first assay, reflects a real group difference at a different phase of the circadian cycle. There is evidence for an IL-6 circadian rhythm in humans (110-113). However, this does not explain the contradictory results obtained by Maes *et al* and Wong *et al.*

In addition to the higher early morning IL-6 plasma levels in subjects with PTSD, our research group found a strong positive correlation between plasma IL-6 and norepinephrine in the symptomatic patients, in contrast to that in normal controls. This correlation is consistent with the theorized relation between increased SNS transmission and increased IL-6 production, and it provides further evidence of altered neuroimmune dynamics in PTSD.

An alternative explanation for the higher morning IL-6 levels found in our PTSD cohort would be increased morning IL-6 production, due to a failure of NE to diminish production overnight, as is seen during it's normal circadian cycle (170,171). Subjects with PTSD consistently show enhanced peripheral catecholamine secretion in reaction to acute stress, but studies of basal plasma or 24 hour urine NE levels are variable and inconsistent (6,26,172-178). However, Mellman *et al* while finding normal urinary catecholamine secretion in PTSD patients over a full 24 hour period, found a relative excess of nocturnal secretion of MHPG, the NE metabolite, leading the authors to hypothesize a failure of NE to diminish to a normal nadir overnight (26,170). Thus, intermittent greater NE production may lead to higher plasma IL-6 levels following acute stress, or during early morning hours, with possible untoward health effects.

3.5 Peripheral health effects

Life stress, and in particular traumatic stress in a given cohort is followed by a statistical increase in somatic complaints and health care use (15,179,180). Data from some recent studies suggest a relation between traumatic stress, PTSD and long term morbidity and, possibly, mortality (181-186).

Although these relations are not unequivocally established, long-term health effects, if they do occur, may in large part be the result of hyperactivity of the central stress circuitry accompanied by intermittent excess catecholamine discharge into the peripheral compartment, either via the peripheral sympathetic nervous system or the adrenal gland. The increase in sympathetic transmission could have direct deleterious end-organ effects. Examples of such direct end organ effects may be at the inner ear, or in the cardiovascular system. For example, excess sympathetic outflow may result in hyperacousis, cochlear damage and ultimate hearing loss (187). In the cardiovascular system, excess SNS transmission may explain the increased nighttime heart rate and the abnormal heart rate variability found in PTSD (188-190). Alternatively, the excess SNS drive might manifest through the peripheral immune system as changes in immune responsivity and cytokine production, causing indirect health effects (10).

The dual immune systems: innate and acquired immunity, if functioning adequately, serve to protect the individual from pathogens and abnormal cells, at the same time delimiting their activity so as not to cause harm to "self". The innate immune system, comprised of phagocytic granulocytes and macrophages is phylogenetically the oldest and is the first, rapid line of defense against invaders. The acquired immune system, takes days or weeks to develop full efficacy. It is mediated by lymphocytes that have evolved an enormous number of recombinant receptors – the immunoglobulins and T-cell receptors- that are capable of recognizing and amplifying the immune response against an invader (191). The glucocorticoids and catecholamines, at physiologic levels seen in stress, influence the innate and acquired immune systems in a complex interplay, in conjunction with other modulators such as histamine and adenosine (35).

Cortisol has a dual effect: It prepares the signaling system for innate immunity in inactivated cells; however, it inhibits innate immunity in activated cells and suppresses the cellular immune response. Catecholamines suppress innate immunity as well as cellular immunity (55). Natural killer Cells (NKC), a subpopulation of lymphocytes active in the innate immune response against certain viruses, intracellular bacteria and parasites, are very sensitive to catecholamine suppression, but less so to cortisol suppression (192-194).

In addition to modulating immune cell activation, cortisol and catecholamines mediate immune cell redistribution (195). An acute *versus* chronic stressor acting through differences in cortisol and catecholamine production can result in differential effects on cell redistribution, T-cell activation, and cytokine secretion, thereby producing opposite effects on delayed-type hypersensitivity (DTH), an antigen-specific skin cell-mediated immune response (196).

Cytokines, modulated by cortisol and catecholamines, also determine the balance between cellular and humoral (B cell) immunity. Interleukin-12 and tumor necrosis factor (TNF)-alpha promote T-helper (Th) 1 responses and cellular immunity, whereas IL-10 suppresses Th1 activities and stimulates Th2 and humoral immune responses (35). Cortisol and catecholamines inhibit production of IL-12 and TNF-α, thereby shifting the balance from cellular to humoral immunity, enhancing the humoral immune response (197-199).

Some aspects of both the innate and acquired immune systems have been measured in traumatized subjects (Table 2). In addition to the measurement of peripheral cytokines, discussed previously, parameters measured include NKC activity, cell-mediated (delayed –type) immunity, and immune activation, as determined by circulating immune cell counts and subtype. Wong *et al* also measured in-vitro immune cell responsivity to LPS challenge (Table 1). In some of these studies concurrent measurements of stress hormone levels and immune parameters were obtained, and relationships between these two variables were sought.

Both natural killer cell (NKC) activity and number were measured in individuals exposed to traumatic events, with variable results (98,99,102, 200). In three of these studies, measurements of plasma hormones (catecholamines, cortisol, DHEA) or cytokines (IL-4, interferon-γ (IFN-γ), were also obtained. Some studies showing increased and the others decreased NKC activity. This inconsistency can, in part, be attributed to study design and subject selection. For example, two studies reported higher NKC activity in subjects with chronic, active PTSD. However, one of these investigators compared the PTSD subjects to recently sober alcoholic controls, likely to show immunosuppression and the other (Kawamura *et al*) analyzed data from only three PTSD subjects (98,102). Of twelve additional subjects in the manuscript by Kawamura *et al,* all had a history of PTSD, but failed to meet criteria for active illness, and all of these had decreased NKC activity and decreased IFN-γ in comparison with the three subjects who had acute PTSD. In the other two studies, the investigators measured NKC activity in acutely traumatized subjects less than six months post-trauma (99,200). Despite having similar cohorts, the investigative groups reported opposite findings; Ironson *et al* reported an association between intrusive thoughts, PTSD diagnosis and diminished NKC activity, whereas Delahanty *et al* reported an positive association between intrusive thoughts represented by higher Impact of Events Scale scores (IES) and increased NKC activity.

No significant correlations between cortisol or catecholamine levels and NKC activity or between stress hormone levels and group (PTSD versus control) status were found in any of the studies in which hormone levels were obtained (98,200). There is some indication, based on the study by Kurakawa *et al*, that nuances, such as the chronicity or acuity of symptoms may have significant effects on the immune alterations observed, possibly due an evolution HPA and SNS responsivity from time the trauma occurs until the disorder ultimately consolidates. However, further studies of NKC activity in PTSD are needed to clarify discrepancies. In the studies discussed, differences in subject recruitment, assessment and study design preclude a definitive understanding of alterations in NKC activity in PTSD.

Aspects of the acquired immune response in chronic PTSD were assessed *in vivo* by measurement of circulating immune cell number and subtype and cell-mediated immunity, and in-vitro, by LPS lymphocyte challenge (97,101,107,108,184). Although none of these studies has been replicated, the evidence for immune activation in PTSD is consistent across the three published *in vivo* studies (97,101,184). In two of the studies, the authors reported either normal or increased cell-mediated (delayed type) immunity (97,184). Additionally, PTSD positive subjects show an increased number of circulating and activated lymphocytes (T-cells, CD_4 cells, CD45RO/CD45RA) (101,184). In contrast, the *in vitro* LPS lymphocyte challenge by Wong *et al* revealed a_blunted response in lymphocytes from PTSD patients, as assessed by IL-6 secretion (107,201). These data are surprising, given evidence for immune activation in the three cited studies, and the positive correlation between NE and plasma IL-6 concentrations in the study by Baker *et al* (106). Replication of current data, and additional studies will be needed to resolve these discrepancies.

Also, although data from Wong *et al*, fail to show an increase in plasma IL-6, the investigators report high plasma IL-10 concentrations in, at least, the civilian participants of their PTSD cohort. These data suggest of a shift toward a humoral immune balance, consistent with the normal or low plasma cortisol and intermittently high plasma catecholamine concentrations found in PTSD (108).

The increase in circulating and activated lymphocytes, high humoral immunity, and increased plasma concentrations of IL-1 and IL-6 suggest a smoldering peripheral inflammation in PTSD that could cause a number of untoward health effects, including a risk for atherosclerotic cardiovascular disease. Immune inflammation and altered cytokine expression are cardiac risk factors, and are associated with atherosclerotic plaque formation and unstable angina (202,203). Although a large epidemiologic study of Vietnam veterans conducted 15 years after the war (National Vietnam Veterans Research Study) showed no increase in heart disease associated with PTSD, a recent survey demonstrates an excess of abnormal EKGs, with evidence of AV conduction defects and infarcts in Vietnam combat veterans with PTSD (184).

Unfortunately, despite an indication of increased morbidity and mortality associated with PTSD, few studies of cytokine mediators or of associated indices of cardiac risk have been published. One, a study of lipid levels in veterans with PTSD reported higher lipid levels in veterans with PTSD than in veterans without (183). However, this study has not been replicated, and there is no indication that the investigators controlled for group differences in BMI or smoking status. Another, a pilot study of 23 subjects with PTSD, and 8 normal control subjects, reported finding no group differences in single time point plasma concentrations of either IL-6sR, or C-reactive protein, a liver derived, acute phase protein stimulated by IL-6 (204). However, despite finding no group differences in the two biologic measures, the researchers reported a high correlation between sIL6-sR, C-reactive protein, and PTSD avoidant and intrusive symptoms respectively.

The accumulated data are consistent with pre-clinical studies showing that HPA axis and SNS abnormalities result in altered cytokine expression and an enhanced peripheral immune response. However, these data are at times contradictory, and are as yet, unreplicated. Replication of current work and new studies will be needed to fully clarify the relationships between the mental events, CNS adaptations, and peripheral health effects in PTSD.

4. SUMMARY

In summary, PTSD is unique among psychiatric disorders, in that a state of threatened homeostatis, or stress, is a known direct precipitant for the disorder. Symptoms, once established, often persist for years. The disorder is associated with well-documented persistent, physiologic abnormalities in the major stress-response systems, the HPA axis and SNS. Evidence for specific abnormalities in the HPA axis and SNS introduces the possibility that the components of the immune system, such as the cytokines, participate in the maladaptive response, since the brain and immune systems are functionally linked. Consequently, altered cytokine physiology has been hypothesized in PTSD, as it has in a number of other mental health disorders.

Pre-clinical investigations are consistent with a hypothesis of cytokine involvement in PTSD. The data indicate that stress induces CNS cytokines and that these cytokines participate, via sensitization and cross-sensitization, in the behavioral and physiological changes in PTSD. Additionally, these data suggest that certain HPA axis feedback dynamics, similar to those of family members or of individuals with PTSD, may be permissive to stress sensitization.

Within the past three years, there have been a number of publications and presentations that have provided basal measurements of cytokines and cytokine receptors in PTSD. None has been replicated, but in total, they suggest abnormalities in CNS cytokine (IL-6) levels, as well as in peripheral cytokine physiology, a likely result of the altered HPA axis and SNS physiology in the disorder. These abnormalities may have consequences not only for the PTSD symptomotology, but in addition, may cause secondary health problems.

REFERENCES

1. True, W.R., Rice, J., Eisen, S.A., et al. A twin study of genetic and environmental contributions to liability for posttraumatic stress symptoms. *Arch.Gen.Psychiatry* 1993; 50:257-264
2. Yehuda, R., Bierer, L.M., Schmeidler, J., et al. Low cortisol and risk for PTSD in adult offspring of holocaust survivors. *Am. J. Psychiatry* 2000; 157:1252-1259
3. McFarlane, A.C. Posttraumatic stress disorder: a model of the longitudinal course and the role of risk factors. *J. Clin. Psychiatry* 2000; 6(5):15-20; discussion 21-3.:15-20
4. Yehuda, R., Boisoneau, D., Lowy, M.T., et al. Dose-response changes in plasma cortisol and lymphocyte glucocorticoid receptors following dexamethasone administration in combat veterans with and without posttraumatic stress disorder. *Arch.Gen.Psychiatry* 1995; 52:583-593

5. Baker, D.G., West, S.A., Nicholson, W.E., et al. Serial CSF corticotropin-releasing hormone levels and adrenocortical activity in combat veterans with posttraumatic stress disorder. *Am. J. Psychiatry* 1999; 156:585-588
6. Geracioti, T.D., Jr., Baker, D.G., Ekhator, N.N., et al. Cerebrospinal fluid norepinephrine concentrations in post-traumatic stress disorder. *Am. J. Psychiatry* 2001; 158:1227-1230
7. Friedman, M.J., Charney, D.S. and Deutch, A.Y. Neurobiological and Clinical Consequences of Stress: From *Normal Adaptation to Posttraumatic Stress Disorder* Philadelphia, Lippincott-Raven Publishers, 1995.
8. Besedovsky, H.O. and del Rey, A. Immune-neuro-endocrine interactions: facts and hypotheses. *Endocr.Rev.* 1996; 17:64-102
9. Chrousos, G.P. Stressors, stress, and neuroendocrine integration of the adaptive response. The 1997 Hans Selye Memorial Lecture. *Ann.N.Y.Acad.Sci* 1998; 851:311-335
10. Elenkov, I.J., Wilder, R.L., Chrousos, G.P., et al. The sympathetic nerve--an integrative interface between two supersystems: the brain and the immune system. *Pharmacol. Rev.* 2000; 52:595-638
11. Crofford, L.J. and Demitrack, M.A. Evidence that abnormalities of central neurohormonal systems are key to understanding fibromyalgia and chronic fatigue syndrome. *Rheum.Dis.Clin North Am.* 1996; 22:267-284
12. Shalev, A.Y., Freedman, S., Peri, T., et al. Prospective study of posttraumatic stress disorder and depression following trauma. *Am. J. Psychiatry* 1998; 155:630-637
13. Heim, C., Newport, D.J, Heit, S., et al. Pituitary-adrenal and autonomic responses to stress in women after sexual and physical abuse in childhood. *JAMA* 2000; 284:592-597
14. Fairbank, J.A., Schlenger, W.E., Saigh, P.A., et al. An epidemiologic profile of post-traumatic stress disorder. In *Neurobiological and Clinical Consequences of Stress: From Normal Adaptation to PTSD* Edited by Friedman MJ, Charney DS, Deutch AY. Philadelphia, Lippincott-Raven Publishers, 1995, pp. 415-427.
15. Baker, D.G., Mendenhall, C.L., Simbartl, L.A., et al. Relationship between posttraumatic stress disorder and self-reported physical symptoms in Persian Gulf War veterans. *Arch.Intern.Med* 1997; 157:2076-2078
16. Amir, M., Kaplan, Z., Neumann, L., et al. Posttraumatic stress disorder, tenderness and fibromyalgia. *J. Psychosom.Res.* 1997; 42:607-613
17. Baker, D.G., McQuarrie, I.G., Murray, M.G., et al. Diagnostic status and treatment recommendation for gulf war veterans with multiple non-specific symptoms. *Mil.Med* . 2001; 166: 972-981.
18. Wong, M.L., Kling, M.A., Munson, P.J., et al. Pronounced and sustained central hypernoradrenergic function in major depression with melancholic features: relation to hypercortisolism and corticotropin-releasing hormone. *Proc.Natl.Acad.Sci.* U.S.A 2000; 97:325-330
19. Legangneux, E., Mora, J.J., Spreux-Varoquaux, O., et al. Cerebrospinal fluid biogenic amine metabolites, plasma-rich platelet serotonin and [3H]imipramine reuptake in the primary fibromyalgia syndrome. *Rheumatology.(Oxford)* 2001; 40:290-296
20. Torpy, D.J., Papanicolaou, D.A., Lotsikas, A.J., et al. Responses of the sympathetic nervous system and the hypothalamic-pituitary-adrenal axis to interleukin-6: a pilot study in fibromyalgia. *Arthritis Rheum.* 2000; 43:872-880
21. Neeck, G. and Crofford, L.J. Neuroendocrine perturbations in fibromyalgia and chronic fatigue syndrome. *Rheum.Dis.Clin. North Am.* 2000; 26:989-1002
22. Cleare, A.J., Blair, D., Chambers, S., et al. Urinary free cortisol in chronic fatigue syndrome. *Am. J. Psychiatry* 2001; 158:641-643

23. Roy, A., Guthrie, S., Pickar, D., et al. Plasma norepinephrine responses to cold challenge in depressed patients and normal controls. *Psychiatry Res.* 1987; 21:161-16834.
24. Siever, L.J., Uhde, T.W., Jimerson, D.C., et al. Indices of noradrenergic output in depression. *Psychiatry Res.* 1986; 19:59-73
25. Yunus, M.B., Dailey, J.W., Aldag, J.C., et al. Plasma and urinary catecholamines in primary fibromyalgia: a controlled study. *J. Rheumatol.* 1992; 19:95-97
26. Mellman, T.A., Kumar, A., Kulick-Bell, R., et al. Nocturnal/daytime urine noradrenergic measures and sleep in combat-related PTSD. *Biol. Psychiatry* 1995; 38:174-179
27. McFall, M.E., Murburg, M.M., Ko, G.N., et al. Autonomic responses to stress in Vietnam combat veterans with posttraumatic stress disorder. *Biol. Psychiatry* 1990; 27:1165-1175
28. Laye, S., Parnet, P., Goujon, E., et al. Peripheral administration of lipopolysaccharide induces the expression of cytokine transcripts in the brain and pituitary of mice. *Brain Res. Mol. Brain Res.* 1994; 27:157-162
29. Miller, A.H., Pariante, C.M., Pearce BD Effects of cytokines on glucocorticoid receptor expression and function. Glucocorticoid resistance and relevance to depression. *Adv. Exp. Med. Biol.* 1999; 461:107-16.:107-116
30. Path, G., Scherbaum, W.A. and Bornstein, S.R. The role of interleukin-6 in the human adrenal gland. *Eur. J. Clin. Invest.* 2000; 30:91-95
31. McCann, S.M., Kimura, M., Karanth, S., et al. The mechanism of action of cytokines to control the release of hypothalamic and pituitary hormones in infection. *Ann.N.Y.Acad.Sci.* 2000; 917:4-18.:4-18
32. Chrousos, G.P. The hypothalamic-pituitary-adrenal axis and immune-mediated inflammation. *N.Engl.J. Med.* 1995; 332:1351-1362
33. Maes, M., Lin, A.H., Verkerk, R., et al. Serotonergic and noradrenergic markers of post-traumatic stress disorder with and without major depression. *Neuropsychopharmacology* 1999; 20:188-197
34. Licinio, J. and Wong, M.L. The role of inflammatory mediators in the biology of major depression: central nervous system cytokines modulate the biological substrate of depressive symptoms, regulate stress-responsive systems, and contribute to neurotoxicity and neuroprotection. *Mol. Psychiatry* 1999; 4:317-327
35. Elenkov, I.J., Chrousos, G.P. and Wilder, R.L. Neuroendocrine regulation of IL-12 and TNF-alpha/IL-10 balance. Clinical implications. *Ann.N.Y.Acad.Sci.* 2000; 917:94-105, 94-105
36. Gold, P.W., Chrousos, G.P. The endocrinology of melancholic and atypical depression: relation to neurocircuitry and somatic consequences. *Proc.Assoc.Am. Physicians* 1999; 111:22-34
37. Asensio, V.C. and Campbell, I.L. Chemokines in the CNS: plurifunctional mediators in diverse states. *Trends Neurosci.* 1999; 22:504-512
38. Schmidt, E.D., Schoffelmeer, A.N., De Vries, T.J., et al. A single administration of interleukin-1 or amphetamine induces long-lasting increases in evoked noradrenaline release in the hypothalamus and sensitization of ACTH and corticosterone responses in rats. *Eur. J. Neurosci.* 2001; 13:1923-1930
39. Reichenberg, A., Yirmiya, R., Schuld, A., et al. Cytokine-associated emotional and cognitive disturbances in humans. *Arch.Gen.Psychiatry* 2001; 58:445-452
40. Bluthe, R.M., Crestani, F., Kelley, K.W., et al. Mechanisms of the behavioral effects of interleukin 1. Role of prostaglandins and CRF. *Ann.N.Y.Acad.Sci.* 1992; 650:268-75.:268-275

41. Turrin, N..P and Plata-Salaman, C.R. Cytokine-cytokine interactions and the brain. *Brain Res. Bull.* 2000; 51:3-9
42. Ek, M., Engblom, D., Saha, S., et al. Inflammatory response: pathway across the blood-brain barrier. *Nature* 2001; 410:430-431
43. Konsman, J.P., Luheshi, G.N., Bluthe, R.M., et al. The vagus nerve mediates behavioural depression, but not fever, in response to peripheral immune signals; a functional anatomical analysis. *Eur. J. Neurosci.* 2000; 12:4434-4446
44. Hansen, M.K., O'Connor, K.A., Goehler, L.E., et al. The contribution of the vagus nerve in interleukin-1beta-induced fever is dependent on dose. *Am. J. Physiol. Regul.Integr.Comp. Physiol.* 2001; 280:R929-R934
45. Chen, G., Castro, W.L., Chow, H.H., et al. Clearance of 125I-labeled interleukin-6 from brain into blood following intracerebroventricular injection in rats. *Endocrinology* 1997; 138:4830-4836
46. Banks, W.A. Enhanced leptin transport across the blood-brain barrier by alpha1-adrenergic agents. *Brain Res.* 2001; 899:209-217
47. Banks, W.A., Kastin, A.J. and Broadwell, R.D. Passage of cytokines across the blood-brain barrier. *Neuroimmunomo.* 1995; 2:241-248
48. Raju, T.N. The Nobel chronicles. 1927: Julius Wagner-Jauregg (1857-1940). *Lancet* 1998; 352:1714
49. Vale, W., Rivier, C., Brown, M.R., et al. Chemical and biological characterization of corticotropin releasing factor. *Recent Prog.Horm.Res.* 1983; 39:245-70.:245-270
50. Von Euler, U.S. The presence of a substance with sympathin E properties in spleen extracts. *Acta Physiol. Scand.* 1946; 11:168-173
51. Mastorakos, G., Chrousos, G.P. and Weber, J.S. Recombinant interleukin-6 activates the hypothalamic-pituitary-adrenal axis in humans. *J. Clin. Endocrinol. Metab.* 1993;77:1690-1694
52. Maier, S.F. and Watkins, L.R. Cytokines for psychologists: implications of bidirectional immune-to-brain communication for understanding behavior, mood, and cognition. *Psychol.Rev.* 1998; 105:83-107
53. Castanon, N., Bluthe, R.M. and Dantzer, R. Chronic treatment with the atypical antidepressant tianeptine attenuates sickness behavior induced by peripheral but not central lipopolysaccharide and interleukin-1beta in the rat. *Psychopharmacology (Berl)* 2001; 154:50-60
54. Papez, J.W. A proposed mechanism of emotion. 1937. *J. Neuropsychiatry Clin. Neurosci.* 1995; 7:103-112
55. Downing, J.EG. and Kendall, M.D. Peripheral and central neural mechanisms for immune regulation through the innervation of immune effector sites. In *The Physiology of Immunity* Edited by Marsh JA, Kendall MD. Boca Raton, CRC Pres, 1996, pp. 103-125.
56. Smagin, G.N., Heinrichs, S.C. and Dunn, A.J. The role of CRH in behavioral responses to stress. *Peptides* 2001; 22:713-724
57. Herman, J.P. and Cullinan, W.E. Neurocircuitry of stress: central control of the hypothalamo-pituitary-adrenocortical axis. *Trends Neurosci.* 1997; 20:78-84
58. Turrin, N.P., Gayle, D., Ilyin, S.E., et al. Pro-inflammatory and anti-inflammatory cytokine mRNA induction in the periphery and brain following intraperitoneal administration of bacterial lipopolysaccharide. *Brain Res. Bull.* 2001; 54:443-453
59. Vallieres, L. and Rivest, S. Regulation of the genes encoding interleukin-6, its receptor, and gp130 in the rat brain in response to the immune activator lipopolysaccharide and the pro-inflamatory cytokine interleukin-1beta. *J. Neurochem.* 1997; 69:1668-1683

60. Westerhaus, M.J. and Loewy, A.D. Central representation of the sympathetic nervous system in the cerebral cortex. *Brain Res.* 2001; 903:117-127
61. Dunn, A.J. Systemic interleukin-1 administration stimulates hypothalamic norepinephrine metabolism parallelling the increased plasma corticosterone. *Life Sci.* 1988; 43:429-435
62. Kabiersch, A., del Rey, A., Honegger, C.G., et al. Interleukin-1 induces changes in norepinephrine metabolism in the rat brain. *Brain Behav.Immun.* 1988; 2:267-274
63. Weidenfeld, J., Abramsky, O. and Ovadia, H. Evidence for the involvement of the central adrenergic system in interleukin 1-induced adrenocortical response. *Neuropharmacology* 1989; 28:1411-1414
64. Day, T.A., Ferguson, A.V. and Renaud, L.P. Noradrenergic afferents facilitate the activity of tuberoinfundibular neurons of the hypothalamic paraventricular nucleus. *Neuroendocrinology* 1985; 41:17-22
65. Feldman, S., Conforti, N. and Melamed, E. Hypothalamic norepinephrine mediates limbic effects on adrenocortical secretion. *Brain Res. Bull.* 1988; 21:587-590
66. Bhatnagar, S. and Dallman, M. Neuroanatomical basis for facilitation of hypothalamic-pituitary-adrenal responses to a novel stressor after chronic stress. *Neuroscience* 1998; 84:1025-1039
67. Dayas, C.V., Buller, K.M. and Day, T.A. Neuroendocrine responses to an emotional stressor: evidence for involvement of the medial but not the central amygdala. Eur J *Neurosci.* 1999; 11:2312-2322
68. Minami, M., Kuraishi, Y., Yabuuchi, K., et al. Induction of interleukin-1 beta mRNA in rat brain after transient forebrain ischemia. *J. Neurochem.* 1992; 58:390-392
69. Nguyen, K.T., Deak, T., Will, M.J., et al. Timecourse and corticosterone sensitivity of the brain, pituitary, and serum interleukin-1beta protein response to acute stress. *Brain Res.* 2000; 859:193-201
70. Shintani, F., Nakaki, T., Kanba, S., et al. Involvement of interleukin-1 in immobilization stress-induced increase in plasma adrenocorticotropic hormone and in release of hypothalamic monoamines in the rat. *J. Neurosci.* 1995; 15:1961-1970
71. Shintani, F., Nakaki, T., Kanba, S., et al. Role of interleukin-1 in stress responses. A putative neurotransmitter. *Mol. Neurobiol.* 1995; 10:47-71
72. Maier, S.F., Nguyen, K.T., Deak, T., et al. Stress, learned helplessness, and brain interleukin-1 beta. *Adv. Exp. Med. Biol.* 1999; 461:235-49.:235-249
73. Pugh, R., Nguyen, K.T., Gonyea, J.L., et al. Role of interleukin-1 beta in impairment of contextual fear conditioning caused by social isolation. *Behav.Brain Res.* 1999; 106:109-118
74. Plata-Salaman, C.R., Ilyin, S.E., Turrin, N.P., et al. Neither acute nor chronic exposure to a naturalistic (predator) stressor influences the interleukin-1beta system, tumor necrosis factor-alpha, transforming growth factor-beta1, and neuropeptide mRNAs in specific brain regions. *Brain Res. Bull.* 2000; 51:187-193
75. Peri, T., Ben Shakhar, G., Orr, S.P., et al. Psychophysiologic assessment of aversive conditioning in posttraumatic stress disorder. *Biol. Psychiatry* 2000; 47:512-519
76. Shalev, A.Y., Peri, T., Brandes, D., et al. Auditory startle response in trauma survivors with posttraumatic stress disorder: a prospective study. *Am. J. Psychiatry* 2000; 157:255-261
77. Orr, S.P., Metzger, L.J., Lasko, N.B., et al. De novo conditioning in trauma-exposed individuals with and without posttraumatic stress disorder. *J. Abnorm. Psychol.* 2000; 109:290-298

78. Papanicolaou, D.A., Petrides, J.S., Tsigos, C., et al. Exercise stimulates interleukin-6 secretion: inhibition by glucocorticoids and correlation with catecholamines. *Am. J. Physiol.* 1996; 271:E601-E605
79. Bullman, T.A., Kang, H.K. and Thomas, T.L. Posttraumatic stress disorder among Vietnam veterans on the Agent Orange Registry. A case-control analysis. Ann. *Epidemiol.* 1991; 1:505-512
80. Ehde, D.M., Patterson, D.R., Wiechman, S.A., et al. Post-traumatic stress symptoms and distress 1 year after burn injury. *J. Burn Care Rehabil.* 2000; 21:105-111
81. van Dijken, H.H., de Goeij, D.C., Sutanto, W., et al. Short inescapable stress produces long-lasting changes in the brain-pituitary-adrenal axis of adult male rats. *Neuroendocrinology* 1993; 58:57-64
82. Schmidt, E.D., Binnekade, R., Janszen, A.W., et al. Short stressor induced long-lasting increases of vasopressin stores in hypothalamic corticotropin-releasing hormone (CRH) neurons in adult rats. *J. Neuroendocrinol.* 1996; 8:703-712
83. Plotsky, P.M., Cunningham, E.T., Jr. and Widmaier, E.P. Catecholaminergic modulation of corticotropin-releasing factor and adrenocorticotropin secretion. *Endocr. Rev.* 1989; 10:437-458
84. Buwalda, B., de Boer, S.F., Schmidt, E.D., et al. Long-lasting deficient dexamethasone suppression of hypothalamic-pituitary-adrenocortical activation following peripheral CRF challenge in socially defeated rats. *J. Neuroendocrinol.* 1999; 11:513-520
85. McQuade, R. A microdialysis study of the noradrenergic response in rat frontal cortex and hypothalamus to a conditioned cue for aversive, naturalistic environmental stimuli. *Psychopharmacology (Berl)* 2000; 148:201-208
86. Marti, O., Garcia, A., Velles, A., et al. Evidence that a single exposure to aversive stimuli triggers long-lasting effects in the hypothalamus-pituitary-adrenal axis that consolidate with time. *Eur. J. Neurosci.* 2001; 13:129-136
87. Pinnock, S.B. and Herbert, J. Corticosterone differentially modulates expression of corticotropin releasing factor and arginine vasopressin mRNA in the hypothalamic paraventricular nucleus following either acute or repeated restraint stress. *Eur. J. Neurosci.* 2001; 13:576-584
88. Delahanty, D.L., Raimonde, A.J. and Spoonster, E. Initial posttraumatic urinary cortisol levels predict subsequent PTSD symptoms in motor vehicle accident victims. *Biol. Psychiatry* 2000; 48:940-947
89. Rothwell, N.J. CRF is involved in the pyrogenic and thermogenic effects of interleukin 1 beta in the rat. *Am. J. Physiol.* 1989; 256:E111-E115
90. Fisher, L.A. Central actions of corticotropin-releasing factor on autonomic nervous activity and cardiovascular functioning. *Ciba Found. Symp.* 1993; 172:243-253; discussion 253-7:243-253
91. Kasckow, J.W., Regmi, A., Seasholtz, A.F., et al. Regulation of corticotropin-releasing factor-binding protein expression in amygdalar neuronal cultures. J *Neuroendocrinology* 1999; 11:959-966
92. Friedman, W.J. Cytokines regulate expression of the type 1 interleukin-1 receptor in rat hippocampal neurons and glia. *Exp. Neurol.* 2001; 168:23-31
93. Schneider, H., Pitossi, F., Balschun, D., et al. A neuromodulatory role of interleukin-1b0ta + 012.0.30. in the hippocampus. *Proc. Natl. Acad. Sci.* U.S.A 1998; 95:7778-7783
94. Katsuki, H., Nakai, S., Hirai, Y., et al. Interleukin-1 beta inhibits long-term potentiation in the CA3 region of mouse hippocampal slices. *Eur. J. Pharmacol.* 1990; 181:323-326

95. Reyes, T.M. and Coe, C.L. The pro-inflamatory cytokine network: interactions in the CNS and blood of rhesus monkeys. *Am. J. Physiol.* 1998; 274:R139-R144
96. Arend, W.P. Cytokine imbalance in the pathogenesis of rheumatoid arthritis: The role of interleukin-1 receptor antagonist. *Semin. Arthritis Rheum.* 2001; 30:1-6
97. Burgess-Watson, I.P., Muller, H.K., Hoffma, L., et al. Cell-mediated immunity in combat veterans with post-traumatic stress disorder. *Med. J. Aust.* 1995; 162:55-56
98. Laudenslager, M.L., Aasal, R., Adler, L., et al. Elevated cytotoxicity in combat veterans with long-term post-traumatic stress disorder: preliminary observations. *Brain Behav. Immun.* 1998; 12:74-79
99. Ironson, G., Wynings, C., Schneiderman, N., et al. Posttraumatic stress symptoms, intrusive thoughts, loss, and immune function after Hurricane Andrew. *Psychosom. Med.* 1997; 59:128-141
100. Boscarino, J.A. and Chang, J. Higher abnormal leukocyte and lymphocyte counts 20 years after exposure to severe stress: research and clinical implications. *Psychosom. Med.* 1999; 61:378-386
101. Wilson, S.N., van der, K.B., Burbridge J., et al. Phenotype of blood lymphocytes in PTSD suggests chronic immune activation. *Psychosomatics* 1999; 40:222-225
102. Kawamura, N., Kim, Y. and Asukai, N. Suppression of cellular immunity in men with a past history of posttraumatic stress disorder. *Am. J. Psychiatry* 2001; 158:484-486\
103. Spivak, B., Shohat, B., Mester, R., et al. Elevated levels of serum interleukin-1 beta in combat-related posttraumatic stress disorder. *Biol. Psychiatry* 1997; 42:345-348
104. Hagan, M.M., Havel, P.J., Seeley, R.J., et al. Cerebrospinal fluid and plasma leptin measurements: covariability with dopamine and cortisol in fasting humans. *J. Clin. Endocrinol. Metab.* 1999; 84:3579-3585
105. Maes, M., Lin, A.H., Delmeire, L., et al. Elevated serum interleukin-6 (IL-6) and IL-6 receptor concentrations in posttraumatic stress disorder following accidental man-made traumatic events. *Biol. Psychiatry* 1999; 45:833-839
106. Baker, D.G., Ekhator, N.N., Kasckow, J.W., et al. Plasma and cerebrospinal fluid interleukin-6 concentrations in posttraumatic stress disorder. *Neuroimmunomo.* 2001; 9: 209-217.
107. Wong, C.M., Rapaport, M.H., Golier, J.A., et al. Cytokine function in PTSD: Serum and LPS whole blood responses to the DST. Poster presented at the annual meeting of the ACNP, Puerto Rico, 2000;
108. Wong, C.M., Rapaport, M.H., Golier, J.A., et al. IL-10 alteration by trauma types in posttraumatic stress disorder. Poster presented at the annual meeting of the ADAA, Atlanta, GA, 2001.
109. Cava, F., Gonzalez, C., Pascual, M.J., et al. Biological variation of interleukin 6 (IL-6) and soluble interleukin 2 receptor (sIL2R) in serum of healthy individuals. *Cytokine* 2000; 12:1423-1425
110. Kanabrocki, E.L., Sothern, R.B., Messmore, H.L., et al. Circadian interrelationships among levels of plasma fibrinogen, blood platelets, and serum interleukin-6. *Clin. Appl. Thromb. Hemost.* 1999; 5:37-42
111. Vgontzas, A.N., Papanicolaou, D.A., Bixler, E.O., et al. Circadian interleukin-6 secretion and quantity and depth of sleep. *J. Clin. Endocrinol. Metab.* 1999; 84:2603-2607
112. Lissoni, P., Rovelli, F., Brivio, F., et al. Circadian secretions of IL-2, IL-12, IL-6 and IL-10 in relation to the light/dark rhythm of the pineal hormone melatonin in healthy humans. *Nat. Immun.* 1998; 16:1-5

113. Petrovsky, N., McNair, P. and Harrison, L.C. Diurnal rhythms of pro-inflammatory cytokines: regulation by plasma cortisol and therapeutic implications. *Cytokine* 1998; 10:307-312
114. Redwine, L., Hauger, R.L., Gillin, J.C., et al. Effects of sleep and sleep deprivation on interleukin-6, growth hormone, cortisol, and melatonin levels in humans. *J. Clin. Endocrinol. Metab.* 2000; 85:3597-3603
115. Bastard, J.P., Jardel, C., Bruckert, E., et al. Elevated levels of interleukin 6 are reduced in serum and subcutaneous adipose tissue of obese women after weight loss. *J. Clin. Endocrinol. Metab.* 2000; 85:3338-3342
116. Vgontzas, A.N., Bixler, E.O., Papanicolaou, D.A., et al. Chronic systemic inflammation in overweight and obese adults. *JAMA* 2000; 283:2235
117. Orban, Z., Remaley, A.T., Sampson, M., et al. The differential effect of food intake and beta-adrenergic stimulation on adipose-derived hormones and cytokines in man. *J. Clin. Endocrinol. Metab.* 1999; 84:2126-2133
118. Wei, J., Xu, H., Davies, J.L., et al. Increase of plasma IL-6 concentration with age in healthysubjects. *Life Science* 1992; 51:1953-1956
119. Daynes, R.A., Araneo, B.A., Ershler, W.B., et al. Altered regulation of IL-6 production with normal aging. Possible linkage to the age-associated decline in dehydroepiandrosterone and its sulfated derivative. *J. Immunol.* 1993; 150:5219-5230
120. Dugue, B. and Leppanen, E. Short-term variability in the concentration of serum interleukin-6 and its soluble receptor in subjectively healthy persons. *Clin. Chem. Lab. Med.* 1998; 36:323-325
121. Ershler, W.B. and Keller, E.T. Age-associated increased interleukin-6 gene expression, late-life diseases, and frailty. *Annu. Rev. Med.* 2000; 51:245-70.:245-270
122. Yoshioka, M., Itoh, Y., Mori, K., et al. Effects of an interleukin-1beta analogue [Lys-D-Pro-Thr], on incomplete cerebral ischemia-induced inhibition of long-term potentiation in rate hippocampal neurons in vivo. *Neuroscience Letters* 1999; 261:171-174
123. Haack, M., Hinze-Selch, D., Fenzel, T., et al. Plasma levels of cytokines and soluble cytokine receptors in psychiatric patients upon hospital admission: effects of confounding factors and diagnosis. *J. Psychiatr. Res.* 1999; 33:407-418
124. Puder, J.J., Freda, P.U., Goland, R.S., et al. Estrogen modulates the hypothalamic-pituitary-adrenal and inflammatory cytokine responses to endotoxin in women. *J. Clin. Endocrinol. Metab.* 2001; 86:2403-2408
125. Chiu, K.M., Arnaud, C.D., Ju, J., et al. Correlation of estradiol, parathyroid hormone, interleukin-6, and soluble interleukin-6 receptor during the normal menstrual cycle. *Bone* 2000; 26:79-85
126. Konecna, L., Yan, M.S., Miller, L.E., et al. Modulation of IL-6 production during the menstrual cycle in vivo and in vitro. *Brain Behav. Immun.* 2000; 14:49-61
127. Angstwurm, M.W., Gartner, R. and Ziegler-Heitbrock, H.W. Cyclic plasma IL-6 levels during normal menstrual cycle. *Cytokine* 1997; 9:370-374
128. Marz, P., Otten, U. and Rose-John, S. Neural activities of IL-6-type cytokines often depend on soluble cytokine receptors. *Eur. J. Neurosci.* 1999; 11:2995-3004
129. Gaillard, J., Pugniere, M., Tresca, J., et al. Interleukin-6 receptor signaling. II. Bio-availability of interleukin-6 in serum. *Eur. Cytokine Netw.* 1999; 10:337-344
130. Akira, S., Taga, T. and Kishimoto, T. Interleukin-1 in biology and medicine. *Advanced Immunology* 1993; 54:1-78
131. Van Wagoner, N.J. and Benveniste, E.N. Interleukin-6 expression and regulation in astrocytes. *J. Neuroimmunol.* 1999; 100:124-139

132. Their, M., Marz, P., Otten, U., et al. Interleukin-6 (IL-6) and its soluble receptor support survival of sensory neurons. *J. Neurosci. Res.* 1999; 55:411-422
133. Gao, Y., Ng, Y.K., Lin, J.Y., et al. Expression of immunoregulatory cytokines in neurons of the lateral hypothalamic area and amygdaloid nuclear complex of rats immunized against human IgG. *Brain Res.* 2000; 859:364-368
134. Keller, E.T., Wanagat, J. and Ershler, W.B. Molecular and cellular biology of interleukin-6 and its receptor. *Front Biosci.* 1996; 1:d340-57.:d340-d357
135. Yeager, M.P., Lunt, P., Arruda, J., et al. Cerebrospinal fluid cytokine levels after surgery with spinal or general anesthesia. *Reg. Anesth. Pain Med.* 1999; 24:557-562
136. Chen, G., McCuskey, R.S. and Reichlin, S. Blood interleukin-6 and tumor necrosis factor-alpha elevation after intracerebroventricular injection of Escherichia coli endotoxin in the rat is determined by two opposing factors: peripheral induction by LPS transferred from brain to blood and inhibition of peripheral response by a brain-mediated mechanism. *Neuroimmunomodulation* 2000; 8:59-69
137. Peltola, J., Palmio, J., Korhonen, L., et al. Interleukin-6 and interleukin-1 receptor antagonist in cerebrospinal fluid from patients with recent tonic-clonic seizures. *Epilepsy Res.* 2000; 41:205-211
138. Maier, B., Schwerdtfeger, K., Mautes, A., et al. Differential release of interleukines 6, 8, and 10 in cerebrospinal fluid and plasma after traumatic brain injury. *Shock* 2001; 15:421-426
139. Levine, J., Barak, Y., Chengappa, K.N., et al. Cerebrospinal cytokine levels in patients with acute depression. *Neuropsychobiology* 1999; 40:171-176
140. Stubner, S., Schon, T., Padberg, F., et al. Interleukin-6 and the soluble IL-6 receptor are decreased in cerebrospinal fluid of geriatric patients with major depression: no alteration of soluble gp130. *Neurosci. Lett.* 1999; 259:145-148
141. Murphy, P.G., Borthwick, L.A., Altares, M., et al. Reciprocal actions of interleukin-6 and brain-derived neurotrophic factor on rat and mouse primary sensory neurons. *Eur. J. Neurosci.* 2000; 12:1891-1899
142. Biber, K., Lubrich, B., Fiebich, B.L., et al. Interleukin-6 enhances expression of adenosine A(1) receptor mRNA and signaling in cultured rat cortical astrocytes and brain slices. *Neuropsychopharmacology* 2001; 24:86-96
143. Himms-Hagen, J. Physiological roles of the leptin endocrine system: differences between mice and humans. *Crit. Rev. Clin. Lab. Sci.* 1999; 36:575-655
144. Ahima, R.S., Flier, J.S. Leptin. *Annu. Rev. Physiol.* 2000; 62:413-37.:413-437
145. Buyse, M., Viengchareun, S., Bado, A., et al. Insulin and glucocorticoids differentially regulate leptin transcription and secretion in brown adipocytes. *FASEB J.* 2001; 15:1357-1366
146. Burguera, B., Couce, M.E., Long, J., et al. The long form of the leptin receptor (OB-Rb) is widely expressed in the human brain. Neuroendocrinology 2000; 71:187-195
147. Fei, H., Okano, H.J., Li, C., et al. Anatomic localization of alternatively spliced leptin receptors (Ob-R) in mouse brain and other tissues. *Proc. Natl. Acad. Sci.* U.S.A 1997; 94:7001-7005
148. Fantuzzi, G. and Faggioni, R. Leptin in the regulation of immunity, inflammation, and hematopoiesis. *J. Leukoc. Biol.* 2000; 68:437-446
149. Gaillard, R.C., Spinedi, E., Chautard, T., et al. Cytokines, leptin, and the hypothalamo-pituitary-adrenal axis. *Ann. N.Y. Acad. Sci.* 2000; 917:647-57:647-657
150. Rayner, D.V. and Trayhurn, P. Regulation of liptin production: sympathetic nervous system interaction. *J. Mol. Med.* 2001; 79:8-20

151. Tanida, M., Iwashita, S., Ootsuka, Y., et al. Leptin injection into white adipose tissue elevates renal sympathetic nerve activity dose-dependently through the afferent nerves pathway in rats. *Neurosci. Lett.* 2000; 293:107-110
152. Caro, J.F., Kolaczynski, J.W., Nyce, M.R., et al. Decreased cerebrospinal-fluid/serum leptin ratio in obesity: a possible mechanism for leptin resistance. *Lancet* 1996; 348:159-161
153. Reichlin, S., Chen, G. and Nicolson, M. Blood to brain transfer of leptin in normal and interleukin-1beta-treated male rats. *Endocrinology* 2000; 141:1951-1954
154. Luheshi, G.N., Gardner, J.D., Rushforth, D.A., et al. Leptin actions on food intake and body temperature are mediated by IL-1. *Proc. Natl. Acad. Sci.* USA 1999; 96:7047-7052
155. Nye, E.J., Bornstein, S.R., Grice, J.E., et al. Interactions between the stimulated hypothalamic-pituitary-adrenal axis and leptin in humans. *J. Neuroendocrinol.* 2000; 12:141-145
156. Figlewicz, D.P., Higgins, M.S., Ng-Evans, S.B., et al. Leptin reverses sucrose-conditioned place preference in food-restricted rats. *Physiol. Behav.* 2001; 73:229-234
157. Dinarello, C.A. Pro-inflamatory cytokines. *Chest* 2000; 118:503-508
158. Born, T.L., Smith, D.E., Garka, K.E., et al. Identification and characterization of two members of a novel class of the interleukin-1 receptor (IL-1R) family. Delineation of a new class of IL-1R-related proteins based on signaling. *J. Biol. Chem.* 2000; 275:29946-29954
159. Rothwell, N.J. and Luheshi, G.N. Interleukin 1 in the brain: biology, pathology and therapeutic target. *Trends Neurosci.* 2000; 23:618-625
160. Alheim, K. and Bartfai, T. The interleukin-1 system: receptors, ligands, and ICE in the brain and their involvement in the fever response. *Ann. N.Y. Acad. Sci.* 1998; 840:51-8.:51-58
161. Krueger, J.M., Fang, J., Taishi, P., et al. Sleep. A physiologic role for IL-1 beta and TNF-alpha. *Ann. N.Y. Acad. Sci.* 1998; 856:148-59.:148-159
162. Elenkov, I. and Vizi, E.S. Negative feedback modulation of noradrenaline release from the spleen. *Acta Physiol. Hung.* 1990; 75(Suppl):89-90.:89-90
163. Linthorst, A.C., Flachskamm, C., Hopkins, S., et al. Long-term intracerebroventricular infusion of corticotropin-releasing hormone alters neuroendocrine, neurochemical, autonomic, behavioral, and cytokine responses to a systemic inflammatory challenge. *J. Neurosci.* 1997; 17:4448-4460
164. Maes, M., Lin, A.H., Verkerk, R., et al. Serontonergic and noradrenergic markers of post-traumatic stress disorder with and without major depression. *Neuropsychopharmacology* 1999; 20:188-197.
165. Sondergaard, S.R., Ostrowski, K., Ullum, H., et al. Changes in plasma concentrations of interleukin-6 and interleukin-1 receptor antagonists in response to adrenaline infusion in humans. *Eur. J. Appl. Physiol.* 2000; 83:95-98
166. Yehuda, R., Siever, L.J., Teicher, M.H., et al. Plasma norepinephrine and 3-methoxy-4-hydroxyphenylglycol concentrations and severity of depression in combat posttraumatic stress disorder and major depressive disorder. *Biological Psychiatry* 1998; 44:56-63
167. Kubera, M., Maes, M., Holan, V., et al. Prolonged desipramine treatment increases the production of interleukin-10, an anti-inflammatory cytokine, in C57BL/6 mice subjected to the chronic mild stress model of depression. *J. Affect. Disord.* 2001; 63:171-178

168. Maes, M., Song, C., Lin, A.H., et al. Negative immunoregulatory effects of antidepressants: inhibition of interferon-gamma and stimulation of interleukin-10 secretion. *Neuropsychopharmacology* 1999; 20:370-379
169. Suzuki, E., Shintani, F., Kanba, S., et al. Immobilization stress increases mRNA levels of interleukin-1 receptor antagonist in various rat brain regions. *Cell. Mol. Neurobiol.* 1997; 17:557-562
170. Linsell, C.R., Lightman, S.L., Mullen, P.E., et al. Circadian rhythms of epinephrine and norepinephrine in man. *J. Clin. Endocrinol. Metab.* 1985; 60:1210-1215
171. Hansen, A.M., Garde, A.H., Skovgaard, L.T., et al. Seasonal and biological variation of urinary epinephrine, norepinephrine, and cortisol in healthy women. *Clin. Chim..Acta* 2001; 309:25-35
172. Blanchard, E.B., Kolb, L.C., Prins, A., et al. Changes in plasma norepinephrine to combat-related stimuli among Vietnam veterans with posttraumatic stress disorder. *J. Nerv. Ment. Dis.* 1991; 179:371-373
173. McFall, M.E., Murburg, M.M., Ko, G.N., et al. Autonomic responses to stress in Vietnam combat veterans with posttraumatic stress disorder. *Biol. Psychiatry* 1990; 27:1165-1175
174. Pitman, R.K. and Orr, S.P. Twenty-four hour urinary cortisol and catecholamine excretion in combat-related posttraumatic stress disorder. *Biol. Psychia.y* 1990; 27:245-247
175. Murburg, M.M., McFall, M.E., Lewis, N., et al. Plasma norepinephrine kinetics in patients with posttraumatic stress disorder. *Biological Psychiatry* 1995; 38:819-825
176. Kosten, T.R., Mason, J.W., Giller, E.L., et al. Sustained urinary norepinephrine and epinephrine elevation in post-traumatic stress disorder. *Psychoneuroendocrinology* 1987; 12:13-20
177. Mason, J.W., Giller, E.L., Kosten, T.R., et al. Elevation of urinary norepinephrine/cortisol ratio in posttraumatic stress disorder. *J. Nerv. Ment. Dis.* 1988; 176:498-502
178. Yehuda, R., Southwick, S., Giller, E.L., et al. Urinary catecholamine excretion and severity of PTSD symptoms in Vietnam combat veterans. *J. Nerv. Ment. Dis.* 1992; 180:321-325
179. McFarlane, A.C., Atchison, M., Rafalowicz, E., et al. Physical symptoms in post-traumatic stress disorder. *J. Psychosom. Res.* 1994; 38:715-726
180. Solomon, S.D. and Davidson, J.R. Trauma: prevalence, impairment, service use, and cost. *J. Clin. Psychiatry* 1997; 58(9):5-11.:5-11
181. Beckham, J.C., Moore, S.D., Feldman, M.E., et al. Health status, somatization, and severity of posttraumatic stress disorder in Vietnam combat veterans with posttraumatic stress disorder. *Am. J. Psychiatry* 1998; 155:1565-1569
182. Felitti, V.J., Anda, R.F., Nordenberg, D., et al. Relationship of childhood abuse and household dysfunction to many of the leading causes of death in adults. The Adverse Childhood Experiences (ACE) Study. *Am. J. Prev. Med.* 1998; 14:245-258
183. Kagan, B.L., Leskin, G., Haas, B., et al. Elevated lipid levels in Vietnam veterans with chronic posttraumatic stress disorder. *Biol. Psych.* 1999; 45:374-377
184. Boscarino, J.A. and Chang, J. Electrocardiogram abnormalities among men with stress-related psychiatric disorders: implications for coronary heart disease and clinical research. *Ann. Behav. Med.* 1999; 21:227-234
185. Lee, K.A., Vaillant, G.E., Torrey, W.C., et al. A 50-year prospective study of the psychological sequelae of World War II combat. *Am. J. Psychiatry* 1995; 152:516-522
186. Elder, G.H., Jr., Shanahan, M.J., Clipp, E.C. Linking combat and physical health: the legacy of World War II in men's lives. *Am. J. Psychiatry* 1997; 154:330-336

187. Horner, K.C., Giraudet, F., Lucciano, M., et al. Sympathectomy improves the ear's resistance to acoustic trauma--could stress render the ear more sensitive? *Eur. J. Neurosci.* 2001; 13:405-408
188. Muraoka, M.Y., Carlson, J.G. and Chemtob, C.M. Twenty-four-hour ambulatory blood pressure and heart rate monitoring in combat-related posttraumatic stress disorder. *J. Trauma. Stress.* 1998; 11:473-484
189. Beckham, J.C., Feldman, M.E., Barefoot, J.C., et al. Ambulatory cardiovascular activity in Vietnam combat veterans with and without posttraumatic stress disorder. *J. Consult Clin. Psychol.* 2000; 68:269-276
190. Cohen, H., Benjamin, J., Geva, A.B., et al. Autonomic dysregulation in panic disorder and in post-traumatic stress disorder: application of power spectrum analysis of heart rate variability at rest and in response to recollection of trauma or panic attacks. *Psychiatry Res.* 2000; 96:1-13
191. Kimbrell, D.A. and Beutler, B. The evolution and genetics of innate immunity. *Nat. Rev. Genet.* 2001; 2:256-267
192. Lanier, L.L. The origin and functions of natural killer cells. *Clin. Immunol.* 2000; 95:S14-S18
193. Bodner, G., Ho, A. and Kreek, M.J. Effect of endogenous cortisol levels on natural killer cell activity in healthy humans. *Brain Behav. Immun.* 1998; 12:285-296
194. Irwin, M., Hauger, R.L., Jones, L., et al. Sympathetic nervous system mediates central corticotropin-releasing factor induced suppression of natural killer cytotoxicity. *J. Pharmacol. Exp. Ther.* 1990; 255:101-107
195. Dhabhar, F.S., Miller, A.H., McEwen, B.S., et al. Effects of stress on immune cell distribution. Dynamics and hormonal mechanisms. *J. Immunol.* 1995; 154:5511-5527
196. Dhabhar, F.S., Satoskar, A.R., Bluethmann, H., et al. Stress-induced enhancement of skin immune function: A role for gamma interferon. *Proc. Natl. Acad. Sci.* USA 2000; 97:2846-2851
197. Elenkov, I.J., Papanicolaou, D.A., Wilder, R.L., et al. Modulatory effects of glucocorticoids and catecholamines on human interleukin-12 and interleukin-10 production: clinical implications. *Proc. Assoc. Am. Physicians* 1996; 108:374-381
198. Blotta, M.H., DeKruyff, R.H. and Umetsu, D.T. Corticosteroids inhibit IL-12 production in human monocytes and enhance their capacity to induce IL-4 synthesis in CD4+ lymphocytes. *J. Immunol.* 1997; 158:5589-5595
199. Panina-Bordignon, P., Mazzeo, D., Lucia, P.D., et al. Beta2-agonists prevent Th1 development by selective inhibition of interleukin 12. *J. Clin. Invest.* 1997; 100:1513-1519
200. Delahanty, D.L., Dougall, A.L., Craig, K.J., et al. Chronic stress and natural killer cell activity after exposure to traumatic death. *Psychosom, Med.* 1997; 59:467-476
201. DeRijk, R.H., Petrides, J., Deuster, P., et al. Changes in corticosteroid sensitivity of peripheral blood lymphocytes after strenuous exercise in humans. *J. Clin. Endocrinol. Metab.* 1996; 81:228-235
202. Ridker, P.M., Rifai, N., Stampfer, M.J., et al. Plasma concentration of interleukin-6 and the risk of future myocardial infarction among apparently healthy men. *Circulation* 2000; 101:1767-1772
203. Yudkin, J.S., Kumari, M., Humphries, S.E., et al. Inflammation, obesity, stress and coronary heart disease: is interleukin-6 the link? *Atherosclerosis* 2000; 148:209-214
204. Miller, R.J., Sutherland, A.G., Hutchison, J.D., et al. C-reactive protein and interleukin 6 receptor in post-traumatic stress disorder: a pilot study. *Cytokine* 2001; 13:253-255

Chapter 15

CYTOKINES AND SCHIZOPHRENIA

Thomas Pollmächer[1], Andreas Schuld[1] and Dunja Hinze-Selch[2]
[1]Max Planck Institute of Psychiatry, Munich, and [2]Department of Psychiatry and Psychotherapy, Christian Albrechts University, Kiel, Germany

1. INTRODUCTION

Despite tremendous scientific efforts made during the past 100 years, the causes of schizophrenia remain unknown, and despite the important methodological advances achieved within the field of clinical neurosciences during the last 2 decades, the definite knowledge about one of the most frequent and debilitating psychiatric disorders can still be summarized quickly(1): Schizophrenia is a brain disorder affecting about 1% of the global population. There is neither a consistent gender preference nor does schizophrenia occur at divergent rates in various parts of the world. Schizophrenia impairs numerous complex brain functions including those thought to be constitutive for human beings. Particularly thought, perception and cognition are impaired, challenging in many patients the integrity of the individual's personality. Although schizophrenia runs a favorable course in one quarter of the patients affected, the remaining cases experience a chronic fluctuating course leading to permanent disability. There is no cure for schizophrenia but pharmacological and psychosocial treatments yield considerable benefit for the majority of patients.(2) Still, however, the burden of the disease for the individual and for society is tremendous.

On a functional level, the present pathophysiological concepts assume that the symptoms of schizophrenia are due to disturbances in various neurotransmitter systems. In fact, most of these concepts have been derived from the effects of antipsychotic drugs. The blocking actions of antipsychotic drugs on dopamine receptors led to the concept of dopaminergic overactivity as a pivotal functional disturbance in schizophrenia. However, this view has been broadened by the pharmacological properties of some novel antipsychotic compounds. These more recent data suggest the involvement of other neurotransmitters, in particular those of the serotoninergic and the glutamatergic systems(3) as well. It is unclear, however, whether an abnormal neurotransmitter function is a primary pathophysiological feature of schizophrenia or secondary to other, more basic dysfunctions. Moreover, it remains to be determined whether or not the effects of antipsychotic drugs on neurotransmitter systems reflect their primary and pivotal mechanism of action.

The causes of schizophrenia are unknown but 2 basic pathogenetic facts are established(4): (a) Genetic factors play an outstanding role but probably interacting weak effects of multiple genes are pivotal rather than strong effects of single genes. (b) Various environmental factors act on this genetic background which are assumed to be effective already during pregnancy and childhood. It is a matter of ongoing debate, whether the interaction between genetic and environmental factors leads to a disturbance of neuro-development or to a neurodegenerative process starting in late childhood or early adolescence.(5) Subtle abnormalities in brain morphology are consistent with one or both of these concepts which, of course, are not mutually exclusive.

This chapter will outline the current knowledge about the potential role of cytokines in schizophrenia. The idea that cytokines might play a role in schizophrenia is primarily fueled by 2 scientific areas. The historically first one is a body of empirical and epidemiological studies about the inter-relationship of host defense activation and schizophrenia, the second one is the increasing evidence that the immune and central nervous systems are intimately linked. On the one hand, studies going back to the end of the 19th century suggest that schizophrenia is linked somehow to host defense processes.(6) The respective evidence comes from very different areas (see Table 1) and includes findings as diverse as a positive response of schizophrenia to fever therapy,(7) a reduced incidence of rheumatoid arthritis in patients with schizophrenia,(8) an increased rate of various infections during pregnancy of the mothers of individuals who later on develop schizophrenia,(9) and the detection of autoantibodies against brain tissue.(10) It is beyond the scope of the present paper to discuss all of this highly variable evidence but it is clear from Table 1 that it is not possible to integrate all or even the majority of these findings in one conclusive model of immunological abnormalities in schizophrenia.

Nevertheless, these unsystematic findings have drawn the psychiatrists' attention to the developments in the second research area suggesting that cytokines might be important for the understanding of schizophrenia. This area encompasses the rapidly growing field of psychoneuroimmunology, investigating the interaction between the immune system and the central nervous system. It is now well established that inflammatory cytokines represent the major link between systemic host defense processes and the

- Increased frequency of viral infections during pregnancy[6]
- Benefits from fever therapy[7]
- Reduced response rates to hepatitis B vaccination[49]
- Reduced incidence of rheumatoid arthritis[8]
- Increased antibody titers to foreign antigens in patients (e.g. toxoplasma gondii[46])
- Increased immunoglobulin levels and HSV type 2 antibodies in patients' mothers during pregnancy[48]
- Presence of autoantibodies to various tissues including brain[10]
- Retroviral RNA in the brain post mortem[50]
- Disturbances of the blood brain barrier[47]
- T- or B cell abnormalities[9]
- Effects of psychotropic drugs on the immune system[18]

Table 1: Findings suggesting an involvement of the immune system in the pathophysiology and pathogenetics (in the broadest sense) of schizophrenia.

CNS.(11) Moreover, cytokines are produced within the CNS by glial cells and neurons and have neuromodulatory functions.(12) They are involved in neurodevelopment, neurodegeneration and neuroregeneration.(13,14) Finally, there is clear-cut evidence from animal and human studies that cytokines are able to influence complex brain functions such as emotion, cognition, perception, and sleep-wake behavior.(15-17) Therefore, cytokines might indeed contribute to symptoms of schizophrenia and disturbances in neurodevelopment and/or neurodegeneration. But they might also be able to have beneficial effects on neurodevelopmental scars and might counteract neurodegeneration suggesting that effects of psychotropic drugs on cytokine networks(18) are of therapeutic relevance and that in the future administration of certain cytokines to schizophrenic patients might turn out to be beneficial.

In this book chapter we will first outline how, from a theoretical perspective, cytokines might be involved in schizophrenia. Then we will present and discuss the studies that indeed have investigated cytokine networks in patients with schizophrenia. This part will also consider methodological aspects such as the problems arising from the difficulty to assess cytokine production and action within the brain of schizophrenic patients in vivo and the numerous confounding factors which alter systemic cytokine production and release. Finally, we will summarize the state of the art and outline some personal conclusions and future perspectives.

2. THEORETICAL CONSIDERATIONS CONCERNING THE INVOLVEMENT OF CYTOKINES IN THE PATHOPHYSIOLOGY OF SCHIZOPHRENIA

2.1 Basics on cytokines and their receptors

The most recent edition of a leading text book on cytokine physiology proposes the following definition: 'Cytokines are regulatory proteins secreted by white blood cells and a variety of other cells in the body; the pleiotropic actions of cytokines include numerous effects on cells of the immune system and modulation of inflammatory responses'.(19) This definition emphasizes the traditional view of cytokines as humoral mediators within the immune system. However, recent developments in cytokine research clearly show that their physiological role is much broader and extends to the signaling from the immune system to the brain, and to the regulation of brain functions even in the absence of an infection or inflammation. Moreover, numerous regulatory proteins using similar signaling pathways, although derived from non-immune cells and mainly targeting physiological systems other than the immune system, have been discovered. In contrast to most hormones, cytokines are not secreted by endocrine organs such as the pituitary or thyroid gland but rather by certain cell types distributed across different organs. Moreover, they act to a large extent in an autocrine or paracrine manner; the circulating levels of most cytokines are very low during baseline conditions, if they are present at all in the blood stream. During infection and inflammation, however, some cytokines may circulate in considerable amounts.

- Interleukins
- Chemokines
- Tumor necrosis factors
- Colony stimulating factors
- Growth factors
- Neurotrophins
- Neuropoeietins

Table 2: Major cytokine systems

The number of known cytokines is high and diverse (for cytokine families see Table 2). Moreover, the nomenclature is confusing and frequently misleading. In many cases the names are derived from the one function that was historically described first but later on has turned out to characterize only very incompletely the diverse physiological role. As an example, tumor necrosis factor-α is a pivotal pro-inflamatory cytokine involved in primary host responses to infections, the activation of the specific immune system, the regulation of hematopoietic systems, cellular survival and differentiation, the regulation of glucose metabolism and numerous other systems(20) but was initially thought to inhibit tumor growth under certain conditions.

Cytokines act through specific receptors which are present in most cases on a huge number of different cells. With some cytokines, such as TNF-α and IL-6, the activity is regulated among other factors through the extracellular part of the receptors which can be shed from the cell surface. These so-called soluble receptors, for example soluble TNF receptor p55 and p75, enter the circulation and bind to the corresponding cytokine. However, the net effect of this binding to cytokine activity depends on a variety of factors and can result in both inhibition and augmentation of cytokine action. For other cytokines, endogenous antagonistic receptor ligands exist. This is particularly important for the IL-1 cytokine system, in which the IL-1 receptor antagonist (IL-1RA) can effectively block IL-1ß actions.(21)

2.2 Physiological functions of cytokines and their receptors

The physiological functions of cytokines are extremely diverse. However, for the actual context, it is sufficient to outline the major functional principles:

a) Cytokine networks show a high degree of *redundancy*. This can easily be demonstrated with respect to acute infection and inflammation. For example, in the case of gram-negative bacterial infections endotoxins induce fever, an activation of the hormonal stress system and behavioral changes through the release of IL-1, IL-6 and TNF-α from monocytes and macrophages.(22) Administering either of these cytokines to animals or humans induces the same symptoms as an endotoxin administration in a similar, rather indistinguishable manner. As a consequence, strategies to specifically inhibit actions of one of these cytokines have no or very limited effects in models of infection or in bacterial sepsis in humans. Similarly, knockout animals do not show major impairments in coping with such challenges although other specific changes in immune functions are present. It is important to take this considerable redundancy into account when interpreting the physiological importance of changes in one particular cytokine system.

b) Most cytokines are involved in the *regulation of cell differentiation, proliferation and survival.* This is particularly evident regarding the sequence of events occurring during an infection in which cytokines released during the early unspecific phase of the host defense are pivotal to the differentiation and clonal expansion of B- and T-lymphocytes. However, cytokines are also of prominent importance during pre- and postnatal development of all organs including the brain. Moreover, cytokines are indispensable to host responses to injury including tissue repair and controlled cell death (apoptosis). In this context, it is noteworthy that one and the same cytokine, for example TNF-α, may be involved in apoptosis and in the protection of cells against apoptotic stimuli, depending on the particular circumstances.(23) Hence, cytokines are indispensable to both neurodegeneration and neuroregeneration. The involvement of cytokines in neurodevelopment, neurodegeneration and neuroregeneration is of crucial importance for the present topic because, as mentioned above, disturbances in these processes are hypothesized to be pivotal to the pathogenesis of schizophrenia.

c) It is generally accepted that cytokine *actions are limited with respect to both space and time*. Probably, this is not a law without exceptions but under most circumstances cytokines indeed act locally and for a restricted period of time only. They act locally because they normally do not enter circulation but develop paracrine and autocrine activity. Their action is temporally restricted as expression and release are strictly controlled and for most cytokines baseline release rates are very low. These temporal and spacial restrictions of cytokine actions are of importance for the present topic since they considerably limit research on altered cytokine networks in schizophrenic patients. If, for example, due to an intrauterine infection, altered cytokine production was involved in neurodevelopmental disturbances causing schizophrenia, it would be very unlikely that it could still be detected in young schizophrenic adults. Consequently, it might be easier to detect altered cytokine production related to a neurodegenerative process in schizophrenia because such neurodegenerative processes are likely to occur in closer temporal relationship to the onset of the disease. However, the local character of the cytokine production imposes additional difficulties to investigations of schizophrenic patients. Blood and CSF which are relatively easy to obtain for research purposes might yield non-informative results, and brain tissue can only be obtained from very few patients after their death and almost never shortly before or after the onset of the disease.

d) A last feature of cytokine actions which is important to the present topic is that they *mediate interactions between the peripheral immune system and the CNS*. As already mentioned above, inflammatory cytokines are pivotal to all those host responses occurring during an acute infection or inflammation which are mediated by the CNS including fever, neuroendocrine activation and changes in behavior. The fascinating details of the underlying mechanisms have recently been reviewed by various other researchers.(11,12) The major pathways include the activation of autonomous afferents (particularly the vagus nerve) by peripheral cytokines, which via the brain stem leads to the production of cytokines within the CNS. Moreover, at least occurring during acute infection, some cytokines, e.g. IL-6 and TNF-α, reach circulation to a considerable extent. These circulating cytokines are able to enter the brain by active transport mechanisms and in areas without a tight blood brain barrier, for example the organum vasculosum laminae terminalis. It is noteworthy that therapeutical administration of inflammatory cytokines to humans and the experimental induction of endogenous cytokine release have powerful effects on emotion, anxiety and cognition.(16) However, it is only in extreme situations that such conditions go along with symptoms characterizing schizophrenia, and actually there is no evidence at present suggesting that the disorder may be caused by systemic host response processes (see below).

As outlined above, cytokine production and secretion is very strictly regulated, and under baseline conditions only very low amounts, if any, are produced. This led many researchers to the opinion that in adulthood cytokines do not play any physiological role under baseline conditions. However, there is a number of very recent studies suggesting that at least with respect to some cytokines and physiological systems this opinion might be incorrect (for review see (24). Mentioning those 2 results which are important for neuroscience and psychiatry, there is growing evidence that inflammatory cytokines such as TNF-α and IL-1ß are involved in physiological sleep regulation and memory processes. Hence, the idea that even very subtle changes in cytokine production not related to infection and inflammation might affect complex brain functions cannot be discarded at present.

2.3 Putative involvement of cytokines in the pathophysiology of schizophrenia

Based on these principles of cytokine actions we will now describe a theoretical framework of the potential involvement of cytokines in the pathogenesis and pathophysiology of schizophrenia. This framework is graphically represented in Figure 1. Presumed alterations in cytokine networks may either be genetically determined or functional. In both situations, albeit by means of different mechanisms, cytokines might contribute to the pathophysiology of schizophrenia by impairing neurodevelopment or neurodegeneration resulting in disturbed brain function but they might also directly impair functions of the CNS. Conversely, cytokines might be protective with respect to the development of schizophrenia at certain developmental stages, or they might even be beneficial to counteract neurodegeneration or functional impairments due to neurodevelopmental disturbances in patients suffering from schizophrenia.

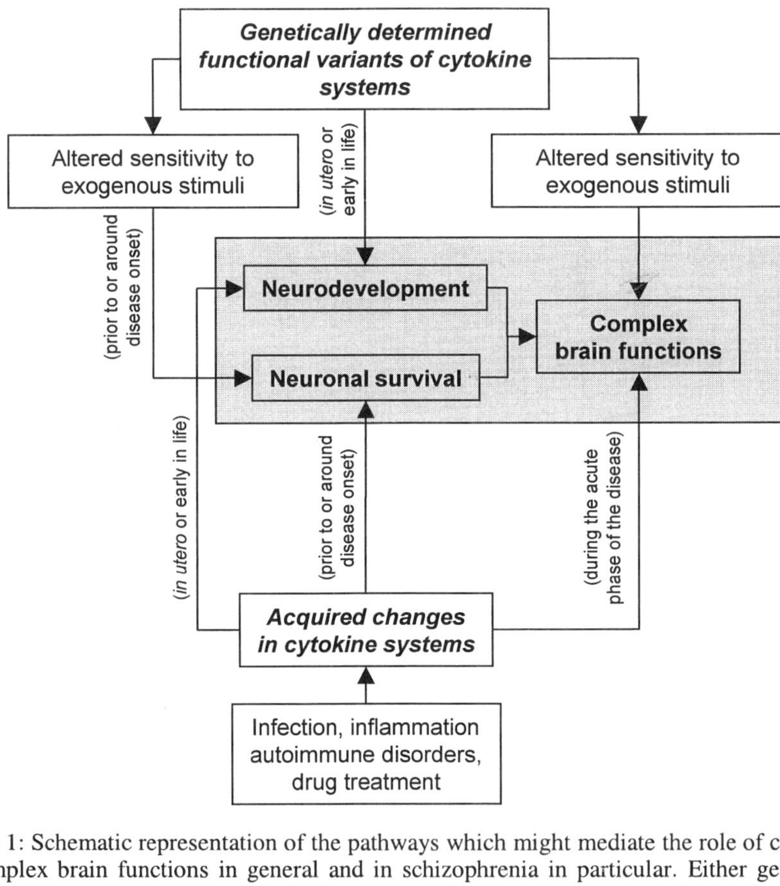

Figure 1: Schematic representation of the pathways which might mediate the role of cytokines in complex brain functions in general and in schizophrenia in particular. Either genetically determined or acquired (functional) alterations of cytokine networks might interact with neurodevelopment, neuronal survival or directly with those complex brain functions involved in the pathophysiology of schizophrenia. For details see text.

Genetically determined alterations in cytokine networks such as functionally relevant mutations of classic inflammatory cytokines or neurotrophic factors could, for example, directly affect neurodevelopment *in utero* or early in live. Alternatively, such genetic abnormalities might not affect brain development directly but sensitize the organism to stimuli in early adulthood, thus affecting neuronal survival. It is also possible that genetic factors contribute indirectly to the pathogenesis of schizophrenia by altering the sensitivity of the organism towards infections, inflammatory or autoimmune processes.

Independent of genetic factors, cytokines released during infection or inflammation *in utero* or early in live could affect neurodevelopment, or, if the infection occurs in early adulthood, influence neuronal survival. Moreover, cytokines released under such conditions might directly affect brain functions. Finally, it may be possible that genetically and acquired abnormalities in cytokine networks interact with respect to various aspects of the pathophysiology of schizophrenia.

Although this framework is complex and theoretical, it provides a solid basis for the design of clinical research on schizophrenia and for the interpretation of results. The hypothesis, for example, that schizophrenia occurs due to an autoimmune process would fit with the idea that cytokines mediate neurodegeneration. To test this hypothesis, research strategies must be directed to the identification of cytokine production abnormalities typical of autoimmunity, and to other signs of an autoimmune process which are independent of the assessment of cytokine networks. In contrast, studies performed to scrutinize the hypothesis that intra-uterine infections via a cytokine-dependent neurodevelopmental disturbance cause schizophrenia, would try to relate signs of maternal or fetal infection to morphological or histological signs of impaired brain development in schizophrenic patients. Since, in this case, cytokines would play only an indirect role restricted to the prenatal period of time, one would not expect persisting abnormalities in cytokine networks in adulthood.

Below, we will summarize studies that have investigated the involvement of cytokine networks in schizophrenia. First, we will summarize the few studies addressing the issue of genetic alterations related to cytokines, followed by those studies investigating functional abnormalities.

3. GENETIC STUDIES ON CYTOKINES IN SCHIZOPHRENIA

So far, only few studies have investigated cytokine or cytokine receptor genes in schizophrenic patients, and the results are mostly negative. As can be seen from Table 3, schizophrenia has not been found to be consistently associated with known variants of genes encoding members of the IL-1 system, the IL-2 receptor, BDNF, CNTF or NT-3. Regarding the TNF-α cytokine system, no association to variants of the TNF p75 receptor has been found but one – so far unconfirmed – study suggests that schizophrenic patients have an increased liability to carry one or two copies of the TNF2(A) allele, the less frequent of 2 known variants at position -308. Interestingly enough, this variant goes along with an increased expression of TNF-α, which fits with some of the studies which have investigated functional aspects of the TNF-α system in schizophrenic patients (see below).

Genetic studies in schizophrenia are particularly difficult due to the complex cross-sectional and longitudinal phenotype and the possibility of phenocopies and heterogeneous etiology.

Therefore, it is reasonable to search for genetic causes of certain subtypes and to relate genetic findings to biological markers of the disease. With regard to cytokines, very little has been done in this field but there is preliminary evidence that an IL-1ß polymorphism might be related to bi-frontotemporal gray matter deficits,(25) and that variants of the BDNF might also influence brain morphology and certain clinical characteristics in schizophrenic patients.(26,27)

Cytokines	Number of Studies	Results
IL-1 system	4	No association of polymorphisms to the disease in all studies; in one study association to brain morphology[51-54]
IL-2 receptor	2	No association of beta chain polymorphisms to schizophrenia[54,55]
TNF-α system	2	Increased frequency of the TNF2(A) allele in schizophrenic patients, no association to TNF type II receptor polymorphisms[34,56]
BDNF	4	No association of polymorphisms to the disease in all studies but to certain clinical characteristics in one, and to brain morphology in another study [26,27,57,58]
CNTF	3	No association of a null mutation of CNTF to the disease[59-61]
NT-3	8	No association in most (5) studies[57,62-68]

Table 3: Genetic studies about cytokines in schizophrenic patients.

4. FUNCTIONAL ALTERATIONS OF CYTOKINE NETWORKS IN SCHIZOPHRENIA

Although schizophrenia is a brain disorder, the vast majority of functional studies about cytokine networks have been performed using blood. We will briefly summarize what is known about cytokines in the brain or in the CSF of schizophrenic patients and then turn to the numerous studies which either investigated cytokine production by peripheral immune cells or cytokine concentrations in the serum or plasma.

4.1 Cytokines in the brain and the CSF of patients with schizophrenia

There is a single study that has investigated the production of specific cytokines in the brain of schizophrenic patients. Takahashi and colleagues(28) reported about increased BDNF levels in the corticolimbic system of schizophrenic patients; in parallel, TrkB receptor expression was reduced. Levels of NGF and NT-3 were not altered. The authors were unable to relate increased BDNF levels to medication, which is important because psychotropic drugs may increase BDNF production.(29)

Cytokines	Number of studies	Results
IL-1	5	No difference compared to controls in all studies[69-73]
IL-2	5	No difference compared to controls in all but one study[69-72,74]
IL-2 receptor	1	Reduced levels compared to controls[70]
IL-4	1	Increased in adolescent patients compared to subjects with OCD or ADHD[74]
IL-6	3	No differences compared to controls in all studies[70,73,75]
TNF-α	2	No difference compared to controls in both studies[70,74]
IFNs	2	Increased activity in one study[76,77]
TGF-ß	1	No difference compared to controls[78]

Table 4. Cytokine levels in the CSF of schizophrenic patients.

Table 4 summarizes those studies which have investigated CSF levels of cytokines in schizophrenic patients. There is quite strong evidence that IL-1 levels are not altered. Four subsequent studies could not confirm an initial report about 10 patients with increased IL-2 levels in the CSF. One single study reports about reduced CSF IL-2 receptor levels, which is in contrast to the vast majority of studies about IL-2 receptors in the blood. IL-6 and TNF-α levels have repeatedly found to be normal, as was the case in a single study concerning TGF-ß. Results on IFNs are inconsistent, and one study reports about increased IL-4 levels.

In summary, the vast majority of studies performed suggested that the levels of major cytokines are not altered in the CSF of schizophrenic patients. But the number of studies is small, and due to the more invasive nature of the procedure to gain CSF as compared to blood, less acutely ill patients have been included.

4.2 Cytokines in peripheral blood of patients with schizophrenia

As already mentioned, the vast majority of studies that have addressed the question whether cytokine networks are altered in schizophrenic patients used blood samples. This is problematic because the regulation of the production and secretion of cytokines within the brain differs considerably from that in the periphery although these 2 compartments interact tightly on a functional level. Hence, it is unclear whether functional investigations of cytokine networks in the periphery allow conclusions regarding the brain. In particular, it is almost impossible to transfer negative findings from the periphery to the CNS. However, even positive findings are difficult to interpret because the respective interactions are highly complex, although cytokine production in the periphery directly affects the brain and induces cytokine production within the CNS.(24)

The studies can be divided into 2 groups investigating either cytokine or soluble cytokine receptor amounts present in plasma or serum, or studying the ex vivo cytokine release by white blood cells upon stimulation with mitogens, bacterial cell wall components or other agents. In all studies cytokines or cytokine receptors are quantified using either bioassays which make use of the dependence of the growth of cell lines on the presence or absence of cytokines, or immunological methods such as RIA or ELISA. It is beyond the scope of this chapter to discuss the methodological aspects of these different approaches [for details see (30)].

Both groups of studies have particular strengths and weaknesses, but most importantly they deliver results which are not comparable at all. Measurements using plasma or serum indeed reflect the amount of a cytokine, circulating in a patient, although it remains very difficult to determine how these amounts relate to cytokine production within the brain and whether and how these peripheral cytokines might directly affect brain function. In contrast, a determination of cytokine secretion by white blood cells upon stimulation ex vivo represents an artificial measurement system mimicking in vitro a presumed response of the immune system to stimulation. Therefore, it is not surprising at all that results from such stimulation experiments in general are not congruent with measurements using plasma or serum. Unfortunately, there is almost no systematic work comparing these 2 approaches. Just to give one example how divergent the results may be, the repeated stimulation of humans with bacterial endotoxin – the major cell-wall component of gram-negative bacteria – has opposite effects on the amounts of TNF-α produced ex vivo or measured in serum. It could be shown in cancer patients that daily administration of endotoxin progressively leads to a reduced amount of circulating TNF-α in response to the stimuli and to a reduced degree of TNF-α mediated host-responses, in particular fever. In contrast, the ability of PBMC to secrete TNF-α in vitro increased progressively.(31)

To account for these fundamental methodological differences, results of studies about circulating cytokines in patients with schizophrenia and studies concerning the ex vivo secretion will be presented separately (Tables 5 and 6). With respect to some findings the evidence that patients with schizophrenia differ from controls is quite consistent, and we will discuss the reasons for these differences and their physiological and clinical significance after the presentation of the results.

4.2.1 Studies on circulating levels of cytokines and soluble cytokine receptors

Since the 1980s more than 50 studies have investigated the levels of cytokines or soluble cytokine receptors in serum or plasma from schizophrenic patients (see Table 5). The vast majority of these studies has investigated four major cytokine systems, namely the IL-1, IL-2, IL-6 and TNF-α systems. As outlined already in part 2 of this chapter, those inflammatory cytokines, for a very long time, have been thought not to be produced in the absence of infection, inflammation or autoimmune processes. Refined analytical techniques have shown in the last 10 to 15 years that at least IL-6 and TNF-α are produced in healthy people as well and circulate in very small quantities of up to some pg/ml of blood or serum. The respective evidence for IL-1 and IL-2 is conflicting. In contrast, soluble cytokine receptors for IL-2, IL-6 and TNF-α are found under baseline conditions in quite considerable quantities of some 100 pg/ml to even some ng/ml.

Cytokines	Number of Studies	Results
IL-1	8	No difference to controls in the majority (6) of studies[73,79-85]
IL-1Ra	5	Increased in three studies[86-89]
IL-2	4	Inconsistent[80,84,85,90]
IL-2 receptor	12	Increased in the majority (9) of studies [81,83,87,91-99]
IL-4	1	Unchanged[100]
IL-6	21	Increased in about half (11) of the studies[75,79-84,86-89,95,96,99,101-107]
IL-6 receptor	5	Inconsistent[86,95,96,104,108]
IL-8	2	Unchanged[83,101]
IL-18	1	Increased[109]
TNF-α	8	Increased in about half (5) of the studies[81,83,85,89,99,101,106]
TNF-receptor I	1	Decreased[99]
TNF-receptor II	1	Unchanged[99]
IFNs	2	No difference in both studies[90,110]
NGF	2	Decreased in both studies[111]

Table 5. Results of studies about circulating levels of cytokines in the serum or plasma of schizophrenic patients.

Regarding the IL-1 system, the majority of studies reports that the circulating levels of IL-1ß are not altered in schizophrenia. This is in line with consistent evidence that IL-1ß hardly appears in the circulation even during experimental host defense activities.(32,33) Interestingly enough, three of five studies report about a slightly increased circulating amount of the endogenous IL-1 receptor antagonist, although the function of IL-1Ra in the virtual absence of IL-1ß is unclear.

Increased serum levels of the soluble IL-2 receptor were the most exciting immunological finding in schizophrenia during the late 1980s. Numerous studies of different groups have confirmed this result, although studies concerning the circulating amounts of IL-2 itself have remained inconsistent.

Regarding IL-6 and TNA-α, results have been inconsistent, too. Slightly more than half of the studies concerning each of these cytokines reported about a minor increase in schizophrenic patients as compared to controls. So far, there does not exist a sufficient number of studies to draw any conclusions regarding the circulating levels of soluble IL-6 and TNF-α receptors in schizophrenia.

Regarding other cytokines, there are only very few studies. These tentatively suggest that IL-4, IL-8 and interferon levels are unchanged. One study suggests that IL-18 levels are increased in patients with schizophrenia, and 2 studies found that NGF levels are decreased.

4.2.2 Studies on in vitro secretion of cytokines and shedding of soluble cytokine receptors

Studies investigating the secretion of cytokines by white blood cells in vitro are much more complicated and involve much more methodological degrees of freedom than studies using serum and plasma. We have recently summarized these problems in detail(30) and will restrict ourselves to the bottom line of results (see Table 6).

Cytokines	Number of studies	Results
IL-1	3	No difference compared to controls in 2 studies[80,112,113]
IL-2	18	Reduced secretion in the majority of studies[42,43,80,113-127]
IL-2 receptor	3	No differences compared to controls in all studies[118,120,128]
IL-3	2	Inconsistent results[112,113]
IL-6	2	No differences compared to controls in both studies[80,118]
IL-10	4	Inconsistent results[119,121,123,124]
TNF-α	1	No difference compared to controls[129]
IFN	11	Reduced secretion in the majority of studies[43,118-121,123,124,130-133]

Table 6. Results of studies about the stimulated secretion of cytokines by white blood cells from schizophrenic patients.

Two of 3 studies did not discover an alteration in stimulated IL-1ß secretion. The vast majority of studies about IL-2 found a reduced stimulated secretion in schizophrenic patients compared to controls, whereas those few studies investigating stimulated sIL-2R shedding found no difference. The majority of studies performed about the secretion of interferon suggests that white blood cells from schizophrenic patients secrete less than cells from controls upon stimulation. Only few studies have looked at other cytokines suggesting a normal stimulated secretion of IL-6 and TNF-α and yielding inconsistent results for IL-3 and IL-10.

5. SYNOPSIS AND INTERPRETATION OF THE MAJOR FINDINGS ON CYTOKINE NETWORKS IN SCHIZOPHRENIA

Prior to attempting an interpretation of the aforementioned results concerning cytokine levels and production in schizophrenic patients we will summarize the results of genetic and functional studies for those cytokine systems in which any kind of approach suggests reproducible differences from controls (see Table 7). For most cytokine systems (IL-4, IL-8, IL-10, IL-18, TGF, BDNF, NT-3) functional data are rare and either inconsistent or to be considered preliminary. Only for the IL-1 system enough studies have been performed to conclude that alterations in schizophrenia are very unlikely. Reproducible differences between patients suffering from schizophrenia and controls have been reported for the IL-2 system and the interferons and, although somewhat less convincingly, for the IL-6 and TNF-α systems.

Table 7 shows that for all these cytokine systems there has been no confirmed evidence to date that alterations occurring on the genome level are present in schizophrenia. Moreover, consistent evidence affirming a difference between patients and controls in CSF cytokine levels has been lacking. It must be stressed, however, at this point that this negative evidence is not very strong because genetic and CSF studies have both been rare (see Tables 3 and 4).

The pattern of confirmed abnormalities of cytokine networks in schizophrenia with respect to blood levels and ex vivo stimulated production is quite complex. Whereas IL-6 and TNF-α blood levels are increased, the stimulated secretion of white blood cells seems not to be altered consistently. In contrast, the circulating amounts of interferons have been reported to be normal, whereas the in vitro secretion has quite consistently been found to be decreased. Concerning the IL-2 system, IL-2 blood levels are normal, whereas soluble IL-2 receptor levels are increased. In contrast, stimulated IL-2 production is reduced, whereas stimulated soluble IL-2 receptor production was found to be unchanged.

It is not surprising that results obtained with respect to circulating cytokine levels differ from those on a stimulated production because as outlined above baseline cytokine production in vivo is regulated differently as compared to ex vivo secretion upon stimulation. Nevertheless, the interpretation of the causes and the significance of the findings summarized

Cytokines	Genetic Studies	CSF levels	Plasma levels	Production by white blood cells
IL-2	Not investigated	Unchanged	Inconsistent	**Reduced**
IL-2 receptor	No effect on the disease	Reduced (one study)	**Increased**	Unchanged
IL-6	Not investigated	Unchanged	**Increased**	Unchanged
IFNs	Not investigated	Inconsistent	Unchanged	**Reduced**
TNF system	Increased frequency of the TNFA(2) allele (one study)	Unchanged	**Increased**	Unchanged

Table 7. Overview of cytokine systems for which consistent alterations have been found in schizophrenia (for details see text and Tables 3-6). Findings which have been replicated and reflect more than half of the studies on this particular topic are printed in bold.

in Table 7 are currently quite difficult and must remain tentative. First of all, it is likely at present that altered cytokine regulation observed in the blood of schizophrenic patients is an acquired phenomenon because genetic findings which could be causative are not at hand. The TNF-α system might be an exception as the TNFA(2) allele which in one study unreplicated to date was found to be more frequent in schizophrenic patients as compared to controls indeed goes along with an increased TNF-α expression.(34) The majority of studies found TNF-α levels to be increased in schizophrenia but unfortunately TNF-α levels have not been related to the TNF polymorphism so far, neither in healthy people nor in schizophrenic patients.

If altered regulation of some cytokine systems in schizophrenia is an acquired phenomenon, 2 major questions arise: what are the causes of these alterations and what is their significance for the pathogenetics and the pathophysiology of the disorder? For both questions there is no clear-cut answer at present but we will discuss the two major options:

Altered cytokine network regulation in patients with schizophrenia might either be related to the cause of the disorder, and cytokines might play a genuine pathophysiological role, or some aspects of the disorder itself might secondarily affect cytokine systems. We will shortly discuss both of these alternatives.

5.1 Does a chronic or relapsing-remitting kind of infection, inflammation or an autoimmune process cause both schizophrenia and the observed abnormalities in cytokine networks?

As outlined already in the early 1990s in a brilliant article hypothesizing that schizophrenia may be caused by a combined overactivation of macrophages and T-lymphocytes,(35) situations of a chronic immune system stimulation might indeed go along with increased levels of IL-6, TNF-α and sIL-R2, and with a reduced stimulated IL-2 production. However, so far no consistent direct evidence in favor of an autoimmune process or a chronic infection has been demonstrated in schizophrenia. It is important to note at this point that numerous authors have argued that the quite robust evidence affirming an increased frequency of infections during the pregnancy of mothers of subjects who later on developed schizophrenia together with the cytokine abnormalities seen in adult patients supports an immunological cause of schizophrenia. This reasoning is problematic since although cytokines released during prenatal infections might well disturb neurodevelopment and via this pathway contribute to the causes of schizophrenia it can almost be excluded that these acute changes in cytokine production will persist into adulthood. Hence, if infection, inflammation or autoimmunity caused the observed peripheral dysregulations of cytokines, the underlying immunological process must be assumed to be still active in schizophrenic patients and should be detectable by future research.

The question arises how, in the context of this presumed immunological process, cytokines could be involved in the pathophysiology of schizophrenia. Cytokines could either directly affect brain functions leading to symptoms of schizophrenia or contribute indirectly to a basic pathophysiological process such as neurodegeneration.

There is abundant evidence that IL-6 and TNF-α are involved in neurodegenerative processes. However, increased levels of these cytokines have been found only in the blood of schizophrenic patients so far. In addition, the idea that neurodegeneration takes place in schizophrenia is derived from neuroimaging studies suggesting a progressive brain atrophy in certain regions.(36) Direct neuropathological evidence has been lacking for both a neurodegenerative process and an activation of cytokine systems in the brain of schizophrenic patients. Hence, at present it is quite unlikely that the alterations in cytokine networks found in the blood of schizophrenic patients either directly contribute to or indicate an ongoing cytokine-driven neurodegenerative process in the brain of schizophrenic patients.

Cytokines could also directly affect brain functions leading to symptoms of schizophrenia. This could even be true for cytokines produced exclusively in the periphery. A large body of evidence clearly shows that inflammatory cytokines such as IL-1, IL-6 and TNF-α are pivotal to those behavioral changes which occur during acute infection and inflammation.(15,16)

This set of behavioral alterations includes reduced locomotion, reduced food intake, reduced sexual activity, anhedonia and disturbed sleep. Moreover, the therapeutic administration of cytokines to patients with certain infections or cancer induces comparable symptoms, sometimes even suicidal ideation and disturbances of perception and thought.(14) Finally, even in healthy subjects modest increases in endogeneously produced IL-6 and TNF-α in response to host defense stimulation with endotoxin temporarily deteriorate mood, increase anxiety and worsen memory.(17) However, although the circulating amounts of IL-6 and TNF-α might be slightly increased in patients with schizophrenia as compared to healthy controls, these increases are very subtle compared to settings such as experimental endotoxinemia or the therapeutic administration of cytokines (for review see (24)). Moreover, although only few studies compared cytokine levels between different psychiatric disorders, there is evidence suggesting that IL-6 and TNF-α levels in schizophrenia and depression are very similar.(99,102)

Despite increasing evidence that circulating cytokines may indeed induce psychopathology, it is unclear whether this is the case for very low amounts. In contrast, there is preliminary evidence discussed elsewhere in more detail(24) suggesting that very low amounts of inflammatory cytokines might influence brain functions in the opposite direction as compared to high amounts. One example is sleep, which is enhanced by very low amounts and disturbed by high amounts of endotoxin.(37) Particularly core symptoms of schizophrenia such as disturbances of thought and perception and delusions are unlikely to be caused by low amounts of circulating inflammatory cytokines.

In summary, the is no conclusive evidence at present that infection, inflammation or autoimmune processes cause the altered regulation of the IL-6, TNF-α, IL-2 and IFN cytokine networks observed in the peripheral blood of patients with schizophrenia. Moreover, the available data do not convincingly support the idea that cytokines contribute to the changes of brain functions or neurodegeneration in schizophrenia. To render such a participation probable, alterations of cytokine networks within the patients' brain will have to be shown in future studies.

5.2 Is cytokine system dysregulation in schizophrenia a secondary phenomenon?

Hence, the second alternative to explain the findings discussed in the present chapter must be considered, which is that cytokine network dysregulation is caused by some aspects of schizophrenia itself. Again, it must be stressed that increases in the circulating amounts of cytokines and changes in stimulated cytokine production found in patients with schizophrenia are very subtle. Actually, the most recent findings could only be obtained because of a dramatic refinement of the analytical techniques making the detection of cytokine amounts of sometimes even below 1 pg/ml possible. In general, analytical refinements make methods more sensitive but often also more vulnerable to artifacts. It was only during the last 5 years that, for example, the methods for the detection of IL-6 became sensitive enough to detect stable baseline values in healthy people. Therefore, it is now possible to detect very small intersubject and intrasubject differences. To give one example, there are numerous studies suggesting slight, sleep-related increases in the circulating levels of IL-6. However, methodological studies show that these findings are flawed by the fact that repeated long-term blood sampling by means of an intravenous catheter induces a local (rather than systemic) IL-6 production which artificially increases the estimated blood levels.(38,39) Similarly, circulating cytokine levels and ex vivo cytokine production by white blood cells are sensitive to numerous other influences (see Table 8), some of which might be of particular

External Factor	Examples of cytokine systems affected
Age	Increased circulating levels of TNF-α and TNF receptors
Gender	Increased IL-1Ra levels in females
Obesity	Increased IL-6 and TNF-α levels
Stress	Increased IL-6 levels
Smoking	IL-2 production decreased, sIL-2R levels increased
Recent infections	Increased IL-6 and TNF-α levels
Psychotropic medication (e.g. clozapine)	Increased circulating levels of TNF-α and TNF receptors, and of sIL-2R

Table 8. Major external influences on cytokine networks. This list is not comprehensive, neither with respect to the possible factors nor the cytokine systems affected (for reviews see 18,30,99).

importance for schizophrenia. It is not possible to discuss all these influences with respect to every cytokine system that is crucial in the context of this chapter; this is impossible not only due to the complexity of this topic but also because several of these possible influences have been investigated for some but not for all cytokines of interest. The major point here is that most studies about cytokines in schizophrenia have not appropriately taken into account all of these possible influences. Some of these factors, such as age and gender, most studies have accounted for. Obesity, although induced by numerous psychotropic drugs, has almost never been accounted for; fortunately, however, in cross-sectional samples, patients suffering from schizophrenia are not obese suggesting that samples used for studies about cytokine systems were not biased by increased weight in the experimental group.

However, other factors known to influence cytokine systems are much more prevalent in patients with schizophrenia as compared to healthy control populations. These include acute stress, smoking and medication. To our knowledge there are no studies comparing stress levels in schizophrenic patients to those of controls but most studies about cytokine networks have been performed in acutely ill, hospitalized patients suggesting a considerable amount of unspecific acute stress in addition to the specific schizophrenia-associated symptomatology. There is clear-cut evidence that smoking is much more prevalent in schizophrenic patients than in healthy controls.(40) Moreover, there is no doubt that smoking affects cytokine production, in particular the IL-2 cytokine system.(41) However, even very recent studies failed to account for this confounding variable.(42,43)

The last very important factor that has to be taken into account when attempting to understand cytokine network regulation in schizophrenia is medication. Immunomodulatory effects of psychotropic agents, which we have recently reviewed in depth elsewhere,(18,30) have been known since the 1950s when chlorpromazine, the first antipsychotic agent, was discovered. Chlorpromazine was also the first agent which has been shown already 15 years ago to affect stimulated cytokine secretion in vitro (134). Although it is clear now that the IFN, IL-2, IL-6 and TNF-α cytokine systems are all affected by some antipsychotic drugs, comprehensive studies regarding the effects on cytokine blood levels and in vitro cytokine production have been lacking for most compounds. So far, clozapine, the prototype of second generation antipsychotics, has been investigated most intensively suggesting clear-cut effects on the IL-2, IL-6 and TNF-α systems in vivo and complex effects in vitro.(44,45)

As outlined above, cytokines are not only involved in neurodegenerative but also neuroprotective processes and under certain circumstances in neuronal repair. Moreover, there is preliminary evidence that inflammatory cytokines can not only compromise complex brain functions but also enhance them (e.g. sleep, see above). Hence, cytokines might not only play a role in the pathogenetics of schizophrenia by, for example, impairing brain development *in utero* but could also exhibit beneficial effects in schizophrenic patients, e.g. by counteracting an ongoing neurodegenerative process. Therefore, we would like to speculate that modulatory effects of psychotropic drugs on cytokine networks might not only be an unwanted confounding factor but point to a new mode of action in the framework of their therapeutic activities. It is to be noted that this is at present a speculative hypothesis but it seems worth to be tested.

To summarize the role of external influences on cytokine networks in schizophrenia it can be concluded that there are at least three important factors: stress, smoking and medication. These are important confounding factors which so far have not sufficiently been taken into consideration in studies about cytokine networks in schizophrenia. Therefore, these factors might account to a considerable extent for the differences from controls which have been observed for cytokine levels or secretion in patients with schizophrenia. The effects of psychotropic agents on cytokine networks, in addition to confounding studies aimed at unraveling disease-related abnormalities in cytokine networks, might contribute to their beneficial effects in schizophrenia, although this is an hypothesis that still awaits experimental confirmation.

6. SUMMARY AND OUTLOOK

We would like to summarize this chapter concerning the presumed role of cytokines in the pathogenetics and pathophysiology of schizophrenia by stressing the following points:

a) Cytokines are humoral immune mediators but they also play an outstanding role in cell development and survival of almost every tissue including the brain. Hence, cytokines are very likely to be directly or indirectly involved in neurodevelopmental and neurodegenerative processes in schizophrenia.
b) However, on the genomic level, in the brain and the CSF there are at present no convincing data demonstrating that cytokine networks are indeed disturbed in schizophrenic patients. This negative evidence, though, is based on a small number of studies.
c) There are quite robust findings obtained from peripheral blood that IL-2, IL-6, TNF-α and IFN cytokine systems are regulated differently as compared to controls in schizophrenic patients. However, these findings

are not specific to schizophrenia, their pathophysiological relevance for processes within the CNS is undetermined, and the respective results are confounded by numerous intervening variables such as stress, smoking and medication.

d) Medication in particular has a profound influence on cytokine networks which might not only be important as a confounding factor but also for the understanding of the drugs' beneficial actions on the CNS.

Future research on the involvement of cytokines in the pathogenetics, pathophysiology and treatment of schizophrenia should consider the following directions:

a) Although studies about the genetics of cytokine systems have mostly yielded negative results, further efforts are warranted to clarify whether cytokines in the broadest sense (see Table 1) contribute to the multigenetic background of schizophrenia. In this context it also has to be considered that altered upstream genes regulating cytokine networks indirectly affect cytokine function, and that cytokine actions might be modified by altered downstream effector genes.
b) It will be crucial to gain information on the regulation of cytokine networks within the brain of schizophrenic patients. Probably genomic and proteomic approaches will considerably accelerate the generation of the respective knowledge but the limiting factor is the low availability of post-mortem brain tissue from schizophrenic patients. Moreover, since disease onset and death are far apart in the majority of schizophrenic patients it will be difficult to separate secondary processes from primary ones.
c) Prospective, longitudinal, large-scale multicenter studies should clarify the reproducibility and biological significance of altered cytokine production and secretion in peripheral blood. These studies should investigate blood levels and stimulated ex vivo cytokine secretion in parallel and carefully take the large number of confounding factors into account. Although investigations in blood are relatively easy to perform, one should face the possibility that these studies might be unable to clarify the role of cytokines in a brain disorder such as schizophrenia.
d) Immunomodulatory effects of psychotropic agents, which have been discovered already in the 1950s and include potent effects on cytokine networks, should be investigated in much more detail because they might be important for the drugs' effects on the CNS. These studies when taking into account neuroprotective effects of various cytokines might ultimately lead to new strategies for the treatment of schizophrenia.

REFERENCES

1. Bukanan, R. Schizophrenia: introduction and overwiew. In: *Comprehensive textbook of Psychiatry*. Benjamin and Virginia Saddock, editors. 7th edition. New York: Lippincott Williams & Wilkins 2000; 1096-1110.
2. Pollmächer, T. Treatment of schizophrenia. In: *International encyclopedia of the social and behavioral sciences*. Smelser NJ and Balters PB, editors. 1st edition. Kidlington: Pergamon 2002; 13546-13552.
3. Pickar, D. Prospects for pharmacotherapy of schizophrenia. *Lancet* 1995; 345:557-562.
4. Torrey, E.F. and Yolken, R.H. Familial and genetic mechanisms in schizophrenia. *Brain Res. Rev.* 2000; 31:113-117.
5. Lieberman, J.A. Is schizophrenia a neurodegenerative disorder? A clinical and neurobiological perspective. *Biol. Psychiatry* 1999; 46:729-739.
6. Yolken, R.H. and Torrey, E.F. Viruses, schizophrenia, and bipolar disorder. *Clin. Microbiol. Rev.* 1995; 8:131-145.
7. Wagner, J. Über die Einwirkung fieberhafter Erkrankungen auf Psychosen. *Allgemeine Zeitschrift für Psychiatrie* 1887; 27:93-131.
8. Oken, R.J. and Schulzer, M. At issue: schizophrenia and rheumatoid arthritis: the negative association revisited. *Schizophr. Bull.* 1999; 25:625-638.
9. Kirch, D.G. Infection and autoimmunity as etiologic factors in schizophrenia: a review and reappraisal. *Schizophr. Bull.* 1993; 19:355-370.
10. Ganguli, R., Brar, J.S, Chengappa, K.N., et al. Autoimmunity in schizophrenia: a review of recent findings. *Ann. Med.* 1993; 25:489-496.
11. Rothwell, N.J. and Hopkins, S.J. Cytokines and the nervous system II: actions and mechanisms of action. *Trends Neurosci.* 1995; 18:130-136.
12. Hopkins, S.J. and Rothwell, N.J. Cytokines and the nervous system I: expression and recognition. *Trends Neurosci.* 1995; 18:83-88.
13. Malek-Ahmadi, P. Neuropsychiatric aspects of cytokines research: an overview. *Neurosci. Biobehav. Rev.* 1996; 20:359-365.
14. Kronfol, Z. and Remick, D.G. Cytokines and the brain: implications for clinical psychiatry. *Am. J. Psychiatry* 2000; 157:683-694.
15. Dantzer, R. Cytokine-induced sickness behavior: where do we stand? *Brain Behav. Immun.* 2001; 15:7-24.
16. Yirmiya, R., Pollak, Y., Morag, M., et al. Illness, cytokines and depression. *Ann. N. Y. Acad. Sci.* 2000; 917:478-487.
17. Reichenberg, A., Yirmiya, R., Schuld, A., et al. Cytokine-associated emotional and cognitive disturbances in humans. *Arch. Gen. Psychiatry* 2001; 58:445-452.
18. Pollmächer, T., Haack, M., Schuld, A., et al. Effects of antipsychotic drugs on cytokine networks. *J. Psychiatr. Res.* 2000; 34:369-382.
19. Vilček, J. The cytokines: an overview. In: *The cytokine handbook.* Angus W. Thomson, editor. 3rd edition. Pittsburgh: Academic Press 1998; 1-21.
20. Tracey, K.J. and Cerami, A. Tumor necrosis factor: a pleiotropic cytokine and therapeutic target. *Annu. Rev. Med.* 1994; 45:491-503.
21. Dinarello, C.A. The interleukin-1 family: 10 years of discovery. *FASEB J.* 1994; 8:1314-1325.
22. Burrell, R. Human responses to bacterial endotoxin. *Circ. Shock* 1994; 43:137-153.
23. Zhang, M. and Tracey, K. Tumor necrosis factor. In: *The cytokine handbook.* Angus W.Thomson, editor. 3rd edition. Pittsburgh: Academic Press 1998; 517-549.

24. Pollmächer, T., Haack, M., Schuld, A., et al. Low levels of circulating inflammatory cytokines - do they affect human brain functions? *Brain, Behav.Immun.*, in press
25. Meisenzahl, E.M., Rujescu, D., Kirner, A., et al. Association of an interleukin-1beta genetic polymorphism with altered brain structure in patients with schizophrenia. *Am. J. Psychiatry* 2001; 158:1316-1319.
26. Krebs, M.O., Guillin, O., Bourdell, M.C., et al. Brain derived neurotrophic factor (BDNF) gene variants association with age at onset and therapeutic response in schizophrenia. *Mol. Psychiatry* 2000; 5:558-562.
27. Wassink, T.H., Nelson, J.J., Crowe, R.R., et al. Heritability of BDNF alleles and their effect on brain morphology in schizophrenia. *Am. J. Med. Genet* 1999; 88:724-728.
28. Takahashi, M., Shirakawa, O., Toyooka, K., et al. Abnormal expression of brain-derived neurotrophic factor and its receptor in the corticolimbic system of schizophrenic patients. *Mol. Psychiatry* 2000; 5:293-300.
29. Chen, B., Dowlatshahi, D., MacQueen, G.M., et al. Increased hippocampal BDNF immunoreactivity in subjects treated with antidepressant medication. *Biol. Psychiatry* 2001; 50:260-265.
30. Hinze-Selch, D. and Pollmächer, T. In vitro cytokine secretion in individuals with schizophrenia: results, confounding factors, and implications for further research. *Brain Behav. Immun.* 2001; 15:282-318.
31. Mackensen, A., Galanos, C., Wehr, U., et al. Endotoxin tolerance: regulation of cytokine production and cellular changes in response to endotoxin application in cancer patients. *Eur. Cytokine Netw.* 1992; 3:571-579.
32. Pollmächer, T., Schreiber, W., Gudewill, S., et al. Influence of endotoxin on nocturnal sleep in humans. *Am. J. Physiol.* 1993; 264:R1077-R1083
33. Pollmächer, T., Korth, C., Mullington, J., et al. Effects of granulocyte colony-stimulating factor on plasma cytokine and cytokine receptor levels and on the in vivo host response to endotoxin in healthy men. *Blood* 1996; 87:900-905.
34. Boin, F., Zanardini, R., Pioli, R., et al. Association between -G308A tumor necrosis factor alpha gene polymorphism and schizophrenia. *Mol. Psychiatry* 2001; 6:79-82.
35. Smith, R.S. A comprehensive macrophage-T-lymphocyte theory of schizophrenia. *Med. Hypotheses* 1992; 39:248-257.
36. Andreasen, N.C., Paradiso, S. and O'Leary, D.S. 'Cognitive dysmetria' as an integrative theory of schizophrenia: a dysfunction in cortical-subcortical-cerebellar circuitry? *Schizophr. Bull.* 1998; 24:203-218.
37. Mullington J, Korth C, Hermann DM, Orth A, Galanos C, Holsboer F, et al. Dose-dependent effects of endotoxin on human sleep. *Am J Physiol Regul Integr Comp Physiol* 2000; 278:R947-R955
38. Haack M, Reichenberg A, Kraus T, et al. Effects of an intravenous catheter on the local production of cytokines and soluble cytokine receptors in healthy men. *Cytokine* 2000; 12:694-698.
39. Haack M, Kraus T, Schuld A, Dalal M, Koethe D, Pollmächer T. Diurnal variations of interleukin-6-plasma levels during sleep and wakefulness are confounded by particularities of blood drawing. *Psychoneuroendocrinology*, in press
40. de Leon, J., Dadvand, M., Canuso, C., et al. Schizophrenia and smoking: an epidemiological survey in a state hospital. *Am. J. Psychiatry* 1995; 152:453-455.
41. Pollmächer, T., Hinze-Selch, D., Mullington, J., et al. Smoking affects plasma-soluble interleukin-2 receptor levels in patients with schizophrenia. *Arch. Gen. Psychiatry* 1997; 54:89

42. Arolt, V., Rothermundt, M., Wandinger, K.P, et al. Decreased in vitro production of interferon-gamma and interleukin-2 in whole blood of patients with schizophrenia during treatment. *Mol. Psychiatry* 2000; 5:150-158.
43. Rothermundt, M., Arolt, V., Leadbeater, J., et al. Cytokine production in unmedicated and treated schizophrenic patients. *Neuroreport* 2000; 11:3385-3388.
44. Pollmächer, T., Hinze-Selch, D. and Mullington, J. Effects of clozapine on plasma cytokine and soluble cytokine receptor levels. *J. Clin. Psychopharmacol.* 1996; 16:403-409.
45. Hinze-Selch, D., Becker, W., Stein, G., et al. Effects of clozapine on in vitro immune parameters: A longitudinal study in clozapine-treated schizophrenic patients. *Neuropsychopharmacology* 1998; 2:114-122.
46. Yolken RH, Bachmann S, Rouslanova I, et al. Antibodies to Toxoplasma gondii in individuals with first-episode schizophrenia. *Clin Infect Dis* 2001; 32:842-844.
47. Müller, N. The role of the cytokine network in the CNS and psychic disorders. *Der Nervenarzt* 1997; 68:11-20.
48. Buka, S.L. Tsuang, M.T., Torrey, E.F., et al. Maternal infections and subsequent psychosis among offspring. *Arch Gen Psychiatry* 2001; 58 1032-1037.
49. Russo R, Ciminale M, Ditommaso S, et al. Hepatitis B vaccination in psychiatric patients. *Lancet* 1994; 343:356
50. Karlsson H., Bachmann, S., Schröder, J., et al. Retroviral RNA identified in the cerebrospinal fluids and brains of individuals with schizophrenia. *Proc Natl Acad Sci U S A* 2001; 98: 4634-4639.
51. Laurent C, Thibaut F, Ravassard P, et al. Detection of two new polymorphic sites in the human interleukin-1 beta gene: lack of association with schizophrenia in a French population. *Psychiatr Genet* 1997; 7:103-105.
52. Chowdari KV, Xu K, Zhang F, et al. Immune related genetic polymorphisms and schizophrenia among the Chinese. *Hum Immunol* 2001; 62:714-724.
53. Katila, H., Hanninen, K. and Hurme, M. Polymorphisms of the interleukin-1 gene complex in schizophrenia. *Mol. Psychiatry* 1999; 4:179-181.
54. Tatsumi, M., Sasaki, T., Sakai, T., et al. Genes for interleukin-2 receptor beta chain, interleukin-1 beta, and schizophrenia: no evidence for the association or linkage. *Am. J. Med. Genet.* 1997; 74:338-341.
55. Nimgaonkar, V.L., Yang, Z.W., Zhang, X.R., et al. Association study of schizophrenia and the IL-2 receptor beta chain gene. *Am. J. Med. Genet.* 1995; 60:448-451.
56. Wassink, T.H., Crowe, R.R., and Andreasen, N.C. Tumor necrosis factor receptor-II: heritability and effect on brain morphology in schizophrenia. *Mol. Psychiatry* 2000; 5:678-682.
57. Virgos, C., Martorell, L., Valero, J., et al. Association study of schizophrenia with polymorphisms at six candidate genes. *Schizophr. Res.* 2001; 49:65-71.
58. Hawi, Z., Straub, R.E., O'Neill, A., et al. No linkage or linkage disequilibrium between brain-derived neurotrophic factor (BDNF) dinucleotide repeat polymorphism and schizophrenia in Irish families. *Psychiatry Res.* 1998; 81:111-116.
59. Tanaka, Y., Ujike, H., Fujiwara, Y., et al. Schizophrenic psychoses and the CNTF null mutation. *Neuroreport* 1998; 9:981-983.
60. Gelernter, J., Van Dyck, C., van Kammen, D.P., et al. Ciliary neurotrophic factor null allele frequencies in schizophrenia, affective disorders, and Alzheimer's disease. *Am. J. Med. Genet.* 1997; 74:497-500.
61. Sakai, T., Sasaki, T., Tatsumi, M., et al. Schizophrenia and the ciliary neurotrophic factor (CNTF) gene: no evidence for association. *Psychiatry Res.* 1997; 71:7-10.

62. Jonsson, E., Brene, S., Zhang, X.R., et al. Schizophrenia and neurotrophin-3 alleles. *Acta Psychiatr. Scand.* 1997; 95:414-419.
63. Nimgaonkar, V.L., Zhang, X.R., Brar, J.S., et al. Lack of association of schizophrenia with the neurotrophin-3 gene locus. *Acta Psychiatr. Scand.* 1995; 92:464-466.
64. Dawson, E., Powell, J.F., Sham, P.C., et al. An association study of a neurotrophin-3 (NT-3) gene polymorphism with schizophrenia. *Acta Psychiatr. Scand.* 1995; 92:425-428.
65. Hattori, M. and Nanko, S. Association of neurotrophin-3 gene variant with severe forms of schizophrenia. *Biochem. Biophys. Res. Commun.* 1995; 209:513-518.
66. Thome, J., Durany, N., Palomo, A., et al. Variants in neurotrophic factor genes and schizophrenic psychoses: no associations in a Spanish population. *Psychiatry Res.* 1997; 71:1-5.
67. Gill, M., Hawi, Z., O'Neill, F.A., et al. Neurotrophin-3 gene polymorphisms and schizophrenia: no evidence for linkage or association. *Psychiatr. Genet.* 1996; 6:183-186.
68. Nanko, S., Hattori, M., Kuwata, S., et al. Neurotrophin-3 gene polymorphism associated with schizophrenia. *Acta Psychiatr. Scand.* 1994; 89:390-392.
69. el-Mallakh, R.S., Suddath, R.L. and Wyatt, R.J. Interleukin-1 alpha and interleukin-2 in cerebrospinal fluid of schizophrenic subjects. *Prog. Neuropsychopharmacol. Biol. Psychiatry* 1993; 17:383-391.
70. Barak, V., Barak, Y., Levine, J., et al. Changes in interleukin-1 beta and soluble interleukin-2 receptor levels in CSF and serum of schizophrenic patients. *J. Basic. Clin. Physiol. Pharmacol.* 1995; 6:61-69.
71. Licinio, J., Seibyl, J.P., Altemus, M., et al. Elevated CSF levels of interleukin-2 in neuroleptic-free schizophrenic patients. *Am. J. Psychiatry* 1993; 150:1408-1410.
72. Rapaport, M.H., McAllister, C.G., Pickar, D., et al. CSF IL-1 and IL-2 in medicated schizophrenic patients and normal volunteers. *Schizophr. Res.* 1997; 25:123-129.
73. Katila, H., Hurme, M., Wahlbeck, K., et al. Plasma and cerebrospinal fluid interleukin-1β and interleukin-6 in hospitalized schizophrenic patients. *Neuropsychobiology* 1994; 30:20-23.
74. Mittleman, B.B., Castellanos, F.X., Jacobsen, L.K., et al. Cerebrospinal fluid cytokines in pediatric neuropsychiatric disease. *J. Immunol.* 1997; 159:2994-2999.
75. van Kammen, D.P., McAllister-Sistilli, C.G., Kelley, M.E., et al. Elevated interleukin-6 in schizophrenia. *Psychiatry Res.* 1999; 87:129-136.
76. Libikova, H., Breier, S., Kocisova, M., et al. Assay of interferon and viral antibodies in the cerebrospinal fluid in clinical neurology and psychiatry. *Acta Biol. Med. Ger.* 1979; 38:879-893.
77. Roy, A., Pickar, D., Ninan, P., et al. A search for interferon in the CSF of chronic schizophrenic patients. *Am. J. Psychiatry* 1985; 142:269.
78. Vawter, M.P., Dillon-Carter, O., Issa, F., et al. Transforming growth factors beta 1 and beta 2 in the cerebrospinal fluid of chronic schizophrenic patients. *Neuropsychopharmacology* 1997; 16:83-87.
79. Katila, H., Appelberg, B., Hurme, M., et al. Plasma levels of interleukin-1β and interleukin-6 in schizophrenia, other psychoses, and affective disorders. *Schizophr. Res.* 1994; 12:29-34.
80. Kim, Y.K., Lee, M.S. and Suh, K.Y. Decreased interleukin-2 production in Korean schizophrenic patients. *Biol. Psychiatry* 1998; 43:701-704.
81. Baker, I., Masserano, J. and Wyatt, R.J. Serum cytokine concentrations in patients with schizophrenia. *Schizophr. Res.* 1996; 20:199-203.

82. Naudin, J., Capo, C., Giusano, B., et al. A differential role for interleukin-6 and tumor necrosis factor-alpha in schizophrenia? *Schizophr. Res.* 1997; 26:227-233.
83. Erbagci, A.B., Herken, H., Koyluoglu, O., et al. Serum IL-1beta, sIL-2R, IL-6, IL-8 and TNF-alpha in schizophrenic patients, relation with symptomatology and responsiveness to risperidone treatment. *Mediators Inflamm.* 2001; 10:109-115.
84. Kim, Y.K., Kim, L. and Lee MS. Relationships between interleukins, neurotransmitters and psychopathology in drug-free male schizophrenics. *Schizophr. Res.* 2000; 44:165-175.
85. Theodoropoulou, S., Spanakos, G., Baxevanis, C.N., et al. Cytokine serum levels, autologous mixed lymphocyte reaction and surface marker analysis in never medicated and chronically medicated schizophrenic patients. *Schizophr. Res.* 2001; 47:13-25.
86. Maes, M., Bosmans, E., Kenis, G., et al. In vivo immunomodulatory effects of clozapine in schizophrenia. *Schizophr. Res.* 1997; 26:221-225.
87. Akiyama, K. Serum levels of soluble IL-2 receptor alpha, IL-6 and IL-1 receptor antagonist in schizophrenia before and during neuroleptic administration. *Schizophr. Res.* 1999; 37:97-106.
88. Maes, M., Bocchio, C.L., Bignotti. S., et al. Effects of atypical antipsychotics on the inflammatory response system in schizophrenic patients resistant to treatment with typical neuroleptics. *Eur. Neuropsychopharmacol.* 2000; 10:119-124.
89. Naudin, J., Mege, J.L., Azorin, J.M., et al. Elevated circulating levels of IL-6 in schizophrenia. *Schizophr. Res.* 1996; 20:269-273.
90. Gattaz, W.F., Dalgalarrondo, P. and Schroder, H.C. Abnormalities in serum concentrations of interleukin-2, interferon-alpha and interferon-gamma in schizophrenia not detected. *Schizophr. Res.* 1992; 6:237-241.
91. Ganguli, R. and Rabin, B.S. Increased serum interleukin 2 receptor concentration in schizophrenic and brain-damaged subjects. *Arch. Gen. Psychiatry* 1989; 46:292
92. Rapaport, M.H., McAllister, C.G., Pickar, D., et al. Elevated levels of soluble interleukin 2 receptors in schizophrenia. *Arch. Gen. Psychiatry* 1989; 46:291-292.
93. Rapaport, M.H. and Lohr JB. Serum-soluble interleukin-2 receptors in neuroleptic-naive schizophrenic subjects and in medicated schizophrenic subjects with and without tardive dyskinesia. *Acta Psychiatr. Scand.* 1994; 90:311-315.
94. Gaughran, F., O'Neill, E., Cole, M., et al. Increased soluble interleukin 2 receptor levels in schizophrenia. *Schizophr. Res.* 1998; 29:263-267.
95. Maes, M., Meltzer, H.Y. and Bosmans E. Immune-inflammatory markers in schizophrenia: comparison to normal controls and effects of clozapine. *Acta Psychiatr. Scand.* 1994; 89:346-351.
96. Maes, M., Bosmans, E., Calabrese, J., et al. Interleukin-2 and interleukin-6 in schizophrenia and mania: effects of neuroleptics and mood stabilizers. *J. Psychiatr. Res.* 1995; 29:141-152.
97. Maes, M., Meltzer, H.Y., Buckley, P., et al. Plasma-soluble interleukin-2 and transferrin receptor in schizophrenia and major depression. *Eur. Arch. Psychiatry Clin. Neurosci.* 1995; 244:325-329.
98. Müller, N., Empl, M., Riedel, M., et al. Neuroleptic treatment increases soluble IL-2 receptors and decreases soluble IL-6 receptors in schizophrenia. *Eur. Arch. Psychiatry Clin. Neurosci.* 1997; 247:308-313.
99. Haack, M., Hinze-Selch, D., Fenzel, T., et al. Plasma levels of cytokines and soluble cytokine receptors in psychiatric patients upon hospital admission: effects of confounding factors and diagnosis. *J. Psychiat.r Res* 1999; 33: 407-418.
100. Ramchand, R., Wei, J., Ramchand, C.N., et al. Increased serum IgE in schizophrenic patients who responded poorly to neuroleptic treatment. *Life Sci.* 1994; 54:1579-1584.

101. Kudoh, A., Sakai, T., Ishihara, H., et al. Plasma cytokine response to surgical stress in schizophrenic patients. *Clin. Exp. Immunol.* 2001; 125:89-93.
102. Frommberger, U.H., Bauer, J., Haselbauer, P. , et al. Interleukin-6-(IL-6) plasma levels in depression and schizophrenia: comparison between the acute state and after remission. *Eur. Arch. Psychiatry Clin. Neurosci.* 1997; 247-228-233.
103. Ganguli, R., Yang, Z., Shurin, G., et al. Serum interleukin-6 concentration in schizophrenia: elevation associated with duration of illness. *Psychiatry Res.* 1994; 51:1-10.
104. Lin, A., Kenis, G., Bignotti, S., et al. The inflammatory response system in treatment-resistant schizophrenia: increased serum interleukin-6. *Schizophr. Res.* 1998; 32:9-15.
105. Maes, M., Meltzer, H.Y., Bosmans, E., et al. Increased plasma concentrations of interleukin-6, soluble interleukin-6, soluble interleukin-2 and transferrin receptor in major depression. *J. Affect. Disord.* 1995; 34:301-309.
106. Monteleone, P., Fabrazzo, M., Tortorella, A., et al. Plasma levels of interleukin-6 and tumor necrosis factor alpha in chronic schizophrenia: effects of clozapine treatment. *Psychiatry Res.* 1997; 71:11-17.
107. Shintani, F., Kanba, S., Maruo, N., et al. Serum interleukin-6 in schizophrenic patients. *Life Sci.* 1991; 49:661-664.
108. Müller, N., Empl, M., Riedel, M., et al. Neuroleptic treatment increases soluble IL-2 receptors and decreases soluble IL-6 receptors in schizophrenia. *Eur. Arch. Psychiatry Clin. Neurosci.* 1997; 247:308-313.
109. Tanaka, K.F., Shintani, F., Fujii, Y., et al. Serum interleukin-18 levels are elevated in schizophrenia. *Psychiatry Res.* 2000; 96:75-80.
110. Rimon, R., Ahokas, A., Hintikka, J., et al. Serum interferon in schizophrenia. *Ann. Clin. Res.* 1985; 17:139-140.
111. Perez-Polo, J.R., Dy P., Westlund, K., et al. Levels of serum nerve growth factor in schizophrenia. *Birth Defects Orig. Artic. Ser.* 1978; 14:311-321.
112. Sirota, P., Fishman, P., Elizur, A., et al. Lymphokine production in schizophrenic patients. *Biol. Psychiatry* 1998; 27:118A.
113. Bessler, H., Levental, Z., Karp, L., et al. Cytokine production in drug-free and neuroleptic-treated schizophrenic patients. *Biol. Psychiatry* 1995; 38:297-302.
114. Ganguli ,R., Rabin, B.S. and Belle, S.H. Decreased interleukin-2 production in schizophrenic patients. *Biol. Psychiatry* 1989; 26:427-430.
115. Ganguli, R., Brar, J.S., Solomon, W., et al. Altered interleukin-2 production in schizophrenia: association between clinical state and autoantibody production. *Psychiatry Res.* 1992; 44:113-123.
116. Yang, Z.W., Chengappa, K.N., Shurin, G., et al. An association between anti-hippocampal antibody concentration and lymphocyte production of IL-2 in patients with schizophrenia. *Psychol. Med.* 1994; 24:449-455.
117. Ganguli, R., Brar, J.S., Chengappa, K.R., et al. Mitogen-stimulated interleukin-2 production in never-medicated, first-episode schizophrenic patients. The influence of age at onset and negative symptoms. *Arch. Gen. Psychiatry* 1995; 52:668-672.
118. Hornberg, M., Arolt, V., Wilke, I., et al. Production of interferons and lymphokines in leukocyte cultures of patients with schizophrenia. *Schizophr. Res.*1995; 15:237-242.
119. Rothermundt, M., Arolt, V., Weitzsch, C., et al. Production of cytokines in acute schizophrenic psychosis. *Biol. Psychiatry* 1996; 40:1294-1297.
120. Wilke, I., Arolt, V., Rothermundt, M., et al. Investigations of cytokine production in whole blood cultures of paranoid and residual schizophrenic patients. *Eur. Arch. Psychiatry Clin. Neurosci.* 1996; 246:279-284.

121. Rothermundt, M., Arolt, V., Weitzsch, C., et al. Immunological dysfunction in schizophrenia: a systematic approach. *Neuropsychobiology* 1998; 37:186-193.
122. O'Donnell, M.C., Catts, S.V., Ward, P.B., et al. Increased production of interleukin-2 (IL-2) but not soluble interleukin-2 receptors (sIL-2R) in unmedicated patients with schizophrenia and schizophreniform disorder. *Psychiatry Res.* 1996; 65:171-178.
123. Cazzullo, C.L., Scarone, S., Grassi, B., et al. Cytokine production in chronic schizophrenia patients with or without paranoid behaviour. *Prog. Neuropsychopharmacol. Biol. Psychiatry* 1998; 22:947-957.
124. Cazzullo, C.L., Sacchetti, E., Galluzzo, A., et al. Cytokine profiles in drug-naive schizophrenic patients. *Schizophr. Res.* 2001; 47:293-298.
125. Villemain, F., Chatenoud, L., Guillibert, E., et al. Decreased production of interleukin-2 in schizophrenia. *Ann. N. Y. Acad. Sci.* 1987; 496:669-675.
126. Villemain, F., Chatenoud, L., Galinowski, A., et al. Aberrant T cell-mediated immunity in untreated schizophrenic patients: deficient interleukin-2 production. *Am. J. Psychiatry* 1989; 146:609-616.
127. Sekirina, T.P., Voronkova, T.L., Tsutsu'lkovskaia, M.I., et al. [Disorders of interleukin 2 biosynthesis by peripheral blood lymphocytes in patients with schizophrenia]. *Zh. Nevropatol. Psikhiatr. Im. S. S. Korsakova* 1989; 89:95-97.
128. O'Donnell, C.P., Schaub, C.D., Haines, A.S., et al. Leptin prevents respiratory depression in obesity. *Am. J. Respir. Crit. Care Med.* 1999; 159:1477-1484.
129. Schattner, A., Cori, Y., Hahn, T., et al. No evidence for autoimmunity in schizophrenia. *J. Autoimmun.* 1996; 9:661-666.
130. Moises, H.W., Schindler, L., Leroux, M., et al. Decreased production of interferon alpha and interferon gamma in leucocyte cultures of schizophrenic patients. *Acta Psychiatr. Scand.* 1985; 72:45-50.
131. Arolt, V., Weitzsch, C., Wilke, I., et al. Production of interferon-gamma in families with multiple occurrence of schizophrenia. *Psychiatry Res.* 1997; 66:145-152.
132. Naidenova, N.N., Miriutova, T.L. and Vasil'eva, O.A. [The interferon system of patients with schizophrenia during therapy]. *Zh. Nevropatol. Psikhiatr. Im. S. S. Korsakova* 1988; 88:96-100.
133. Katila, H., Cantell, K. and Hirvonen, S. Production of interferon-alpha and gamma by leukocytes from patients with schizophrenia. *Schizophr. Res.* 1989; 2:361-365.
134. Johnson, H.M., Vassallo, T., Torres, B.A. Interleukin 2-mediated events in gamma-interferon production are calcium dependent at more than one site. *J. Immunol.* 1985; 2:967-970.

Chapter 16

CYTOKINE REGULATION IN AUTISM

Vijendra K. Singh
Department of Biology & Biotechnology Center, Utah State University, Logan, Utah

INTRODUCTION

Autism is a biological disorder of unknown cause and etiology. The disorder impacts the functioning of the central nervous system (CNS), manifesting devastating psychiatric as well as neurological problems in the affected person. Autism is commonly described as a neurodevelopmental disorder, which implies that the structure and function of the developing brain is impaired. Today, autism is known to affect more than one-half million American people, by and large children but a significant number of adults also. The diagnosis is made during early childhood before the age of 34 months but the affliction continues well into the adulthood, becoming a life-long disability.

Currently, autism is defined not by etiology or pathology but by the presence of a particular pattern of behavioral characteristics that follow a particular developmental course with evidence of delay or deviant development within the first three years of life. Most commonly, the autistic behaviors that characterize the disorder include "qualitative deficits" in four main categories: (i) deficits of developmental rates and/or sequences; (ii) deficits of responses to sensory stimuli; (iii) deficits of speech, language, and cognitive capacity; and (iv) deficits of ways in relating to other people. The clinical features of autism may vary with age and include disturbances of mood, thought, and behavior. Until recently, the incidence of autism was 4-5 in every 10,000 births but the number of autistic cases is rising sharply. Today, the incidence of autism is estimated to be 20-80 for every 10,000 births (1 in 500 or 1 in 125) in the United States, as well as in the United Kingdom (1).

Autism is a complex heterogeneous disorder, the cause of which is not known. Causally speaking, multiple factors have been implicated in the pathogenesis of the disorder (2,3). Although ten or more genes may be linked to autistic spectrum disorders, no single gene has so far been identified specifically for autism. Moreover, while the search for genetic factors is heavily supported, genetic factors are expected to account for no more than 10% of the entire autistic population. This means that approximately 90% of the autistic population will be explained by factors other than the genetic factors. They possibly include environmental factors, immune factors, neurochemical factors and other as yet unknown factors. Several years ago, we hypothesized that an autoimmune reaction to brain, in particular myelin sheath, may play an important role in the pathogenesis of autism (4). Furthermore, we postulated that an immune insult, secondary to a viral infection or vaccination, to developing myelin may cause "nicks" or small changes in the myelin sheath, modifying neuronal functions and thereby ultimately leading to life-long disturbances of higher mental functions such as learning and memory, social interaction and communication (4,5). This article concerns with immune factors especially cytokines and their role in the induction of autoimmunity in autism.

1. AUTOIMMUNE HYPOTHESIS IN AUTISM

Autoimmunity is an abnormal immune reaction in which the immune system becomes primed to react against body organs, and the net result is an autoimmune disease. Diseases are commonly labeled as autoimmune when the etiology and pathogenesis is not well known or established. However, only some diseases are 'classical' autoimmune diseases while others are only suspected of having autoimmune pathology. As recently reviewed by Ermann and Fathman (6), the clinical presentation of autoimmune diseases involves several factors: environmental factors, genetic link especially of immune response (IR) genes, immune abnormalities of thymus-derived immunoregulatory T cells, autoantibodies especially organ-specific autoantibodies, gender factor for greater prevalence in either males or females, hormonal factors, and response to immune therapy. Although the evidence is circumstantial, viruses are commonly considered as trigger agents for autoimmune diseases, which are also regarded as the commonest manifestations of chemical toxicity and/or vaccine immunizations. Autoimmune diseases are generally linked to IR genes, for example HLA, haplotypes or Gm isotypes, located on chromosome 6 in man. For last several years, we studied autism as an autoimmune disorder (2,4,5) and together with other researchers we have identified several autoimmune factors in people with autism. The autoimmune factors in autism are:

1. Autism shows microbial associations of certain viruses such as measles, rubella and CMV (7-9).

2. Autism displays increased frequency of immune response (IR) genes, for example HLA, C4B null allele, extended haplotype B44-SC30-DR4 and hypervariable region of HLA-DRb1 (10-12).
3. Autistic patients have impairments of cellular and humoral immunity: decrease of IgA; increase of IgG3, antinuclear antibodies and immune complexes; decrease of lymphocyte count, CD4+ T helper cells and natural killer (NK) cells; and suppressed cellular immunity as reflected by decreased mitogen-induced lymphocyte stimulation and reduced NK cell activity (13-19).
4. Autistic children show inappropriate cellular immunity to myelin basic protein, which induces an antigen-specific macrophage migration-inhibition factor (MMIF) (20).
5. Autism involves a gender factor, i.e., it affects males about four times more than females.
6. Autism often occurs in conjunction with a family history of autoimmune diseases, e.g., multiple sclerosis, rheumatoid, type II diabetes, arthritis, etc. (21,22).
7. Autism also involves hormonal factors, e.g., secretin, beta-endorphin, etc.
8. Autistic patients have organ-specific autoantibodies to brain antigens such as myelin basic protein (MBP), neuron-axon filament proteins (NAFP), serotonin receptor proteins (SRP), galactocerebrosides (GC), and caudate nucleus proteins (CP) (4,5,23-27).
9. Autistic patients show immune activation as reflected by T cell activation (28) and enhanced levels of circulating cytokines (29, 30).
10. Autistic patients respond well to autoimmune therapy with oral autoantigen (5), leukocyte-derived transfer factor (31), and intravenous immunoglobulin (19).

In autism, one of the salient features of autoimmune hypothesis is the brain-specific autoantibodies, in particular myelin basic protein (MBP) autoantibodies (4,5). MBP is the main protein constituent of a fatty coating called "myelin sheath" that protects nerve fibers in the brain. We recently showed that up to 85% of autistic children have autoantibodies specifically to MBP, but not to two other key constituents (2,3-cyclic nucleotide 3'-phosphohydrolase and galactocerebrosides) of the myelin sheath (5). This led us to postulate that an autoimmune response to MBP is involved in autism (2,5). Moreover, we think that autism in many cases may be explained by increased incidence of autoimmunity in autistic children. Since viruses are commonly associated with autoimmune disease, we hypothesized that a virus-induced autoimmune response may play a causal role in autism. Indeed, the rapidly accumulating evidence supports an etiological link for measles virus in autoimmunity to brain in autism (32). To that end, we recently measured virus serology for measles virus (MV), mumps virus (MuV), rubella virus (RV), cytomegalovirus (CMV), and human herpes virus-6 (HHV-6). We found that autistic children harbor significantly higher than normal levels of antibodies to measles virus. However, the level of antibodies to other four viruses did not significantly differ between autistic and normal children. Furthermore, we found that a vast majority (greater than 90%) of autistic children showed a serological (antibody) association between measles virus and MBP autoantibodies, i.e., the higher the measles virus antibody level the greater the chance of MBP autoantibody. However, this association was not found for other viruses and/or other brain autoantibodies that we studied (33). This observation led us to speculate that a measles infection may be etiologically linked to autoimmunity in autism (5,32,33).

The source of measles virus in children with autism is not known. Autistic children do not have a history of a German measles rash hence a wild type measles infection is rather unlikely. However, there is another possibility of an "atypical" measles infection in the absence of a typical measles rash. Such an infection could occur either by a variant measles infection or it could be acquired from immunization with measles-mumps-rubella (MMR) vaccine. Indeed, an atypical measles infection in the absence of a rash and unusual neurological symptoms was recently described to suggest the existence of a variant measles virus in man (34). In our own laboratory, we have recently gathered evidence that many autistic children have abnormal or inappropriate antibodies to MMR vaccine but not to other vaccines such as diptheria-tetanus-pertussis (DPT), diptheria-tetanus (DT) or hepatitis B (Hep B), and these antibodies are specifically directed against the measles subunit of this multivalent vaccine (35). Furthermore, there was a strong serological correlation (greater than 90%) between MMR antibodies and MBP autoantibodies, suggesting a causal association of MMR vaccine with autism or autistic regression that has been described post-MMR immunization in a considerable number of children (33,35). While more research is necessary on this topic, we speculate that an atypical measles infection may etiologically be linked to brain autoimmunity in autism. To that end, the studies of autoimmunity-producing cytokines are also quite relevant: (i) We previously reported that autistic children have significant increases of autoimmunity-specific cytokines such as interleukin-12 (IL-12) and interferon-gamma (IFN-γ) in favor of a Th-1 immune response (30); and (ii) measles vaccination with MMR vaccine essentially induces IFN-γ cytokine for Th-1 type of immune response (36), which could indirectly explain a potential connection between MMR and autism. Clearly, these findings are important for understanding a fundamental mechanism of autoimmunity in autism but more research is necessary to understand their precise role in the pathogenesis of the disorder.

Although there is virtually no research, there appears to be a connection between autism and myelin (5). It is derived from such observations as: (i) Autism is a neurodevelopmental disorder with an early-onset before the age of 36 months, a period when myelin formation is particularely sensitive to various stimuli; (ii) Autism involves a myelin defect based on brainstem auditory evoked potential (BAEP) measurements (37); (iii) An autoimmune reaction to myelin has been suggested in children with Landau-Kleffner syndrome (38), a disorder in which the early symptoms are very much autistic-like especially the deficits of auditory and language skills; (iv) delayed or incomplete myelinization in the corpus callosum (which is the largest myelinated area of the brain) has been suggested as the basis of auditory processing problems in some children with learning disabilities (39) and auditory processing deficits are found in autistic children also; (v) Autistic children respond abnormally to myelin basic protein (MBP) in a cellular immune response assay (20) and many have circulating autoantibodies to MBP (4,5); more importantly, the magnetic resonance imaging (MRI) shows brain abnormalities of diffused myelin, dysmyelination or delayed myelinization in autistic children (5); and, albeit anecdotal, the oral administration of Sphingolin (a natural product containing MBP autoantigen) results in clinical improvement of behavioral characteristics in autistic children (5). Thus the CNS myelin in developing brain appears to be a prime target of immune assault, which via MBP autoantibodies and/or antigen-primed T lymphocytes may lead to autoimmune pathology in autism (4,5). Myelin is a covering sheath surrounding the neuronal-axons to insulate them and to facilitate the conduction of nerve impulses. In central nervous system (CNS), the myelin is produced by oligodendrocytes whereas Schwann cells produce myelin in the peripheral nervous system. MBP is a major protein (30% of total protein) of the CNS myelin. The function of MBP is not well established except that it compacts immature myelin lamellae into mature myelin sheath. Myelination is essentially a postnatal event with rapid accumulation of myelin between 7 to 30 days after birth but it continues well into the adulthood.

2. CYTOKINE PROFILES IN AUTISM

Cytokine studies in autistic disorder are scarce, in part due to the lack of funding for this type of research in autism. Cytokine studies can be performed by three different approaches: (i) Cytokines can be measured in biological fluids such as serum, plasma or cerebrospinal fluid, which represents endogenously (or *in vivo*) produced circulating cytokines; (ii) Cytokines production can be studied by peripheral blood mononuclear cells (PBMNC) after mitogenic or antigenic stimulation *in vitro*; and (iii) Cytokine-specific mRNA expression can be measured in PBMNC after *in vitro* stimulation with mitogens. In our laboratory, we initially took the first approach because it represents an *in vivo* state, and measured circulating levels of cytokines in children with autism (29,30). As summarized in Table 1, the serum or plasma level of only three cytokines (IL-2, IL-12 and IFN-γ) was significantly elevated in autistic children. The serum level of six other cytokines such as IL-1, IL-4, IL-6, IL-10, IFN-α and TNF-α did not significantly differ between normal children and autistic children.

Cytokine	**Normal children (Mean±S.E.)**	**Autistic children (Mean±S.E.)**	**P value***
IL-1 (pg/ml)	96±40 (n=15)	151±164 (n=19)	0.222
IL-2 (Unit/ml)	5±7 (n=33)	39±85 (n=23)	0.028*
IL-4 (pg/ml)	12.4±4.1 (n=21)	19.2±11.7 (n=23)	0.598
IL-6 (pg/ml)	3.3±1.2 (n=16)	4.1±0.8 (n=15)	0.673
IL-10 (pg/ml)	6.43±4 (n=21)	8.26±5 (n=23)	0.782
IL-12 (pg/ml)	0.2±0.1 (n=16)	5.2±2.2 (n=16)	0.05*
IFN-γ (pg/ml)	0.5±0.3 (n=20)	15±6 (n=20)	0.04*
IFN-α (pg/ml)	274±108 (n=18)	318±111 (n=18)	0.832
TNF-α (pg/ml)	0.14±0.1 (n=18)	0.2±0.14 (n=18)	0.738

Table 1. Serum Level of Cytokines in Autistic Children

*The P value of ≤0.05 indicates a significant difference between two groups of subjects: normal children versus autistic children.

We also conducted a pilot study of IL-2, IL-6 and TNF production by PBMNC. We found that the IL-2 production was significantly higher in autistic children as compared to normal children (Fig. 1). Although the

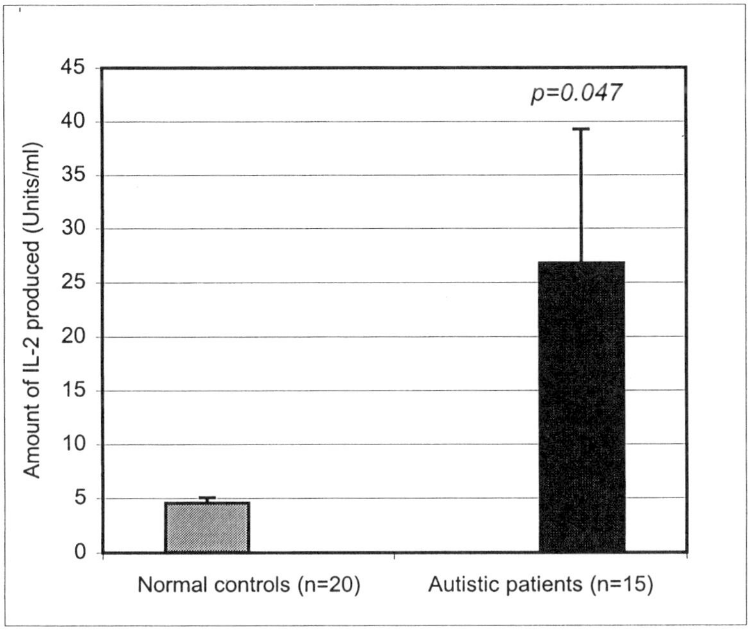

Figure 1.IL-2 Production by PBMNC of Autistic and Normal Children. Two million PBMNC of either autistic children or normal children were incubated in the absence or presence of 1% (w/v) phytohemagglutinin (PHA) for 48 hrs, followed by collection of supernatants that were tested in HT-2 cell bioassay according to Singh and Leu (53). The data, as given in terms of Units/ml of bioactivity in reference to a recombinant IL-2 standard, were significantly (p=0.047) different between autistic and normal children.

production of IL-6 (Fig. 2) and TNF (Fig. 3) by PBMNC of autistic children was slightly higher than the production of these cytokines by PBMNC of normal children, the difference between these two groups of subjects did not attain statistical significance. Our result of TNF production in autistic children is consistent with a previous report (40). More recently, two other groups of researchers took alternative approaches. They found that PBMNC of autistic children produce elevated levels of IL-12 and IFN-γ (41) or express higher than normal levels of mRNA for IFN-γ (42). Collectively, these observations demonstrate the existence of Th-1 type of the immune response in autistic children and that would also be consistent with an autoimmune pathology, simply because IL-2, IFN-γ and IL-12 are well known to induce autoimmune diseases (43,44).

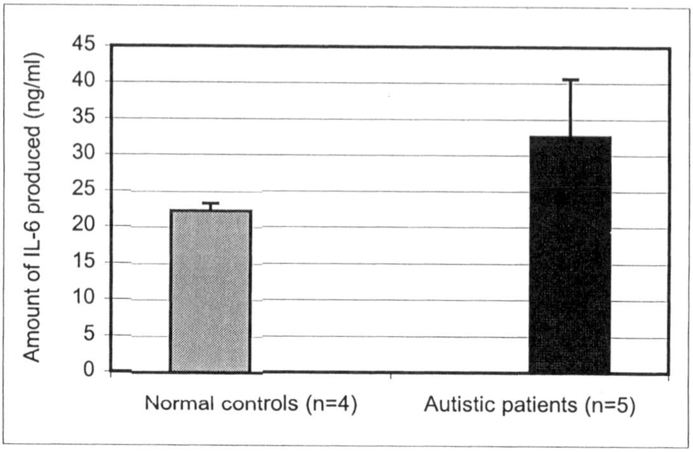

Figure 2.IL-6 Production of PBMNC of Autistic and Normal Children. One million PBMNC of either autistic or normal children were incubated in the absence or presence of 1 μg/ml of lipopolysaccharide (LPS) for 24 hrs, followed by collection of supernatants that were tested in MH60.BSF2 cell bioassay according to Leu and Singh (54). The data, as given in terms of ng/ml of bioactivity in reference to a recombinant IL-6 standard, were not significantly different between autistic and normal children.

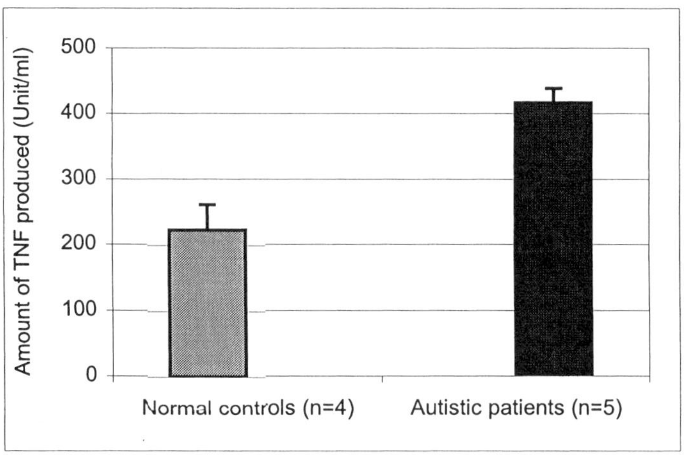

Figure 3.TNF Production of PBMNC of Autistic and Normal Children. Monocytes from 5 million PBMNC of either autistic or normal children were incubated in the absence or presence of 5 μg/ml of lipopolysaccharide (LPS) for 24 hrs, followed by collection of supernatants that were tested in L929 fibroblasts bioassay according to Hogan and Vogel (55). The data, as given in terms of Units/ml of bioactivity in reference to a recombinant TNF standard, were not significantly different between autistic and normal children.

Since IL-2, IL-12 and IFN-γ are well known to induce Th-1 type of immune response in autoimmune diseases and since their levels are selectively and significantly elevated in autistic children we postulated that a Th-1 type of immune response plays a pathogenic role in autism (30). Afterwards, other researchers (45,46) described the existence of Th-2 type of immune response in children with autistic spectrum disorders; however, the downside of these reports is that the researchers did not assay for actual cytokines. Instead, they only analyzed the phenotype of T cells (CD4 or CD8) in the blood – a measurement that does not specify the nature of either the Th-1 or Th-2 type of immune response. More recently, however, new evidence from three different lines of research supports the natural existence of Th-1 type of immune response in autistic children. First, as shown in Table 1, we did not find a significant increase of IL-4 and IL-10, which are the cytokines that characterize Th-2 cells (47-49). Secondly, the peripheral blood mononuclear cells of autistic children produce significantly higher levels of IL-12 and IFN-γ (41), which are the two key cytokines that characterize Th-1 cells. Thirdly, the IFN-γ mRNA expression in the peripheral blood mononuclear cells of autistic children is significantly higher (42), which is also consistent with Th-1 type of immune response in autistic children.

It is now generally believed that the blood-brain barrier is much more permeable to circulating autoantibodies, activated T cells and immune activation antigens (e.g. IL-12 and IFN-γ) than conventionally thought, enabling these molecules to cross the blood-brain barrier especially after IFN-γ-induced MHC class II expression (2,43). Furthermore, IL-12 specifically induces IFN-γ in Th-1 cells both *in vivo* and *in vitro* (50-52). This activity in autism might be reflected by the plasma increase of IFN-γ in autistic patients. Since brain is the affected organ in autism and IFN-γ is the most distinct inducer of MHC class II (HLA-DR, DP, and DQ) in brain, the increase of IFN-γ in autism may be associated with abnormal expression of class II. The increased expression of MHC class II antigens is a well-known risk factor in many autoimmune diseases including those of the nervous system (43). The class II activation permits enhanced permeability across the blood-brain barrier for circulating brain autoantibodies and/or antigen-activated T cells. We suggested that the brain MBP serves as a candidate autoantigen since it induces macrophage migration-inhibition factor (20), MBP autoantibodies (2,5), and lymphocyte proliferation (our unpublished data) in autistic children. Because IL-12 induces autoimmune diseases, its selective increase *in vivo* may provide a new clue to the autoimmune basis of autism (30). We believe that this is a novel finding for understanding a pathogenic role of autoimmunity in autism.

3. TH-1/TH-2 PARADIGM IN AUTISM

In recent years, one of the most important paradigms in immunology is the nature of a cellular immune response to a foreign antigen. Indeed, this is largely determined by the pattern of cytokines produced by the activated CD+ T helper cells. The naïve T helper cells (Th-o) cells differentiate into at least two types of polarized effector cells, namely Th-1 and Th-2 cells depending on the cytokine they synthesize (Fig. 4). The antigen presenting cells such as macrophages and/or monocytes produce a cytokine known as IL-12 that regulates the development of Th-1/Th-2 immune responses (50,51). As a general rule, upon antigenic stimulation, Th-1 cells are known to promote cellular and autoimmune immune responses whereas Th-2 cells are presumed to direct humoral immune responses (47-52). At the onset, foreign antigens activate antigen-presenting cells to produce IL-12 that via IFN-γ production promotes development of naïve helper cells (Th-0) into Th-1 cells. IL-12 is also known as a growth factor for NK cells and T cells as it induces NK cell- and T cell-mediated cytotoxicity. More importantly, IL-12 plays a key role in the induction of autoimmune diseases (43).

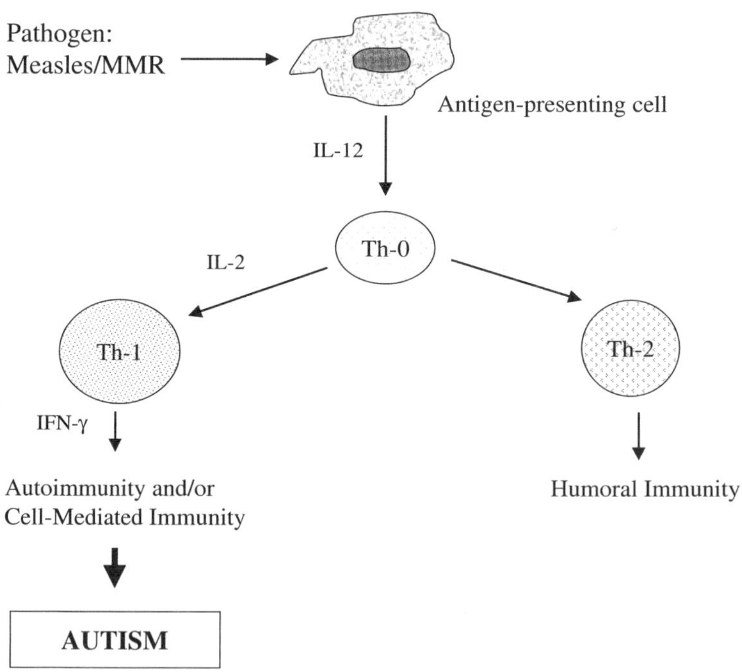

Figure 4.Diagrammatic Relationship between Cytokines and Autoimmunity in Autism

Regarding the pathogenesis of immune-mediated diseases, immune activation is one of the primary events in autoimmunity, inflammation and viral infections. Immune activation leads to spontaneous proliferation of peripheral blood mononuclear cells, increased expression of activation markers on peripheral blood mononuclear cells, and increased accumulation of blood mononuclear cell-derived soluble antigens, mainly cytokines, cytokine receptors, and adhesion molecules. Based on these considerations, immune activation occurs naturally in autistic children because they have elevated levels of immune activation antigens such as sCD8, IL-2, IL-12 and IFN-γ (28,30) and their blood contains activated T cells (28). Thus it is reasonable to infer that the increase of IL-12 in autistic children points to antigenic stimulation of Th-1 cells, which via INF-γ may induce autoimmunity as diagrammatically shown in Figure 4. The IL-12 cytokine selectively promotes the development of Th-1 cells (50,51) and Th-1 cells initiate the pathogenesis of organ-specific autoimmune diseases (43).

4. RESEARCH METHODS IN OUR STUDIES

We studied two groups of subjects: Autistic children and normal children. As far as possible, the subjects were matched for age and gender and no one was denied participation in the study because of race, age or gender factors, except those beyond the scope of our research study. The clinical diagnosis of autism was essentially according to the Diagnostic and Statistical Manual of Mental Disorders, Fourth Edition (DSM-IV) and only those with a diagnosis of autism were included; this excluded children with Asperger's syndrome, childhood disintegrative disorder and related pervasive developmental disorders (PDD). The normal children were those having firm normal physical health without any sign of brain disease or mental illness or any other known medical condition. Before drawing blood samples of human subjects, we obtained permission of the Institutional Review Board (IRB) at Utah State University and formerly at the University of Michigan. By and large, we employed previously collected serum samples that were stored frozen in a freezer at –20°C while keeping the freezing-thawing cycle to a minimum.

For cytokine studies, we measured serum levels of interleukin-1 (IL-1), interleukin-2 (IL-2), interleukin-4 (IL-4), interleukin-6 (IL-6), interleukin-10 (IL-10), interleukin-12 (IL-12), interferon-gamma (IFN-γ), interferon-alpha (IFN-α), and tumor necrosis factor-alpha (TNF-α). The cytokine levels were measured according to our published reports (29,30), but by following the strict instructions of the manufacturer of cytokine EIA kits that were purchased commercially. The IL-4 and IL-10 EIA kits were purchased from

Immunotech, a Beckman Coulter Company (Miami, Florida), whereas all other kits were purchased from sources as previously described (29,30).

Furthermore we studied production of three cytokines, namely IL-2, IL-6 and TNF. The assay procedure for IL-2 production by PBMNC was essentially according to our published report (53). The IL-6 production by PBMNC was performed with MH60.BSF2 cell bioassay as described previously (54). The TNF production by monocytes, separated from PBMNC by adherance to plastic surface, was measured by following a cell-cytotoxicity assay of Hogan and Vogel (55).

5. CONCLUSION

The rapidly accumulating evidence suggests that autism may be an autoimmune disorder that involves organ-specific autoimmune reaction to brain. In autistic brain, the most likely target of autoimmunity is the myelin sheath, but other neural targets such as brain serotonin receptor should be investigated. Currently, the cytokine studies in autism are scarce but nevertheless they yielded some very important information for a causative role of autoimmunity in the disorder. This specifically relates to IL-12 cytokine, which via Th-1 immune response plays a key role in inducing autoimmune diseases. Indeed, this also appears to be the case in autism. The factors that regulate cytokine production and induce shifts in Th-1/Th-2 paradigm are not known but some possibilities are environmental factors such as chemical/metal toxicity, immunization factors such as MMR vaccine, dietary factory such as fat-rich and/or carbohydrate-rich diets or hormonal factors such as corticosteroids and catecholamines. Clearly, the rationale design of cytokine-based therapy, in particular involving IL-12 as a target, would be highly desirable for autism. Immune therapy at IL-12 level is already being developed for other autoimmune diseases (2,56). We believe that autoimmunity, especially at the cytokine level, should be exploited as a viable target of drug development for autism. In a much broader sense, the cytokine and autoimmunity research will have a global impact, especially in terms of treatment for autism in America, as well as in other countries worldwide. Considering an estimated population of 500,000 autistic people in the United States, it is tempting to suggest that the autoimmunity treatment may benefit between 250,000 - 350,000 Americans today. Simply put, there is tremendous hope for autistic people through cytokine and autoimmunity research.

ACKNOWLEDGMENTS

The author would like to extend sincere thanks to Courtney Nelson, Crystal Brock, Sheren Lin, and Elizabeth Newell for their assistance in part with laboratory studies described herein. He greatly appreciates and thanks the participation of families in the study. The study was supported without any conflict of interest by private contributions from the James Dougherty Jr Foundation, Forrest Lattner Jr Foundation, Autism Autoimmunity Project, Autism Research Institute, BHARE Foundation, Yorio family, Unanue family, and several other families.

REFERENCES

1. Rimland, B. Huge increase of autism incidence reported in California; autism cluster investigated in Brick, New Jersey. *Autism Research Review International* 1999; 3:1.
2. Singh, V.K. Immunotherapy for brain disease and mental illnesses. *Progr. Drug Res.* 1997; 48:129-146.
3. Trottier, G., Srivastava, L. and Walker, C.D. Etiology and infantile autism: a review of recent advances in genetic and neurobiological research. *J. Psychiat. Neurosci.* 1999; 24:95-96.
4. Singh, V.K., Warren, R.P., Odell, J.D., et al. Antibodies to myelin basic protein in children with autistic disorder. *Brain Beha. Immun.* 1993; 7:97-103.
5. Singh, V.K., Neuro-immunopathogenesis in autism. In *New Foundation of Biology*, Istvan, Berczi, Reginald, M., Gorczynski, R.M, ed. The Netherlands, Elsevier B.V. Publishing, 2001; 1:443-454.
6. Ermann, J.., Fathman, C.G. Autoimmune diseases: genes, bugs and failed regulation. *Nature Immunol.* 2001; 2:759-761.
7. Chess, S., Fernandez, P. and Korn, S. Behavioral consequences of congenital rubella. *J. Pediatrics* 1978; 93:699-703.
8. Stubbs, E.G., Ash, E., Williams, P.S. Autism and congenital cytomegalovirus. *J. Aut. Dev. Disord.* 1984; 14:183-89.
9. Ivarsson, S.A., Bjerre, I., Vegfors, P., et al. Autism as one of several disabilities in two children with congenital cytomegalovirus infection. *Neuropediatrics* 1989; 21:102-103.
10. Warren, R.P., Singh, V.K., Cole, P., et al. Increased frequency of the null allele at the complement C4B locus in autism. *Clin. Exptl. Immunol.* 1991; 83:438-440.
11. Warren, R.P., Singh, V.K., Averett, R.E., et al. Immunogenetic studies in autism and related disorders. *Mol. Chem. Neuropathol.* 1996; 28:77-81.
12. Warren, R.P., Odell, J.D., Warren, W.L., et al. Strong association of the third hypervariable region of HLA-DRb1 with autism. *J. Neuroimmunol.* 1996; 67:97-102.
13. Stubbs, E.G., Crawford, M.L., Burger, D.R., et al. Depressed lymphocyte responsiveness in autistic children. *J. Aut. Childh. Schizophr.* 1977; 7:49-55.
14. Warren, R.P., Foster, A., Margaretten. N.C., et al. Immune abnormalities in patients with autism. *J. Aut. Develop. Disord.* 1986; 16:189-197.
15. Warren, R.P., Foster, A. and Margaretten, N.C. Reduced natural killer cell activity in autism. *J. Am. Acad. Child Adol. Psychol.* 1987; 26:333-35.
16. Singh, V.K., Fudenberg, H.H., Emerson, D., et al. Immunodiagnosis and immunotherapy in autistic children. *Ann. N.Y. Acad. Sci.* 1988; 540:602-604.

17. Yonk, L.J., Warren, R.P., Burger, R.A., et al. CD4+ helper T cell depression in autism. *Immunol. Lett.* 1990; 25:341-346.
18. Menage, P., Thibault, G., Barthelemy, C., et al. CD4+ CD45RA+ T lymphocyte deficiency in autistic children: effect of a pyridoxine-magnesium treatment. *Brain Dysfunct.* 1992; 5:326-333.
19. Gupta, S., Aggarwal, S. and Heads, C. Dysregulated immune system in children with autism: Beneficial effects of intravenous immune globulin on autistic characteristics. *J Aut Develop Dis* 1996; 26:439-452.
20. Weizman, A., Weizman, R., Szekely, G.A., et al. Abnormal immune response to brain tissue antigen in the syndrome of autism. *Am. J. Psychiat.* 1982; 139:1462-1465.
21. Money, J., Bobrow, N.A. and Clarke, F.C. Autism and autoimmune disease: a family study. *J. Aut. Child Schizo.* 1971; 1:146-160.
22. Comi, A.M., Zimmerman, A.W., Frye, V.H., et al. Familial clustering of autoimmune disorders and evaluation of medical risk factors in autism. *J. Child Neurol.* 1999; 14:388-394.
23. Singh, V.K., Warren, R.P., Averett, R., et al. Circulating autoantibodies to neuronal and glial filament proteins in autism. *Pediatric Neurology* 1997; 16:88-90.
24. Todd, R.D., Ciaranello, R.D. Demonstration of inter- and intraspecies differences in serotonin binding sites by antibodies from an autistic child. *Proc. Natl. Acad. Sci.* USA 1985; 82:612-616.
25. Singh, V.K, Singh, E.A. and Warren, R.P. Hyperserotoninemia and serotonin receptor antibodies in children with autism but not mental retardation. *Biol. Psychiat.* 1997; 41:753-755.
26. Connelly, A.M., Chez, M.G., Pestronk, A., et al. Serum antibodies to brain in Landau-Kleffner variant, autism, and other neurological disorders. *J. Pediatrics* 1999; 134:607-613.
27. Singh, V.K., Wu, E. and Tsui, B. Autoantibodies to neural proteins of caudate nucleus in autism. Proceedings of the International Meeting for Autism Research; San Diego, California, November 9-10, 2001
28. Plioplys, A.V., Greaves, A., Kazemi, K., et al. Lymphocyte function in autism and Rett syndrome. *Neuropsychobiology* 1994; 29:12-16.
29. Singh, V.K., Warren, R.P., Odell, J.D., et al. Changes of soluble interleukin-2, interleukin-2-receptor, T8 antigen, and interleukin-1 in the serum of autistic children. *Clin. Immunol. Immunopath.* 1991; 61:448-455.
30. Singh, V.K. Plasma increase of interleukin-12 and interferon-γ: pathological significance in autism. *J. Neuroimmunol.* 1996; 66:143-145.
31. Fudenberg, H.H. Dialyzable lymphocyte extract (DlyE) in infantile onset autism: A pilot study. *Biotherapy* 1996; 9:143-147.
32. Singh, V.K., Lin, S.Y. and Yang, V.C. Serological association of measles virus and human herpesvirus-6 with brain autoantibodies in autism. *Clin. Immunol. Immunopathol.* 1998; 89:105-108.
33. Singh, V.K. Immunological link between measles-mumps-rubella and autoimmune autism. Proceedings of the 11th International Congress for Immunology (ICI); 2001; A9; Stockholm, Sweden.
34. Madhur. G. Indian scientists warn of "mutant measles" virus. *Brit. Med. J.* 2001; 322:693.
35. Singh, V.K. and Nelson, C. Abnormal measles serology and autoimmunity in autistic children. *J. Allergy Clin. Immunol.* (Supplement), 2002; 109: Abs. # 702.
36. Pabst, H.F., Spady, D.W., Carson, MM, et al. Kinetics of immunologic responses after primary MMR vaccination. *Vaccine* 1997; 15:10-14.

37. McClelland, R.J., Eyre, D.G., Watson, D., et al. Central conduction time in childhood autism. *Brit. J. Psychiat.* 1992; 160:659-663.
38. Nevsimalova, S., Taubaerova, A., Doutlik, S., et al. A role of autoimmunity in the etiopathogenesis of Landau-Kleffner syndrome? *Brain Dev.* 1992; 14:342-350.
39. Musiek, F.E., Gollegly, K.M. and Baran, J.A. Myelination of the corpus callosum and autditory processing problems in children: theoretical and clinical correlates. *Seminar Hearing* 1989; 5:231-241.
40. Gupta, S., Rimland, B. and Shilling, P.D. Pentoxifylline: brief review and rationale for its possible use in the treatment of autism. *J. Child Neurol.* 1996; 11:501-504.
41. Jyonouchi, H., Sun, S. and Le, H. Innate and adaptive immune responses in children with regression autism: evaluation of the effects of environmental factors including vaccination. *J. Allergy Clin. Immunol.* (Supplment) 2001; 107:Part 2, Abstract #897.
42. Hu, Y. Altered T cell-mediated immunity and infectious factors in autism. Doctor of Philosophy dissertation, Utah State University, Logan, Utah, 2000.
43. Liblau, R.S., Singer, S.M. and McDevitt, H.O. Th1 and Th2 CD4+ cells in the pathogenesis of organ-specific diseases. *Immunol. Today* 1995; 16:34-38.
44. Singh, V.K., Mehrotra, S. and Agarwal, S.S. The paradigm of Th1 and Th2 cytokines: its relevance to autoimmunity and allergy. *Immunol. Res.* 1999; 20:147-161.
45. Gupta, S., Aggarwal, S., Rashanravan, B., et al. Th1- and Th2-like cytokines in CD4+ and CD8+ cells in autism. *J. Neuroimmunol.* 1998; 85:106-109.
46. Furlano, R.I., Anthony, A., Day, R., et al. Colonic CD8 and γδ T-cell infiltration with epithelial damage in children with autism. *J. Pediatrics* 2001; 138:366-372.
47. Mosmann, T.R. and Sad, S. The expanding universe of T cell subsets: Th1, Th2 and more. *Immunol. Today* 1996; 17:138-146.
48. Romagnani, S. Human Th1 and Th2 subsets. *Immunol. Today* 1997; 18:256-257.
49. Abbas, A.K., Murphy, K.M. and Sher, A. Functional diversity of helper T lymphocytes. *Nature* 1996; 383:787-793.
50. Trinchieri, G. Interleukin-12 and its role in the generation of T_H1 cells. *Immunol. Today* 1993; 14:335-338.
51. McNight, A.J., Zimmer, G.J., Fogelman, I., et al. Effects of IL-12 on helper T cell-dependent immune responses in vivo. *J. Immunol.* 1994; 152:2172-2179.
52. Das, G., Sheridan, S. and Janeway Jr, C.A. The source of early IFN-γ that plays a role in Th1 priming. *J. Immunol.* 2001; 167:2004-2010.
53. Singh, V.K. and Leu, S.J.C. Enhancing effect of corticotropin-releasing neurohormone on production of interleukin-1 and interleukin-2. *Neurosci. Lett.* 1990; 120:151-154.
54. Leu, S.C. and Singh, V.K., Stimulation of interleukin-6 production by corticotropin-releasing factor. *Cell. Immunol.* 1992; 143:220-227.
55. Hogan, M.M. and Vogel, S.N., Measurement of tumor necrosis factor α and β. *Current Protocols in Immunology* 1991; Unit 6.10.1-6.10.5.
56. Ichikawa, M., Koh, C.S., Inoue, A., et al. Anti-IL-12 antibody prevents the development and progression of multiple sclerosis-like relapsing-remitting demyelinating disease in NOD mice induce with myelin oligodendrocyte glycoprotein peptide. *J. Neuroimmunol.* 2000; 102:56-66.

Chapter 17

CYTOKINE REGULATION IN ALZHEIMER'S DISEASE

Robert E. Mrak
University of Arkansas for Medical Sciences, Little Rock, Arkansas

1. INTRODUCTION

The idea that Alzheimer's disease involves an 'inflammation-like' process, involving orchestrated cytokine and cellular responses that culminate in neuronal injury and destruction, has become so widely accepted that it is now difficult to remember how radical this idea was less than a decade ago. Alzheimer's disease, at that time, was seen as a 'neuro-degenerative' disease, in which accumulations of Aβ protein damaged neurons through direct and, we thought, soon-to-be elucidated, toxic mechanisms. Any associated glial cells were considered to be passive or 'reactive' late-comers, of no significance toward understanding the disease or its progression.

Griffin et al. were the first to challenge these assumptions, in a landmark 1989 paper that described intense microglial immunoreactivity for the potent immunomodulatory and pro-inflammatory cytokine interleukin-1 (IL-1) in tissue sections from brains of Alzheimer patients (1). These authors also found, at that time, increased astrocytic expression of S100B (formerly known as S100β), an important astrocyte-derived, secreted, neurotrophic molecule that was at that time little appreciated for its cytokine-like activities. Within a few years, Alzheimer brain had been shown to also contain elevated levels of the cytokines interleukin-6 (2), transforming growth factor β_1 (3), interferon α (4,5), and interleukins-2 and -3 (6). More recently, increased concentrations of other cytokines, growth factors, and trophic molecules, or of their associated synthesizing enzymes or receptors, have been described for Alzheimer brain. These include heparin binding growth-associated molecule (7), nitric oxide synthase (8), macrophage-colony stimulating factor (9), interleukin-8 receptor B (10), monocyte chemoattractant protein-1 (11), the β-chemokine receptors CCR3 and CCR5 and macrophage inflammatory protein-1β (12), fibroblast growth factor-9

(13), vascular endothelial growth factor (14), and the interferon γ-inducible chemokine IP-10 (15). In addition to these 'classical' cytokines, the α-secretase proteolytic fragment of the Alzheimer-associated Aβ precursor protein, sAPPα, has been shown to manifest important intercellular signaling functions (16-18). This raises the possibility that altered (or β-secretase-mediated) cleavage of the Aβ precursor protein in Alzheimer's disease may, in addition to generating pathogenic Aβ peptides, disturb normal brain homeostasis through loss of sAPPα cytokine functions (19).

Genetic studies have recently provided substantial new support for causal associations between the overexpression of cytokines, especially IL-1, and Alzheimer's disease. For the IL-1A gene (that encodes IL-1α), there is a 2-to-3-fold increased risk for AD associated with homozygosity for a common polymorphism located -889 bp upstream to the transcription initiation site, within a negative regulatory element. When this is coupled with homozygosity for a polymorphism in exon 5 of the IL-1B gene (that encodes IL-1β), this increased risk rises to 10-fold (20,21). These associations have now been confirmed by several other studies (22-24) although one group found no association (25). Two groups have now found increased risk for Alzheimer's disease associated with certain polymorphisms in the tumor necrosis factor gene (26,27), and a modest increase in risk has been reported for a -509 polymorphism in the transforming growth factor-β_1 gene (28). There is, in addition, one report of increased risk for Alzheimer's disease associated with a polymorphism in the 3' flanking region of the interleukin-6 gene (29), but a second study did not confirm this (30).

These studies and other studies of cytokines in Alzheimer's disease, together with considerable experimental work implicating various cytokines in the pathophysiological processes that underlie Alzheimer's disease, have shaped a new understanding of Alzheimer pathogenesis. An understanding in which activated, 'inflammatory' glial elements, and glia-derived cytokines, are key initiating and orchestrating elements in a self-propagating cascade of molecular and cellular events that ultimately proves injurious and toxic to neurons and their processes in Alzheimer brain.

2. CYTOKINE LINKS TO ALZHEIMER PATHOLOGY: INTERLEUKIN-1

The demonstration of microglial IL-1 overexpression in Alzheimer brain in 1989 (1) coincided with the discovery that IL-1 regulates the synthesis of the Aβ precursor protein (APP) from which the pathologic and diagnostic amyloid plaques of Alzheimer's disease originate (31). These twin findings suggested immediately that glial cytokine overexpression and actions could be important pathogenic events in the initiation or progression of the disease. Further work confirmed trophic effects of IL-1 on APP expression in neurons (32) and also showed IL-1 effects on the processing of APP and secretion of APP fragments (33). IL-1 enhances neuronal acetylcholinesterase expression and activity, possibly explaining in part the cholinergic dysfunction characteristic of this disease (34). In addition to these neurotrophic actions, IL-1 has autocrine, gliotrophic actions that include promotion of microglial activation and further promotion of IL-1 expression (34-36). IL-1 activates astrocytes, and promotes astrocytic expression of α_1-antichymotrypsin (37) – a protein that is found co-deposited with Aβ protein in the amyloid plaques of Alzheimer's disease – and of S100B (38), a neurotrophic and gliotrophic cytokine that is itself of potential importance in Alzheimer's disease. Recent demonstrations of genetic associations between certain IL-1 gene polymorphisms and increased risk for Alzheimer's disease have further strengthened the idea that IL-1 plays an important and causal role in this disease (39,40).

The functions of IL-1 suggest that this cytokine plays a key role in the pathogenesis of Alzheimer's disease, and there is now abundant evidence for this from studies of Alzheimer's disease itself. Brain tissue levels of IL-1 are markedly elevated in Alzheimer's disease, and the numbers of IL-1-immunoreactive microglia are markedly increased in tissue sections of Alzheimer brain (1). These microglia show a close association with Aβ plaques, and the distribution of these microglia across brain regions reflects the distribution of Aβ plaque pathology (41). Even more interesting is the observation that the pattern of Aβ distribution within cerebral cortical layers in Alzheimer brain mirrors the distribution of microglia within cerebral cortical layers in brains of normal controls (42), suggesting that the normal distribution of microglia within the brain determines in part the distribution of Aβ plaques in Alzheimer's disease.

Microglia overexpressing IL-1 have been shown to be present at all stages of Aβ plaque formation and progression in Alzheimer's disease, and are also found associated with Aβ plaques in transgenic animal models of Alzheimer's disease (43,44). In Alzheimer's disease, these microglia show patterns of association with different plaque types that correlate with evidence of neuritic injury (45). The diagnostic neuritic amyloid plaques of Alzheimer's disease are thought to evolve from early, 'pre-amyloid' or 'diffuse' deposits of Aβ peptide that are not associated with discernible evidence of neuritic or neuronal damage. These pre-amyloid deposits are found in Alzheimer brain concurrent with more 'advanced' or 'classical' plaques that show, in addition, clumping or 'condensation' of Aβ peptides to form histologically identifiable 'congophilic' amyloid. Within these latter plaques there are abnormal, swollen, 'dystrophic' neurites reflecting local neurotoxicity associated with the deposit. Further condensation of Aβ yields a dense core of congophilic amyloid, and this progression is accompanied by decreasing evidence of neuritic injury, culminating in a putative 'end-stage' or 'burnt-out' plaque that consists solely of dense compact amyloid, without associated dystrophic neurites.

Aβ promotes microglial production of IL-1 in vitro (46), and even the early, diffuse pre-amyloid deposits of Alzheimer's disease contain activated microglia, overexpressing IL-1 (45). With plaque progression, these plaque-associated, IL-1-overexpressing microglia increase in number and state of activation, but then their numbers decline as the volume of non-compacted amyloid declines and as the degree of associated neritic injury declines. End stage plaques are devoid of non-compacted amyloid, devoid of microglia and devoid of damaged neurites. It is noteworthy that diffuse 'pre-amyloid' deposits of Aβ peptide are sometimes found in brains of elderly, non-demented individuals, without clinical or pathological evidence of Alzheimer's disease. These apparently benign deposits do not have associated activated microglia (47), in contrast to the histologically similar deposits in Alzheimer brain. These observations collectively suggest that there is some chemical or physical difference between the early, diffuse 'pre-amyloid' deposits in Alzheimer's disease and the similar-appearing plaques that are sometimes seen in cognitively normal, aged individuals, and that this difference renders the diffuse deposits of Alzheimer's disease both immunogenic and pathogenic. They further suggest that microglia-derived IL-1 may be an important factor in neuritic pathology and neurotoxicity associated with plaque progression in Alzheimer's disease.

IL-1 may have direct toxic effects on cortical neurons (48), but IL-1-associated neurotoxicity may also arise indirectly from a number of 'downstream' cascade actions. As noted, IL-1 promotes astrocytic expression of the neurite growth-promoting factor S100B, which in turn promotes potentially toxic increases in intraneuronal free calcium concentration (49). S100B also induces astrocytic expression of nitric oxide synthase with release of potentially neurotoxic nitric oxide (50), and increased nitric oxide production is seen in response to IL-1 in vivo (51).

Evidence of progressive neurotoxicity can be found in the reactions of neurons found within or adjacent to Aβ plaques. These neurons frequently show evidence of DNA damage, demonstrated using the TUNEL technique (52). Furthermore, the extent of this injury increases progressively with plaque stage and, presumably, with continued exposure to locally elevated levels of IL-1. End-stage plaques show markedly diminished numbers of plaque-associated neurons, and all of those neurons label with TUNEL, suggesting that there is also neuronal loss during plaque evolution (52). This is an important observation, as such Aβ-plaque-associated neuronal loss may far exceed that associated with the other major form of neuropathological change in Alzheimer's disease – the neurofibrillary tangle – and may be the major proximate cause of neurodegeneration and cognitive decline in this disease. IL-1 may also play a role in neurofibrillary tangle pathology in Alzheimer's disease, as microglia overexpressing IL-1 also show progressive association with neurons bearing neurofibrillary tangles (53), and intrathecal IL-1 levels correlate with those of the tangle-associated protein tau (54). Indeed, IL-1 has been shown to promote expression of both neurofilament proteins and of tau both in vitro (55) and in vivo (56), suggesting that microglia-derived IL-1 may be important in tangle formation and progression.

In summary, a large body of data derived from pathological analysis of Alzheimer brain, from genetic studies, from observations in vitro, and from experimental animal studies implicate interleukin-1 as a key initiating and orchestrating cytokine in Alzheimer's disease. According to these ideas, a *cytokine cycle* (57) of IL-1-driven cascades forms a glia-derived brain 'inflammatory' response that drives plaque progression (and possibly tangle progression) in Alzheimer's disease, and that underlies progressive neurotoxicity and loss associated with disease progression (40,57,58).

3. CYTOKINE LINKS TO ALZHEIMER PATHOLOGY: S100β

A second intercellular signaling molecule that was identified early in Alzheimer lesions is S100B (1). S100B was first described as a calcium-binding protein, and is now recognized as a member of a large family of such proteins (59). However, unlike other members of this family, S100B is secreted from cells (60,61) and has important intercellular signaling (i.e., cytokine) functions that are demonstrable at nanomolar concentrations. For instance, S100B promotes neuronal survival and neurite outgrowth (62,63), enhances neuronal survival during development and after injury (64), and stimulates astrocytic proliferation in vitro (65). S100B has been shown to act on a variety of central nervous system cell types to increase intracellular free calcium levels through a mechanism that involves activation of phospholipase C and formation of IP3 (49). Such S100B-mediated increases in free calcium levels are toxic in cultured PC-12 neuroblastoma cells (66).

S100B is overexpressed by activated astrocytes associated with the Aβ plaques of Alzheimer's disease (1,67,68), and tissue levels of biologically active, homodimeric S100B are markedly increased in Alzheimer brain (69). Astrocyte S100B expression is promoted by IL-1, and there are correlations between microglial IL-1 overexpression and astrocytic S100B overexpression in Alzheimer's disease that are consistent with a causal association. The activated astrocytes in Alzheimer brain are frequently found at the periphery of Aβ plaques, where they envelope the amyloid deposits with thick processes and extend finer processes deep into the lesion core (70,68). Moreover, the distribution of these astrocytes across brain regions (71), like the distribution of IL-1-overexpressing microglia (41), mirrors recognized distribution patterns for Aβ plaques (72).

S100B-overexpressing astrocytes, like IL-1-overexpressing microglia, show characteristic patterns of association with different stages of Aβ plaque formation. These patterns mirror the patterns seen with microglia, and suggest a concerted waxing and waning of an 'inflammation-like' process in response to the Aβ deposits (67). Thus activated, S100B-overexpressing astrocytes are found in most early 'pre-amyloid' deposits in Alzheimer's disease, and are found in greater numbers and with more prominent S100B overexpression in early neuritic plaques lacking dense amyloid cores. The numbers of these astrocytes correlates well with the degree of neuritic pathology in all types of neuritic plaques, consistent with a role for S100B's neurite-growth-promoting activity in the genesis of neuritic overgrowth and injury in Aβ plaques. End-stage, 'burnt out' Aβ plaques are devoid of activated astrocytes, as they are devoid injured neurites and devoid of associated microglia (67). Activated, S100B-overexpressing astrocytes also show a progressive association with neurons at various stages of neurofibrillary tangle formation that mirrors, but lags slightly behind, the progressive association of activated, IL-1-overexpressing microglia with these same neurons (53). This is consistent with the trophic and activating effects of IL-1 on astrocytes, and is also consistent with a role for S100B-derived (i.e., calcium-induced) neurotoxicity (66) in the progressive neuronal injury that accompanies tangle formation (53).

In summary, S100B is an important intercellular signaling molecule that is overexpressed by astrocytes in response to stimulation by IL-1. S100B has several important actions, including promotion of neurite outgrowth and promotion of increased intracellular calcium levels in target cells. S100B overexpression in neuritic plaques of Alzheimer's disease appears to correlate with neuritic pathologic changes, and the toxic and trophic effects of S100B may be responsible in part for these changes (73).

4. THE ROLE OF TRANSFORMING GROWTH FACTOR Β IN ALZHEIMER'S DISEASE

Transforming growth factor β (TGFβ) is an important cytokine that manifests both pro-inflammatory and anti-inflammatory properties. TGFβ is expressed in brain by astrocytes, microglia, and oligodendrocytes (74). Expression of TGFβ, unlike that of IL-1, does not appear to be induced by Aβ protein (75), but TGFβ expression is induced by IL-1 (74), and brain tissue levels of TGFβ (76) and TGFβ mRNA (77,78) are increased in Alzheimer's disease. TGFβ immunoreactivity is found in neurons bearing neurofibrillary tangles and in some Aβ plaques (3,76,79). There appears to be some isoform specificity to these increases, with TGFβ1 preferentially overexpressed in dystrophic neurites of Aβ plaques, TGFβ3 overexpressed in Hirano bodies, and TGFβ2 showing more general expression in plaque neurites, tangles, and activated glia (79). Cerebrospinal fluid levels of TGFβ are also increased in Alzheimer's disease (80).

A number of observations suggest that TGFβ1 may be important in Aβ deposition, at least in cerebral blood vessels, in Alzheimer's disease. TGFβ1 acts on astrocytes to promote expression of mRNA for the Aβ precursor protein (81), and intraventricular infusion of Aβ protein, combined with neuropil injections of TGFβ1, produces Aβ-immunoreactive plaque-like deposits in the brains of elderly rats. In contrast, infusion of Aβ alone produces no discernible pathologic change (82). Transgenic mice overexpressing TGFβ1 in astrocytes develop Aβ deposits in cerebral vasculature and in meninges (78), along with small vessel sclerosis (83) and alterations in brain glucose utilization (84). Moreover, these changes are accelerated in double-transgenic mice overexpressing TGFβ1 and human APP (78). A most interesting observation is that these double-transgenic mice show an apparent shift in Aβ deposition from parenchymal plaques to cerebral blood vessels, relative to single-transgenic mice overexpressing only APP (85). This finding is supported by the parallel observation that human Alzheimer brain shows a positive correlation between tissue TGFβ1 mRNA levels and cerebral vascular Aβ deposition, and a negative correlation between tissue TGFβ1 mRNA levels and parenchymal Aβ deposition. These observations suggest that TGFβ1 might promote removal or redistribution of Aβ in Alzheimer brain, and this suggestion is further supported by the findings that double-transgenic, TGFβ1/APP mice show increased activation of microglia, and that TGFβ1 promotes microglial removal of Aβ in vitro (85). In humans, there is a TGFβ1 gene polymorphism that is associated with a small increase in gene transcriptional activity, and this polymorphism appears to confer modestly-increased risk for Alzheimer's disease (28).

In summary, TGFβ appears to have important modulating effects on the development and progression of Aβ pathology in Alzheimer's disease. TGFβ may promote vascular Aβ deposition, small vessel sclerosis, and consequent metabolic abnormalities in brain, but TGFβ also appears to have ameliorating effects on parenchymal Aβ deposition.

5. OTHER CYTOKINES IMPLICATED IN ALZHEIMER'S DISEASE

5.1 Interleukin-6

Interleukin-6 (IL-6) is a major pro-inflammatory cytokine, expressed by several cell types in brain, that can be induced by either IL-1 (35,36) or S100B (86). IL-6 manifests several functions of potential importance in Alzheimer's disease. Like IL-1, IL-6 promotes neuronal expression of neurofilaments and of tau protein (55) and of APP (87,88). Like IL-1, IL-6 promotes astrocytic expression of the Aβ plaque-associated molecule α_1-antichymotrypsin (89). IL-1-induced astrocytic IL-6 expression is markedly potentiated by aged (i.e., fibrillar) Aβ protein (90).

In Alzheimer's disease, IL-6 immunoreactivity is found within extracellular Aβ deposits (2). This phenomenon appears to be restricted to certain plaques (91), particularly early plaque types (92,93), suggesting a role for IL-6 in the initiation of plaque progression. The cellular origin of this extracellular immunoreactivity is not clear. One study found IL-6 overexpression by plaque-associated astrocytes in an animal model of Alzheimer's disease (43), but a second group was unable to reproduce this finding (44). In Alzheimer's disease itself, astrocytic IL-6 overexpression has not been reported but IL-6 immunoreactivity has been found in microglia and in neurofibrillary tangle-bearing neurons (94). Brain tissue levels of IL-6 are not increased in Alzheimer's disease (6,95), although increased levels of IL-6 mRNA have been reported (77,96). Cerebrospinal fluid IL-6 levels have been reported to be increased (97,98) normal (2,99,100), or even decreased (101) in Alzheimer's disease. One study found a correlation between cerebrospinal levels of IL-6 and of tau in Alzheimer patients (54). As noted above, the possibility of increased risk for Alzheimer's disease associated with certain IL-6 gene polymorphisms remains controversial (30,29).

5.2 Tumor Necrosis Factor α

Tumor necrosis factor α (TNFα) is a third major pro-inflammatory cytokine, which frequently manifests synergistic effects with IL-1 in inflammatory processes (102). TNFα is expressed by astrocytes and microglia (103), and this expression is promoted by Aβ protein (104) or by damaged or dying neurons (105). TNFα, in combination with interferon γ, promotes production of Aβ peptides in human neuronal and nonneuronal cells in vitro (106). One group found elevated cerebrospinal fluid TNFα levels in Alzheimer's disease (54), but a second group did not (98). Similarly, one study found TNFα overexpression by plaque-associated astrocytes in an animal model of Alzheimer's disease (43), but a second group was unable to reproduce this finding (44). As noted earlier, two groups have reported increased risk for Alzheimer's disease associated with certain polymorphisms in the TNF gene (26,27).

5.3 Secreted α fragments of the Aβ precursor protein

The Aβ precursor protein, APP, as the name suggests, was discovered as the molecule of origin for the Aβ peptides that accumulate in extracellular amyloid deposits in Alzheimer's disease. This transmembrane protein is cleaved by an α-secretase to yield a secreted α fragment, sAPPα. This cleavage occurs in the middle of the Aβ sequence and thus precludes formation of pathogenic Aβ peptides. A second cleavage pathway involves a β-secretase and, following further processing, yields the Aβ fragment. This pathway is active in Alzheimer's disease, and precludes generation of sAPPα fragments. Secreted sAPPα fragments have important trophic and signaling functions. They can induce proliferation of non-neuronal cells, and they promote survival and neurite outgrowth in neurons (107). These observations suggest the interesting possibility that loss of sAPPα activity, due to shunting of APP processing to the alternative β-secretase pathway, may be responsible in part for neuronal toxicity and loss in Alzheimer's disease. Another possible role for sAPPα in Alzheimer's disease is suggested by the observation that sAPPα promotes expression of IL-1 and inducible nitric oxide synthase in microglia, resulting in toxicity to co-cultured neurons (19). This effect is inhibited by the ε3 isoform of apolipoprotein E (the gene for which is protective against Alzheimer's disease), but not by the ε4 isoform (the gene for which is an established risk factor for Alzheimer's disease), suggesting that differential modulation of sAPPα functions may explain in part the protective and risk-enhancing effects of these apolipoprotein E alleles.

5.4 Macrophage colony-stimulating factor

Macrophage colony-stimulating factor (M-CSF) is released by neurons in response to Aβ protein (9). M-CSF dramatically augments Aβ-induced microglial production of IL-1, IL-6, and nitric oxide (108). In Alzheimer's disease, M-CSF immunoreactivity is found in neurons adjacent to Aβ deposits, and M-CSF levels are increased in cerebrospinal fluid (9). Transgenic mice, overexpressing a human mutant APP that is associated with familial Alzheimer's disease, show increased brain tissue levels of M-

CSF and M-CSF mRNA; and show M-CSF immunoreactivity in neuritic structures near Aβ plaques (109).

5.5 Chemokines

Chemokines (*chemo*tactic cyto*kines*) are a recently-recognized family of cytokines, the primary function of which is recruitment of leukocytes to inflammatory sites. Many, but not all, of these molecules have been detected in brain (110,111). In Alzheimer's disease, there is increased expression of the macrophage inflammatory protein MIP-1β and of the γ-interferon-inducible protein IP-10 by activated astrocytes, and of the chemokine receptors CCR3 and CCR5 by activated microglia, adjacent to Aβ deposits in Alzheimer's disease (12,15). Within the Aβ plaques of Alzheimer's disease, dystrophic neurites have been shown to express the IL-8 receptor CXCR2 (10). The monocyte chemoattractant protein-1 (MCP-1) is found in activated microglia and within neuritic Aβ plaques, but not in early plaque forms, in Alzheimer's disease (11). Aβ promotes production of MCP-1, of MIP-1α, of MIP-1β, and of IL-8 by monocytes and microglia (75,112,113); promotes expression of MCP-1 and of the T-cell attractant RANTES ("*Regulated upon Activation, Normal T cells, Expressed and Secreted*") by astrocytes (114); and promotes expression of IL-8 by human astrocytoma cells (90). These observations collectively suggest that chemokines may play a role in the pathogenesis of Alzheimer's disease (111).

5.6 Other cytokines and growth factors

Enhanced immunoreactivity for vascular endothelial growth factor is present in clusters of reactive astrocytes, and in parenchymal vascular walls and perivascular deposits, in Alzheimer brain (14). Immunoreactivity for fibroblast growth factor-9 is present in dystrophic neurites within Aβ plaques, and in activated astrocytes (13). Aβ promotes expression of basic fibroblast growth factor by cultured glial cells (46); and fibroblast growth factor, in turn, increases the expression of APP by cultured PC-12 neuroblastoma cells (115).

Interferon γ potentiates TNFα-induced Aβ production (106) and Aβ-induced microglial activation (104,116) in vitro, but there is, at present, no good evidence for increased expression of interferon γ in Alzheimer brain (29,67). In contrast, microglial immunoreactivity for interferon α has been reported for Alzheimer's disease (4,5).

Some plaque-associated astrocytes in Alzheimer's disease express inducible nitric oxide synthase (iNOS) (8). IL-1 induces iNOS and NO production in vivo (51), and Aβ in combination with various pro-inflammatory cytokines induces iNOS expression and NO production in cultured astrocytes (116). Increased IL-3 immunoreactivity has been reported for Alzheimer brain (6), and heparin binding growth-associated molecule is present in elevated quantities in Alzheimer brain, where it is found co-deposited with Aβ protein (7).

6. SUMMARY

In the decade since the first demonstration of increased cytokine (i.e., IL-1 and S100B) expression in Alzheimer brain (1), a number of additional cytokines and growth factors have been shown to be upregulated in this disease. Interleukin-1, in particular, is a key orchestrating immunomodulatory cytokine in peripheral immune responses, and the topographical distribution (within Aβ deposits, across brain regions, and across cortical layers) and apparent temporal distribution (among different plaque types in a hypothesized sequence of plaque evolution) suggest a key orchestrating function for this cytokine in the 'neuroinflammatory' processes of Alzheimer's disease as well (40). The known actions of IL-1 suggest that IL-1 initiates amplifying cytokine cascades, with potential 'feedback' arms involving neuronal injury, further microglial activation, and further overexpression of IL-1. These amplifying, self-propagating cascades have been conceptualized as a pathogenic *cytokine cycle* that drives plaque progression and neuronal injury in Alzheimer's disease (50). Recent genetic studies have demonstrated increased risk for Alzheimer's disease associated with certain polymorphisms in cytokine genes, especially those encoding IL-1α and IL-1β (39), adding further weight to the pathogenic importance of these intercellular signaling molecules in Alzheimer's disease.

ACKNOWLEDGMENTS

This work was supported in part by NIH AG 12411.

REFERENCES

1. Griffin, W.S.T., Stanley, L.C., Ling, C., et al. Brain interleukin 1 and S-100 immunoreactivity are elevated in Down syndrome and Alzheimer disease. *Proc Natl Acad Sci USA* 1989; 86: 7611-7615.
2. Bauer, J., Strauss, S., Schreiter-Gasser, U., et al. Interleukin-6 and α-2-macroglobulin indicate an acute-phase state in Alzheimer's disease cortices. *FEBS Lett* 1991; 285:111-4.

3. van der Wal, E.A., Gomez-Pinilla, F., Cotman, C.W. Transforming growth factor-β1 is in plaques in Alzheimer and Down pathologies. *Neuroreport* 1993; 4: 69-72.
4. Akiyama, H., Ikeda, K., Katoh, M., et al. Expression of MRP14, 27E10, interferon-α and leukocyte common antigen by reactive microglia in postmortem human brain tissue. *J Neuroimmunol* 1994; 50:195-201.
5. Yamada T., Horisberger, M.A., Kawaguchi, et al. Immunohistochemistry using antibodies to α-interferon and its induced protein, MxA, in Alzheimer's and Parkinson's disease brain tissues. *Neurosci Lett* 1994; 181: 61-4.
6. Araujo, D.M., Lapchak, P.A. Induction of immune system mediators in the hippocampal formation in Alzheimer's and Parkinson's diseases: selective effects on specific interleukins and interleukin receptors. *Neuroscience* 1994; 61:745-54.
7. Wisniewski, T., Lalowski, M., Baumann, M., et al. HB-GAM is a cytokine present in Alzheimer's and Down's syndrome lesions. *Neuroreport* 1996; 7: 667-71.
8. Wallace, M.N., Geddes, J.G., Farquhar, D.A., et al. Nitric oxide synthase in reactive astrocytes adjacent to β-amyloid plaques. *Exp Neurol* 1997; 144: 266-72.
9. Du Yan, S., Zhu, H., Fu, J., et al. Amyloid-β peptide-receptor for advanced glycation endproduct interaction elicits neuronal expression of macrophage-colony stimulating factor: a pro-inflamatory pathway in Alzheimer disease. *Proc Natl Acad Sci U S A* 1997; 94: 5296-301.
10. Xia, M., Qin, S., McNamara, M., et al. Interleukin-8 receptor B immunoreactivity in brain and neuritic plaques of Alzheimer's disease. *Am J Pathol* 1997; 150: 1267-74.
11. Ishizuka, K., Kimura, T., Igata-yi, R., et al. Identification of monocyte chemoattractant protein-1 in senile plaques and reactive microglia of Alzheimer's disease. *Psychiatry. Clin Neurosci* 1997; 51: 135-8.
12. Xia, M.Q., Qin, S.X., Wu, L.J., et al. Immunohistochemical study of the β-chemokine receptors CCR3 and CCR5 and their ligands in normal and Alzheimer's disease brains. *Am J Pathol* 1998; 153: 31-7.
13. Nakamura, S., Arima, K., Haga, S., et al. Fibroblast growth factor (FGF)-9 immunoreactivity in senile plaques. *Brain Res* 1998; 814: 222-5.
14. Kalaria, R.N., Cohen, D.L., Premkumar, D.R., et al. Vascular endothelial growth factor in Alzheimer's disease and experimental cerebral ischemia. *Brain Res Molec Brain Res* 1998; 62: 101-5.
15. Xia, M.Q., Bacskai, B.J., Knowles, R.B., et al. Expression of the chemokine receptor CXCR3 on neurons and the elevated expression of its ligand IP-10 in reactive astrocytes: in vitro ERK1/2 activation and role in Alzheimer's disease. *J Neuroimmunol* 2000; 108: 227-35.
16. Barger, S.W., Mattson, M.P. Secreted form of the Alzheimer's amyloid precursor protein stimulates a membrane-associated guanylate cyclase. *Biochem J* 1995; 311:45-47.
17. Barger, S.W., Mattson, M.P. Secreted form of the Alzheimer's amyloid precursor protein induces neuroprotective κB-dependent transcription. *Mol Brain Res* 1996; 40:116-126.
18. Mattson, M.P., Barger, S.W., Furukawa, K., et al. Cellular signaling roles of TGFβ, TNF and βAPP in brain injury responses and Alzheimer's disease. *Brain Res - Brain Res Rev* 1997; 23: 47-61
19. Barger, S.W., Harmon, A.D. Microglial activation by Alzheimer amyloid precursor protein and modulation by apolipoprotein E. *Nature* 1997; 388:878-81.
20. Grimaldi, L.M., Casadei, V.M., Ferri, C., et al. Association of early-onset Alzheimer's disease with an interleukin-1α gene polymorphism. *Ann Neurol* 2000; 47: 361-5.

21. Nicoll, J.A., Mrak, R.E., Graham ,D.I., et al. Association of interleukin-1 gene polymorphisms with Alzheimer's disease. *Ann Neurol* 2000; 47: 365-8.
22. Du, Y., Dode,l R.C., Eastwood, B.J., et al. Association of an interleukin 1α polymorphism with Alzheimer's disease. *Neurology* 2000; 55: 480-3, 2000
23. Minster, R.L., DeKosky, S.T., Ganguli, M., Belle S, et al. Genetic association studies of interleukin-1 (IL-1A and IL-1B) and interleukin-1 receptor antagonist genes and the risk of Alzheimer's disease. *Ann Neurol* 2000; 48: 817-9.
24. Rebeck, G.W. Confirmation of the genetic association of interleukin-1A with early onset sporadic Alzheimer's disease. *Neurosci Lett* 2000; 293: 75-7.
25. Bertram, L., Blacker, D., Crystal, A., et al. Candidate genes showing no evidence for association or linkage with Alzheimer's disease using family-based methodologies. *Exp Gerontol* 2000; 35: 1353-61, 2000
26. Collins, J.S., Perry, R.T., Watson Jr, B., et al. Association of a haplotype for tumor necrosis factor in siblings with late-onset Alzheimer disease: the NIMH Alzheimer Disease Genetics Initiative. *Am J Med Gen* 2000; 96: 823-30.
27. McCusker, S.M., Curran, M.D., Dynan, K.B., et al. Association between polymorphism in regulatory region of gene encoding tumour necrosis factor α and risk of Alzheimer's disease and vascular dementia: a case-control study. *Lancet* 2001; 357: 436-9.
28. Luedecking, E.K., DeKosky, S.T., Mehdi, H., et al. Analysis of genetic polymorphisms in the transforming growth factor-β1 gene and the risk of Alzheimer's disease. *Human Genet* 2000; 106: 565-9.
29. Papassotiropoulos, A., Bagli, M., Jessen, F., et al. A genetic variation of the inflammatory cytokine interleukin-6 delays the initial onset and reduces the risk for sporadic Alzheimer's disease. *Ann Neurol* 1999; 45: 666-8.
30. Bhojak, T.J., DeKosky, S.T., Ganguli, M., et al. Genetic polymorphisms in the cathepsin D and interleukin-6 genes and the risk of Alzheimer's disease. *Neurosci Lett* 2000; 288: 21-4.
31. Goldgaber, D., Harris, H.W., Hla, T., et al. Interleukin 1 regulates synthesis of amyloid β-protein precursor mRNA in human endothelial cells. *Proc Natl Acad Sci USA* 1989; 86: 7606-10.
32. Forloni, G., Demicheli, F., Giorgi, S., et al. Expression of amyloid precursor protein mRNAs in endothelial, neuronal and glial cells: modulation by interleukin-1. *Brain Res Molec Brain Res* 1992; 16: 128-34.
33. Buxbaum, J.D., Oishi, M., Chen, H.I., et al. Cholinergic agonists and interleukin 1 regulate processing and secretion of the Alzheimer β/A4 amyloid protein precursor. *Proc Natl Acad Sci USA* 1992; 89: 10075-10078.
34. Li, Y., Liu, L., Kang, J., et al. Neuronal-glial interactions mediated by interleukin-1 enhance neuronal acetylcholinesterase activity and mRNA expression. *J Neurosci* 2000; 20: 149-155.
35. Lee, S.C., Liu, W., Dickson, D.W., et al. Cytokine production by human fetal microglia and astrocytes. Differential induction by lipopolysaccharide and IL1β. *J Immunol* 1993; 150, 2659-67.
36. Sebire, G., Emilie, D., Wallon, C., Hery C, et al. In vitro production of IL6, IL1β, and tumor necrosis factor β by human embryonic microglial and neural cells. *J Immunol* 1993; 150: 1517-23.
37. Das, S., Potter, H. Expression of the Alzheimer amyloid-promoting factor antichymotrypsin is induced in human astrocytes by IL-1. *Neuron* 1995; 14: 447-56.
38. Sheng, J.G., Ito, K., Skinner, R.D., et al. In vivo and in vitro evidence supporting a role for the inflammatory cytokine interleukin-1 as a driving force in Alzheimer pathogenesis. *Neurobiol Aging* 1996; 17: 761-6.

39. Mrak, R.E., Griffin, W.S.T. Interleukin-1 and the immunogenetics of Alzheimer disease. *J Neuropathol Exp Neurol* 2000; 59: 471-6.
40. Mrak, R.E., Griffin, W.S.T. Interleukin-1, neuroinflammation, and Alzheimer's disease. *Neurobiol Aging* 2001; 22: 903-908.
41. Sheng, J.G., Mrak, R.E., Griffin, W.S.T. Microglial interleukin-1α expression in brain regions in Alzheimer's disease: Correlation with neuritic plaque distribution. *Neuropathol Appl Neurobiol* 1996; 21: 290-301.
42. Sheng, J.G., Griffin, W.S.T., Royston, C.M., et al. Distribution of IL-1-immunoreactive microglia in cerebral cortical layers: Implications for neuritic plaque formation in Alzheimer's disease. *Neuropathol Appl Neurobiol* 1998; 24: 278-83.
43. Benzing, W.C., Wujek, J.R., Ward, E.K., et al. Evidence for glial-mediated inflammation in aged APP(SW) transgenic mice. *Neurobiol Aging* 1999; 20: 581-9.
44. Mehlhorn, G., Hollborn, M., Schliebs, R. Induction of cytokines in glial cells surrounding cortical β-amyloid plaques in transgenic Tg2576 mice with Alzheimer pathology. *Int J Develop Neurosci* 2000; 18: 423-31.
45. Griffin, W.S.T., Sheng, J.G., Roberts, G.W., et al. Interleukin-1 expression in different plaque types in Alzheimer's disease: significance in plaque evolution. *J Neuropathol Exp Neurol* 1995; 54: 276-281, 1995
46. Araujo, D.M., Cotman, C.W.. β-Amyloid stimulates glial cells in vitro to produce growth factors that accumulate in senile plaques in Alzheimer's disease. *Brain Res* 1992; 569:141-5.
47. Mackenzie IR, Hao C, Munoz DG. Role of microglia in senile plaque formation. *Neurobiol Aging* 16: 797-804, 1995
48. Allan, S.M., Parker, L.C., Collins, B., et al. Cortical cell death induced by IL-1 is mediated via actions in the hypothalamus of the rat. *Proc Natl Acad Sci USA* 2000; 97:5580-5.
49. Barger SW, Van Eldik LJ. S100B stimulates calcium fluxes in glial and neuronal cells. *J Biol Chem* 1992; 267:9689-9694.
50. Hu, J., Castets, F., Guevara, J.L., et al. S100B stimulates inducible nitric oxide synthase activity and mRNA levels in rat cortical astrocytes. J *Biol Chem* 1996; 271: 2543-7.
51. Casamenti, F., Prosperi, C., Scali, C., et al. Interleukin-1β activates forebrain glial cells and increases nitric oxide production and cortical glutamate and GABA release in vivo: implications for Alzheimer's disease. *Neurosci.* 1999; 91: 831-42.
52. Sheng, J.G., Zhou, X.Q., Mrak, R.E., et al. Progressive neuronal injury associated with amyloid plaque formation in Alzheimer's disease. *J Neuropathol Exp Neurol* 1998; 57: 714-7.
53. Sheng, J.G., Mrak, R.E., Griffin, W.S.T. Glial-neuronal interactions in Alzheimer disease: progressive association of IL-1α$^+$ microglia and S100β$^+$ astrocytes with neurofibrillary tangle stages. *J Neuropathol Exp Neurol* 1997; 56: 285-90.
54. Tarkowski, E., Blennow, K., Wallin, A., et al. Intracerebral production of tumor necrosis factor-α, a local neuroprotective agent, in Alzheimer disease and vascular dementia. *J Clin Immunol* 1999; 19: 223-30.
55. Wolozin, B., Lesch, P., Lebovics, R., et al. A.E. Bennett Research Award 1993. Olfactory neuroblasts from Alzheimer donors: studies on APP processing and cell regulation. *Biol Psychiatry* 1993; 34: 824-38.
56. Sheng, J.G., Zhu, S.G., Jones, R.A., et al. Interleukin-1 promotes expression and phosphorylation of neurofilament and tau proteins in vivo. *Exp Neurol* 2000; 163: 388-391.

57. Griffin, W.S.T., Sheng, J.G., Royston, M.C., et al. Glial-neuronal interactions in Alzheimer's disease: the potential role of a 'cytokine cycle' in disease progression. *Brain Pathol* 1998; 8: 65-72.
58. Mrak, R.E., Sheng, J.G., Griffin, W.S.T. Glial cytokines in Alzheimer's disease: review and pathogenic implications. *Hum Pathol* 1995; 26: 816-823.
59. Donato, R. Functional roles of S100 proteins, calcium-binding proteins of the EF-hand type. Biochim Biophys Acta 1450: 191-231, 1999
60. Shashoua, V.E., Hesse, G.W., Moore, B.W. Proteins of the brain extracellular fluid: evidence for release of S100 protein. *J Neurochem* 1984; 42: 1536-1541.
61. Van Eldik, L.J., Zimmer, D.B. Secretion of S-100 from rat C6 glioma cells. *Brain Res* 1987; 436: 367-370.
62. Kligman, D., Marshak, D.R. Purification and characterization of a neurite extension factor from bovine brain. *Proc Natl Acad Sci USA* 1985; 82: 7136-7139.
63. Winningham-Major, F., Staecher, J.L., Barger, S.W., et al. Neurite extension and neuronal survival activities of recombinant S100B proteins that differ in the content and position of cysteine residues. *J Cell Biol* 1989; 109: 3036-71.
64. Barger, S.W., Van Eldik, L.J., Mattson, M.P. S100β protects hippocampal neurons from damage induced by glucose deprivation. *Brain Res* 1995; 677:167-70.
65. Selinfreund, R.H., Barger, S.W., Pledger, W.J., et al. Neurotrophic protein S100B stimulates glial cell proliferation. *Proc Nat Acad Sci USA* 1991; 88: 3554-3558.
66. Fano, G.. Mariggio, M.A.. Angelella, P., et al. The S-100 protein causes an increase of intracellular calcium and death of PC12 cells. *Neurosci.* 1993; 53: 919-25.
67. Mrak, R.E., Sheng, J.G., Griffin WST. Correlation of astrocytic S100B expression with dystrophic neurites in amyloid plaques of Alzheimer's disease. *J Neuropathol Exp Neurol* 1996; 55: 273-279.
68. Sheng, J.G., Mrak, R.E., Griffin, W.S.T. S100B protein expression in Alzheimer disease: potential role in the pathogenesis of neuritic plaques. *J Neurosci Res* 1994; 39: 398-404.
69. Marshak, D.R., Pesce, S.A., Stanley, L.C., et al. Increased S100 neurotrophic activity in Alzheimer disease temporal lobe. *Neurobiol Aging* 1992; 13: 1-7.
70. Mandybur, T.I., Chuirazzi, C.C. Astrocytes and the plaques of Alzheimer's disease. *Neurology* 1990; 40: 635-639.
71. Van Eldik, L.J., Griffin, W.S.T. S100B expression in Alzheimer's disease: Relation to neuropathology in brain regions. *Biochim Biophys Acta* 1994; 1223: 398-403.
72. Braak, H., Braak, E. Neuropathological stageing of Alzheimer-related changes. *Acta Neuropathol* 1991; 82: 239-259.
73. Mrak, R.E., Griffin, W.S.T. The role of activated astrocytes and of the neurotrophic cytokine S100B in the pathogenesis of Alzheimer's disease. *Neurobiol Aging* 2001; 22: 915-922.
74. Da Cunha, A., Jefferson, J.A., Jackson, R.W., et al. Glial cell-specific mechanisms of TGF-β1 induction by IL-1 in cerebral cortex. *J Neuroimmunol* 1993; 42: 71-86.
75. Meda, L., Baron, P., Prat, E., et al. Pro-inflamatory profile of cytokine production by human monocytes and murine microglia stimulated with β-amyloid(25-35). *J Neuroimmunol* 1999; 93: 45-52.
76. Flanders, K.C., Lippa, C.F., Smith, T.W., et al. Altered expression of transforming growth factor-β in Alzheimer's disease. *Neurology* 1995; 45: 1561-9.
77. Luterman, J.D., Haroutunian, V., Yemul, S., et al. Cytokine gene expression as a function of the clinical progression of Alzheimer disease dementia. *Arch Neurol* 2000; 57: 1153-60.

78. Wyss-Coray, T., Masliah, E., Mallory, M., et al. Amyloidogenic role of cytokine TGF-β1 in transgenic mice and in Alzheimer's disease. *Nature* 1997; 389: 603-6.
79. Peress, N.S., Perillo, E. Differential expression of TGF-β 1, 2 and 3 isotypes in Alzheimer's disease: a comparative immunohistochemical study with cerebral infarction, aged human and mouse control brains. *J Neuropathol Exp Neurol* 1995; 54: 802-11.
80. Chao, C.C., Ala, T.A., Hu, S., et al. Serum cytokine levels in patients with Alzheimer's disease. *Clin Diagn Lab Immunol* 1994; 1: 433-6.
81. Gray, C.W., Patel, A.J. Regulation of β-amyloid precursor protein isoform mRNAs by transforming growth factor-β1 and interleukin-1β in astrocytes. *Brain Res Molec Brain Res* 1993; 19: 251-6.
82. Frautschy, S.A., Yang, F., Calderon, L., et al. Rodent models of Alzheimer's disease: rat Aβ infusion approaches to amyloid deposits. *Neurobiol Aging* 1996; 17: 311-21.
83. Wyss-Coray, T., Lin, C., Sanan, D.A., et al. Chronic overproduction of transforming growth factor-β1 by astrocytes promotes Alzheimer's disease-like microvascular degeneration in transgenic mice. *Am J Pathol* 2000; 156: 139-50.
84. Wyss-Coray, T., Lin, C., von Euw, D., et al. Alzheimer's disease-like cerebrovascular pathology in transforming growth factor-β1 transgenic mice and functional metabolic correlates. *Ann N Y Acad Sci 2000;* 903: 317-23.
85. Wyss-Coray, T., Lin, C., Yan, F., et al. TGF-β1 promotes microglial amyloid-β clearance and reduces plaque burden in transgenic mice. *Nature Med* 2001; 7: 612-8.
86. Li, Y., Barger, S.W., Liu, L., et al. S100B induction of the pro-inflammatory cytokine interleukin-6 in neurons: Implications for Alzheimer pathogenesis. *J Neurochem* 2000; 74: 143-50.
87. Del Bo, R., Angeretti, N., Lucca, E., et al. Reciprocal control of inflammatory cytokines, IL-1 and IL-6, and β-amyloid production in cultures. *Neurosci Lett* 1995; 188: 70-4.
88. Ringheim, G.E., Szczepanik, A.M., Petko, W., et al. Enhancement of β-amyloid precursor protein transcription and expression by the soluble interleukin-6 receptor/interleukin-6 complex. *Brain Res Molec Brain Res* 1998; 55: 35-44.
89. Kordula, T., Rydel, R.E., Brigham, E.F., et al. Oncostatin M and the interleukin-6 and soluble interleukin-6 receptor complex regulate α_1-antichymotrypsin expression in human cortical astrocytes. *J Biol Chem* 1998; 273: 4112-8.
90. Gitter, B.D., Cox, L.M., Rydel, R.E., et al. Amyloid β peptide potentiates cytokine secretion by interleukin-1β-activated human astrocytoma cells. *Proc Natl Acad Sci USA* 1995; 92: 10738-41.
91. Strauss, S., Bauer, J., Ganter, U., et al. Detection of interleukin-6 and α_2-macroglobulin immunoreactivity in cortex and hippocampus of Alzheimer's disease patients. *Lab Invest* 1992; 66: 223-30.
92. Huell, M., Strauss, S., Volk, B., et al. Interleukin-6 is present in early stages of plaque formation and is restricted to the brains of Alzheimer's disease patients. *Acta Neuropathol* 1995; 89: 544-51.
93. Hull, M., Berger, M., Volk, B., et al. Occurrence of interleukin-6 in cortical plaques of Alzheimer's disease patients may precede transformation of diffuse into neuritic plaques. *Ann N Y Acad Sci* 1996; 777: 205-12.
94. Thal, D.R., Schober, R., Birkenmeier, G. The subunits of α_2-macroglobulin receptor/low density lipoprotein receptor-related protein, native and transformed α_2-macroglobulin and interleukin 6 in Alzheimer's disease. *Brain Res* 1997; 777: 223-7.
95. Lanzrein, A.S., Johnston, C.M., Perry, V.H., et al. Longitudinal study of inflammatory factors in serum, cerebrospinal fluid, and brain tissue in Alzheimer disease: interleukin-1β, interleukin-6, interleukin-1 receptor antagonist, tumor necrosis factor-α, the soluble

tumor necrosis factor receptors I and II, and α_1-antichymotrypsin. *Alzh Dis Assoc Dis* 1998; 12: 215-27.
96. Wood, J.A., Wood, P.L., Ryan, R., et al. Cytokine indices in Alzheimer's temporal cortex: no changes in mature IL-1β or IL-1RA but increases in the associated acute phase proteins IL-6, α_2-macroglobulin and C-reactive protein. *Brain Res* 1993; 629: 245-52.
97. Blum-Degen, D., Muller, T., Kuhn, W., et al. Interleukin-1β and interleukin-6 are elevated in the cerebrospinal fluid of Alzheimer's and de novo Parkinson's disease patients. *Neurosci Lett* 1995; 202: 17-20.
98. Martinez, M., Fernandez-Vivancos, E., Frank, A., et al. Increased cerebrospinal fluid fas (Apo-1) levels in Alzheimer's disease. Relationship with IL-6 concentrations. *Brain Res* 2000; 869: 216-9.
99. Marz, P., Heese, K., Hock, C., et al. Interleukin-6 (IL-6) and soluble forms of IL-6 receptors are not altered in cerebrospinal fluid of Alzheimer's disease patients. *Neurosci Lett* 1997; 239: 29-32.
100. Engelborghs, S., De Brabander, M., De Cree, J., et al. Unchanged levels of interleukins, neopterin, interferon-γ and tumor necrosis factor-α in cerebrospinal fluid of patients with dementia of the Alzheimer type. *Neurochem Int* 1999; 34: 523-30.
101. Yamada, K., Kono, K., Umegaki, H., et al. Decreased interleukin-6 level in the cerebrospinal fluid of patients with Alzheimer-type dementia. *Neurosci Lett* 1995; 186: 219-21.
102. Dinarello, C.A. Pro-inflamatory cytokines. *Chest* 2000; 118: 503-508.
103. Benveniste EN, Benos DJ. TNF-α- and IFN-γ-mediated signal transduction pathways: effects on glial cell gene expression and function. *FASEB J* 1995; 9: 15577-1584.
104. Meda, L., Cassatella, M.A., Szendrei, G.I., et al. Activation of microglial cells by β-amyloid protein and interferon-γ. *Nature* 1995; 374: 647-50.
105. Viviani, B., Corsini, E., Galli, C.L., et al. Dying neural cells activate glia through the release of a protease product. *GLIA* 2000; 32: 84-90.
106. Blasko, I., Marx, F., Steiner, E., et al. TNFα plus IFNγ induce the production of Alzheimer β-amyloid peptides and decrease the secretion of APPs. *FASEB J* 1999; 13: 63-8.
107. Mattson, M.P., Barger, S.W., Cheng, B., et al. β-Amyloid precursor protein metabolites and loss of neuronal calcium homeostasis in Alzheimer's disease. *Trends Neurosci* 1993; 16: 409-415.
108. Murphy, G.M., Jr, Yang, L., Cordell, B.. Macrophage colony-stimulating factor augments β-amyloid-induced interleukin-1, interleukin-6, and nitric oxide production by microglial cells. *J Biol Chem* 1998; 273: 20967-71.
109. Murphy, G.M., Jr, Zhao, F., Yang, L., et al. Expression of macrophage colony-stimulating factor receptor is increased in the AβPP(V717F) transgenic mouse model of Alzheimer's disease. *Am J Pathol* 2000; 157: 895-904.
110. Mennicken, F., Maki, R., de Souza, E.B., et al. Chemokines and chemokine receptors in the CNS: a possible role in neuroinflammation and patterning. *Trends Pharmacol Sci* 1999; 20: 73-8.
111. Xia, M.Q., Hyman,, B.T. Chemokines/chemokine receptors in the central nervous system and Alzheimer's disease. *J NeuroVirol* 1999; 5: 32-41.
112. Fiala, M., Zhang, L., Gan, X., et al. Amyloid-β induces chemokine secretion and monocyte migration across a human blood--brain barrier model. *Molec Med* 1998; 4: 480-9.
113. Szczepanik, A.M., Funes, S., Petko, W., et al. IL-4, IL-10 and IL-13 modulate Aβ(1--42)-induced cytokine and chemokine production in primary murine microglia and a human monocyte cell line. *J Neuroimmunol* 2001; 113: 49-62

114. Johnstone, M., Gearing, A.J., Miller, K.M.. A central role for astrocytes in the inflammatory response to β-amyloid; chemokines, cytokines and reactive oxygen species are produced. *J Neuroimmunol* 1999; 93: 182-93.
115. Lahiri, D.K., Nall, C. Promoter activity of the gene encoding the β-amyloid precursor protein is up-regulated by growth factors, phorbol ester, retinoic acid and interleukin-1. *Brain Res Molec Brain Res* 1995; 32: 233-40.
116. Rossi, F., Bianchini E. Synergistic induction of nitric oxide by β-amyloid and cytokines in astrocytes. *Biochem Biophys Res Commun* 1996; 225: 474-8.

Chapter 18

CYTOKINES, IMMUNITY AND DISORDERED SLEEP

Michael Irwin
Cousins Center for Psychoneuroimmunology, Neuropsychiatric Institute, University of California, Los Angeles, California

1. INTRODUCTION

Sleep is hypothesized to have a restorative function on immune processes (1). In turn, disordered sleep and sleep loss are thought to impair host defense mechanisms and impact susceptibility to viral- and bacterial pathogens (2,3). Few studies have evaluated the association between disordered sleep and immunity in psychiatric patient populations, even though depressed- and alcoholic patients show prominent disturbances of sleep (4) and are at risk for infectious and other immune-related diseases (5,6).

Increasing evidence also suggests a bi-directional communication between sleep and the immune system. In addition to the effects of sleep on cytokine expression, the complex cytokine network also influences sleep and sleep depth (7) with inflammatory cytokines having both somnogenic and inhibitory effects depending on the cytokine, dose, and circadian phase (3,8,9). Translation of these basic mechanisms into the clinic has implications for answering why sleep may be disordered in psychiatric populations and for the development of novel therapies for disordered sleep (10).

Thus, the objectives of this review are to examine the relationships between disturbances of sleep and immunity in humans by assessing the association between disordered sleep and immune alterations that coincide in depression and alcoholism. Second, to further evaluate the impact of sleep on immunity by using an experimental approach, namely partial night sleep loss, that tests the role of sleep- vs. circadian processes on measures of natural and cellular immunity. Finally, to examine the sympathetic and neuroendocrine mechanisms that might link disordered sleep, sleep loss, and immunity.

2. SLEEP AND HOST DEFENSE MECHANISMS

What evidence is available regarding the impact of sleep on host defense mechanisms and infectious disease risk? Decreased sleep amounts are associated with increased morbidity and mortality in humans (11). However, but there is scant clinical information about whether this relationship is due to changes in host resistance to infectious disease (3).

In animals, several studies support the notion that sleep loss impacts host defense mechanisms. For example in rats, sleep deprivation lasting 7 hours impairs influenza viral clearance, suppresses specific influenza antibody production and is associated with the presence of viral particles in lung homogenates (12). Further work by Brown and colleagues suggests that sleep deprivation impairs secondary antibody responses (13). In rats that are immunized with sheep red blood cells and then rechallenged with the same antigen 14 days later, ex vivo immune responses are decreased in the sleep-deprived as compared to the sham treated animals (13). The effects of sustained sleep loss have also been investigated. Under conditions in which sleep loss is sustained for days, rats show a breakdown in host defense mechanisms and ultimately die (14). In addition to increased metabolism, hyperphagia, weight loss, and hypothermia, Everson et al. (14) found that sleep-deprived animals show evidence of systemic infection and septicemia with opportunistic bacterial microorganisms (i.e., Pseudomonas aeruginosa, streptococcus) during the final stages of life. Finally, Toth and colleagues (15) have suggested that failure to exhibit increases of slow wave sleep following infection with E. coli, S. aureus, or C. albicans correlates with increased morbidity as well as rates of mortality in rabbits. Nevertheless, not all data have supported the hypothesis that sleep influences host defenses or vulnerability to infections. Benca and colleagues (16) did not find that sleep deprived rats are at increased risk for bacteremia or changes in cellular and humoral immune responses.

In humans, even less is known about the effects of sleep and sleep loss on host defense mechanisms and only one study has evaluated changes in infectious disease risk and/or disease specific immune measures in relationship to sleep. In this instance, circadian shift workers who commonly experience disordered sleep were found to exhibit depressed cellular immune function and increased rates of respiratory tract infections (17,18). In contrast, most clinical investigations have evaluated the effects of sleep deprivation and relied on ex vivo assessments of immune parameters including numbers of immune cells and function of isolated peripheral blood mononuclear cells (2). Inference about disease risk from these studies requires cautious interpretation since many of the immune measures are nonspecific (e.g. mitogen-induced lymphocyte proliferation).

In addition, changes in peripheral blood immune measures represent alterations in only one immunological compartment and function of immune cells in other compartments must be simultaneously assessed to determine changes in host immune competence (3).

Studies of the immunological effects of sleep deprivation in humans have not yielded consistent results and both enhancing and suppressive effects have been observed depending upon the immune measures and the duration of the sleep deprivation. For example, Dinges and colleagues (19) found in 20 subjects that 64 h of sleep deprivation was associated with leukocytosis and increased NK cell activity, but with no changes in lymphocyte counts or nonspecific pokeweed mitogen-, phytohemagglutinin (PHA)-, or concanavalin-A (Con-A)- lymphocyte proliferation. In contrast, Moldofsky et al. (20) reported in 10 men that 40 h of sleep loss induced a prolonged decline of NK activity and a delayed nocturnal rise of lymphocyte proliferation to pokeweed mitogen but not to PHA. Palmblad and colleagues found decreased PHA-induced lymphocyte proliferation following 48 h of wakefulness in 12 men and increased interferon production and decreased phagocytic activity in 8 women sleep deprived for 77 hours (21,22). Finally, Born et al. examined the effects of 40 h of sustained wakefulness on measures of both lymphocyte numbers and function (23). When changes in cytokine production were expressed in reference to numbers of circulating lymphocyte subsets, sleep was found to be associated with enhanced production of IL-2 by T cells but not to have effects on IL-1, TNF or IL-6.

The varying effects of sleep deprivation on immune function and the lack of consistent results is not surprising in view of the substantial differences in duration of sleep deprivation between the various studies. Sleep loss that varies in duration may differentially impact circadian- and stress-dependent processes and such differences may contribute to variability in the immune results (3). For example, immune parameters show a circadian cycle yet most studies only assessed immune function at one or two time points (2,24). Prolonged wakefulness (>40 hours) may be more likely to produce a shift in circadian phase as compared to sleep loss periods of shorter duration (14,25). Similarly, prolonged wakefulness lasting for days at a time may induce stress with elevations in glucocorticoids and other typical stress responses, and the immunological effects may be due to stress rather than sleep loss per se. However, as reviewed by Dinges et al (2), studies of total sleep deprivation have not generally been found to elevate plasma levels of glucocorticoids or urinary catecholamines.

In addition to these limitations, it is useful to note that these experimental studies have uniformly used total sleep deprivation paradigms. Clinical samples are likely to show sleep loss during only part of the night (4), as opposed to total sleep deprivation, and the relevance of total sleep deprivation to understanding the relationships between disordered sleep and immune function under naturalistic conditions remains unclear. Moreover, total sleep deprivation fails to assess the relationships between disordered sleep architecture, decreases of sleep depth and immune processes. Studies of sleep architecture in relation to autonomic functioning have shown that impairments of sleep depth, not merely sleep amounts, have substantial effects on sympathetic tone and predict risk for hypertension and cardiovascular disease. Thus, in an effort to address these issues, we have evaluated the influence of sleep, sleep depth, and sleep intensity on immunity in clinical samples of depressed- and alcoholic subjects. In addition, partial night sleep deprivation has been used as a naturalistic probe to evaluate the effects of sleep- vs circadian dependent processes on immune processes.

3. DEPRESSION, SLEEP AND IMMUNITY

Depression serves as an excellent model to learn more about sleep-immune interactions as depressed patients exhibit prominent abnormalities of sleep (4), dysregulation of the neuroendocrine- and sympathetic nervous systems (26,27), and alterations of immunity such as a reduction of lymphocyte responses and natural killer (NK) activity (5,28). Peripheral blood NK cell activity is an in vitro measure of cellular immune function that correlates with host resistance to viral illness (29).

Prior studies have found that severity of sleep disturbance is key in identifying those depressed patients at risk for immune alterations. In depressed patients, NK activity is negatively correlated with severity of insomnia ($r=-0.33$, $p<0.05$) but not with other depressive symptoms including anxiety/somatization, weight loss, cognitive disturbance or diurnal variation. In addition, the relationships between quality of sleep and immunity remain robust even when sleep is objectively assessed by EEG. For example, in depressed patients, total sleep time, sleep efficiency, and duration of nonREM sleep positively correlate with lytic activity (30). In other words, those depressed subjects who had the lowest amounts of EEG sleep and sleep continuity showed the lowest NK activity. Whether the secretory pattern of cytokines, or some other peptide with sleep-immune effects, is altered in depressed patients to produce coincident changes of sleep and NK activity is not known.

Similar relationships between sleep and NK activity have also been found in non-depressed groups which further supports the notion that disordered sleep affects immunity independent of a mood disorder. Even in controls who have no lifetime- or current history of a major psychiatric disorder, total sleep time, sleep efficiency, and duration of nonREM sleep are positively correlated with NK activity.

Sleep is regulated by a multitude of substances interwoven with each other which are also capable of affecting immune function (8), and we speculate that both sleep processes and natural cytotoxicity might be modulated by a common central factor. For example, the neuropeptide corticotropin releasing factor is increased in the cerebrospinal fluid of depressed patients (27), inhibits normal sleep and interleukin-1 induced sleep (31), and acts within the central nervous system to activate sympathetic outflow and reduce cellular immunity (32). Conversely, interleukin-1 induces slow wave sleep in animals (8), is increased in the plasma of humans following the onset of slow wave sleep (33), and acts on NK cells to stimulate cytolytic activity (29). Future studies are needed to evaluate whether the secretory pattern of cytokines, or some other peptide with sleep-immune effects, is altered in depressed patients to produce coincident changes of sleep and immunity.

4. ALCOHOLISM, SLEEP AND IMMUNITY

Alcoholics are another patient population who deserve attention in the study of sleep-immune interactions. Sleep disturbance in alcoholics is marked by a severe loss of sleep continuity and depth. Moreover, disordered sleep in this population persists for months and years after abstinence. Given the substantial evidence that alcoholics are at increased risk for a number of infectious illnesses including HIV, tuberculosis, and hepatitis C and that this risk is especially increased in African American alcoholics (as reviewed by Irwin& Miller, 2000) (34), the consequences of disordered sleep on immunity in this population was evaluated.

Severity of alcohol dependence, age, and ethnicity are all predictors of abnormal sleep in alcoholics. Recent data, for example, indicate that African American ethnicity and alcohol dependence interact to produce a more profound loss of delta sleep as measured by polysomnographic and spectral sleep analyses than that found in Euro-American alcoholics (35). Indeed, none of the African American alcoholics had measurable amounts of Stage 4 sleep when scored by raters blind to the group status of the EEG records. Spectral analyses of the sleep EEG further confirmed the loss of delta sleep in the alcohol dependent patients. As compared to the controls, alcoholics showed significantly lower delta (0.75-4.5 Hz) activity over the whole night and especially during the first NREM period (35).

Given the hypothesis that sleep contributes to the maintenance of the immune system, (1,2), African American alcoholics who show the most severe sleep disturbance were hypothesized to have the greatest immunological disturbances as compared to other groups. Assessment of immune function included assay of natural immunity and evaluation of levels of pro-inflammatory cytokine (IL-6) and Th-1 like (IL-2, IL-12) and Th2 like (IL-10) cytokines. In humans, two polar categories of T helper (Th) cells are described: Th1 cells secrete IL-2, IL-12 and interferon-γ (IFN) which predominantly stimulate cell-mediated responses, whereas Th2 cells secrete IL-4 and IL-10 which mainly stimulate antibody-mediated responses (36).

In parallel with the increased risk of disordered sleep, African American alcoholics had significantly lower levels of NK activity and IL-2 stimulated NK activity than the other three groups (34). Furthermore, production of the pro-inflammatory cytokine, IL-6, was significantly lower in the African American alcoholics as compared to the other three groups (Figure 1),

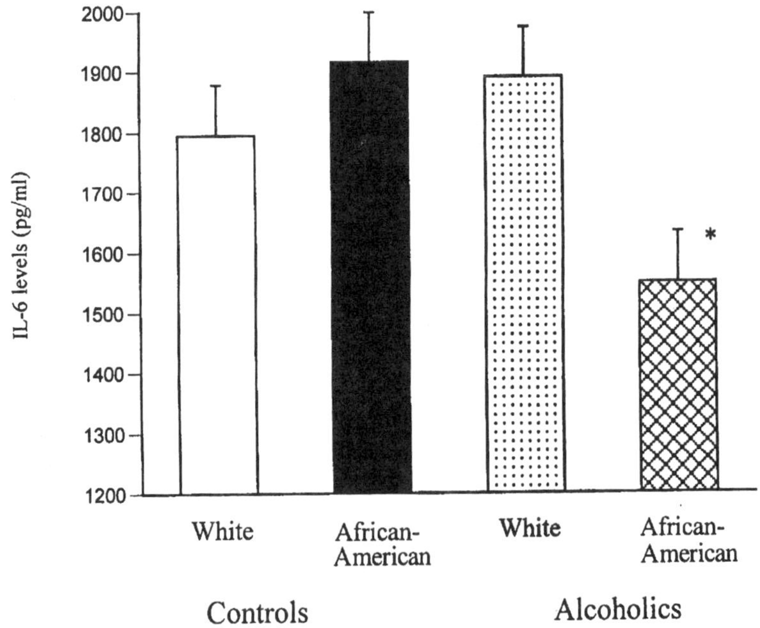

Figure 1: Con-A stimulated production of IL-6 by peripheral blood mononuclear cells in control and alcoholic patients stratified as white- and African American subjects. Data are presented as mean $\pm$ SEM. There are 16 white controls, 14 African-American controls, 16 white alcoholics, and 14 African American alcoholics. * Supernatant levels of IL-6 were significantly lower in the African American alcoholics as compared to the other three groups ($F=7.2$, $p<0.01$). (Reprinted from ref #34)

whereas, production of the inhibitory cytokine IL-10 was significantly increased in the African American alcoholics as compared to controls and white alcoholics (Figure 2).

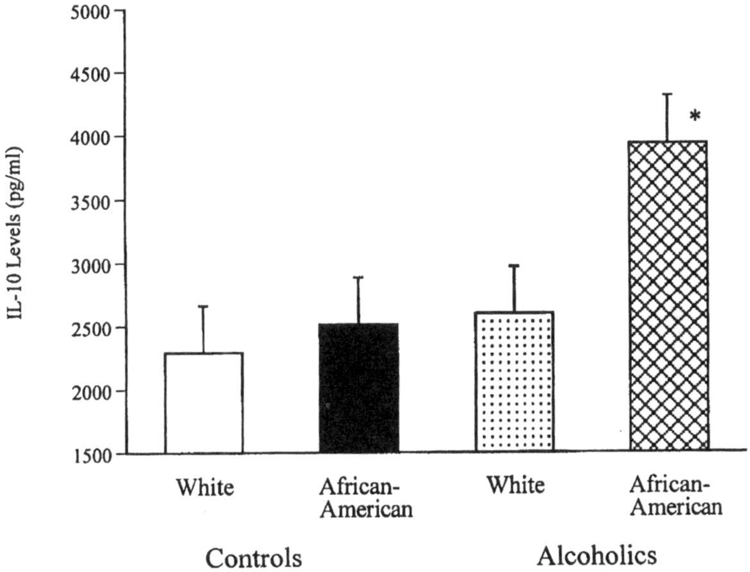

Figure 2: Con-A stimulated production of IL-10 by peripheral blood mononuclear cells in control and alcoholic patients stratified as white- and African American subjects. Data are presented as mean ± SEM. There are 16 white controls, 14 African-American controls, 16 white alcoholics, and 14 African American alcoholics. * Supernatant levels of IL-10 were significantly increased in the African American alcoholics as compared to controls and white alcoholics. ($F=4.9$, $p<0.05$). (Reprinted from ref #34)

Changes in the relative expression of Th1 to Th2 cytokines, as found in the present study, may have important clinical implications in alcoholics. For example, alcoholics show a high incidence of hepatitis C (37), and the chronicity of this viral infection is related to a lack of Th1 cytokine production and/or an increase in Th2 release. For example, Woitas et al. (38) have found that hepatitis C seropositive blood donors without viremia show increased Th1 cytokine production (IFN, IL-2) in response to hepatitis C core protein, whereas seropositive donors with viremia show increases in IL-10 producing T cells.

The model put forward in this review posits an inter-relationship between sleep, and cytokine expression in which feedback systems are operating. Not only is disordered sleep thought to impact sleep, but alternatively cytokines are hypothesized to alter sleep depth and intensity. In animals, IL-1, TNF, IFN and IL-10 are involved in the physiological regulation of sleep (8,39). For example, pro-inflammatory cytokines such as IL-1 and TNF increase delta sleep (40,41), whereas IL-10 inhibits slow wave sleep in animals (42). IL-6 also decreases non-REM sleep, although this action was small due to the use of a cross-species form of IL-6 (43). However, in humans, acute peripheral administration of IL-6 reduces delta sleep in the first half of the night when circulating levels of IL-6 are elevated (44). Vgontzas have further found that daytime elevations of circulating IL-6 are associated with severity of sleepiness and fatigue in sleep disordered patients (45), and that sleep deprivation leads to daytime over-secretion of IL-6 (46) in which daytime plasma levels of IL-6 negatively correlate with amounts of slow wave sleep. In view of this converging evidence, it is tempting to speculate about a link between cytokines and disordered sleep in the African American alcoholics in which cytokine abnormalities contribute to the severity of sleep disturbance in these patients.

5. PARTIAL NIGHT SLEEP DEPRIVATION

5.1 Effects on natural and cellular immunity

While the findings in depressed- and alcoholic subjects studies together indicate an association between disordered sleep and immune processes, their interpretation remains obscure regarding the precise role of sleep and sleep loss. Rather than simply relying on correlational approaches, experimental strategies such as sleep deprivation are needed to evaluate the effects of sleep on immunity. However, previous studies have almost exclusively used prolonged, total sleep deprivation strategies (2,3), even though clinical samples often report symptoms of early- or late insomnia, rather than sleep loss throughout an entire night.

It is important to examine the effects of sleep loss using a naturalistic probe, namely partial deprivation of sleep during the late- or early part of the night, as a way of estimating further the role of sleep in altering immunity in depressed- and alcoholic populations. Indeed, partial night sleep deprivation was found to produce a significant reduction of cellular immunity as measured by NK activity, IL-2 stimulated NK activity, and stimulated production of IL-2 (47,48) in which declines of IL-2 production was driven by deficits in the function of both lymphocyte and monocyte cell populations. Born and colleagues have also shown that IL-2 production is affected by sleep; stimulated ex vivo production of interleukin-2 (IL-2) is enhanced during sleep as compared to wake and this effect is dependent on sleep, rather than circadian processes (23).

These findings extend previous investigations in humans on the immunological effects of prolonged- or total night sleep deprivation and show that even a rather modest disturbance of sleep produces an acute reduction of NK cell, T-cell and monocyte function. The clinical implications of reduced immune cell function following sleep loss is not known, although NK cells mediate protection against primary herpes virus infections and compromised natural immunity may be a prognostic factor for recurrence in patients with malignant disease (29).

However, to interpret these findings in relationship to host defense function, it is useful to keep in mind that the immune parameters tested were obtained from peripheral blood. In the case of an acute stress paradigm in rats, Dhabhar and McEwen (49) have shown that decreases of NK activity in the peripheral blood are accompanied by a redistribution of NK cells to tissues and an increase in NK function in these sites. In the present studies, changes of NK- and LAK activity were due to decreases in lytic activity per cell, not simply related to changes in the distribution of these cell types in the peripheral blood. Nevertheless, the question about whether sleep deprivation compromises host defense function remains unanswered. Dinges and colleagues have found that prolonged sleep deprivation induces leucocytosis and enhancement of NK cell function in humans (19). Moreover, in a murine model of influenza viral infection, sleep deprivation was associated with increased levels of IL-1 and interferon-α and decreases of viral shedding in bronchioalveolar lavages (50).

Although the immediate effects of modest loss of sleep on NK cell function are robust, an additional caveat requires consideration. These alterations of natural and cellular immune function following sleep deprivation are transient and return to basal levels following a night of recovery sleep. Indeed, rebound increases of slow wave sleep that typically occur following a night of sleep deprivation may underlie recovery increases of immunity (48). Slow wave sleep has been associated with increased serum concentrations of IL-2 in humans that in turn might stimulate NK cells during the recovery night (33). Alternatively, endogenous interleukin-1 is implicated in the homeostatic recovery of slow wave sleep following sleep loss. Opp and Krueger (51) have shown that the administration of antibodies to interleukin-1 antagonizes rebound increases of slow wave sleep in animals following sleep deprivation. Thus, release of pro-inflamatory cytokines such as interleukin-1 may occur before recovery sleep and coordinate the induction of increases of slow wave sleep as well as increases of NK responses and T helper cytokine production.

5.2 Effects on circulating cytokines

Most studies have relied on ex vivo measures of immune function, and information about the effects of sleep on *in vivo* measures of immune function, particularly in relation to sleep activity and specific sleep stages, is rather limited. In six volunteers, Moldofsky (33) reported that the onset of sleep and slow wave sleep was associated with increases in circulating concentrations of IL-1 followed by elevations of IL-2. Another study has found that circulating concentrations of IL-2 change across the nocturnal period, with IL-2 plasma levels increasing in the one hour period after sleep (52). Other measures of IL-2 taken during the night were comparable to awake values. In addition, sleep deprivation had no effect on circulating levels of IL-2 and there was no association between sleep activity and change of circulating levels of IL-2 (52). The lack of a consistent influence of sleep on circulating concentrations of IL-2, as opposed to IL-2 production, is most parsimoniously explained by the local release of IL-2 at cellular sites where it is rapidly utilized, degraded, or bound to soluble IL-2 receptors for uptake. Thus, in vivo circulating concentrations of IL-2 may not reliably demonstrate the impact of sleep on the expression and release of this cytokine.

In contrast with the rather limited data linking sleep and IL-2, accumulating data suggest a robust interaction between sleep, circadian rhythms, and IL-6. Circulating concentrations of IL-6 show a periodicity with low values during daytime and maxima during the night (53). Other studies have found that patients with nocturnal sleep disturbance show altered circulating concentrations of TNFα and IL-6 (45,46) and that exogenous intravenous doses of IL-6 induce decreases of slow wave sleep and REM sleep (44). However, two other studies failed to detect changes in circulating levels of IL-6 in relation to sleep deprivation and/or circadian rhythms(2,23), and it is not known whether nocturnal IL-6 concentrations are related to sleep activity or sleep stages.

To address these questions regarding the effects of nocturnal sleep and sleep loss on the nocturnal secretion of IL-6, circulating levels of IL-6 were measured during two nights, baseline and partial sleep deprivation, early night (PSD-E). Sleep EEG was monitored along with sampling of circulating concentrations of growth hormone, cortisol and melatonin (25). Growth hormone is thought to be sleep-dependent (54,55) whereas cortisol and melatonin are proposed to driven primarily by circadian processes (56,57). If IL-6 is sleep-dependent, the secretory profile of IL-6 would parallel that of growth hormone but not cortisol or melatonin. Change scores were calculated for each time point in relation to awake levels to evaluate the effects of sleep and sleep deprivation on IL-6, growth hormone, cortisol and melatonin levels.(Figure 3)

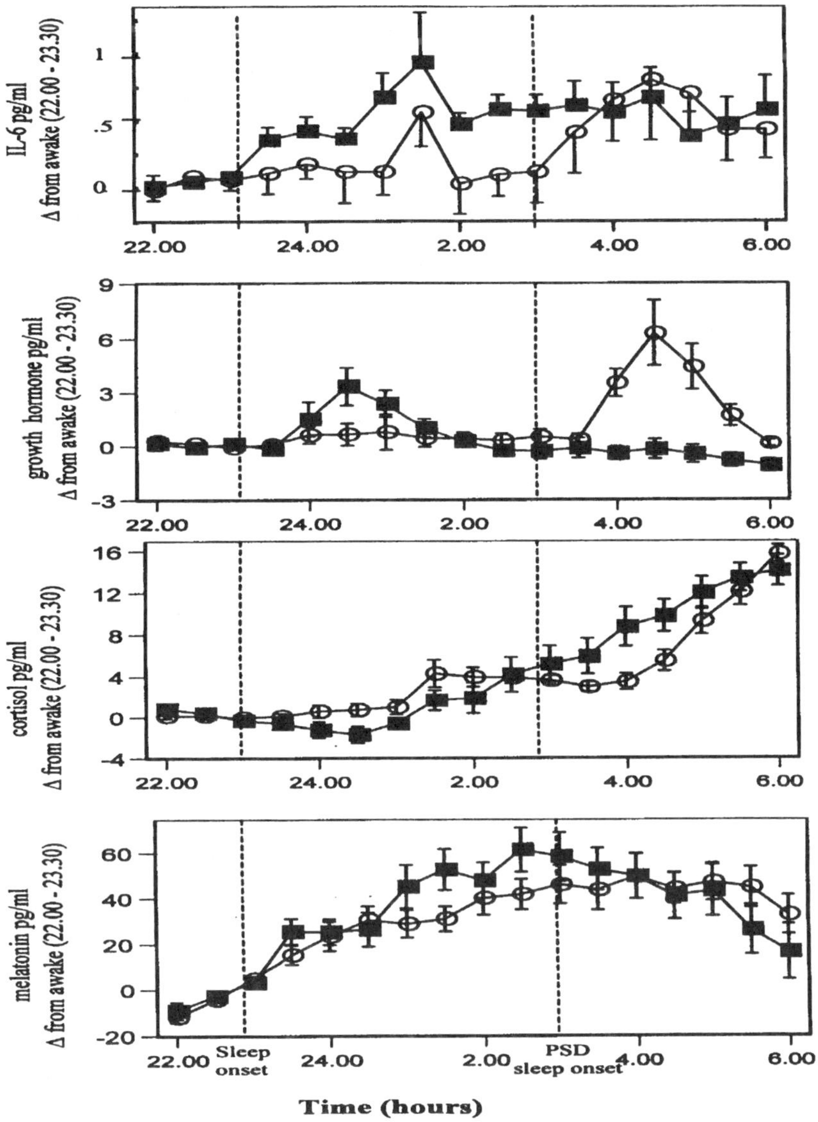

Figure 3. Averaged change scores from awake (± SEM) for circulating levels of IL-6 (A), growth hormone (B), cortisol (C) and melatonin (D) in subjects during baseline (■) and and PSD-E nights (O). The *vertical dashed line* at 2300 h indicates the average time that the subjects were asleep on the baseline night; the *vertical dashed line* at 0300 h indicates the time that the subjects were asleep on PSD-E night. For IL-6, change in the early part of the night was elevated during the baselines- as compared to the PSD-E night ($t=2.4$, $p<0.05$). For growth hormone, late night change scores were elevated in the PSD-E night as compared to baseline ($t=-3.9$, $p<0.01$). For cortisol , late night change scores were decreased in the PSD-E as compared to baseline ($t=6.7$, $p<0.001$). For melatonin, there were no differences between the PSD-E and baseline nights. (Reprinted from ref #25)

IL-6 showed a nocturnal profile that was similar to that found for the sleep-dependent hormone, growth hormone. Sleep onset was associated with an increase in circulating levels of IL-6, with peak values occurring 2.5 h after sleep onset. When sleep onset was delayed during partial night sleep deprivation, the nocturnal increase of IL-6 did not occur until after sleep onset at 3.00h. In contrast, sleep deprivation had no effect on the nocturnal secretory profiles of cortisol or melatonin, which taken together suggest that sleep, rather than a circadian pacemaker, influences nocturnal IL-6 and growth hormone secretion (25).

Levels of IL-6 were also compared between awake and different sleep stages over the nocturnal period (Figure 4). Circulating concentrations of IL-6 were different across sleep stages in which IL-6 was higher during stages 1-2 sleep and REM sleep as compared to average levels during the awake period. Levels of IL-6 were similar between the awake period and stages 3-4.

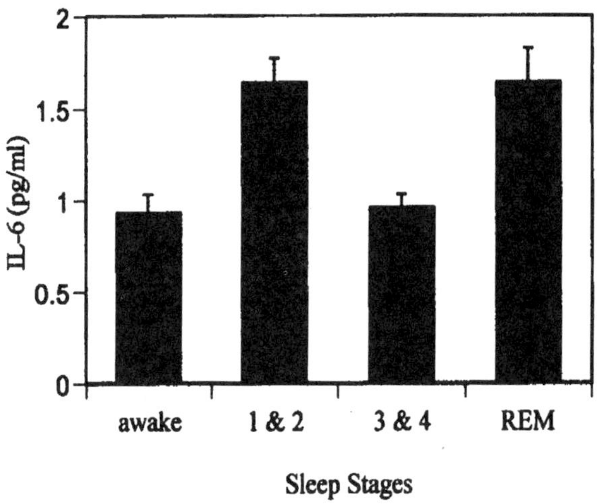

Figure 4. Mean IL-6 serum levels across the various sleep stages. Concentrations of IL-6 were different across sleep stages ($F = 8.9$, $p<0.001$) during the baseline night. Post-hoc comparisons demonstrated that circulating concentrations of IL-6 were higher during stages 1-2 sleep ($t = 3.2$, $p< 0.01$) and REM sleep ($t = 3.6$, $p<0.01$) as compared to average levels during the awake period. Levels of IL-6 were similar between the awake period and stages 3-4 ($t = -0.4$, $p=0.72$).(Reprinted from ref #25)

The present observations linking decreases of nocturnal IL-6 to the onset of slow wave sleep are consistent with those of Vgontzas et al. who found a negative correlation between slow wave sleep and IL-6 release (46). Moreover, Vgontzas and colleagues found that IL-6 was decreased during a night of recovery sleep; a night of recovery sleep is typically associated with increased relative amounts of slow wave sleep(46). Alternatively, as already discussed, cytokines such as IL-6 are hypothesized to have regulatory influence on sleep. Exogenous doses of IL-6 are associated with decreased REM sleep and with decreased amounts of slow wave sleep in the first part of the night followed by increases in the second half (44).

There are several implications regarding the association between sleep and IL-6. Consistent with the notion that sleep has a restorative function, normal sleep is associated with enhancement of IL-6 release as well as other immune processes (23). Thus, loss of sleep during part of the night may be one factor exacerbating the immunological alterations found in persons who experience psychological stress or suffer from a psychiatric disorder and commonly complain of insomnia (4,58). In addition, given the association between increases of IL-6 and REM sleep, disturbances of sleep architecture may lead to abnormalities in the sleep-related secretion of IL-6. For example in conditions such as alcoholism, depression and possibly aging where there is a relative increase of REM sleep at the expense of slow wave sleep (4), it is possible that abnormal increases of IL-6 may occur during sleep with implications for inflammatory- and cardiovascular disease risk in these populations (59,60).

6. SLEEP DEPRIVATION AND IMMUNITY: POSSIBLE MECHANISMS

Multiple pathways involving activation of the hypothalamic pituitary adrenal axis, secretion of melatonin, or release of sympathetic neurotransmitters have been postulated to mediate the relationship between depression, sleep and immune function. Both growth hormone and melatonin are secreted during sleep, and act in vitro to induce monocyte secretion of IL-1 and to alter cellular immune responses (61,62). ACTH and cortisol release, also related to circadian rhythm, increase or suppress in vitro production of Th1 and Th2 cytokines depending on the stage of T cell activation(63). Finally, sympathetic tone, generally thought to decrease during the night especially during delta sleep (52), has suppressive effects on immune function depending on the cellular target, the concentration available at the cellular, tissue, or system level, binding of α- or ß adrenergic receptors, and/or competing effects on peripheral blood immune cell traffic and cell activation (32). Production of both Th1 (IL-2) and Th2 (IL-4) is increased by sympathetic denervation (32).

Studies have begun to examine the mechanisms that mediate the effects of PSD on NK activity and have initially focused on adrenergic mechanisms. Substantial evidence has shown that the release of catecholamines and β-adrenergic receptor activation results in a reduction of ex vivo levels of NK activity in humans (64). The results showed that PSD was associated with increases in circulating concentrations of epinephrine and norepinephrine and that late night elevations of norepinephrine during PSD negatively correlated with change of NK activity from baseline to PSD night (rho=0.48, p=0.05) (52). There was no association between late night levels of epinephrine and change of NK activity, and early night levels of catecholamines were not correlated with change of NK activity.

Among the mechanisms underlying the relation between sleep architecture and nocturnal IL-6 secretion, the release of catecholamines also needs to be considered. Catecholamines stimulate IL-6 secretion via the β-adrenergic receptor (65,66), and exercise-induced release of epinephrine and norepinephrine correlates with increases of IL-6 (67). Sympathetic neural activity increases during REM sleep (68) and decreases during slow wave sleep (52). Thus, increases of norepinephrine during REM sleep may account for the increase of IL-6 during this sleep stage. Nevertheless, as compared to awake, average levels of circulating catecholamines decrease during sleep and it not known what accounts for the average increase of IL-6 during sleep. Growth hormone and melatonin have been reported to promote pro-inflammatory and T cell function (62,69,70), but there was no correlation between IL-6 levels and these neuroendocrine hormones in our studies.

7. CONCLUSIONS

Sleep loss is endemic in society, and loss of sleep during only part of the night is one of the most common complaints of persons who experience environmental- or psychological stress, travel across time meridians, engage in shift work, or suffer from a psychiatric disorder. The present studies indicate that only a modest amount of sleep loss impacts measures of natural- and cellular immunity and alters the nocturnal secretion of the pro-inflammatory cytokines such as IL-6. These data, together with findings in clinical samples that correlate disordered sleep with impairments in immune functioning, indicate that sleep disturbance has to be taken into account as a cause of immune alterations or as a factor exacerbating them. However, understanding the physiological- and clinical significance of these acute immunological changes observed in healthy adults requires additional research. The ultimate question about whether sleep compromises subjects' health remains unanswered. None of the studies have employed disease specific immune measures nor intensively monitored changes in health status in relationship to administration of a sleep deprivation paradigm. Experimental approaches that involve administration of a viral challenge or biologically relevant immunization that is coupled with sleep deprivation are needed to test the impact of sleep on clinically relevant endpoints.

Partial night sleep deprivation serves as naturalistic probe to evaluate the immune changes engendered by sleep loss with minimal confounding effects due to changes in stress hormones (e.g., cortisol) or circadian phase (e.g. melatonin). Thus, partial sleep deprivation is exceedingly useful in testing the biological- and immunological responses that are linked to homeostatic drive for sleep and in exploring the reciprocal interactions between sleep and cytokines. Given the responsivity of cytokines to disordered sleep and compelling basic evidence that pro-inflammatory cytokines are somnogenic, studies that permit more precise estimates of the temporal relationships between recovery sleep and these cytokines are needed. If changes of pro-inflammatory cytokines impact sleep and mediate deficits of sleep (e.g., loss of delta sleep) in humans, this finding could have important clinical implications in the potential use of these cytokines or their antagonists for evaluating and possibly treating disordered sleep in clinical populations.

ACKNOWLEDGEMENTS

This work was supported in part by NIAAA grants AA10215 and AA13239, NIMH 5T32-18399, NIMH 2P30-MH30914, NIH M01 RR00827, NIMH M4-55253.

Parts of this chapter were previously published in the forms of a review of conference proceedings and journal articles.

REFERENCES

1. Horne, J., *Why We Sleep: The Function of Sleep in Humans and Other Mammals.* Oxford: Oxford University Press, 1988; 1-319.
2. Dinges, D. F., Douglas, S. D., Hamarman, S., et al. Sleep deprivation and human immune function. *Adv. Neuroimmunol.* 1995; 5: 97-110.
3. Benca, R. M. and Quintas, J. Sleep and host defenses: a review. *Sleep* 1997; 20: 1027-37.
4. Benca, R. M., Obermeyer, W. H., Thisted, R. A., et al. Sleep and psychiatric disorders: a meta analysis. *Arch. Gen. Psychiatry* 1992; 49: 651-668.
5. Irwin, M. Immune correlates of depression. *Adv. Exp. Med. Biol.* 1999; 461: 1-24.
6. Cook, R. T. Alcohol abuse, alcoholism, and damange to the immune system - A review. *Alc: Clin Exp Res* 1998; 22: 1927-1942.
7. Krueger, J. M. and Majde, J. A. Short Analytical Review. Sleep as a host defense: its regulation by microbial products and cytokines. *Clin. Immunol. Immunopathol.* 1990; 57: 188-199.
8. Krueger, J. M. and Toth, L. A. Cytokines as regulators of sleep. *Ann. N.Y. Acad. Sci.* 1994; 739: 299-310.

9. Krueger, J. M., Takahashi, S., Kapás, L., et al. Cytokines in sleep regulation. *Adv. Neuroimmunol.* 1995; 5: 171-88.
10. Pollmächer, T., Schuld, A., Kraus, T., et al. Experimental immunomodulation, sleep, and sleepiness in humans. *Ann. N.Y. Acad. Sci.* 2000; 917: 488-99.
11. Kripke, D. F. and Wingard, D. Mortality risk factors in the elderly [letter]. *Am. J. Pub. Health* 1985; 75: 788-9.
12. Brown, R., Pang, G., Husband, A. J., et al. Suppression of immunity to influenza virus infection in the respiratory tract following sleep disturbance. *Regional Immunol.* 1989; 2: 321-325.
13. Brown, R., Price, R. J., King, M. G., et al. Interleukin-1 beta and muramyl dipeptide can prevent decreased antibody response associated with sleep deprivation. *Brain Behav. Immun.* 1989; 3: 320-330.
14. Everson, C. A. Functional consequences of sustained sleep deprivation in the rat. *Behav. Brain Res.* 1995; 69: 43-54.
15. Toth, L. A., Tolley, E. A. and Krueger, J. M. Sleep as a prognostic indicator during infectious disease in rabbits. *Proc. Soc. Exp. Bio. Med.* 1993; 203: 179-192.
16. Benca, R. M., Kushida, C. A., Everson, C. A., et al. Sleep deprivation in the rat: VII. Immune function. *Sleep* 1989; 12: 47-52.
17. Nakano, Y., Miura, T., Mara, I., et al. The effect of shift work on cellular immune function. *J. Human Ergol.* 1982; 11: 131-137.
18. Curti, R., Radice, G., C., Cesana, R., et al. Work, stress and immune system: lymphocyte reactions during rotating shift work. *Med. Lav.* 1982; 6: 564-570.
19. Dinges, D. F., Douglas, S. D., Zaugg, L., et al. Leukocytosis and natural killer cell function parallel neurobehavioral fatigue induced by 64 hours of sleep deprivation. *J. Clin. Invest.* 1994; 93: 1930-1939.
20. Moldofsky, H., Lue, F. A., Davidson, J. R., et al. Effects of sleep deprivation on human immune function. *FASEB J.* 1989; 3: 1972-1977.
21. Palmblad, J., Cantell, K., Strander, H., et al. Stressor exposure and immunological responses in man: interferon producing capacity and phagocytosis. *J. Psychosom. Res.* 1976; 20: 193-199.
22. Palmblad, J., Petrini, B., Wasserman, J., et al. Lymphocyte and granulocyte reactions during sleep deprivation. *Psychosom. Med.* 1979; 41(4): 273-278.
23. Born, J., Lange, T., Hansen, K., et al. Effects of sleep and circadian rhythm on human circulating immune cells. *J. Immunol.* 1997; 158: 4454-4464.
24. Kronfol, Z., Nair, M., Zhang, Q., et al. Circadian immune measures in healthy volunteers: relationship to hypothalamic-pituitary-adrenal axis hormones and sympathetic neurotransmitters. *Psychosom. Med.* 1997; 59: 42-50.
25. Redwine, L., Hauger, R., Gillin, J., et al. Effects of sleep and sleep deprivation on interleukin-6, growth hormone, cortisol and melatonin levels in humans. *J. Clin. Endocrin. Metab.* 2000; 85: 3597-3603.
26. Irwin, M., Brown, M., Patterson, T., et al. Neuropeptide Y and natural killer cell activity: findings in depression and Alzheimer caregiver stress. *FASEB J.* 1991; 5: 3100-3107.
27. Owens, M. J. and Nemeroff, C. B. Physiology and pharmacology of corticotropin-releasing factor. *Pharm. Rev.* 1991; 91: 425-473.
28. Irwin, M., Caldwell, C., Smith, T. L., et al. Major depressive disorder, alcoholism, and reduced natural killer cell cytotoxicity: Role of severity of depressive symptoms and alcohol consumption. *Arch. Gen. Psychiat.* 1990; 47: 713-719.
29. Trinchieri, G. Biology of natural killer cells. *Adv. Immunol.* 1989; 47: 187-376.

30. Irwin, M., Smith, T. L. and Gillin, J. C. Electroencephalographic sleep and natural killer activity in depressed patients and control subjects. *Psychosom. Med.* 1992; 54: 10-21.
31. Ehlers, C. L., Somes, C., Seifritz, E., et al. CRF/NPY interactions: a potential role in sleep dysregulation in depression and anxiety. *Dep. Anxiety* 1997; 6: 1-9.
32. Friedman, E. M. and Irwin, M. R. Modulation of immune cell function by the autonomic nervous system. *Pharmacol. Ther.* 1997; 74: 27-38.
33. Moldofsky, H., Lue, F. A., Eisen, J., et al. The relationship of interleukin-1 and immune functions to sleep in humans. *Psychosom. Med.* 1986; 48(5): 309-318.
34. Irwin, M. and Miller, C. Decreased natural killer cell responses and altered interleukin-6 and interleukin-10 production in alcoholism: an interaction between alcohol dependence and African-American ethnicity. *Alc. Clin. Exp. Res.* 2000; 24: 560-9.
35. Irwin, M., Miller, C., Gillin, J.C., et al. Polysomnographic and spectral sleep EEG in primary alcoholics: an interaction between alcohol dependence and African American ethnicity. *Alc. Clin. Exp. Res.* 2000; 24: 1376-1384.
36. Street, N. E. and Mosmann, T. R. Functional diversity of T lymphocytes due to secretion of different cytokine patterns. *FASEB J.* 1991; 5: 171-177.
37. Mendenhall, C. L., Moritz, T., Rouster, S., et al. Epidemiologyof hepatitis C among veterans with alcoholic liver disease. The VA Cooperative Study Group 275. *Am. J. Gastroenterol.* 1993; 88: 1022-1026.
38. Woitas, R. P., Lechmann, M., Jung, G., et al. CD30 induction and cytokine profiles in hepatitis C virus core-specific pheripheral blood T lymphocytes. *J. Immunol.* 1997; 159: 1012-1018.
39. Opp, M. R. and Imeri, L. Sleep as a behavioral model of neuro-immune interactions. *Acta Neurobiologiae Experimentalis* 1999; 59: 45-53.
40. Opp, M. R., Obal, F., Jr. and Krueger, J. M. Interleukin 1 alters rat sleep: temporal and dose-related effects. *Am. J. Physiol.* 1991; 260(29): R52-R58.
41. Kapás, L., Hong, L., Cady, A. B., et al. Somnogenic, pyrogenic, and anorectic activities of tumor necrosis factor-alpha and TNF-alpha fragments. *Am. J. Physiol.* 1992; 263: R708-15.
42. Opp, M. R., Smith, E. M. and Hughes, T. K. Interleukin-10(cytokine synthesis inhibitory factor) acts in the central nervous system of rats to reduce sleep. *J. Neuroimmunol.* 1995; 60: 165-168.
43. Opp, M., Obal, F., Jr., Cady, A. B., et al. Interleukin-6 is pyrogenic but not somnogenic. *Physiol. Behav.* 1989; 45: 1069-72.
44. Späth-Schwalbe, E., Hansen, K., Schmidt, F., et al. Acute effects of recombinant human interleukin-6 on endocrine and central nervous sleep functions in healthy men. *J. Clin. Endocrin. Metab.* 1998; 83: 1573-9.
45. Vgontzas, A. N., Papanicolaou, D. A., Bixler, E. O., et al. Elevation of plasma cytokines in disorders of excessive daytime sleepiness: role of sleep disturbance and obesity [see comments]. *J. Clin. Endocrin. Metab.* 1997; 82: 1313-6.
46. Vgontzas, A. N., Papanicolaou, D. A., Bixler, E. O., et al. Circadian interleukin-6 secretion and quantity and depth of sleep. *J. Clin. Endocrin. Metab.* 1999; 84: 2603-7.
47. Irwin, M., Mascovich, A., Gillin, J. C., et al. Partial sleep deprivation reduces natural killer cell activity in humans. *Psychosom. Med.* 1994; 56: 493-498.
48. Irwin, M., McClintick, J., Costlow, C., et al. Partial night sleep deprivation reduces natural killer and cellular immune responses in humans. *FASEB J.* 1996; 10: 643-653.
49. Dhabhar, F. S., Miller, A. H., McEwen, B. S., et al. Stress-induced changes in blood leukocyte distribution. *J. Immunol.* 1996; 157: 1638-1644.

50. Renegar, K. B., Crouse, D., Floyd, R. A., et al. Progression of influenza viral infection through the murine respiratory tract: the protective role of sleep deprivation. *Sleep* 2000; 23: 859-63.
51. Opp, M. R. and Krueger, J. M. Anti-interleukin-1β reduces sleep and sleep rebound after sleep deprivation in rats. *Am. J. Physiol.* 1994; 266: R688-R695.
52. Irwin, M., Thompson, J., Miller, C., et al. Effects of sleep and sleep deprivation on catecholamine and interleukin-2 levels in humans: clinical implications. *J. Clin. Endocrin. Metab.* 1999; 84: 1979-85.
53. Bauer, J., Hohagen, F., Ebert, T., et al. Interleukin-6 serum levels in healthy persons correspond to the sleep-wake cycle. *Clin. Invest.* 1994; 72: 315.
54. Jarrett, D. B., Greenhouse, J. B., Miewald, J. M., et al. A reexamination of the relationship between growth hormone secretion and slow wave sleep using delta wave analysis. *Biol. Psychiat.* 1990; 27: 497-509.
55. Perras, B., Marshall, L., Köhler, G., et al. Sleep and endocrine changes after intranasal administration of growth hormone-releasing hormone in young and aged humans. *Psychoneuroendocrinol.* 1999; 24: 743-57.
56. Gronfier, C., Chapotot, F., Weibel, L., et al. Pulsatile cortisol secretion and EEG delta waves are controlled by two independent but synchronized generators. *Am. J. Physiol.* 1998; 275: E94-100.
57. Youngstedt, S. D., Kripke, D. F., Elliott, J. A. Melatonin excretion is not related to sleep in the elderly. *J. Pineal. Res.* 1998; 24: 142-5.
58. Hall, M., Baum, A., Buysse, D. J., et al. Sleep as a mediator of the stress-immune relationship. *Psychosom. Med.* 1998; 60: 48-51.
59. Papanicolaou, D. A., Wilder, R. L., Manolagas, S. C., et al. The pathophysiologic roles of interleukin-6 in human disease. *Ann. Intern. Med.* 1998; 128: 127-37.
60. Musselman, D. L., Evans, D. L. and Nemeroff, C. B. The relationship of depression to cardiovascular disease: epidemiology, biology, and treatment. *Arch. Gen. Psychiatry* 1998; 55: 580-92.
61. Weigent, D. A. and Blalock, J. E. Growth hormone and the immune system. *Prog. Neuroendocrinimmunol.* 1990; 3: 231.
62. Garcia-Mauriño, S., Gonzalez-Haba, M. G., Calvo, J. R., et al. Melatonin enhances IL-2, IL-6, and IFN-gamma production by human circulating CD4+ cells: a possible nuclear receptor-mediated mechanism involving T helper type 1 lymphocytes and monocytes. *J. Immunol.* 1997; 159: 574-81.
63. Chiappelli, F., Manfrini, E., Franceschi, C., et al. Steroid regulation of cytokines. Relevance for TH1-to-TH2 shift? *Ann. N.Y. Acad. Sci.* 1994; 746: 204-15.
64. Murray, D. R., Irwin, M., Rearden, C. A., et al. Sympathetic and immune interactions during dynamic exercise: Mediation via Beta2-adrenergic dependent mechanism. *Circulation* 1992; 83: 203-213.
65. van Gool, J., van Vugt, H., Helle, M., et al. The relation among stress, adrenalin, interleukin 6 and acute phase proteins in the rat. *Clin. Immunol. Immunopathol.* 1990; 57: 200-10.
66. DeRijk, R. H., Boelen, A., Tilders, F. J., et al. Induction of plasma interleukin-6 by circulating adrenaline in the rat. *Psychoneuroendocrinol.* 1994; 19: 155-63.
67. Papanicolaou, D. A., Petrides, J. S., Tsigos, C., et al. Exercise stimulates interleukin-6 secretion: inhibition by glucocorticoids and correlation with catecholamines. *Am. J. Physiol.* 1996; 271: E601-5.
68. Somers, V. K., Phil, D., Dyken, M. E., et al. Sympathetic-nerve activity during sleep in normal subjects. *N. Engl. J. Med.* 1993; 328: 303-307.

69. LeRoith, D., Yanowski, J., Kaldjian, E. P., et al. The effects of growth hormone and insulin-like growth factor I on the immune system of aged female monkeys. *Endocrinol.* 1996; 137: 1071-1079.
70. Lissoni, P., Rovelli, F., Brivio, F., et al. Circadian secretions of IL-2, IL-12, IL-6 and IL-10 in relation to the light/dark rhythm of the pineal hormone melatonin in healthy humans. *Nat. Immun.* 1998; 16: 1-5.

INDEX

11β-hydroxysteroid dehydrogenase, 87, 105, 106
Acetylcholine, 62, 124, 160, 381
Adrenocorticotropic hormone (ACTH), 58-69, 85-111, 188, 225-234, 238, 244-247, 267, 283, 285, 302, 415
Adenosine, 147, 148, 153, 156, 318, 324
Aging, 158, 167, 178, 214, 266, 365, 415
Alcoholism, 403, 407, 415
Amyloid, 386-390, 393
Antidepressant, 40, 64, 71, 72, 123, 125, 140, 141, 249, 267-273, 290-294, 321, 322
see also MAOI, SSRI, TCA
Anti-inflammatory cytokines, 6, 18, 131, 141, 151, 170, 178, 179, 292, 321
Anxiety, 15, 33, 231, 266, 281-294, 302, 312-318, 345, 357, 406
Appetite, 124, 167-196, 265-274, 304, 305
Acquired Immune Deficiency Syndrome (AIDS), 124, 138, 149, 262-265,
Area postrema, 24, 25, 31, 43, 44, 50, 66, 91, 92, 134, 188, 232
Arginine vasopressin (AVP), 86, 95, 228, 232-236, 310, 311
Autism, 369-381
Basal forebrain, 153-159
B cell, 4-6, 178, 260, 288, 290, 324, 341
Benzodiazepine, 46, 52, 56
Brain derived neurotrophic factor (BDNF), 151, 156, 348, 349, 354
Cachexia, 3, 5, 138, 167-179, 196
Cancer, 12, 13, 138, 140, 149, 167, 176, 177, 185, 193-196, 211, 214, 260-268, 351, 357
Cardiovascular, 16, 155, 180, 225, 232, 328, 330, 412, 421
Catecholamine, 46, 47, 56-73, 91, 100, 101, 125, 126, 133, 192, 229, 285, 302-310, 319-326, 380, 405, 415
CD_4 cells, 289, 290, 314, 325, 377
CD_8 cells, 260, 261, 283, 286, 287, 290, 314
CD_{14}, 19, 49, 50, 173, 189
c-Fos, 26, 30, 31, 44-46, 94, 95, 132, 153, 154, 185-189, 226-240
Chemokine, 5, 124, 179, 342, 385, 386, 394
Choroid plexus, 22-33, 42-50, 62, 67, 92-97, 123, 133, 134, 184
Chronic fatigue, 110, 138, 139, 149, 302, 310
Circumventricular, 22-33, 43-50, 66, 91, 93, 133, 184, 227-232, 303
Corticosterone, 56-73, 86-90, 105-107, 132, 225-248, 310, 311
Corticotropin releasing hormone (CRH), 46, 85-111, 126, 150, 228-244, 266-274, 283, 285, 302-312, 318-321
Cortisol, 71, 86, 106, 107, 259-267, 283-290, 302-326, 423-427
Cross-sensitization, 233, 245, 309-313, 327
Cyclooxygenase, 44, 46, 50, 67, 70, 124, 133, 153, 189-191, 220
Depression, 33, 40, 41, 71, 72, 109, 110, 123-126, 134, 138-141, 186, 249, 259-274, 284, 287, 302-321, 357, 403, 406, 415
Dopamine, 56, 57, 62, 125, 126, 155, 193, 229, 233, 266, 267, 283, 286, 307, 319, 339
Eating, 72, 129-136, 167-194, 268, 417
Electroencephalogram (EEG), 147-158, 266-268, 273, 274, 406-412
see also sleep EEG
Endothelium, 22, 24, 33, 67, 227
Enkephalin, 124, 126
Exercise, 8, 179, 416
Fever, 4, 5, 18, 24-32, 39, 40, 70, 123, 124, 129-138, 167-171, 186-194, 213, 244, 289, 321, 340-351

Gamma-aminobutyric acid (GABA), 62, 65, 126, 152-157, 283
Genetic, 40, 41, 50, 102, 170-185, 218, 220, 266, 281-294, 303, 310-323, 340-348, 154-361, 369, 370, 386-395
Glossopharyngeal nerve, 27, 28
Glucocorticoid, 101, 105, 108, 310
Glutamate, 31, 62, 65, 133, 339
Growth factor, 4-8, 124, 148-160, 179, 215, 342, 378, 385-395
  see also brain derived neurotrophic factor (BDNF)
  see also nerve growth factor (NGF)
  see also transforming growth factor (TGF)
Growth hormone releasing hormone (GHRH), 147-157, 283
Hepatitis, 12, 268, 273, 341, 372, 407, 409
Hippocampus, 48, 57-67, 92-104, 125, 133, 149, 151, 182, 212-220, 229-247, 270, 306-320,
Histamine, 19, 62, 70, 195, 196, 236, 238, 239, 244, 248, 256
Human immunodeficiency virus (HIV), 124, 149, 263, 264, 407
Hypothalamic-pituitary adrenal axis (HPA),
  see adrenocorticoptropic hormone (ACTH)
  see corticotropin releasing hormone (CRH)
  see glucocorticoid
Hypothalamus, 42-43, 53-69, 85, 86, 91-104, 124-126, 133, 136, 151-157, 176-196, 226-245, 270, 284, 285, 306-310
Infection, 3, 8, 11, 15-33, 39-51, 56-73, 85-107, 129-140, 149, 167-196, 213, 220, 225-249, 260-268, 285-291, 305, 340-358, 370, 379, 404, 409, 411
Inflammation, 1-9, 17-33, 39-50, 87-110, 129, 168-173, 243, 244, 264, 303, 317-326, 342-357, 379, 385, 390
Interferon (IFN)
  autism, and 376-381
  brain effects, 22, 32, 64, 65, 73, 90, 103, 167-196, 230
  depression, and 72, 124-126, 140, 259-274
  disordered sleep, and 408, 409
  general properties, 5, 17
  schizophrenia, and 351-362
  sickness behavior, and 130, 137-141, 168-176, 268, 269
IFNα see Interferon
IFNβ see Interferon
IFNγ see Interferon
Interleukin-1 (IL-1),
  Alzheimer's disease, and 386-390
  anxiety disorder, and 286-289
  autism and, 373-377, 379
  brain expression, 41-51, 226-228, 239, 303, 304
  brain effects, 15-31, 59-74, 124, 126, 229-235, 243-248, 284, 306-316
  depression, and 33, 40, 41, 71, 72, 110, 123-126, 137-141, 249, 265-274, 284, 321
  general properties, 4-13, 320
  PTSD, and 289, 290, 315, 320-322
  schizophrenia, and 348-356
  sickness behavior, and 132-134, 168-196, 212-218
  sleep disorders, and 405, 409-412
Interleukin-1 receptor (IL-1R), 26, 43, 47, 97, 132-134, 182, 213, 320
Interleukin-1 Receptor Antagonist (IL-1Ra), 9, 18, 40, 42, 65-68, 90, 131-134, 141, 172, 231, 270, 271, 290, 307, 320-322, 343, 352, 358
Interleukin-2 (IL-2), 5, 9, 13, 17-20, 63, 73, 103, 109, 123-125, 140, 167, 196, 211-220, 267-272, 285-292, 312, 315, 349-360, 374-379, 408-412
Interleukin-2 Receptor (IL-2R), 288, 292, 316, 349
  soluble IL-2 receptor (sIL-2R), 260, 270, 271, 289, 290, 315, 353, 358
Interleukin-3 (IL-3), 7, 270, 272, 288-292, 353, 355
Interleukin-4 (IL-4), 6, 18, 20, 32, 129, 131, 171, 291, 314, 324, 349-354, 374-379, 408, 415
Interleukin-6 (IL-6)
  Alzheimer's disease, and 392, 393
  anxiety disorders, and 284-290
  autism and, 374-380
  brain effects, 15, 27, 49, 50, 60-73, 85-110, 124, 125, 130-132, 167, 181, 195, 211-216, 227-248

depression, and 109, 110, 123, 124, 260-272, 284, 321
general properties, 2, 5, 8, 343-345, 358
PTSD, and 315-327
schizophrenia, and 349-357, 360
sleep disorders, and 405-416
soluble IL-6 Receptor (sIL-6R), 270-271, 289-290, 315, 316, 322
Interleukin-8 (IL-8), 5, 6, 49, 167, 170, 179, 181, 196, 264, 352-354, 394
Interleukin-10 (IL-10), 6, 18, 20, 42, 104, 131, 171, 179, 218, 271, 284-292, 313, 325, 353, 374-379, 408, 410
Interleukin-12 (IL-12), 5, 89, 174-177, 230, 324, 372-380, 408
Interleukin-18 (IL-18), 5, 10, 104, 174, 175, 352-354, 385
Intracerebroventricular, 21, 24, 26, 58, 134, 141, 168, 193, 195, 267
Learning, 33, 138, 140, 156, 211-217, 268, 284, 285, 370, 373
Leptin, 184, 191, 192, 317, 319, 320, 322-324
Lipopolysaccharide (LPS)
behavioral effects, 72, 73, 130-141, 173-176, 184-196, 213, 227-249, 318
blood cell stimulation with, 271, 272, 315, 324, 325, 336
brain effects, 26-31, 41-50, 58-74, 88, 93-104
Lymphocyte, 3, 19, 20, 65, 128, 129, 139, 178, 182, 311, 219, 260-265, 273, 285-291, 313-326, 344, 356, 373, 404-410
Macrophages, 4, 5, 19, 20, 23, 28, 85, 88, 96-98, 102, 124, 130, 132, 176-175, 184, 189, 238, 291, 304, 320, 343, 356, 378
Monoamine oxidase inhibitor (MAOI), 270
Memory, 33, 125, 156, 211-220, 265, 268, 269, 285, 301, 312, 318, 345, 357, 370
Meninges, 22, 23, 25, 33, 42, 48, 50, 92, 93, 96, 97, 184, 188, 391
Molecular, 2, 5, 6, 130, 132, 141, 151, 215, 218, 220, 386
Natural Killer (NK), 4-6, 89, 174, 175, 260, 266, 283, 291, 314, 324, 325, 371, 378, 405-416
Nerve growth factor (NGF), 150-159, 231, 349-353
Neuropeptide, 100, 123-126, 195, 196, 211, 232-234, 262, 283, 285, 303-307, 407
Neurotransmitter, 31, 55, 58-72, 124, 140, 153, 216, 225-231, 243, 262-269, 283-293, 303-306, 339, 415
Nitric oxide, 41, 43, 50, 62, 91, 92, 124, 147, 150, 156, 189, 248, 385-394
Non-REM-sleep, 406, 407, 410
Norepinephrine, 57, 59, 72, 74, 93, 270, 271, 273, 308, 327, 422
Obesity, 180, 181
Obsessive Compulsive Disorder (OCD), 289, 293, 349
Opioid peptide, 124, 125
Panic, 288
Paraventricular nucleus (PVN), 46, 47, 55-69, 85-99, 185-195, 226-248, 306-312
Phobia, 288, 292
Post Traumatic Stress Disorder (PTSD), 109, 110, 289, 290, 293, 301-327
Pro-inflammatory cytokines, 4-8, 18, 50, 91, 123, 167-196, 264-271, 292, 305, 131, 395, 409, 416, 417
Prostaglandins, 22, 24, 30, 34, 48, 67, 91, 92, 101, 189, 191, 248
Psychological stress, 58, 243, 263, 281-286, 305, 415, 416
Psychosis, 40
Rapid eye movement (REM) sleep, 147, 148, 151, 156, 268, 412-416
Schizophrenia, 71, 219, 220, 339-361
Sensitization, 99, 169, 225, 226, 232-248, 284, 303, 309-313, 327
Sexual behavior, 40, 130 230, 265, 268, 285
Sickness behavior, 40, 69-72, 110, 123, 129-137, 213, 235, 236, 238, 268, 285, 305, 312, 321
Sleep, 16-32, 40, 69, 72, 124, 135, 147-160, 167, 168, 182, 265-274, 301, 304, 320, 341, 345, 357-360, 403-417
Sleep EEG, 266, 268, 273, 274, 407, 412
see also REM sleep

see also non-REM sleep
Sympathetic nervous system, 57, 60, 285, 301, 302, 318, 323, 406
Selective serotonin reuptake inhibitors (SSRI), 269, 270, 290, 291
TH1/TH2, 378, 380
Transforming growth factor (TGF), 4, 6, 8, 123, 150, 349, 350, 354, 390, 391
Trauma, 8, 17, 101, 109, 220, 301, 303 308, 315, 325
see also PTSD
Tricyclic antidepressant (TCA), 269, 270
Tryptophan, 56-73, 140, 229, 230
Tumor necrosis factor (TNF)
Alzheimer's disease, and 385-394
anxiety, and 284-289
autism, and 374-380
behavioral effects, 123, 130-132, 147-156, 167-196, 211, 215
brain effects, 17-30, 64-73, 85-110, 227-249
brain expression, 43, 50, 227
depression, and 109, 123, 265-272, 284
general properties, 3-12, 305, 343-345
schizophrenia, and 347-360
Vagus nerve, 27-29, 66, 93, 132, 133, 226, 345
Ventricular ependyma, 22, 26
Virus, 6, 56-73, 85, 89, 103, 129, 138, 172, 173, 175, 263, 265, 267, 289, 324, 370, 372, 411